Elliptic & Parabolic Equations

Elliptic & Parabolic Equations

Zhuoqun Wu, Jingxue Yin & Chunpeng Wang

Jilin University, China

World Scientific

NEW JERSEY · LONDON · SINGAPORE · BEIJING · SHANGHAI · HONG KONG · TAIPEI · CHENNAI

Published by

World Scientific Publishing Co. Pte. Ltd.
5 Toh Tuck Link, Singapore 596224
USA office: 27 Warren Street, Suite 401-402, Hackensack, NJ 07601
UK office: 57 Shelton Street, Covent Garden, London WC2H 9HE

British Library Cataloguing-in-Publication Data
A catalogue record for this book is available from the British Library.

ELLIPTIC AND PARABOLIC EQUATIONS

ISBN-13 978-981-270-025-4
ISBN-10 981-270-025-0
ISBN-13 978-981-270-026-1 (pbk)
ISBN-13 981-270-026-9 (pbk)

Printed in Singapore

Preface

Elliptic equations and parabolic equations are two important branches in the field of partial differential equations. These two kinds of equations arise frequently at the same time in many applications: elliptic equations arise from the stationary case and parabolic equations from the nonstationary case. In history, theories of them have been developed almost simultaneously. Many methods applied to elliptic equations are also available to parabolic equations, although some new methods are required to be developed for the latter. So far there have been numerous monographs focusing separately on each kind of equations, see [Ladyženskaja and Ural'ceva (1968)], [Ladyženskaja, Solonnikov and Ural'ceva (1968)], [Gilbarg and Trudinger (1977)], [Lieberman (1996)], [Chen and Wu (1997)], [Chen (2003)] and [Gu (1995)]. However, there are very few books treating them in combination. In this respect, the book [Oleĭnik and Radkevič (1973)] should be mentioned, in which the equations considered include not only both linear elliptic and parabolic equations, but also all kinds of linear degenerate elliptic equations of second order. However, in the framework of this book, parabolic equations are regarded as degenerate elliptic equations by treating the time variable and space variables equally and thus only the commonalities between these two kinds of equations are presented. As a matter of course, in such a book, it is impossible to discuss deeply the specific properties of each kind.

From our own experiences of teaching and research, we are aware of the necessity of writing a book which merges these two kinds of equations into an organic whole, involving the related basic theories and methods. This book is the result of a try following this idea, which is completed on the basis of lectures for graduate students majored in partial differential equations at Jilin University of China. The lectures have also been used at the summer school for graduate students in China.

The purpose of this book is to provide an introduction to elliptic and parabolic equations of second order for graduate students and young scholars who want to work in the field of partial differential equations. It is our hope that the book will be beneficial not only to stress the commonalities between these two kinds of equations, but also to expose the specific properties of each kind, so that the readers can efficiently learn the related knowledge by observing the relationship and contrasting the similarities and differences. An exhaustive theory of these two kinds of equations is outside the scope of this book. The book covers only the related basic theories and methods in a reasonable volume. More attention is paid to typical equations. In treating each kind of equations, usually we first give a detailed discussion on some typical equations and then discuss general equations in a brief fashion. Our principal intention is to prevent the complicate derivation due to the generality of equations in form from concealing and obscuring substantial of the argument. Emphasis is put on introducing methods and techniques rather than collecting theorems and facts.

The book consists of thirteen chapters.

Some preliminary knowledge needed in the book is collected in Chapter 1, the main part of which is an introduction to the theory of Sobolev spaces and Hölder spaces.

Linear equations are discussed in Chapter 2 through Chapter 9. Chapter 2 and Chapter 3 are devoted to weak solutions and the L^2 theory of linear elliptic equations and parabolic equations respectively.

Properties of weak solutions are discussed in Chapter 4 and Chapter 5. In Chapter 4, we introduce two important methods, the De Giorgi iteration and the Moser iteration which are described only for some typical equations and applied only to the maximum estimates on weak solutions. Chapter 5 discusses Harnack's inequalities.

In Chapter 6 and Chapter 7, we establish Schauder's estimates for elliptic equations and parabolic equations respectively. Based on these estimates, we prove the existence of classical solutions in Chapter 8. In establishing Schauder's estimates, we apply Campanato's approach which is based on the important fact that the Hölder continuity of functions can be described in an equivalent integral form. By means of this approach, the proof is relatively simple.

Chapter 9 is an argument of the L^p estimates which are used to discuss the existence of strong solutions.

The solvability of quasilinear equations is studied in Chapter 10 through

Chapter 12. Three methods are introduced, they are: the fixed point method (Chapter 10), the topology degree method (Chapter 11) and the monotone method (Chapter 12).

The book finishes with Chapter 13, which contains an investigation of elliptic and parabolic equations with degeneracy. The first part of this chapter deals with linear equations, namely, equations with nonnegative characteristic form. Quasilinear equations are discussed in the second part of this chapter.

As space is limited, we are not able to cover the study of fully nonlinear elliptic and parabolic equations in the book.

Wu Zhuoqun
Yin Jingxue
Wang Chunpeng

Jilin University, P. R. China
August, 2006

Contents

Preface   v

1. **Preliminary Knowledge**   1

  1.1   Some Frequently Applied Inequalities and Basic Techniques   1
     1.1.1   Some frequently applied inequalities   1
     1.1.2   Spaces $C^k(\Omega)$ and $C_0^k(\Omega)$ 2
     1.1.3   Smoothing operators   3
     1.1.4   Cut-off functions   5
     1.1.5   Partition of unity   6
     1.1.6   Local flatting of the boundary   6
  1.2   Hölder Spaces   7
     1.2.1   Spaces $C^{k,\alpha}(\overline{\Omega})$ and $C^{k,\alpha}(\Omega)$ 7
     1.2.2   Interpolation inequalities   8
     1.2.3   Spaces $C^{2k+\alpha,k+\alpha/2}(\overline{Q}_T)$ 13
  1.3   Isotropic Sobolev Spaces   14
     1.3.1   Weak derivatives   14
     1.3.2   Sobolev spaces $W^{k,p}(\Omega)$ and $W_0^{k,p}(\Omega)$ 15
     1.3.3   Operation rules of weak derivatives   17
     1.3.4   Interpolation inequality   17
     1.3.5   Embedding theorem   19
     1.3.6   Poincaré's inequality   21
  1.4 t-Anisotropic Sobolev Spaces   24
     1.4.1   Spaces $W_p^{2k,k}(Q_T)$, $\overset{\circ}{W}_p^{2k,k}(Q_T)$, $\overset{\bullet}{W}_p^{2k,k}(Q_T)$, $V_2(Q_T)$ and $V(Q_T)$ 24
     1.4.2   Embedding theorem   26
     1.4.3   Poincaré's inequality   28

1.5   Trace of Functions in $H^1(\Omega)$ 29
      1.5.1   Some propositions on functions in $H^1(Q^+)$ 29
      1.5.2   Trace of functions in $H^1(\Omega)$ 33
      1.5.3   Trace of functions in $H^1(Q_T) = W_2^{1,1}(Q_T)$ 35

2. L^2 Theory of Linear Elliptic Equations   39

    2.1   Weak Solutions of Poisson's Equation   39
          2.1.1   Definition of weak solutions   40
          2.1.2   Riesz's representation theorem and its application . .   41
          2.1.3   Transformation of the problem   43
          2.1.4   Existence of minimizers of the corresponding
                  functional   44
    2.2   Regularity of Weak Solutions of Poisson's Equation   47
          2.2.1   Difference operators   47
          2.2.2   Interior regularity   50
          2.2.3   Regularity near the boundary   53
          2.2.4   Global regularity   56
          2.2.5   Study of regularity by means of smoothing operators   58
    2.3 L^2 Theory of General Elliptic Equations   60
          2.3.1   Weak solutions   60
          2.3.2   Riesz's representation theorem and its application . .   61
          2.3.3   Variational method   62
          2.3.4   Lax-Milgram's theorem and its application   64
          2.3.5   Fredholm's alternative theorem and its application .   67

3. L^2 Theory of Linear Parabolic Equations   71

    3.1   Energy Method   71
          3.1.1   Definition of weak solutions   72
          3.1.2   A modified Lax-Milgram's theorem   73
          3.1.3   Existence and uniqueness of the weak solution   75
    3.2   Rothe's Method   79
    3.3   Galerkin's Method   85
    3.4   Regularity of Weak Solutions   89
    3.5 L^2 Theory of General Parabolic Equations   94
          3.5.1   Energy method   94
          3.5.2   Rothe's method   96
          3.5.3   Galerkin's method   97

4. De Giorgi Iteration and Moser Iteration   105

4.1 Global Boundedness Estimates of Weak Solutions of Poisson's Equation 105

    4.1.1 Weak maximum principle for solutions of Laplace's equation 105

    4.1.2 Weak maximum principle for solutions of Poisson's equation 107

4.2 Global Boundedness Estimates for Weak Solutions of the Heat Equation 111

    4.2.1 Weak maximum principle for solutions of the homogeneous heat equation 111

    4.2.2 Weak maximum principle for solutions of the nonhomogeneous heat equation 112

4.3 Local Boundedness Estimates for Weak Solutions of Poisson's Equation 116

    4.3.1 Weak subsolutions (supersolutions) 116

    4.3.2 Local boundedness estimate for weak solutions of Laplace's equation 118

    4.3.3 Local boundedness estimate for solutions of Poisson's equation 120

    4.3.4 Estimate near the boundary for weak solutions of Poisson's equation 122

4.4 Local Boundedness Estimates for Weak Solutions of the Heat Equation 123

    4.4.1 Weak subsolutions (supersolutions) 123

    4.4.2 Local boundedness estimate for weak solutions of the homogeneous heat equation 123

    4.4.3 Local boundedness estimate for weak solutions of the nonhomogeneous heat equation 126

5. Harnack's Inequalities   131

5.1 Harnack's Inequalities for Solutions of Laplace's Equation . 131

    5.1.1 Mean value formula 131

    5.1.2 Classical Harnack's inequality 133

    5.1.3 Estimate of $\sup_{B_{\theta R}} u$ 133

    5.1.4 Estimate of $\inf_{B_{\theta R}} u$ 135

    5.1.5 Harnack's inequality 141

    5.1.6 Hölder's estimate 143

5.2   Harnack's Inequalities for Solutions of the Homogeneous
       Heat Equation   145
       5.2.1 Weak Harnack's inequality   146
       5.2.2 Hölder's estimate   155
       5.2.3 Harnack's inequality   156

6. Schauder's Estimates for Linear Elliptic Equations   159

     6.1   Campanato Spaces   159
     6.2   Schauder's Estimates for Poisson's Equation   165
            6.2.1 Estimates to be established   165
            6.2.2 Caccioppoli's inequalities   168
            6.2.3 Interior estimate for Laplace's equation   173
            6.2.4 Near boundary estimate for Laplace's equation . . .   175
            6.2.5 Iteration lemma   177
            6.2.6 Interior estimate for Poisson's equation   178
            6.2.7 Near boundary estimate for Poisson's equation . . .   181
     6.3   Schauder's Estimates for General Linear Elliptic Equations   187
            6.3.1 Simplification of the problem   188
            6.3.2 Interior estimate   188
            6.3.3 Near boundary estimate   191
            6.3.4 Global estimate   193

7. Schauder's Estimates for Linear Parabolic Equations   197

     7.1 t-Anisotropic Campanato Spaces   197
     7.2   Schauder's Estimates for the Heat Equation   199
            7.2.1 Estimates to be established   199
            7.2.2 Interior estimate   200
            7.2.3 Near bottom estimate   208
            7.2.4 Near lateral estimate   214
            7.2.5 Near lateral-bottom estimate   227
            7.2.6 Schauder's estimates for general linear parabolic
                   equations   231

8. Existence of Classical Solutions for Linear Equations   233

     8.1   Maximum Principle and Comparison Principle   233
            8.1.1 The case of elliptic equations   233
            8.1.2 The case of parabolic equations   236

8.2 Existence and Uniqueness of Classical Solutions for Linear
Elliptic Equations 240
   8.2.1 Existence and uniqueness of the classical solution for
Poisson's equation 240
   8.2.2 The method of continuity 246
   8.2.3 Existence and uniqueness of classical solutions for
general linear elliptic equations 248
8.3 Existence and Uniqueness of Classical Solutions for Linear
Parabolic Equations 249
   8.3.1 Existence and uniqueness of the classical solution for
the heat equation 250
   8.3.2 Existence and uniqueness of classical solutions for
general linear parabolic equations 251

9. L^p Estimates for Linear Equations and Existence of
Strong Solutions   255

9.1 L^p Estimates for Linear Elliptic Equations and Existence
and Uniqueness of Strong Solutions 255
   9.1.1 L^p estimates for Poisson's equation in cubes 255
   9.1.2 L^p estimates for general linear elliptic equations . . . 260
   9.1.3 Existence and uniqueness of strong solutions for linear
elliptic equations 264
9.2 L^p Estimates for Linear Parabolic Equations and Existence
and Uniqueness of Strong Solutions 266
   9.2.1 L^p estimates for the heat equation in cubes 266
   9.2.2 L^p estimates for general linear parabolic equations . 271
   9.2.3 Existence and uniqueness of strong solutions for linear
parabolic equations 272

10. Fixed Point Method   277

10.1 Framework of Solving Quasilinear Equations via Fixed Point
Method 277
   10.1.1 Leray-Schauder's fixed point theorem 277
   10.1.2 Solvability of quasilinear elliptic equations 277
   10.1.3 Solvability of quasilinear parabolic equations 280
   10.1.4 The procedures of the a priori estimates 282
10.2 Maximum Estimate 282
10.3 Interior Hölder's Estimate 284

10.4 Boundary Hölder's Estimate and Boundary Gradient Esti-
     mate for Solutions of Poisson's Equation 287
10.5 Boundary Hölder's Estimate and Boundary Gradient
     Estimate 289
10.6 Global Gradient Estimate 296
10.7 Hölder's Estimate for a Linear Equation 301
     10.7.1 An iteration lemma 301
     10.7.2 Morrey's theorem 302
     10.7.3 Hölder's estimate 303
10.8 Hölder's Estimate for Gradients 307
     10.8.1 Interior Hölder's estimate for gradients of solutions . 307
     10.8.2 Boundary Hölder's estimate for gradients of solutions 308
     10.8.3 Global Hölder's estimate for gradients of solutions . 310
10.9 Solvability of More General Quasilinear Equations 310
     10.9.1 Solvability of more general quasilinear elliptic
            equations 310
     10.9.2 Solvability of more general quasilinear parabolic
            equations 311

11. Topological Degree Method   313

    11.1 Topological Degree 313
         11.1.1 Brouwer degree 313
         11.1.2 Leray-Schauder degree 315
    11.2 Existence of a Heat Equation with Strong Nonlinear Source 317

12. Monotone Method   323

    12.1 Monotone Method for Parabolic Problems 323
         12.1.1 Definition of supersolutions and subsolutions 324
         12.1.2 Iteration and monotone property 324
         12.1.3 Existence results 327
         12.1.4 Application to more general parabolic equations . . . 330
         12.1.5 Nonuniqueness of solutions 332
    12.2 Monotone Method for Coupled Parabolic Systems 336
         12.2.1 Quasimonotone reaction functions 337
         12.2.2 Definition of supersolutions and subsolutions 337
         12.2.3 Monotone sequences 339
         12.2.4 Existence results 350
         12.2.5 Extension 353

13. Degenerate Equations 355

13.1 Linear Equations 355

13.1.1 Formulation of the first boundary value problem . . 356

13.1.2 Solvability of the problem in a space similar to H^1 . 361

13.1.3 Solvability of the problem in $L^p(\Omega)$ 362

13.1.4 Method of elliptic regularization 365

13.1.5 Uniqueness of weak solutions in $L^p(\Omega)$ and regularity 366

13.2 A Class of Special Quasilinear Degenerate Parabolic Equations – Filtration Equations 368

13.2.1 Definition of weak solutions 369

13.2.2 Uniqueness of weak solutions for one dimensional equations 371

13.2.3 Existence of weak solutions for one dimensional equations 373

13.2.4 Uniqueness of weak solutions for higher dimensional equations 378

13.2.5 Existence of weak solutions for higher dimensional equations 381

13.3 General Quasilinear Degenerate Parabolic Equations 384

13.3.1 Uniqueness of weak solutions for weakly degenerate equations 385

13.3.2 Existence of weak solutions for weakly degenerate equations 393

13.3.3 A remark on quasilinear parabolic equations with strong degeneracy 399

Bibliography 403

Index 405

Chapter 1

Preliminary Knowledge

In this chapter, we provide some preliminary knowledge needed in this book. The central part is a brief introduction to the theory of Sobolev spaces and Hölder spaces. Most results are stated without proof, but references containing detailed proofs are indicated. An exception is that, for the convenience of the reader, a thorough discussion about the trace on the boundary of functions in a class of special Sobolev spaces is presented. The reader is assumed to have some acquaintance with elementary knowledge of functional analysis. Some specific facts in this field will be quoted wherever we need in the following chapters.

1.1 Some Frequently Applied Inequalities and Basic Techniques

This section presents some frequently applied inequalities and basic techniques such as mollifying, cutting off, partition of unity and local flatting of the boundary.

1.1.1 *Some frequently applied inequalities*

Young's inequality *Let $a > 0$, $b > 0$, $p > 1$, $q > 1$ and $\dfrac{1}{p} + \dfrac{1}{q} = 1$. Then*

$$ab \leq \frac{a^p}{p} + \frac{b^q}{q}.$$

It is called Cauchy's inequality when $p = q = 2$.

Replacing a, b by $\varepsilon^{1/p}a$, $\varepsilon^{-1/p}b$ with $\varepsilon > 0$ in the above inequality, we get

Young's inequality with ε *Let $a > 0$, $b > 0$, $\varepsilon > 0$, $p > 1$, $q > 1$ and $\dfrac{1}{p} + \dfrac{1}{q} = 1$. Then*

$$ab \leq \frac{\varepsilon a^p}{p} + \frac{\varepsilon^{-q/p} b^q}{q} \leq \varepsilon a^p + \varepsilon^{-q/p} b^q.$$

In particular, when $p = q = 2$, it becomes

$$ab \leq \frac{\varepsilon}{2} a^2 + \frac{1}{2\varepsilon} b^2,$$

which is called Cauchy's inequality with ε.

The following inequalities for functions in L^p are used frequently:

Hölder's inequality *Let $p > 1$, $q > 1$ and $\dfrac{1}{p} + \dfrac{1}{q} = 1$. If $f \in L^p(\Omega)$, $g \in L^q(\Omega)$, then $f \cdot g \in L^1(\Omega)$ and*

$$\int_\Omega |f(x)g(x)| dx \leq \|f(x)\|_{L^p(\Omega)} \|g(x)\|_{L^q(\Omega)}.$$

In particular, when $p = q = 2$, it becomes

$$\int_\Omega |f(x)g(x)| dx \leq \|f\|_{L^2(\Omega)} \|g\|_{L^2(\Omega)},$$

which is called Schwarz's inequality.

Minkowski's inequality *Let $1 \leq p < +\infty$, $f, g \in L^p(\Omega)$. Then $f + g \in L^p(\Omega)$ and*

$$\|f + g\|_{L^p(\Omega)} \leq \|f\|_{L^p(\Omega)} + \|g\|_{L^p(\Omega)}.$$

Here and below, throughout this chapter, Ω is always assumed to be a domain of $\mathbb{R}^n$, unless stated otherwise, although many propositions presented are valid when Ω is merely an open set or even a measurable set.

1.1.2 *Spaces $C^k(\Omega)$ and $C_0^k(\Omega)$*

Let k be a nonnegative number or ∞.

Definition 1.1.1 $C^k(\Omega)$ and $C^k(\overline{\Omega})$ denote sets of all functions having continuous derivatives up to order k on Ω and $\overline{\Omega}$ respectively. Usually, we simply denote $C^0(\Omega)$ and $C^0(\overline{\Omega})$ by $C(\Omega)$ and $C(\overline{\Omega})$ respectively. Define the norm on $C^k(\overline{\Omega})$ as follows

$$|u|_{k;\Omega} = \sum_{|\alpha| \leq k} \sup_{\Omega} |D^\alpha u|,$$

where $\alpha = (\alpha_1, \cdots, \alpha_n)$ is called a multi-index, $\alpha_1, \cdots, \alpha_n$ are nonnegative integers, $|\alpha| = \alpha_1 + \cdots + \alpha_n$ and

$$D^\alpha u = \frac{\partial^{|\alpha|} u}{\partial x_1^{\alpha_1} \cdots \partial x_n^{\alpha_n}}.$$

It is easy to verify that endowed with the norm defined above, $C^k(\overline{\Omega})$ is a Banach space (see [Chen and Wu (1997)], [Adams (1975)]).

Definition 1.1.2   For a function $u(x)$ on Ω, we define

$$\text{supp} u = \overline{\{x \in \Omega; u(x) \neq 0\}}$$

and call it the support of $u(x)$.

Definition 1.1.3 $C_0^k(\Omega)$ denotes the set of all functions in $C^k(\Omega)$ whose supports are compact in Ω. Usually we simply denote $C_0^0(\Omega)$ by $C_0(\Omega)$.

1.1.3   *Smoothing operators*

Approximating a given function by smooth functions is a basic technique used frequently in the study of partial differential equations. There have been a variety of ways to this purpose, among them is the following method of mollification.

Let $j(x) \in C_0^\infty(\mathbb{R}^n)$ be a nonnegative function, vanishing outside the unit ball $B_1(0) = \{x \in \mathbb{R}^n; |x| < 1\}$ and satisfying $\int_{\mathbb{R}^n} j(x) dx = 1$. A typical example is

$$j(x) = \begin{cases} \dfrac{1}{A} e^{1/(|x|^2 - 1)}, & |x| < 1, \\ 0, & |x| \geq 1, \end{cases}$$

where

$$A = \int_{B_1(0)} e^{1/(|x|^2 - 1)} dx.$$

Obviously for any $\varepsilon > 0$, the function

$$j_\varepsilon(x) = \frac{1}{\varepsilon^n} j\left(\frac{x}{\varepsilon}\right),$$

vanishes outside the ball $B_\varepsilon(0) = \{x \in \mathbb{R}^n; |x| < \varepsilon\}$ and $\int_{\mathbb{R}^n} j_\varepsilon(x)dx = 1$.

Definition 1.1.4   For a function $u \in L^1_{\text{loc}}(\mathbb{R}^n)$, the operator J_ε defined by

$$J_\varepsilon u(x) = (j_\varepsilon \star u)(x) \equiv \int_{\mathbb{R}^n} j_\varepsilon(x - y)u(y)dy$$

is called a smoothing operator, $J_\varepsilon u(x)$ the mollification of u, and $j_\varepsilon(x)$ the mollifier or kernel of radius ε of the operator J_ε.

Here and below, for any open set $\Omega \subset \mathbb{R}^n$, we denote by $L^1_{\text{loc}}(\Omega)$ the set of all locally integrable functions in Ω.

Proposition 1.1.1   *Let u be a function defined on $\mathbb{R}^n$, vanishing outside a bounded domain Ω.*

i) If $u \in L^1(\Omega)$, then $J_\varepsilon u \in C^\infty(\mathbb{R}^n)$.

ii) If $\text{supp}\,u \subset \Omega$ and $\text{dist}(\text{supp}\,u, \partial\Omega) > \varepsilon$, then $J_\varepsilon u \in C_0^\infty(\Omega)$.

iii) If $u \in L^p(\Omega)(1 \le p < +\infty)$, then $J_\varepsilon u \in L^p(\Omega)$ and

$$\|J_\varepsilon u\|_{L^p(\Omega)} \le \|u\|_{L^p(\Omega)}, \quad \lim_{\varepsilon \to 0^+} \|J_\varepsilon u - u\|_{L^p(\Omega)} = 0.$$

iv) If $u \in C(\Omega)$, $G \subset \overline{G} \subset \Omega$, then

$$\lim_{\varepsilon \to 0^+} J_\varepsilon u(x) = u(x)$$

uniformly on G.

v) If $u \in C(\overline{\Omega})$, then

$$\lim_{\varepsilon \to 0^+} J_\varepsilon u(x) = u(x)$$

uniformly on Ω.

Corollary 1.1.1 $C_0^\infty(\Omega)$ *is dense in* $L^p(\Omega)(p \ge 1)$.

For the proof of Proposition 1.1.1 and Corollary 1.1.1, we refer to [Adams (1975)] Chapter 2.

From the definition of smoothing operators, we see that the value of the mollification of a function at a point depends on the value of the function in the ε-neighborhood of this point. So in approximating a given function at the points near the boundary, the method of mollification stated above is not available. In this case, we may mollify the function after supplementing its definition, say, letting it equal zero outside Ω, and use some modified mollifiers. As an example, we consider the mollification of a function near the upper boundary $\{x \in \mathbb{R}^n; x_n = 1, |x_i| < 1, i = 1, \cdots, n-1\}$ and the lower boundary $\{x \in \mathbb{R}^n; x_n = -1, |x_i| < 1, i = 1, \cdots, n-1\}$ of the domain $Q = \{x \in \mathbb{R}^n; |x_i| < 1, i = 1, 2, \cdots, n\}$.

Definition 1.1.5 For $u \in L^1(Q)$, define

$$J_\varepsilon^- u(x) = \int_Q j_\varepsilon(x_1 - y_1) \cdots j_\varepsilon(x_{n-1} - y_{n-1}) j_\varepsilon(x_n - y_n - 2\varepsilon) u(y) dy,$$

$$J_\varepsilon^+ u(x) = \int_Q j_\varepsilon(x_1 - y_1) \cdots j_\varepsilon(x_{n-1} - y_{n-1}) j_\varepsilon(x_n - y_n + 2\varepsilon) u(y) dy,$$

where $j_\varepsilon(\tau)$ is an one dimensional mollifier.

It is easy to verify that $J_\varepsilon^- u(x)$ is well defined on the upper boundary of Q, and so is $J_\varepsilon^+ u(x)$ on the lower boundary of Q.

1.1.4 *Cut-off functions*

Let $\Omega \subset \mathbb{R}^n$ be a bounded domain with suitably smooth boundary, $\Omega' \subset\subset \Omega$ (i.e. Ω' is a subdomain of Ω such that $\overline{\Omega}' \subset \Omega$) and $d = \frac{1}{4}\text{dist}(\Omega', \partial\Omega)$. Then $d > 0$. Set $\Omega'' = \{x \in \Omega; \text{dist}(x, \Omega') < d\}$. Then $\text{dist}(\Omega'', \partial\Omega) = 3d$. Let $\eta(x) = J_d(\chi_{\Omega''}(x))$ be the mollification of the characteristic function $\chi_{\Omega''}(x)$ of Ω'', where d is the radius of the mollifier. It is easy to verify that $\eta(x)$ possesses the following properties:

$$\eta \in C_0^\infty(\Omega), \quad 0 \leq \eta(x) \leq 1, \quad \eta(x) \equiv 1 \text{ in } \Omega', \quad |\nabla\eta(x)| \leq \frac{C}{d},$$

where C is a constant depending only on Ω. The value of $\eta(x)$ outside Ω will be always regarded as zero, unless stated otherwise. Functions having the above properties like η will be called cut-off functions on Ω relative to Ω'.

In later applications, we frequently use the cut-off functions on the ball $B_R(x^0) = \{x \in \mathbb{R}^n; |x - x^0| < R\}$. Let $0 < \rho < R$ and $\eta(x)$ be a cut-off function on $B_R(x^0)$ relative to $B_\rho(x^0)$ defined in the above manner. Then it is easy to verify that $\eta(x)$ satisfies

$$|\nabla\eta(x)| \leq \frac{C}{R - \rho},$$

and

$$|D^k\eta(x)| \leq \frac{C}{|R - \rho|^k}, \quad [D^k\eta]_\alpha \leq \frac{C}{|R - \rho|^{k+\alpha}}, \quad k = 1, 2, \cdots, \alpha \in (0, 1),$$

where C is a universal constant independent of R, ρ, k and α. For the definition of $[D^k \eta]_\alpha$, see §1.2.

In studying the properties such as regularity of solutions, we always confine ourselves to the consideration in a small neighborhood for the moment. An important measure to localize the problem is the usage of the cut-off functions. In this way, all local properties of the given function are retained and no influence outside the small neighborhood has to be considered.

1.1.5 *Partition of unity*

As observed above, we can localize the problem by using cut-off functions. In the study of partial differential equations, we also need frequently to integrate the result obtained by localization to deduce a global one. To this end, we need another measure, called, partition of unity. The following is a basic theorem on the partition of unity, for the proof, we refer to [Adams (1975)] Chapter 2 or [Cui, Jin and Lu (1991)] Chapter 1.

Theorem 1.1.1 *Let $K \subset \mathbb{R}^n$ be a compact subset, $U_1, \cdots, U_N$ be an open covering of K. Then there exist functions $\eta_1 \in C_0^\infty(U_1), \cdots, \eta_N \in C_0^\infty(U_N)$, such that*

i) $0 \leq \eta_i(x) \leq 1, \quad \forall x \in U_i \quad (i = 1, \cdots, N);$

ii) $\displaystyle\sum_{i=1}^N \eta_i(x) = 1, \quad \forall x \in K.$

We call $\eta_1, \cdots, \eta_N$ a partition of unity associated to $U_1, \cdots, U_N$.

1.1.6 *Local flatting of the boundary*

In studying boundary value problems, we have to talk about the smoothness of the boundary. Usually smoothness of the boundary is defined by means of flatting the boundary locally.

Definition 1.1.6 Let $\Omega \subset \mathbb{R}^n$ be a bounded domain. The boundary $\partial\Omega$ of Ω is said to have C^k smoothness, denoted by $\partial\Omega \in C^k$, if for any $x^0 \in \partial\Omega$, there exists a neighborhood U of x^0 and an invertible C^k mapping $\Psi : U \to B_1(0)$, such that

$$\Psi(U \cap \Omega) = B_1^+(0) = \{y \in B_1(0); y_n > 0\},$$
$$\Psi(U \cap \partial\Omega) = \partial B_1^+(0) \cap \{y \in \mathbb{R}^n; y_n = 0\}.$$

As we will see later, to discuss the properties of a function near the boundary, we usually flat the boundary locally in this way to transform the original problem locally into a problem on a domain with a superplane as its lower boundary.

1.2 Hölder Spaces

1.2.1 Spaces $C^{k,\alpha}(\overline{\Omega})$ and $C^{k,\alpha}(\Omega)$

In this section we introduce a class of functions, called Hölder continuous functions, which can be regarded as functions differentiable of fraction order.

Definition 1.2.1 Let $u(x)$ be a function on $\Omega \subset \mathbb{R}^n$. For $0 < \alpha < 1$, define the semi-norm

$$[u]_{\alpha;\Omega} = \sup_{x,y\in\Omega, x\neq y} \frac{|u(x) - u(y)|}{|x - y|^\alpha}.$$

Denote the set of all functions u satisfying $[u]_{\alpha;\Omega} < +\infty$ by $C^\alpha(\overline{\Omega})$ and define on it the norm

$$|u|_{\alpha;\Omega} = |u|_{0;\Omega} + [u]_{\alpha;\Omega},$$

where $|u|_{0;\Omega}$ is the maximum norm of $u(x)$, namely,

$$|u|_{0;\Omega} = \sup_{x\in\Omega} |u(x)|.$$

Furthermore, we may define the function space

$$C^{k,\alpha}(\overline{\Omega}) = \{u; D^\beta u \in C^\alpha(\overline{\Omega}), \text{for any } \beta \text{ such that} |\beta| \leq k\}$$

for any nonnegative integer k and introduce the semi-norm

$$[u]_{k,\alpha;\Omega} = \sum_{|\beta|=k} [D^\beta u]_{\alpha;\Omega},$$

$$[u]_{k,0;\Omega} = [u]_{k;\Omega} = \sum_{|\beta|=k} |D^\beta u|_{0;\Omega}$$

and the norm

$$|u|_{k,\alpha;\Omega} = \sum_{|\beta|\leq k} |D^\beta u|_{\alpha;\Omega},$$

$$|u|_{k,0;\Omega} = |u|_{k;\Omega} = \sum_{|\beta|\leq k} |D^\beta u|_{0;\Omega}.$$

If for any domain $\Omega' \subset\subset \Omega$, $u \in C^{k,\alpha}(\overline{\Omega}')$, then we say that $u \in C^{k,\alpha}(\Omega)$. We always omit the notation Ω in the subscripts of the Hölder semi-norm and norm, if no confusion will be caused.

It is easy to verify that $C^{k,\alpha}(\overline{\Omega})$ is a Banach space.

If $\alpha = 1$, then we obtain the Lipschitz space.

From the definition of the Hölder semi-norm and norm, it follows immediately

Proposition 1.2.1   *Let* $u, v \in C^\alpha(\overline{\Omega})$. *Then*

i) $[uv]_{\alpha;\Omega} \leq |u|_{0;\Omega}[v]_{\alpha;\Omega} + [u]_{\alpha;\Omega}|v|_{0;\Omega}$;

ii) $|uv|_{\alpha;\Omega} \leq |u|_{\alpha;\Omega}|v|_{\alpha;\Omega}$.

1.2.2   *Interpolation inequalities*

The most important property of Hölder spaces is the interpolation inequalities which enable us to concentrate on the key point and thus can be used to simplify the proof.

Theorem 1.2.1   *Let* B_ρ *be a ball of radius* ρ *in* $\mathbb{R}^n$ *and* $u \in C^{1,\alpha}(\overline{B}_\rho)$. *Then for any* $0 < \sigma \leq \rho$,

$$[u]_{1;B_\rho} \leq \sigma^\alpha [u]_{1,\alpha;B_\rho} + \frac{C(n)}{\sigma}|u|_{0;B_\rho}, \tag{1.2.1}$$

$$[u]_{\alpha;B_\rho} \leq \sigma[u]_{1,\alpha;B_\rho} + \frac{C(n)}{\sigma^\alpha}|u|_{0;B_\rho}, \tag{1.2.2}$$

where $C(n)$ *is a positive constant depending only on* n.

Proof.   For any $x \in B_\rho$, choose $x^0 \in B_\rho$, such that $x \in B_{\sigma/2}(x^0) \subset B_\rho$. Integrating $D_i u$ over $B_{\sigma/2}(x^0)$ and using Green's formula yield

$$\int_{B_{\sigma/2}} D_i u\, dx = \int_{\partial B_{\sigma/2}} u \cos(\vec{\nu}, x_i)\, ds,$$

where $\vec{\nu}$ denotes the unit normal vector outward to $\partial B_{\sigma/2}$. By the mean value theorem, for some $\bar{x} \in B_{\sigma/2}$,

$$D_i u(\bar{x}) |B_{\sigma/2}| = \int_{B_{\sigma/2}} D_i u \, dx,$$

and hence

$$|D_i u(\bar{x})| = \frac{1}{|B_{\sigma/2}|} \left| \int_{\partial B_{\sigma/2}} u \cos(\vec{\nu}, x_i) ds \right| \leq \frac{|\partial B_{\sigma/2}|}{|B_{\sigma/2}|} |u|_{0;B_\rho} = \frac{2n}{\sigma} |u|_{0;B_\rho}.$$

If $\bar{x} \neq x$, then

$$|D_i u(x)| \leq |D_i u(x) - D_i u(\bar{x})| + |D_i u(\bar{x})|$$
$$\leq \frac{|D_i u(x) - D_i u(\bar{x})|}{|x - \bar{x}|^\alpha} |x - \bar{x}|^\alpha + \frac{2n}{\sigma} |u|_{0;B_\rho}$$
$$\leq \sigma^\alpha [D_i u]_{\alpha;B_\rho} + \frac{2n}{\sigma} |u|_{0;B_\rho}$$

and (1.2.1) is proved.

Since for any $y \neq x$,

$$\frac{|u(x) - u(y)|}{|x - y|^\alpha} = \frac{|u(x) - u(y)|}{|x - y|} |x - y|^{1-\alpha} < \sigma^{1-\alpha} [u]_{1;B_\rho}$$

when $|x - y| \leq \sigma$ and

$$\frac{|u(x) - u(y)|}{|x - y|^\alpha} \leq \frac{2}{\sigma^\alpha} |u|_{0;B_\rho}$$

when $|x - y| \geq \sigma$, we have, in either case,

$$\frac{|u(x) - u(y)|}{|x - y|^\alpha} \leq \sigma^{1-\alpha} [u]_{1;B_\rho} + \frac{2}{\sigma^\alpha} |u|_{0;B_\rho}.$$

This combined with (1.2.1) leads to (1.2.2). $\square$

Remark 1.2.1 *Special interpolation inequalities can be derived by choosing special values of σ. For example, choosing $\sigma = \varepsilon^{1/\alpha} \rho$ in (1.2.1) gives*

$$[u]_{1;B_\rho} \leq \varepsilon \rho^\alpha [u]_{1,\alpha;B_\rho} + \frac{C(n)}{\varepsilon^{1/\alpha}} \rho^{-1} |u|_{0;B_\rho}.$$

Corollary 1.2.1 *Let B_ρ be a ball of radius ρ in $\mathbb{R}^n$ and $u \in C^{2,\alpha}(\overline{B}_\rho)$. Then for any $0 < \sigma \leq \rho$,*

$$\sigma^\alpha [u]_{\alpha;B_\rho} + \sigma [u]_{1;B_\rho} + \sigma^{1+\alpha} [u]_{1,\alpha;B_\rho} + \sigma^2 [u]_{2;B_\rho}$$

$$\leq \sigma^{2+\alpha}[u]_{2,\alpha;B_\rho} + C(n)|u|_{0;B_\rho}. \tag{1.2.3}$$

Proof.   (1.2.2) implies that for $0 < \sigma \leq \rho$,

$$[u]_{1,\alpha;B_\rho} \leq \sigma[u]_{2,\alpha;B_\rho} + \frac{C(n)}{\sigma^\alpha}[u]_{1;B_\rho}. \tag{1.2.4}$$

Denote $\sigma = \sigma_1$ in (1.2.4) and $\sigma = \sigma_2$ in (1.2.1). Then for $0 < \sigma_1, \sigma_2 \leq \rho$,

$$[u]_{1,\alpha;B_\rho} \leq \sigma_1[u]_{2,\alpha;B_\rho} + \frac{C(n)}{\sigma_1^\alpha}[u]_{1;B_\rho},$$

$$[u]_{1;B_\rho} \leq \sigma_2^\alpha[u]_{1,\alpha;B_\rho} + \frac{C(n)}{\sigma_2}|u|_{0;B_\rho}.$$

Hence

$$\left(1 - C(n)\left(\frac{\sigma_2}{\sigma_1}\right)^\alpha\right)[u]_{1,\alpha;B_\rho} \leq \sigma_1[u]_{2,\alpha;B_\rho} + \frac{C^2(n)}{\sigma_1^\alpha \sigma_2}|u|_{0;B_\rho}.$$

Letting $\sigma_1 = \lambda_1\sigma$, $\sigma_2 = \lambda_2\sigma$ gives

$$\left(1 - C(n)\left(\frac{\lambda_2}{\lambda_1}\right)^\alpha\right)[u]_{1,\alpha;B_\rho} \leq \lambda_1\sigma[u]_{2,\alpha;B_\rho} + \frac{C^2(n)}{\lambda_1^\alpha \lambda_2 \sigma^{\alpha+1}}|u|_{0;B_\rho}.$$

Now we choose $\lambda_1, \lambda_2 \in (0,1)$ such that $1 - C(n)\left(\frac{\lambda_2}{\lambda_1}\right)^\alpha = \lambda_1$. Then

$$\sigma^{1+\alpha}[u]_{1,\alpha;B_\rho} \leq \sigma^{2+\alpha}[u]_{2,\alpha;B_\rho} + C(n)|u|_{0;B_\rho} \tag{1.2.5}$$

with another constant $C(n)$. Since $\lambda_1, \lambda_2 \in (0,1)$, (1.2.5) holds for $0 < \sigma \leq \rho$.

Combining (1.2.5) with (1.2.1) derives

$$\sigma[u]_{1;B_\rho} \leq \sigma^{2+\alpha}[u]_{2,\alpha;B_\rho} + C(n)|u|_{0;B_\rho} \tag{1.2.6}$$

and combining (1.2.5) with (1.2.2) derives

$$\sigma^\alpha[u]_{\alpha;B_\rho} \leq \sigma^{2+\alpha}[u]_{2,\alpha;B_\rho} + C(n)|u|_{0;B_\rho}. \tag{1.2.7}$$

Furthermore, using (1.2.1) and (1.2.6), we obtain

$$\sigma^2[u]_{2;B_\rho} \leq \sigma^{2+\alpha}[u]_{2,\alpha;B_\rho} + C(n)\sigma[u]_{1;B_\rho}$$
$$\leq (1 + C(n))\sigma^{2+\alpha}[u]_{2,\alpha;B_\rho} + C^2(n)|u|_{0;B_\rho}.$$

With $\lambda\sigma$ in place of σ, we are led to

$$\sigma^2[u]_{2;B_\rho} \leq (1 + C(n))\lambda^\alpha \sigma^{2+\alpha}[u]_{2,\alpha;B_\rho} + \left(\frac{C(n)}{\lambda}\right)^2|u|_{0;B_\rho}$$

and hence, by choosing $\lambda \in (0,1)$ such that $(1 + C(n))\lambda^\alpha = 1$, we have, for $0 < \sigma \le \rho$,

$$\sigma^2 [u]_{2;B_\rho} \le \sigma^{2+\alpha} [u]_{2,\alpha;B_\rho} + C(n)|u|_{0;B_\rho} \tag{1.2.8}$$

with another constant $C(n)$.

The conclusion (1.2.3) is just a combination of (1.2.5)–(1.2.8). $\qquad\square$

Theorem 1.2.2 *Let Ω be a bounded domain in $\mathbb{R}^n$ with $\partial\Omega \in C^1$ and $u \in C^{1,\alpha}(\overline{\Omega})$. Then there exists a constant $\rho > 0$, such that for $0 < \sigma \le \rho$,*

$$[u]_{1;\Omega} \le \sigma^\alpha [u]_{1,\alpha;\Omega} + \frac{C(n)}{\sigma}|u|_{0;\Omega}, \tag{1.2.9}$$

$$[u]_{\alpha;\Omega} \le \sigma [u]_{1,\alpha;\Omega} + \frac{C(n)}{\sigma^\alpha}|u|_{0;\Omega}. \tag{1.2.10}$$

Proof. Since $\partial\Omega \in C^1$, by the finite covering theorem, there exist finite number of points $x^j \in \partial\Omega (j = 1, \cdots, N)$ and neighborhoods U_j of x^j, such that $\partial\Omega \subset \bigcup_{j=1}^{N} U_j$ and $U_j \cap \partial\Omega$ can be flatted, namely, there exist invertible C^1 mappings $\Psi_j : U_j \to B_1(0)$ satisfying

$$\Psi_j(U_j \cap \Omega) = B_1^+(0) = \{y \in B_1(0); y_n > 0\},$$
$$\Psi_j(U_j \cap \partial\Omega) = \partial B_1^+(0) \cap \{y \in \mathbb{R}^n; y_n = 0\} \quad (j = 1, \cdots, N).$$

Denote $U_0 = \Omega \setminus \bigcup_{j=1}^{N} U_j$, $d = \text{dist}(U_0, \partial\Omega)$. Suppose $x \in U_0$. Then for any $0 < \sigma \le d$, $B_\sigma(x) \subset \Omega$. Similar to the proof of Theorem 1.2.1, from

$$\int_{B_\sigma(x)} D_i u\, dx = \int_{\partial B_\sigma(x)} u \cos(\vec{\nu}, x_i)\, ds \quad (i = 1, 2, \cdots, n)$$

and the mean value theorem, we may find some $\bar{x} \in B_\sigma(x)$, such that

$$|D_i u(\bar{x})| \le \frac{2n}{\sigma}|u|_{0;\Omega} \quad (i = 1, 2, \cdots, n)$$

and hence

$$|D_i u(x)| \le \sigma^\alpha [D_i u]_{\alpha;\Omega} + \frac{2n}{\sigma}|u|_{0;\Omega}, \quad x \in U_0 \quad (i = 1, 2, \cdots, n). \tag{1.2.11}$$

If $x \notin U_0$, then for some $j = j_0$, $x \in U_{j_0} \cap \Omega$. As indicated above, the mapping $\Psi_{j_0} : U_{j_0} \to B_1(0)$ flats $U_{j_0} \cap \Omega$, namely,

$$\Psi_{j_0}(U_{j_0} \cap \Omega) = B_1^+(0) = \{y \in B_1(0); y_n > 0\},$$
$$\Psi_{j_0}(U_{j_0} \cap \partial\Omega) = \partial B_1^+(0) \cap \{y \in \mathbb{R}^n; y_n = 0\}.$$

$x \in U_{j_0} \cap \Omega$ implies $y = \Psi_{j_0}(x) \in B_1^+(0)$. It is easy to see that for any $0 < \sigma \le 1$, we may choose $y^0 \in B_1^+(0)$, such that $y \in K = \{B_\sigma(y^0); y_n > 0\} \subset B_1^+(0)$. From

$$\int_{\Psi_{j_0}^{-1}(K)} D_i u \, dx = \int_{\partial\Psi_{j_0}^{-1}(K)} u \cos(\vec{\nu}, x_i) \, ds$$

and the mean value theorem, for some $\tilde{x} \in \Psi_{j_0}^{-1}(K)$,

$$|D_i u(\tilde{x})| \int_{\Psi_{j_0}^{-1}(K)} dx \le |u|_{0;\Omega} \int_{\partial\Psi_{j_0}^{-1}(K)} ds \quad (i = 1, 2, \cdots, n).$$

Since the Jacobi determinant $\dfrac{\partial(x_1, \cdots, x_n)}{\partial(y_1, \cdots, y_n)}$ is bounded and has a positive lower bound on K, noticing that $\{B_\sigma(y^0); y_n > y_n^0\} \subset K$ and $|\partial K| \le |\partial B_\sigma(y^0)|$, we have

$$\int_{\Psi_{j_0}^{-1}(K)} dx = \int_K \frac{\partial(x_1, \cdots, x_n)}{\partial(y_1, \cdots, y_n)} dy \ge c_1 \int_K dy \ge c_1 |\{B_\sigma(y^0); y_n > y_n^0\}|,$$

$$\int_{\partial\Psi_{j_0}^{-1}(K)} ds \le c_2 \int_{\partial K} ds = c_2 |\partial K| \le c_2 |\partial B_\sigma(y^0)|,$$

where $c_1 > 0$ and $c_2 > 0$ are constants depending on Ψ_{j_0}. Since the covering is finite, we may choose $c_1 > 0$, $c_2 > 0$ to be universal ones. Combining the above inequalities leads to

$$|D_i u(\tilde{x})| \le \frac{c_2 |\partial B_\sigma(y^0)|}{c_1 |\{B_\sigma(y^0); y_n > y_n^0\}|} = \frac{C(n)}{\sigma} |u|_{0;\Omega} \quad (i = 1, 2, \cdots, n).$$

Hence, if $\tilde{x} \ne x$, then for $0 < \sigma \le 1$,

$$|D_i u(x)| \le \frac{|D_i u(x) - D_i u(\tilde{x})|}{|x - \tilde{x}|^\alpha} \cdot \left| \frac{\Psi_{j_0}^{-1}(y) - \Psi_{j_0}^{-1}(\tilde{y})}{y - \tilde{y}} \right|^\alpha |y - \tilde{y}|^\alpha + |D_i u(\tilde{x})|$$

$$\le c_3 \sigma^\alpha [D_i u]_{\alpha;\Omega} + \frac{C(n)}{\sigma} |u|_{0;\Omega} \quad (i = 1, 2, \cdots, n),$$

where $c_3 > 0$ is a constant independent of σ and j_0, $y = \Psi(x)$ and $\tilde{y} = \Psi(\tilde{x})$. This implies that for $0 < \sigma \le c_3^{1/\alpha}$,

$$|D_i u(x)| \le \sigma^\alpha [D_i u]_{\alpha;\Omega} + \frac{C(n)}{\sigma} |u|_{0;\Omega}, \quad x \notin U_0 \quad (i = 1, 2, \cdots, n) \quad (1.2.12)$$

with another constant $C(n)$.

Let $\rho = \min\{d, c_3^{1/\alpha}\}$. Then (1.2.9) follows from (1.2.11), (1.2.12). We may argue as Theorem 1.2.1 to further prove (1.2.10). □

Corollary 1.2.2 *Let Ω be a bounded domain in $\mathbb{R}^n$ with $\partial\Omega \in C^1$ and $u \in C^{2,\alpha}(\overline{\Omega})$. Then there exists a constant $\rho > 0$, such that for $0 < \sigma \le \rho$,*

$$\sigma^\alpha [u]_{\alpha;\Omega} + \sigma[u]_{1;\Omega} + \sigma^{1+\alpha}[u]_{1,\alpha;\Omega} + \sigma^2 [u]_{2;\Omega}$$
$$\le \sigma^{2+\alpha}[u]_{2,\alpha;\Omega} + C(n)|u|_{0;\Omega}.$$

1.2.3 Spaces $C^{2k+\alpha, k+\alpha/2}(\overline{Q}_T)$

Denote $Q_T = \Omega \times (0, T)$. For any points $P(x, t)$, $Q(y, s) \in Q_T$, define the parabolic distance between them as

$$d(P, Q) = \left(|x - y|^2 + |t - s|\right)^{1/2}.$$

Definition 1.2.2 *Let $u(x, t)$ be a function on Q_T. For $0 < \alpha < 1$, define*

$$[u]_{\alpha, \alpha/2; Q_T} = \sup_{P, Q \in Q_T, P \ne Q} \frac{|u(P) - u(Q)|}{d^\alpha(P, Q)},$$

which is a semi-norm, and denote by $C^{\alpha, \alpha/2}(\overline{Q}_T)$ the set of all functions on Q_T such that $[u]_{\alpha, \alpha/2; Q_T} < +\infty$, endowed with the norm

$$|u|_{\alpha, \alpha/2; Q_T} = |u|_{0; Q_T} + [u]_{\alpha, \alpha/2; Q_T},$$

where $|u|_{0; Q_T}$ is the maximum norm of $u(x, t)$ on Q_T, i.e.

$$|u|_{0; Q_T} = \sup_{(x, t) \in Q_T} |u(x, t)|.$$

Furthermore, for any nonnegative integer k, denote

$$C^{2k+\alpha, k+\alpha/2}(\overline{Q}_T) = \Big\{ u; D^\beta D_t^r u \in C^{\alpha, \alpha/2}(\overline{Q}_T),$$

$$\text{for any } \beta, r \text{ such that } |\beta| + 2r \le 2k \Big\},$$

and define the semi-norm

$$[u]_{2k+\alpha,k+\alpha/2;Q_T} = \sum_{|\beta|+2r=2k} [D^\beta D_t^r u]_{\alpha,\alpha/2;Q_T},$$

$$[u]_{2k,k;Q_T} = \sum_{|\beta|+2r=2k} |D^\beta D_t^r u|_{0;Q_T}$$

and the norm

$$|u|_{2k+\alpha,k+\alpha/2;Q_T} = \sum_{|\beta|+2r\leq 2k} |D^\beta D_t^r u|_{\alpha,\alpha/2;Q_T},$$

$$|u|_{2k,k;Q_T} = \sum_{|\beta|+2r\leq 2k} |D^\beta D_t^r u|_{0;Q_T}.$$

It is not difficult to prove that $C^{2k+\alpha,k+\alpha/2}(\overline{Q}_T)$ is a Banach space. Clearly

$$|u|_{2,1;Q_T} = |u|_{0;Q_T} + |Du|_{0;Q_T} + |D^2 u|_{0;Q_T} + |u_t|_{0;Q_T},$$

$$|u|_{2+\alpha,1+\alpha/2;Q_T} = |u|_{\alpha,\alpha/2;Q_T} + |Du|_{\alpha,\alpha/2;Q_T} + |D^2 u|_{\alpha,\alpha/2;Q_T} + |u_t|_{\alpha,\alpha/2;Q_T},$$

where $|D^2 u|_{0;Q_T}$, $|D^2 u|_{\alpha,\alpha/2;Q_T}$ denote the sums of the corresponded norms of the second order derivatives of u with respect to x. Usually we omit Q_T in the subscripts of the semi-norm and norm if no confusion is caused.

1.3 Isotropic Sobolev Spaces

Now we introduce another kinds of function spaces, i.e. Sobolev spaces which is also of great use in the theory of partial differential equations.

1.3.1 *Weak derivatives*

Definition 1.3.1   Let $u \in L^1_{loc}(\Omega)$, $1 \leq i \leq n$. If there exists $g_i \in L^1_{loc}(\Omega)$, such that

$$\int_\Omega g_i \varphi dx = -\int_\Omega u \frac{\partial \varphi}{\partial x_i} dx, \quad \forall \varphi \in C_0^\infty(\Omega),$$

then g_i is called the weak derivative of u with respect to the variable x_i, denoted by

$$\frac{\partial u}{\partial x_i} = g_i,$$

or $D_i u = g_i$. If for any $i = 1, 2, \cdots, n$, u has weak derivative g_i with respect to x_i, then we call $\vec{g} = (g_1, \cdots, g_n)$ the weak gradient of u, denoted by $\nabla u = \vec{g}$ or $Du = \vec{g}$, and u is said to be weakly differentiable, denoted by $u \in W^1(\Omega)$. Similarly we may define weak derivatives and weak differentiability of higher order. If u is weakly differentiable up to k order, then we denote $u \in W^k(\Omega)$.

1.3.2 Sobolev spaces $W^{k,p}(\Omega)$ and $W_0^{k,p}(\Omega)$

Definition 1.3.2 Let k be a nonnegative integer, $p \geq 1$. The family of functions

$$\left\{ u \in W^k(\Omega); D^\alpha u \in L^p(\Omega), \text{ for any } \alpha \text{ with } |\alpha| \leq k \right\}$$

endowed with the norm

$$\|u\|_{W^{k,p}(\Omega)} = \left(\int_\Omega \sum_{|\alpha| \leq k} |D^\alpha u|^p dx \right)^{1/p} \tag{1.3.1}$$

is called a Sobolev space, denoted by $W^{k,p}(\Omega)$.

It can be proved easily that for $p > 1$, $W^{k,p}(\Omega)$ is a Banach space. We always denote $W^{k,2}(\Omega)$ by $H^k(\Omega)$, which is a Hilbert space with the inner product

$$(u, v)_{H^k(\Omega)} = \int_\Omega \sum_{|\alpha| \leq k} D^\alpha u \cdot D^\alpha v dx, \quad u, v \in H^k(\Omega).$$

Definition 1.3.3 $W_0^{k,p}(\Omega)$ denotes the closure of $C_0^\infty(\Omega)$ in $W^{k,p}(\Omega)$.

Proposition 1.3.1 $W^{k,p}(\mathbb{R}^n) = W_0^{k,p}(\mathbb{R}^n)$, $W^{0,p}(\Omega) = W_0^{0,p}(\Omega) = L^p(\Omega)$. However, for the bounded domain Ω and $k \geq 1$, $W_0^{k,p}(\Omega)$ is a proper subset of $W^{k,p}(\Omega)$.

Proposition 1.3.2 $C^\infty(\Omega) \cap W^{k,p}(\Omega)$ is dense in $W^{k,p}(\Omega)$.

This proposition means that $W^{k,p}(\Omega)$ is the completion of $C^\infty(\Omega)$ with the norm (1.3.1).

It is to be noted that, in general, we can not replace $C^\infty(\Omega)$ by $C^\infty(\overline{\Omega})$ in Proposition 1.3.2. However for a large number of domains Ω including those having Lipschitz continuous boundaries, it is certainly the case.

Definition 1.3.4 A domain Ω is said to have the property of segment, if there exist a finite open cover $\{U_i\}$ of $\partial\Omega$ and corresponding nonzero vectors $\{y^i\}$, such that for any $x \in \overline{\Omega} \cap U_i$ and $t \in (0,1)$, we have $x + ty^i \in \Omega$.

Proposition 1.3.3 *If the domain Ω has the property of segment, then $C^\infty(\overline{\Omega})$ is dense in $W^{k,p}(\Omega)$.*

For the proof of Proposition 1.3.2 and Proposition 1.3.3 see [Gilbarg and Trudinger (1977)] Chapter 7 and [Adams (1975)] Chapter 3.

Proposition 1.3.4 *If $1 < p < +\infty$, then a subset of $L^p(\Omega)$ is (relatively) weakly compact (i.e. each sequence in it contains a weakly convergent subsequence) if and only if the norm of the sequence is bounded.*

Proposition 1.3.5 *If $1 \le p < +\infty$, then a subset X of $L^p(\Omega)$ is (relatively) strongly compact (i.e. each sequence in it contains a strongly convergent subsequence) if and only if:*
i) $\{\|f\|_{L^p(\Omega)}; f \in X\}$ is bounded;
ii) X is globally equicontinuous, i.e. there holds

$$\lim_{h \to 0} \int_\Omega |f(x+h) - f(x)|^p dx = 0$$

uniformly in $f \in X$;
iii) There holds

$$\lim_{R \to +\infty} \int_{\{x \in \Omega; |x| \ge R\}} |f(x)|^p dx = 0$$

uniformly in $f \in X$.

For the proof see [Adams (1975)] Chapter 2. We note that iii) is satisfied automatically if Ω is bounded.

As corollaries of Proposition 1.3.4 and Proposition 1.3.5, we have

Proposition 1.3.6 *For $1 < p < +\infty$, a subset of $W^{k,p}(\Omega)$ is (relatively) weakly compact if and only if it is bounded in $W^{k,p}(\Omega)$.*

Proposition 1.3.7 *Let $\Omega \subset \mathbb{R}^n$ be bounded and $1 \le p < +\infty$. If the subset X of $L^p(\Omega)$ is bounded in $W^{k+1,p}(\Omega)$, then X is (relatively) strongly compact in $W^{k,p}(\Omega)$.*

Here we merely present a sufficient condition for a subset of $W^{k+1,p}(\Omega)$ to be (relatively) strongly compact in the case of bounded domain Ω, which is enough for our later usage.

1.3.3 *Operation rules of weak derivatives*

Some operation rules in calculus can be extended to weak derivatives by the approximation theorem(Proposition 1.1.1).

Proposition 1.3.8   *Let $u, v \in H^1(\Omega)$. Then*

$$\frac{\partial(uv)}{\partial x_i} = u\frac{\partial v}{\partial x_i} + v\frac{\partial u}{\partial x_i}, \quad i = 1, \cdots, n.$$

Proposition 1.3.9   *Let Ω, D be domains of $\mathbb{R}^n$, $u(x) \in W^1(\Omega)$ and $\Phi = (\Phi_1, \cdots, \Phi_n) : D \to \Omega$ be a continuously differential mapping. Then*

$$\frac{\partial u(\Phi(y))}{\partial y_k} = \sum_{i=1}^{n} \frac{\partial \Phi_i}{\partial y_k} \cdot \frac{\partial u}{\partial x_i}, \quad k = 1, \cdots, n.$$

Proposition 1.3.10   *Let $f(s)$ be a continuous function on $\mathbb{R}$ with $f'(s)$ piecewise continuous and bounded. Then $u \in W^1(\Omega)$ implies $f(u) \in W^1(\Omega)$ and*

$$\frac{\partial f(u)}{\partial x_i} = \begin{cases} f'(u)\dfrac{\partial u}{\partial x_i}, & \text{if } u \notin L, \\ 0, & \text{if } u \in L, \end{cases}$$

where L is the set of jump points of $f'(s)$.

1.3.4 *Interpolation inequality*

Definition 1.3.5   A domain Ω is said to have the property of uniform inner cone, if there is a finite cone V, such that every point $x \in \Omega$ is the vertex of a finite cone $V_x \subset \Omega$ congruent with V.

Theorem 1.3.1   *(Ehrling–Nirenberg–Gagliardo's Interpolation Inequality) Let $\Omega \subset \mathbb{R}^n$ be a bounded domain having the property of uniform inner cone. Then for any $\varepsilon > 0$, there exists a constant $C > 0$ depending only on $p \geq 1$, k, ε and Ω, such that for any $u \in W^{k,p}(\Omega)$,*

$$\sum_{|\beta| \leq k-1} \int_\Omega |D^\beta u|^p dx \leq \varepsilon \sum_{|\alpha|=k} \int_\Omega |D^\alpha u|^p dx + C \int_\Omega |u|^p dx.$$

This inequality exposes such an important fact as that the L^p norm of any intermediate derivative of functions in $W^{k,p}(\Omega)$ can be estimated by the L^p norm of the function itself and its derivatives of the highest order.

We refer to [Adams (1975)] Chapter 4 for a detailed proof of this theorem. The basic idea can be shown by the following description for the special case $k = 2$, $n = 1$, $\Omega = (0,1)$, $u \in C^2[0,1]$.

Let $0 < \xi < \dfrac{1}{3}$, $\dfrac{2}{3} < \eta < 1$. By the mean value theorem, we have

$$|u'(\lambda)| = \left| \frac{u(\eta) - u(\xi)}{\eta - \xi} \right| \le 3|u(\xi)| + 3|u(\eta)|$$

for some constant $\lambda \in (\xi, \eta)$. Hence for any $x \in (0,1)$,

$$|u'(x)| = \left| u'(\lambda) + \int_\lambda^x u''(t)dt \right| \le 3|u(\xi)| + 3|u(\eta)| + \int_0^1 |u''(t)|dt.$$

Integrating this inequality with respect to ξ over $(0, 1/3)$ and with respect to η over $(2/3, 1)$ yields

$$\frac{1}{9}|u'(x)| \le \int_0^{1/3} |u(\xi)|d\xi + \int_{2/3}^1 |u(\eta)|d\eta + \frac{1}{9}\int_0^1 |u''(t)|dt$$

$$\le \int_0^1 |u(t)|dt + \frac{1}{9}\int_0^1 |u''(t)|dt.$$

Using Hölder's inequality we further obtain

$$|u'(x)|^p \le 2^{p-1} \cdot 9^p \int_0^1 |u(t)|^p dt + 2^{p-1} \int_0^1 |u''(t)|^p dt.$$

Thus

$$\int_0^1 |u'(t)|^p dt \le K_p \int_0^1 |u''(t)|^p dt + K_p \int_0^1 |u(t)|^p dt,$$

where $K_p = 2^{p-1} \cdot 9^p$. From this, by a change of variables, we are led to the following inequality for an interval (a,b)

$$\int_a^b |u'(t)|^p dt \le K_p(b-a)^p \int_a^b |u''(t)|^p dt$$

$$+ K_p(b-a)^{-p} \int_a^b |u(t)|^p dt. \qquad (1.3.2)$$

Let $\varepsilon \in (0,1)$. Choose a positive integer N, such that

$$\frac{1}{2}\left(\frac{\varepsilon}{K_p}\right)^{1/p} \leq \frac{1}{N} \leq \left(\frac{\varepsilon}{K_p}\right)^{1/p}.$$

Let $a_j = \dfrac{j}{N}(j = 0, 1, \cdots, N)$. Then $a_j - a_{j-1} = \dfrac{1}{N}$. Using (1.3.2) for (a_{j-1}, a_j) and summing on j from 1 to N, we arrive at

$$\int_0^1 |u'(t)|^p dt = \sum_{j=1}^N \int_{a_{j-1}}^{a_j} |u'(t)|^p dt$$

$$\leq K_p \sum_{j=1}^N \left\{ \frac{1}{N^p} \int_{a_{j-1}}^{a_j} |u''(t)|^p dt + N^p \int_{a_{j-1}}^{a_j} |u(t)|^p dt \right\}$$

$$\leq \varepsilon \int_0^1 |u''(t)|^p dt + \frac{2^p K_p^2}{\varepsilon} \int_0^1 |u(t)|^p dt.$$

1.3.5 *Embedding theorem*

Now we proceed to state the most important theorem in the theory of Sobolev spaces – embedding theorem, for its proof, the reader may refer to [Adams (1975)].

Theorem 1.3.2 *(Isotropic Embedding Theorem) Let $\Omega \subset \mathbb{R}^n$ be a bounded domain and $1 \leq p \leq +\infty$.*

i) If Ω has the property of uniform inner cone, then, when $p = n$,

$$W^{1,p}(\Omega) \subset L^q(\Omega), \quad 1 \leq q < +\infty,$$

and for any $u \in W^{1,p}(\Omega)$,

$$\|u\|_{L^q(\Omega)} \leq C(n, q, \Omega) \|u\|_{W^{1,p}(\Omega)}, \quad 1 \leq q < +\infty;$$

when $p < n$,

$$W^{1,p}(\Omega) \subset L^q(\Omega), \quad 1 \leq q \leq p^* = \frac{np}{n-p},$$

and for any $u \in W^{1,p}(\Omega)$,

$$\|u\|_{L^q(\Omega)} \leq C(n, p, \Omega) \|u\|_{W^{1,p}(\Omega)}, \quad 1 \leq q \leq p^*.$$

ii) If $\partial\Omega$ is appropriately smooth, then, when $p > n$,

$$W^{1,p}(\Omega) \subset C^\alpha(\overline{\Omega}), \quad 0 < \alpha \leq 1 - \frac{n}{p},$$

and for any $u \in W^{1,p}(\Omega)$,

$$|u|_{\alpha;\Omega} \le C(n,p,\Omega)\|u\|_{W^{1,p}(\Omega)}, \quad 0 < \alpha \le 1 - \frac{n}{p}.$$

Here p^* is the Sobolev conjugate exponent of p. The constant C in the above three inequalities is called embedding constant.

Remark 1.3.1   *The above embedding theorem can be expressed as*

$$W^{1,p}(\Omega) \hookrightarrow \begin{cases} L^q(\Omega), & 1 \le q \le p^* = \dfrac{np}{n-p}, & p < n, \\[2mm] L^q(\Omega), & 1 \le q < +\infty, & p = n, \\[2mm] C^\alpha(\overline{\Omega}), & 0 < \alpha \le 1 - \dfrac{n}{p}, & p > n. \end{cases}$$

Remark 1.3.2   *The embedding*

$$W^{1,p}(\Omega) \hookrightarrow C^\alpha(\overline{\Omega}),$$

means that one can change the value of any function of $W^{1,p}(\Omega)$ on a set of measure zero so that it becomes a function in $C^\alpha(\overline{\Omega})$.

Applying Theorem 1.3.2 for k times repeatedly leads to

Corollary 1.3.1

$$W^{k,p}(\Omega) \hookrightarrow \begin{cases} L^q(\Omega), & 1 \le q \le \dfrac{np}{n-kp}, & kp < n, \\[2mm] L^q(\Omega), & 1 \le q < +\infty, & kp = n, \\[2mm] C^\alpha(\overline{\Omega}), & 0 < \alpha \le 1 - \dfrac{n}{kp}, & kp > n. \end{cases}$$

Theorem 1.3.3   *(Compact Embedding Theorem) Let $\Omega \subset \mathbb{R}^n$ be a bounded domain and $1 \le p \le +\infty$.*

i) If Ω has the property of uniform inner cone, then the embedding

$$W^{1,p}(\Omega) \hookrightarrow L^q(\Omega)$$

with $1 \le q < p^*$, $p < n$ and the embedding

$$W^{1,p}(\Omega) \hookrightarrow L^q(\Omega)$$

with $1 \le q < +\infty$, $p = n$ are compact.

ii) If $\partial\Omega$ is appropriately smooth, then the embedding

$$W^{1,p}(\Omega) \hookrightarrow C^\alpha(\overline{\Omega})$$

with $p > n$, $0 < \alpha < 1 - \dfrac{n}{p}$ is compact.

Remark 1.3.3   *The same embedding relation and compact embedding relations presented in Theorem 1.3.2 and Theorem 1.3.3 respectively hold for the spaces $W_0^{1,p}(\Omega)$. In this case, no more conditions on the domain are needed.*

Remark 1.3.4   *An embedding is said to be compact, if any bounded sequence in the embedded space contains a subsequence which strongly converges in the embedding space, namely, the embedding operator is compact.*

1.3.6 *Poincaré's inequality*

Theorem 1.3.4   *Let $1 \le p < +\infty$ and $\Omega \subset \mathbb{R}^n$ be a bounded domain.*
   i) If $u \in W_0^{1,p}(\Omega)$, then

$$\int_\Omega |u|^p dx \le C \int_\Omega |Du|^p dx. \tag{1.3.3}$$

ii) If $\partial\Omega$ is locally Lipschitz continuous and $u \in W^{1,p}(\Omega)$, then

$$\int_\Omega |u - u_\Omega|^p dx \le C \int_\Omega |Du|^p dx \tag{1.3.4}$$

with a constant C depending only on n, p and Ω, where

$$u_\Omega = \frac{1}{|\Omega|} \int_\Omega u(x)dx$$

and $|\Omega|$ is the measure of Ω.

Proof.   We first prove (1.3.3). Since $C_0^\infty(\Omega)$ is dense in $W_0^{1,p}(\Omega)$ (see Definition 1.3.3), it suffices to prove (1.3.3) for $u \in C_0^\infty(\Omega)$.
   Choose a cube

$$Q = \{x \in \mathbb{R}^n; a_i < x_i < a_i + d, i = 1, 2, \cdots, n\},$$

to contain Ω where $d = \mathrm{diam}\Omega$ and define $u = 0$ outside Ω. Then $u \in C_0^\infty(Q)$ and for any $x \in Q$,

$$|u(x)|^p = \left| \int_{a_1}^{x_1} D_1 u(s, x_2, \cdots, x_n) ds \right|^p$$
$$\le \left(\int_{a_1}^{a_1+d} |D_1 u(s, x_2, \cdots, x_n)| ds \right)^p$$

$$\leq d^{p-1} \int_{a_1}^{a_1+d} |D_1 u(s, x_2, \cdots, x_n)|^p ds.$$

Integrating over Q leads to

$$\int_\Omega |u(x)|^p dx = \int_Q |u(x)|^p dx$$

$$\leq d^{p-1} \int_Q \int_{a_1}^{a_1+d} |D_1 u(s, x_2, \cdots, x_n)|^p ds dx$$

$$\leq d^p \int_Q |D_1 u(x_1, x_2, \cdots, x_n)|^p dx$$

$$\leq d^p \int_Q |Du|^p dx,$$

which is just (1.3.3) with $C = d^p$.

For simplicity, we merely prove (1.3.4) for $p > 1$; for the proof in the case $p = 1$, we refer to [Maz'ja (1985)].

Since (1.3.4) is unvarying with u replaced by u plus any constant, without loss of generality, we may assume that $u_\Omega = 0$. Suppose (1.3.4) failed, namely, for any positive integer $k \geq 1$, there would exists $u_k \in W^{1,p}(\Omega)$, such that $\int_\Omega u_k(x) dx = 0$, but

$$\int_\Omega |u_k|^p dx > k \int_\Omega |Du_k|^p dx.$$

Set

$$w_k(x) = \frac{u_k(x)}{\|u_k\|_{L^p(\Omega)}}, \quad x \in \Omega \quad (k = 1, 2, \cdots).$$

Then $w_k \in W^{1,p}(\Omega)$ satisfies

$$\int_\Omega w_k(x) dx = 0 \quad (k = 1, 2, \cdots), \tag{1.3.5}$$

$$\|w_k\|_{L^p(\Omega)} = 1 \quad (k = 1, 2, \cdots) \tag{1.3.6}$$

and

$$\int_\Omega |Dw_k|^p dx < \frac{1}{k} \quad (k = 1, 2, \cdots). \tag{1.3.7}$$

(1.3.6) and (1.3.7) imply the boundedness of $\|w_k\|_{W^{1,p}(\Omega)}$. Thus by the weak compactness of the bounded set in $W^{1,p}(\Omega)$ and the compact embedding theorem, we may assert the existence of a subsequence of $\{w_k\}$, assumed to be $\{w_k\}$ itself, and a function $w \in W^{1,p}(\Omega)$, such that

$$w_k \to w \quad (k \to \infty) \quad \text{in } L^p(Q), \tag{1.3.8}$$

$$Dw_k \rightharpoonup Dw \quad (k \to \infty) \quad \text{in } L^p(Q, \mathbb{R}^n). \tag{1.3.9}$$

Here $\rightharpoonup$ denotes the weak convergence. (1.3.7), (1.3.9) imply

$$w(x) = \text{const}, \quad \text{a.e. } x \in \Omega,$$

and (1.3.5), (1.3.8) imply $\int_{\Omega} w(x)dx = 0$. Thus

$$w(x) = 0, \quad \text{a.e. } x \in \Omega. \tag{1.3.10}$$

However, (1.3.6), (1.3.8) imply $\|w\|_{L^p(\Omega)} = 1$ which contradicts (1.3.10). $\square$

Corollary 1.3.2 *Let B_R be a ball of radius R in $\mathbb{R}^n$.*
i) If $u \in W_0^{1,p}(B_R)$, $1 \le p < +\infty$, then

$$\int_{B_R} |u|^p dx \le C(n,p)R^p \int_{B_R} |Du|^p dx.$$

ii) If $u \in W^{1,p}(B_R)$, $1 \le p < +\infty$, then

$$\int_{B_R} |u - u_R|^p dx \le C(n,p)R^p \int_{B_R} |Du|^p dx,$$

where

$$u_R = \frac{1}{|B_R|} \int_{B_R} u(x)dx.$$

Proof. By rescaling, i.e. letting $y = x/R$, we are led to an inequality on B_1, which can be proved easily by Theorem 1.3.4. $\square$

Remark 1.3.5 *From the embedding theorem we see that if $1 \le p < n$, then the exponent p on the left side of the inequalities in Theorem 1.3.4 can be replaced by any q such that $1 \le q \le p^* = \dfrac{np}{n-p}$, namely,*
i) If $u \in W_0^{1,p}(\Omega)$, $1 \le p < n$, then for any $1 \le q \le p^*$,

$$\left(\int_{\Omega} |u|^q dx \right)^{1/q} \le C(n,p,\Omega) \left(\int_{\Omega} |Du|^p dx \right)^{1/p};$$

ii) If $\partial\Omega$ is locally Lipschitz continuous and $u \in W^{1,p}(\Omega)$ with $1 \le p < n$, then for any $1 \le q \le p^*$,

$$\left(\int_\Omega |u - u_\Omega|^q dx \right)^{1/q} \le C(n,p,\Omega) \left(\int_\Omega |Du|^p dx \right)^{1/p}.$$

Remark 1.3.6   *Similarly the exponent p on the left side of the inequalities in Corollary 1.3.2 can be replaced by any q such that $1 \le q \le p^* = \dfrac{np}{n-p}$, namely, we have*

i) If $u \in W_0^{1,p}(B_R)$, $1 \le p < n$, then for any $1 \le q \le p^*$,

$$\left(\int_{B_R} |u|^q dx \right)^{1/q} \le C(n,p) R^{1+n/q-n/p} \left(\int_{B_R} |Du|^p dx \right)^{1/p};$$

ii) If $u \in W^{1,p}(B_R)$, $1 \le p < n$, then for any $1 \le q \le p^*$,

$$\left(\int_{B_R} |u - u_R|^q dx \right)^{1/q} \le C(n,p) R^{1+n/q-n/p} \left(\int_{B_R} |Du|^p dx \right)^{1/p}.$$

To make certain of the dependence of the constant C on R on the right side of the inequalities in applying the embedding theorem, we can use the rescaling technique as we did in the proof of Corollary 1.3.2.

If $p \ge n$, then the exponent q on the left side of the inequalities in Remark 1.3.5 and Remark 1.3.6 can be chosen to be any real number not less than 1. However it is to be noted that in the case $p = n$, the constant C on the right side of the inequalities depends on q in addition to n, Ω.

1.4   *t*-Anisotropic Sobolev Spaces

Since the space variable x and time variable t play different roles in parabolic equations, the function spaces adopted in the study of parabolic equations are different from those in the study of elliptic equations. In this section we introduce the so-called t-anisotropic Sobolev spaces available to parabolic equations.

1.4.1   *Spaces $W_p^{2k,k}(Q_T)$, $\overset{\circ}{W}_p^{2k,k}(Q_T)$, $\overset{\bullet}{W}_p^{2k,k}(Q_T)$, $V_2(Q_T)$ and $V(Q_T)$*

Denote $Q_T = \Omega \times (0, T)$ for $T > 0$.

Definition 1.4.1 Let k be a nonnegative integer and $1 \leq p < +\infty$. The set

$$\left\{u; D^\alpha D_t^r u \in L^p(Q_T), \text{for any } \alpha \text{ and } r \text{ such that } |\alpha| + 2r \leq 2k\right\}$$

endowed with the norm

$$\|u\|_{W_p^{2k,k}(Q_T)} = \left(\iint_{Q_T} \sum_{|\alpha|+2r\leq 2k} |D^\alpha D_t^r u|^p dxdt\right)^{1/p}$$

is denoted by $W_p^{2k,k}(Q_T)$.

It can be proved easily that $W_p^{2k,k}(Q_T)$ is a Banach space. From the definition of $W_p^{2k,k}(Q_T)$ we see that the order of weak derivatives with respect to t of the function in $W_p^{2k,k}(Q_T)$ does not exceed half of the highest order of weak derivatives with respect to x.

We need also the following supplemental definition of the space $W^{m,k}(Q_T)$ with m, k being 0 or 1.

Definition 1.4.2 Let m, k be 0 or 1, and $1 \leq p < +\infty$. The set

$$\left\{u; D^\alpha u, D_t^r u \in L^p(Q_T), \text{for any } \alpha, r \text{ such that } |\alpha| \leq m, r \leq k\right\}$$

endowed with the norm

$$\|u\|_{W_p^{m,k}(Q_T)} = \left(\iint_{Q_T} \left(\sum_{|\alpha|\leq m} |D^\alpha u|^p + \sum_{r\leq k} |D_t^r u|^p\right) dxdt\right)^{1/p}$$

is denoted by $W_p^{m,k}(Q_T)$.

If $p = 2$, then in all spaces defined above, we may define the inner products so that they become Hilbert spaces. In particular, the space with $p = 2$, $m = k = 1$, i.e. the space $W_2^{1,1}(Q_T)$ is just the space $H^1(Q_T)$ defined in §1.3.

We always denote by Du, sometimes by ∇u, the weak gradient of the function u with respect to the space variables and denote by $D_t u$ or u_t the weak derivatives with respect to the time variable t.

Let $\partial_l Q_T$ and $\partial_p Q_T$ be the lateral boundary $\partial\Omega \times (0,T)$ and the parabolic boundary $\partial_l Q_T \cup \{(x,t); x \in \overline{\Omega}, t = 0\}$ of Q_T. Denote by $\overset{\circ}{C}{}^\infty(\overline{Q}_T)$ the set of all functions infinitely differentiable on $\overline{Q}_T$, vanishing near the lateral boundary $\partial_l Q_T$, and by $\overset{\bullet}{C}{}^\infty(\overline{Q}_T)$ the set of all functions infinitely differentiable on $\overline{Q}_T$, vanishing near the parabolic boundary $\partial_p Q_T$.

Definition 1.4.3 Denote by $\overset{\circ}{W}{}^{2k,k}_p(Q_T)$ the closure of $\overset{\circ}{C}{}^{\infty}(\overline{Q}_T)$ in $W^{2k,k}_p(Q_T)$; by $\overset{\circ}{W}{}^{m,k}_p(Q_T)$ with m, k being 0 or 1 the closure of $\overset{\circ}{C}{}^{\infty}(\overline{Q}_T)$ in $W^{m,k}_p(Q_T)$; by $\overset{\bullet}{W}{}^{2k,k}_p(Q_T)$ the closure of $\overset{\bullet}{C}{}^{\infty}(\overline{Q}_T)$ in $W^{2k,k}_p(Q_T)$; by $\overset{\bullet}{W}{}^{m,k}_p(Q_T)$ with m, k being 0 or 1 the closure of $\overset{\bullet}{C}{}^{\infty}(\overline{Q}_T)$ in $W^{m,k}_p(Q_T)$.

Definition 1.4.4 Let $L^{\infty}\left(0,T;L^2(\Omega)\right)$ be the set of all functions u such that for almost all $t \in (0,T)$, $u(\cdot,t) \in L^2(\Omega)$ with $\|u(\cdot,t)\|_{L^2(\Omega)}$ bounded. Denote by $V_2(Q_T)$ the set $L^{\infty}(0,T;L^2(\Omega)) \cap W^{1,0}_2(Q_T)$ endowed with the norm

$$\|u\|_{V_2(Q_T)} = \sup_{0 \leq t \leq T} \|u(\cdot,t)\|_{L^2(\Omega)} + \left(\iint_{Q_T} |Du|^2 dx dt\right)^{1/2}.$$

One may verify that $V_2(Q_T)$ is a Banach space.

Definition 1.4.5 Denote

$$V(Q_T) = \left\{u \in \overset{\bullet}{W}{}^{1,1}_2(Q_T); Du_t \in L^2(Q_T;\mathbb{R}^n)\right\},$$

and define the inner product as

$$(u,v)_{V(Q_T)} = (u,v)_{W^{1,1}_2(Q_T)} + (Du_t, Dv_t)_{L^2(Q_T)}.$$

It is easy to prove

Proposition 1.4.1 $V(Q_T)$ *is dense in* $\overset{\bullet}{W}{}^{1,1}_2(Q_T)$.

Remark 1.4.1 $\overset{\bullet}{W}{}^{0,1}_2(Q_T)$ *can be regarded as the closure in* $W^{0,1}_2(Q_T)$, *of the set of all infinitely differentiable functions on* $\overline{Q}_T$, *vanishing near the bottom* $\overline{\Omega} \times \{t = 0\}$.

1.4.2 Embedding theorem

For the t-anisotropic Sobolev spaces, we also have the embedding theorem, whose proof can be found in [Gu (1995)].

Theorem 1.4.1 *(t-Anisotropic Embedding Theorem) Let* $\Omega \subset \mathbb{R}^n$ *be a bounded domain and* $1 \leq p < +\infty$.

i) *If* Ω *has the property of uniform inner cone, then, when* $p = (n+2)/2$,

$$W^{2,1}_p(Q_T) \subset L^q(Q_T), \quad 1 \leq q < +\infty$$

and for any $u \in W_p^{2,1}(Q_T)$,

$$\|u\|_{L^q(Q_T)} \leq C(n, q, Q_T) \|u\|_{W_p^{2,1}(Q_T)}, \quad 1 \leq q < +\infty;$$

when $p < (n+2)/2$,

$$W_p^{2,1}(Q_T) \subset L^q(Q_T), \quad 1 \leq q \leq \frac{(n+2)p}{n+2-2p}$$

and for any $u \in W_p^{2,1}(Q_T)$,

$$\|u\|_{L^q(Q_T)} \leq C(n, p, Q_T) \|u\|_{W_p^{2,1}(Q_T)}, \quad 1 \leq q \leq \frac{(n+2)p}{n+2-2p}.$$

ii) If $\partial\Omega$ is appropriately smooth, then, when $p > (n+2)/2$,

$$W_p^{2,1}(Q_T) \subset C^{\alpha, \alpha/2}(\overline{Q}_T), \quad 0 < \alpha \leq 2 - \frac{n+2}{p}$$

and for any $u \in W_p^{2,1}(Q_T)$,

$$|u|_{\alpha, \alpha/2; Q_T} \leq C(n, p, Q_T) \|u\|_{W_p^{2,1}(Q_T)}, \quad 0 < \alpha \leq 2 - \frac{n+2}{p}.$$

Remark 1.4.2 *The above embedding theorem can be expressed as*

$$W_p^{2,1}(Q_T) \hookrightarrow \begin{cases} L^q(Q_T), & 1 \leq q \leq \dfrac{(n+2)p}{n+2-2p}, & p < \dfrac{n+2}{2}, \\[2mm] L^q(Q_T), & 1 \leq q < +\infty, & p = \dfrac{n+2}{2}, \\[2mm] C^{\alpha, \alpha/2}(\overline{Q}_T), & 0 < \alpha \leq 2 - \dfrac{n+2}{p}, & p > \dfrac{n+2}{2}. \end{cases}$$

Using Theorem 1.4.1 for k times repeatedly leads to

Corollary 1.4.1

$$W_p^{2k,k}(Q_T) \hookrightarrow \begin{cases} L^q(Q_T), & 1 \leq q \leq \dfrac{(n+2)p}{n+2-2kp}, & kp < \dfrac{n+2}{2}, \\[2mm] L^q(Q_T), & 1 \leq q < +\infty, & kp = \dfrac{n+2}{2}, \\[2mm] C^{\alpha, \alpha/2}(\overline{Q}_T), & 0 < \alpha \leq 2 - \dfrac{n+2}{kp}, & kp > \dfrac{n+2}{2}. \end{cases}$$

The embedding theorem can be established for the space $V_2(Q_T)$. For the convenience of applications, we state it for the standard cylinder

$$Q_\rho = B_\rho \times (-\rho^2, \rho^2), \quad B_\rho = \{x \in \mathbb{R}^n; |x| < \rho\}.$$

Theorem 1.4.2   *Let $u \in V_2(Q_\rho)$. Then*

$$\left(\frac{1}{\rho^{n+2}} \iint_{Q_\rho} |u|^{2q} dx dt \right)^{1/q}$$

$$\leq C(n) \rho^{-n} \left(\sup_{-\rho^2 \leq t \leq \rho^2} \int_{B_\rho} u^2 dx + \iint_{Q_\rho} |Du|^2 dx dt \right),$$

where

$$q = \begin{cases} \dfrac{5}{3}, & \text{when } n = 1, 2; \\[2mm] 1 + \dfrac{2}{n}, & \text{when } n \geq 3. \end{cases}$$

1.4.3   *Poincaré's inequality*

Poincaré's inequality can be also established for the t-anisotropic Sobolev space $W_p^{1,1}(Q_T)$. We state it for the standard cylinder Q_ρ.

Theorem 1.4.3   *(t-Anisotropic Poincaré's Inequality) Let $1 \leq p < +\infty$, $\rho > 0$.*

   i) If $u \in \overset{\bullet}{W}_p^{1,1}(Q_\rho)$, then

$$\iint_{Q_\rho} |u|^p dx dt \leq C(n,p) \left(\rho^p \iint_{Q_\rho} |Du|^p dx dt + \rho^{2p} \iint_{Q_\rho} |D_t u|^p dx dt \right).$$

   ii) If $u \in W_p^{1,1}(Q_\rho)$, then

$$\iint_{Q_\rho} |u - u_\rho|^p dx dt \leq C(n,p) \left(\rho^p \iint_{Q_\rho} |Du|^p dx dt \right.$$
$$\left. + \rho^{2p} \iint_{Q_\rho} |D_t u|^p dx dt \right),$$

where

$$u_\rho = \frac{1}{|Q_\rho|} \iint_{Q_\rho} u(x,t) dx dt.$$

Proof.   First we use the standard Poincaré's inequality (Theorem 1.3.4) to obtain the conclusion for $\rho = 1$ and then deduce the desired result by rescaling. $\square$

1.5 Trace of Functions in $H^1(\Omega)$

In this section, we discuss whether we can and how to define the boundary value for functions in $H^1(\Omega)$ and $H^1(Q_T) = W_2^{1,1}(Q_T)$.

1.5.1 *Some propositions on functions in $H^1(Q^+)$*

Denote

$$Q = \{x \in \mathbb{R}^n; |x_i| < 1, i = 1, \cdots, n\}, \quad Q^+ = \{x \in Q; x_n > 0\},$$
$$x = (x', x_n), \quad x' = (x_1, \cdots, x_{n-1}),$$
$$\Gamma = \{x \in Q; x_n = 0\} = \{x' \in \mathbb{R}^{n-1}; |x_i| < 1, i = 1, \cdots, n-1\}.$$

Proposition 1.5.1 *For any $u \in H^1(Q^+)$, there exists a unique function $w \in L^2(\Gamma)$, such that*

$$\operatorname*{esslim}_{x_n \to 0^+} \int_\Gamma |u(x', x_n) - w(x')|^2 dx' = 0.$$

We call $w(x')$ the trace of u on Γ and denote it by $\gamma u(x', 0)$.

Proof. Uniqueness is obvious. To prove the existence of the trace, we first note that, by Proposition 1.3.3, there exists $u_m \in C^\infty(\overline{Q^+})$, such that

$$\lim_{m \to \infty} \|u_m - u\|_{H^1(Q^+)} = 0.$$

Let $0 < \delta < 1$. Choose a smooth function $\eta(x_n) \in C^1[0,1]$, such that $\eta(x_n) = 1$ for $0 \le x_n \le \delta$ and $\eta(x_n) = 0$ for x_n less than or equal to 1 but close to 1. Clearly

$$\lim_{m \to \infty} \|\eta u_m - \eta u\|_{H^1(Q^+)} = 0. \tag{1.5.1}$$

Since for sufficiently small $\varepsilon \in [0, \delta]$,

$$u_m(x', \varepsilon) = -\int_\varepsilon^1 \frac{\partial(\eta u_m)}{\partial x_n} dx_n,$$

we have

$$|u_m(x', \varepsilon) - u_k(x', \varepsilon)|^2 \le \int_0^1 \left| \frac{\partial(\eta u_m)}{\partial x_n} - \frac{\partial(\eta u_k)}{\partial x_n} \right|^2 dx_n,$$

$$\int_\Gamma |u_m(x', \varepsilon) - u_k(x', \varepsilon)|^2 dx' \le \int_{Q^+} \left| \frac{\partial(\eta u_m)}{\partial x_n} - \frac{\partial(\eta u_k)}{\partial x_n} \right|^2 dx$$

$$\leq \|\eta u_m - \eta u\|^2_{H^1(Q^+)}. \qquad (1.5.2)$$

(1.5.1) and (1.5.2) imply that, for arbitrary fixed small $\varepsilon \in [0, \delta]$, $\{u_m(\cdot, \varepsilon)\}$ is a Cauchy sequence in $L^2(\Gamma)$; we denote its limit function by $v(\cdot, \varepsilon)$. On the other hand, since $\{u_m\}$ converges to u in $L^2(Q^+)$, there exists a set $E \subset (0, \delta)$ of measure zero, such that for any $\varepsilon \in (0, \delta) \backslash E$,

$$v(x', \varepsilon) = u(x', \varepsilon), \quad \text{a.e. } x' \in \Gamma.$$

Thus, from (1.5.2) we obtain, for $\varepsilon \in (0, \delta) \backslash E$,

$$\int_\Gamma |u_m(x', 0) - v(x', 0)|^2 dx' \leq \|\eta u_m - \eta u\|_{H^1(Q^+)}, \qquad (1.5.3)$$

$$\int_\Gamma |u_m(x', \varepsilon) - u(x', \varepsilon)|^2 dx' \leq \|\eta u_m - \eta u\|_{H^1(Q^+)}. \qquad (1.5.4)$$

In addition, clearly

$$\int_\Gamma |u_m(x', \varepsilon) - u_m(x', 0)|^2 dx' \leq \varepsilon \int_{Q^+} \left| \frac{\partial u_m}{\partial x_n} \right|^2 dx. \qquad (1.5.5)$$

Combining (1.5.3), (1.5.4), (1.5.5) with

$$\int_\Gamma |u(x', \varepsilon) - w(x')|^2 dx' \leq \int_\Gamma |u(x', \varepsilon) - u_m(x', \varepsilon)|^2 dx'$$
$$+ \int_\Gamma |u_m(x', \varepsilon) - u_m(x', 0)|^2 dx'$$
$$+ \int_\Gamma |u_m(x', 0) - w(x')|^2 dx' \qquad (1.5.6)$$

yields

$$\lim_{\substack{\varepsilon \in (0,\delta) \backslash E \\ \varepsilon \to 0}} \int_\Gamma |u(x', \varepsilon) - w(x')|^2 dx' = 0,$$

where $w(x') = v(x', 0)$. $\qquad\qquad\qquad\qquad\qquad\qquad\qquad\qquad\qquad\qquad$ □

Remark 1.5.1 *From (1.5.3), we have*

$$\lim_{m \to \infty} \int_\Gamma |u_m(x', 0) - w(x')|^2 dx' = 0. \qquad (1.5.7)$$

By virtue of this fact, we can define the trace of u on Γ as the function $w(x')$ satisfying (1.5.7) for any sequence $\{u_m\} \subset C^\infty(\overline{Q}^+)$ converging to u in $H^1(Q^+)$. It is easy to verify that the trace defined in this manner is equivalent to that defined by Proposition 1.5.1. Since $C^\infty(\overline{Q}^+)$ is dense in $C^1(\overline{Q}^+)$, we can replace the sequence $\{u_m\}$ in formula (1.5.7) by any sequence in $C^1(\overline{Q}^+)$, converging to u in $H^1(Q^+)$.

Remark 1.5.2 *From the proof of Proposition 1.5.1 we see that the conclusion of the proposition still holds, if we replace Q^+ by an arbitrary cylinder $D \times (0,\delta)$ ($\delta > 0$, and D is a bounded domain in $\mathbb{R}^{n-1}$), provided ∂D is Lipschitz continuous, or satisfies more general condition such that $C^\infty(\overline{D \times (0,\delta)})$ is dense in $H^1(D \times (0,\delta))$ (see Proposition 1.3.3).*

Corollary 1.5.1 *If $u \in H^1(Q^+) \cap C(\overline{Q}^+)$, then $\gamma u(x',0) = u(x',0)$ a.e. on Γ.*

Proposition 1.5.2 *Let $u \in H^1(Q^+) \cap C(\overline{Q}^+)$ and $u = 0$ near the upper boundary and the lateral of Q^+. If $u = 0$ on the bottom Γ of Q^+, then $u \in H^1_0(Q^+)$.*

Proof. Extend u to the whole Q by setting $u = 0$ outside Q^+ and denote the new function by $\tilde{u}$. The proof will be proceeded in two steps.

The first step is to prove $\tilde{u} \in H^1(Q)$. Obviously $\tilde{u} \in L^2(Q)$ and the weak derivatives $\dfrac{\partial \tilde{u}}{\partial x_i}$ ($i = 1, \cdots, n-1$) exist with $\dfrac{\partial \tilde{u}}{\partial x_i} = \dfrac{\partial u}{\partial x_i}$ on Q^+ and $\dfrac{\partial \tilde{u}}{\partial x_i} = 0$ on $Q \backslash Q^+$. Hence $\dfrac{\partial \tilde{u}}{\partial x_i} \in L^2(Q)$ for $1 \leq i \leq n-1$. It remains to prove

$$\int_{Q^+} u \frac{\partial \varphi}{\partial x_n} dx = -\int_{Q^+} \frac{\partial u}{\partial x_n} \varphi dx, \quad \forall \varphi \in C^\infty_0(Q). \tag{1.5.8}$$

To this end, for sufficiently small $\varepsilon > 0$, choose a cut-off function $\eta(x_n) \in C^\infty_0(-1,1)$, satisfying the following conditions

$$\eta(x_n) = \begin{cases} 1, & \text{if } |x_n| \leq \varepsilon, \\ 0, & \text{if } |x_n| \geq 2\varepsilon, \end{cases}$$

$$0 \leq \eta(x_n) \leq 1, \quad |\eta'(x_n)| \leq \frac{C}{\varepsilon}, \quad -1 < x_n < 1.$$

Divide $\displaystyle\int_{Q^+} u \frac{\partial \varphi}{\partial x_n} dx$ into

$$\int_{Q^+} u \frac{\partial \varphi}{\partial x_n} dx = \int_{Q^+} u \frac{\partial}{\partial x_n} (\eta(x_n)\varphi + (1 - \eta(x_n))\varphi) dx$$

$$= \int_{Q^+} u \eta'(x_n) \varphi dx + \int_{Q^+} u \eta(x_n) \frac{\partial \varphi}{\partial x_n} dx$$

$$+ \int_{Q^+} u \frac{\partial((1 - \eta(x_n))\varphi)}{\partial x_n} dx$$

$$= I_1^\varepsilon + I_2^\varepsilon + I_3^\varepsilon. \tag{1.5.9}$$

Evidently

$$\lim_{\varepsilon \to 0} I_2^\varepsilon = 0. \tag{1.5.10}$$

Since for $\varepsilon > 0$ small enough, $(1 - \eta(x_n))\varphi \in C_0^\infty(Q^+)$, we have

$$I_3^\varepsilon = - \int_{Q^+} \frac{\partial u}{\partial x_n} ((1 - \eta(x_n))\varphi) dx$$

and hence

$$\lim_{\varepsilon \to 0} I_3^\varepsilon = - \int_{Q^+} \frac{\partial u}{\partial x_n} \varphi dx. \tag{1.5.11}$$

Finally, note that $u \in C(\overline{Q}^+)$ and $u\big|_{x_n=0} = 0$. Thus from

$$|I_1^\varepsilon| \leq \int_{Q^+ \cap \{x; |x_n| \leq 2\varepsilon\}} |u\eta'(x_n)\varphi| dx$$

$$\leq \frac{C}{\varepsilon} \int_{Q^+ \cap \{x; |x_n| \leq 2\varepsilon\}} |u| dx,$$

we arrive at

$$\lim_{\varepsilon \to 0} I_1^\varepsilon = 0. \tag{1.5.12}$$

Letting $\varepsilon \to 0$ in (1.5.9) and using (1.5.10), (1.5.11),(1.5.12) yield (1.5.8).

The second step is to prove $u \in H_0^1(Q^+)$. To this end, we consider the modified mollification of $\tilde{u}$:

$$J_\varepsilon^- \tilde{u}(x) = \int_Q j_\varepsilon(x_1 - y_1) \cdots j_\varepsilon(x_{n-1} - y_{n-1}) j_\varepsilon(x_n - y_n - 2\varepsilon) \tilde{u}(y) dy$$

with $\varepsilon > 0$ small enough, where $j_\varepsilon(\tau)$ is the mollifier in one dimension. Since $j_\varepsilon(\tau) = 0$ for $|\tau| \geq \varepsilon$ and $\tilde{u} = 0$ near the upper boundary and the lateral of Q and on $Q \backslash Q^+$, we have $J_\varepsilon^- \tilde{u} \in C_0^\infty(\overline{Q}^+)$. Since from the first step, $\tilde{u} \in H^1(Q)$, similar to the case of the standard mollification, we can assert

$$\lim_{\varepsilon \to 0} \|J_\varepsilon^- \tilde{u} - u\|_{H^1(Q^+)} = 0.$$

Therefore $u \in H_0^1(Q^+)$. $\qquad\qquad\qquad\qquad\qquad\qquad\qquad\qquad\qquad \square$

Proposition 1.5.3 *Let $u \in H^1(Q^+) \cap C(\overline{Q}^+)$ and $\tilde{u}$ be the extension of u to Q by setting $\tilde{u}(x) = u(x', 0)$ on $Q^- = Q \backslash \overline{Q}^+$. Then $\tilde{u} \in H^1(Q)$.*

Proof. The proof is just the same as the first step of the proof of Proposition 1.5.2, and the only difference is that here we need to require $\eta(x_n)$ to be an even function. $\qquad \square$

1.5.2 Trace of functions in $H^1(\Omega)$

Theorem 1.5.1 *Let $\Omega \subset \mathbb{R}^n$ be a bounded domain with smooth boundary. Then any $u \in H^1(\Omega)$ has trace γu on $\partial\Omega$ and $\gamma u \in L^2(\partial\Omega)$, namely, there exists a unique function $\gamma u \in L^2(\partial\Omega)$ satisfying*

$$\lim_{m \to \infty} \int_{\partial\Omega} |u_m - \gamma u|^2 d\sigma = 0 \qquad (1.5.13)$$

where $\{u_m\} \subset C^1(\overline{\Omega})$ is an arbitrary sequence converging to u in $H^1(\Omega)$.

Proof. Since $\partial\Omega$ is smooth, every point on $\partial\Omega$ has a small neighborhood U, with the following property: there exists a smooth invertible mapping Ψ, transforming $Q = \{y \in \mathbb{R}^n; |y_i| < 1, i = 1, \cdots, n\}$ into U and $Q^+ = \{y \in Q; y_n > 0\}$ into $U \cap \Omega$, such that $\Psi(\Gamma) = U \cap \partial\Omega$, where $\Gamma = \{y \in Q; y_n = 0\}$.

Let $\{u_m\} \subset C^1(\overline{\Omega})$ be an arbitrary sequence converging to u in $H^1(\Omega)$. Denote $v_m = (\eta u_m) \circ \Psi$ with $\eta(x) \in C_0^\infty(U)$. Then $v_m \in C^1(\overline{Q}^+)$ and $v_m = 0$ near the upper boundary and the lateral of Q^+ and

$$\lim_{m \to \infty} \|v_m - (\eta u) \circ \Psi\|_{H^1(Q^+)} = 0.$$

By Proposition 1.5.1, there exists $h \in L^2(\Gamma)$, such that

$$\lim_{m \to \infty} \int_\Gamma |v_m(y', 0) - h(y')|^2 dy' = 0.$$

Obviously, $h = 0$ near the boundary of Γ. Returning to the variable x, we obtain

$$\lim_{m \to \infty} \int_{U \cap \partial\Omega} |\eta u_m - w|^2 d\sigma = 0,$$

where $w = h \circ \Phi\big|_{\Phi_n(x)=0}$, and $\Phi = (\Phi_1, \cdots, \Phi_n)$ is the inverse of Ψ. Note $w = 0$ near the boundary of $U \cap \partial\Omega$. After a zero-extension of w to $\partial\Omega$, the above formula can be written as

$$\lim_{m\to\infty} \int_{\partial\Omega} |\eta u_m - w|^2 d\sigma = 0.$$

By the finite covering theorem, there exist neighborhoods U_i ($i = 1, \cdots, N$) with the property stated above, such that $\partial\Omega \subset \bigcup_{i=1}^{N} U_i$. Let $\eta_i(x)$ ($i = 1, \cdots, N$) be the partition of unity associated to U_i ($i = 1, \cdots, N$) (see §1.1.5), and w_i be the functions corresponding to U_i, η_i obtained as above, namely, $w_i \in L^2(\partial\Omega)$, such that

$$\lim_{m\to\infty} \int_{\partial\Omega} |\eta_i u_m - w_i|^2 d\sigma = 0. \tag{1.5.14}$$

Denote $w = \sum_{i=1}^{N} w_i$. Then $w \in L^2(\partial\Omega)$ and it follows from (1.5.14) and

$$u_m - w = \sum_{i=1}^{N} (\eta_i u_m - w_i), \quad x \in \partial\Omega$$

that

$$\lim_{m\to\infty} \int_{\partial\Omega} |u_m - w|^2 d\sigma = 0.$$

This proves the existence of the trace $\gamma u\big|_{\partial\Omega}$.

By using the local flatting technique to the small neighborhood of any point of $\partial\Omega$, one can prove the uniqueness of the trace. $\square$

Corollary 1.5.2   *Let* $\Omega \subset \mathbb{R}^n$ *be a bounded domain with smooth boundary. If* $u \in H^1(\Omega) \cap C(\overline{\Omega})$, *then* $\gamma u\big|_{\partial\Omega} = u\big|_{\partial\Omega}$.

Proof.   Use the local flatting technique and Corollary 1.5.1. $\square$

Corollary 1.5.3   *Let* $\Omega \subset \mathbb{R}^n$ *be a bounded domain with smooth boundary. If* $u \in H_0^1(\Omega)$, *then* $\gamma u\big|_{\partial\Omega} = 0$.

Proof.   Since $C_0^\infty(\Omega)$ is dense in $H_0^1(\Omega)$, there exists a sequence $\{u_m\} \subset H^1(\Omega)$, which converges to u in $H^1(\Omega)$. Hence $\gamma u\big|_{\partial\Omega} = 0$ follows from (1.5.13). $\square$

Corollary 1.5.4   *Let* $\Omega \subset \mathbb{R}^n$ *be a bounded domain with smooth boundary. If* $u \in H_0^1(\Omega) \cap C(\overline{\Omega})$, *then* $u\big|_{\partial\Omega} = 0$.

Proof. The desired conclusion follows from Corollary 1.5.2 and Corollary 1.5.3. □

Theorem 1.5.2 *Let $\Omega \subset \mathbb{R}^n$ be a bounded domain with smooth boundary. If $u \in H^1(\Omega) \cap C(\overline{\Omega})$ and $u\big|_{\partial\Omega} = 0$, then $u \in H_0^1(\Omega)$.*

Proof. Cut-off u at the small neighborhood of a given point of $\partial\Omega$, namely, consider ηu instead of u with η being a cut-off function, flat $U \cap \partial\Omega$ locally (as we did in the proof of Theorem 1.5.1) and then use Proposition 1.5.2 to conclude the existence of a sequence $u_m \in C_0^1(U \cap \Omega)$ converging to ηu in $H^1(\Omega)$.

Choose such neighborhoods U_i $(i = 1, \cdots, N)$ and an open set U_0, such that $\partial\Omega \subset \bigcup_{i=1}^{N} U_i$, $U_0 \supset \Omega \backslash \bigcup_{i=1}^{N} U_i$. Let $\eta_i(x)$ $(i = 0, 1, \cdots, N)$ be a partition of unity associated to U_i $(i = 0, 1, \cdots, N)$, u_m^i $(i = 1, \cdots, N)$ be the sequences corresponding to U_i, η_i $(i = 1, \cdots, N)$ obtained in the above manner. It is evident that there exists a sequence $\{u_m^0\} \in C_0^\infty(\Omega)$ converging to $\eta_0 u$ in $H^1(\Omega)$. Denote $u_m = \sum_{i=0}^{N} u_m^i$. Then $u_m \in C_0^1(\Omega)$ and $\{u_m\}$ converges to u in $H^1(\Omega)$. Since $C_0^\infty(\Omega)$ is dense in $C_0^1(\Omega)$, this shows that $u \in H_0^1(\Omega)$. □

1.5.3 Trace of functions in $H^1(Q_T) = W_2^{1,1}(Q_T)$

Theorem 1.5.3 *Let $\Omega \subset \mathbb{R}^n$ be a bounded domain with smooth boundary. Then any $u \in H^1(Q_T)$ has trace γu on $\partial_p Q_T$ and $\gamma u \in L^2(\partial_p Q_T)$.*

Proof. By Proposition 1.5.1 and Remark 1.5.1, there exist a sequence $\{u_m\} \in C^\infty(\overline{Q}_T)$ converging to u in $H^1(Q_T)$ and a function $v(x) \in L^2(\Omega)$, such that

$$\lim_{m \to \infty} \int_\Omega |u_m(x, 0) - v(x)|^2 dx = 0.$$

Similar to the proof of Theorem 1.5.1, by using the techniques such as local flatting $\partial_l Q_T$, finite covering and partition of unity, we may assert the existence of a function $w(x, t) \in L^2(\partial_l Q_T)$, such that

$$\lim_{m \to \infty} \int_{\partial_l Q_T} |u_m - w|^2 d\sigma = 0.$$

□

Remark 1.5.3   *Any $u \in H^1(Q_T)$ also has trace on the upper boundary of Q_T. However, such fact is not necessary in later applications.*

Corollary 1.5.5   *Let $\Omega \subset \mathbb{R}^n$ be a bounded domain with smooth boundary. If $u \in H^1(Q_T) \cap C(\overline{Q}_T)$, then $\gamma u\big|_{\partial_p Q_T} = u\big|_{\partial_p Q}$.*

Corollary 1.5.6   *Let $\Omega \subset \mathbb{R}^n$ be a bounded domain with smooth boundary. If $u \in \overset{\bullet}{W}{}^{1,1}_2(Q_T)$, then $\gamma u\big|_{\partial_p Q_T} = 0$; if $u \in \overset{\circ}{W}{}^{1,1}_2(Q_T)$, then $\gamma u\big|_{\partial_l Q_T} = 0$.*

Corollary 1.5.7   *Let $\Omega \subset \mathbb{R}^n$ be a bounded domain with smooth boundary. If $u \in \overset{\bullet}{W}{}^{1,1}_2(Q_T) \cap C(\overline{Q}_T)$, then $u\big|_{\partial_p Q_T} = 0$; if $u \in \overset{\circ}{W}{}^{1,1}_2(Q_T) \cap C(\overline{Q}_T)$, then $u\big|_{\partial_l Q_T} = 0$.*

Theorem 1.5.4   *Let $\Omega \subset \mathbb{R}^n$ be a bounded domain with smooth boundary. If $u \in H^1(Q_T) \cap C(\overline{Q}_T)$, and $u\big|_{\partial_p Q_T} = 0$, then $u \in \overset{\bullet}{W}{}^{1,1}_2(Q_T)$; if $u \in H^1(Q_T) \cap C(\overline{Q}_T)$, and $u\big|_{\partial_l Q_T} = 0$, then $u \in \overset{\circ}{W}{}^{1,1}_2(Q_T)$.*

Proof.   To prove the second part, we first extend u to $t < 0$ and $t > T$ as we did in the proof of Proposition 1.5.3 and then use the techniques such as local flatting, finite covering and partition of unity. Proposition 1.5.2 is applied after local flatting. To prove the first part, we extend u to $t < 0$ by defining $u = 0$ there and extend u to $t > T$ as in the proof of Proposition 1.5.3. The new function is denoted still by u. A modified mollification of u in x_n

$$v_\varepsilon(x,t) = J_\varepsilon^- u(x,t) = \int_{\mathbb{R}} j_\varepsilon(t - s - 2\varepsilon) u(x,s) ds \quad (\varepsilon > 0)$$

is introduced to approximate u in $H^1(Q_T)$. Then the techniques such as local flatting, finite covering and partition of unity are used to construct the smooth approximation of v_ε in $H^1(Q_T)$, which vanishes near $\partial_p Q_T$. $\square$

Exercises

   1. Prove Proposition 1.1.1.

   2. Let U and V be open sets of $\mathbb{R}^n$ with $V \subset \overline{V} \subset U$. Construct a function $\xi \in C_0^\infty(U)$, such that

$$\xi(x) = 1, \quad \forall x \in V.$$

3. Prove Proposition 1.2.1.

4. Prove Corollary 1.2.1.

5. Judge whether $W^{1,1}(\Omega)$ is a Banach space, where $\Omega \subset \mathbb{R}^n$ is an open set.

6. Prove Proposition 1.3.2.

7. Prove Propositions 1.3.8–1.3.10.

8. Let $u \in W^{1,p}((0,1))$ with $p > 1$. Prove

$$|u(x) - u(y)| \le |x - y|^{1-1/p} \left(\int_0^1 |u'(t)|^p dt \right)^{1/p}, \quad \text{for almost all } x, y \in [0, 1].$$

9. Let $\Omega \subset \mathbb{R}^n$ be an open set, $1 \le p \le +\infty$ and $u \in W^{1,p}(\Omega)$.

i) Prove $u^+, u^- \in W^{1,p}(\Omega)$, and

$$Du^+(x) = \begin{cases} Du(x), & \text{whenever } u(x) > 0, \\ 0, & \text{whenever } u(x) \le 0, \end{cases}$$

$$Du^-(x) = \begin{cases} Du(x), & \text{whenever } u(x) < 0, \\ 0, & \text{whenever } u(x) \ge 0, \end{cases}$$

where

$$u^+ = \max\{u, 0\}, \quad u^- = \min\{u, 0\};$$

ii) Prove

$$Du(x) = 0, \quad \text{a.e. } x \in \{x \in \Omega; u(x) = 0\}.$$

10. Prove Corollary 1.3.2 and Remark 1.3.6.

11. Prove Theorem 1.4.3.

12. Prove Proposition 1.5.3.

Chapter 2

L^2 Theory of Linear Elliptic Equations

This chapter is devoted to the L^2 theory of linear elliptic equations. We first present the argument for a typical equation, i.e. Poisson's equation thoroughly and then turn to the general equations in divergence form.

2.1 Weak Solutions of Poisson's Equation

Let $\Omega \subset \mathbb{R}^n$ be a bounded domain with piecewise smooth boundary $\partial\Omega$. Consider the equation

$$-\Delta u = f(x) \qquad (2.1.1)$$

in Ω, where $x = (x_1, \cdots, x_n)$, Δ is Laplace operator in n dimension, i.e.

$$\Delta = \frac{\partial^2}{\partial x_1^2} + \frac{\partial^2}{\partial x_2^2} + \cdots + \frac{\partial^2}{\partial x_n^2}$$

and $f \in L^2(\Omega)$. For simplicity, we merely discuss the Dirichlet problem for equation (2.1.1) with the homogeneous boundary value condition

$$u\Big|_{\partial\Omega} = 0. \qquad (2.1.2)$$

If a nonhomogeneous boundary value condition

$$u\Big|_{\partial\Omega} = g(x) \qquad (2.1.3)$$

is assumed with $g(x)$ appropriately smooth on $\overline{\Omega}$, then we can transform the problem into the one with the homogeneous boundary value condition by considering the equation for $u(x) - g(x)$, which is still a Poisson's equation with another function as its right member.

2.1.1   *Definition of weak solutions*

Assume that $u \in C^2(\Omega)$ is a solution of (2.1.1), $\varphi \in C_0^\infty(\Omega)$ is an arbitrary function. Substituting u into (2.1.1), multiplying the two sides by φ and integrating over Ω yield

$$-\int_\Omega \Delta u \varphi dx = \int_\Omega f\varphi dx. \qquad (2.1.4)$$

By integrating by parts, we can move the operation of derivatives acting on u to φ partially or even completely. In fact, since the support of φ is contained in Ω, we have

$$-\int_\Omega \Delta u \varphi dx = -\int_{\partial\Omega} \varphi \frac{\partial u}{\partial \vec{\nu}} ds + \int_\Omega \nabla u \cdot \nabla \varphi dx = \int_\Omega \nabla u \cdot \nabla \varphi dx$$

$$= \int_{\partial\Omega} u \frac{\partial \varphi}{\partial \vec{\nu}} ds - \int_\Omega u \Delta \varphi dx = -\int_\Omega u \Delta \varphi dx,$$

where $\vec{\nu}$ is the unit normal vector outward to $\partial\Omega$. Thus (2.1.4) can be changed into

$$\int_\Omega \nabla u \cdot \nabla \varphi dx = \int_\Omega f\varphi dx \qquad (2.1.5)$$

or

$$-\int_\Omega u \Delta \varphi dx = \int_\Omega f\varphi dx. \qquad (2.1.6)$$

This shows that if $u \in C^2(\Omega)$ is a solution of (2.1.1), then for any $\varphi \in C_0^\infty(\Omega)$, the integral identities (2.1.5) and (2.1.6) hold.

Conversely, if for any $\varphi \in C_0^\infty(\Omega)$, $u \in C^2(\Omega)$ satisfies (2.1.5) or (2.1.6), then deriving in a contrary way leads to (2.1.4), i.e.

$$\int_\Omega (-\Delta u - f)\varphi dx = 0.$$

Because of the arbitrariness of φ, from this it follows that $-\Delta u - f = 0$ in Ω, i.e. u is a solution of (2.1.1).

It is to be noted that, in order that the integral in (2.1.5) makes sense, it suffices to require $u \in H^1(\Omega)$, and in order that the integral in (2.1.6) makes sense, it even suffices to require $u \in L^2(\Omega)$. In view of this, it is reasonable to regard a function $u \in H^1(\Omega)$ $\left(u \in L^2(\Omega)\right)$ satisfying the integral identity (2.1.5) $\left((2.1.6)\right)$ for any $\varphi \in C_0^\infty(\Omega)$ as a solution of (2.1.1) in a general sense.

Definition 2.1.1 A function $u \in H^1(\Omega)$ is said to be a weak solution of equation (2.1.1), if the integral identity (2.1.5) holds for any $\varphi \in C_0^\infty(\Omega)$.

Remark 2.1.1 *Since $C_0^\infty(\Omega)$ is dense in $H_0^1(\Omega)$, satisfying (2.1.5) for any $\varphi \in C_0^\infty(\Omega)$ implies the same for any $\varphi \in H_0^1(\Omega)$.*

Using the integral identity (2.1.6) we may define another kind of solutions weaker than those stated in Definition 2.1.1, which satisfy (2.1.1) in the sense of distributions. However, this kind of weak solutions will not be concerned in this book.

As weak solutions of the Dirichlet problem (2.1.1), (2.1.2) to be defined, they are required to satisfy, in addition to (2.1.1), the boundary value condition (2.1.2) in certain sense. We have indicated in §1.5.1 that functions in $H_0^1(\Omega)$ take zero boundary value in a general sense. Hence the following definition is reasonable:

Definition 2.1.2 A function $u \in H_0^1(\Omega)$ is said to be a weak solution of the Dirichlet problem (2.1.1), (2.1.2), if the integral identity (2.1.5) holds for any $\varphi \in C_0^\infty(\Omega)$.

Remark 2.1.2 *Weak solutions of the Dirichlet problem (2.1.1), (2.1.3) can be defined as functions u in $H^1(\Omega)$ satisfying the integral identity (2.1.5) for any $\varphi \in C_0^\infty(\Omega)$ and $u - g \in H_0^1(\Omega)$.*

For a few domains of special shape, we may obtain the explicit solutions of the Dirichlet problem (2.1.1), (2.1.2) by constructing Green's functions. However, it is impossible to do for general domains in this way. For general domains there is no alternative but to discuss the solvability theoretically. So far a number of methods have been developed. In this section, we present some of these methods. As we see later what we obtain by means of these methods are weak solutions. However we can further prove their regularity under some additional conditions on $\partial\Omega$ and f, and thus arrive at classical solutions.

2.1.2 *Riesz's representation theorem and its application*

First we present a method which is based on the following theorem:

Riesz's representation theorem *Let $F(v)$ be a bounded linear functional in the Hilbert space H. Then there exists a unique $u \in H$ with $\|u\| = \|F\|$, such that*

$$F(v) = (u, v), \quad \forall v \in H,$$

where $(\cdot, \cdot)$ is the inner product on H.

In order to apply this theorem to the solvability of (2.1.1), (2.1.2), we introduce a new inner product

$$\langle u, v \rangle = \int_\Omega \nabla u \cdot \nabla v dx$$

on $H_0^1(\Omega)$ with a little difference from the one defined before. That $\langle u, v \rangle$ satisfies all properties of inner product can be checked evidently. For example, using Poincaré's inequality (§1.3.6), we see that $\langle u, v \rangle = 0$ in $H_0^1(\Omega)$ implies $u = 0$.

Denote $\|u\| = (u, u)^{1/2}, \|\|u\|\| = \langle u, u \rangle^{1/2}$. Then for $u \in H_0^1(\Omega)$, we have

$$\alpha \|u\| \leq \|\|u\|\| \leq \beta \|u\|$$

where $\alpha > 0, \beta > 0$ are some constants. The second part of the above inequality is trivial and the first part follows from Poincaré's inequality. This means that the new inner product is equivalent to the older one.

Clearly, for any $f \in L^2(\Omega)$,

$$F(v) = \int_\Omega fv dx, \quad v \in H_0^1(\Omega)$$

is a bounded linear functional in $H_0^1(\Omega)$. If we denote the space with the inner product $\langle u, v \rangle$ by $\hat{H}_0^1(\Omega)$, then from the equivalence, $F(v)$ is also a bounded linear functional in $\hat{H}_0^1(\Omega)$. Thus by Riesz's representation theorem, there exists a unique $u \in \hat{H}_0^1(\Omega)$, such that

$$\langle u, v \rangle = F(v) = \int_\Omega fv dx, \quad \forall v \in \hat{H}_0^1(\Omega)$$

i.e.

$$\int_\Omega \nabla u \cdot \nabla v dx = \int_\Omega fv dx, \quad \forall v \in \hat{H}_0^1(\Omega).$$

This shows the unique existence of the weak solution of the Dirichlet problem (2.1.1), (2.1.2).

Theorem 2.1.1 *For any $f \in L^2(\Omega)$, the Dirichlet problem (2.1.1), (2.1.2) admits a unique weak solution.*

2.1.3 Transformation of the problem

Now we turn to another useful method, variational method, which can be applied to not only a wide class of linear elliptic equations, but also a certain kind of quasilinear elliptic equations.

First of all, let us observe the following fact: if $x = (x_1, \cdots, x_n)$ is a minimizer of the quadratic form

$$\Psi(y) = \frac{1}{2} y A y^T - b y^T, \quad y \in \mathbb{R}^n,$$

where A is a positive definite matrix and b is a given vector (existence of the minimizer is obvious), then for any $y \in \mathbb{R}^n$, $F(\varepsilon) = \Psi(x + \varepsilon y)$, as a function of $\varepsilon \in \mathbb{R}$, achieves its minimum at $\varepsilon = 0$ and hence $F'(0) = 0$. Since

$$F'(\varepsilon) = \frac{d}{d\varepsilon} \left(\frac{1}{2}(x + \varepsilon y) A (x + \varepsilon y)^T - b(x + \varepsilon y)^T \right)$$
$$= y A x^T + \varepsilon y A y^T - y b^T,$$

we have

$$F'(0) = y \left(A x^T - b^T \right) = 0, \quad \forall y \in \mathbb{R}^n$$

and hence

$$A x^T = b^T$$

because of the arbitrariness of $y \in \mathbb{R}^n$.

This shows that, to solve a system of linear algebraic equations, it suffices to find a minimizer of its corresponding quadratic form.

The basic idea revealed above is available to differential equations. According to this idea, to solve a given problem for some differential equation, one tries to find its corresponding functional and then to minimize it in a suitable function space. Of course, doing in this way is not always successful for any problem of differential equations. However a wide class of differential equations do have their corresponding functionals. It will be seen soon that the functional corresponding to Poisson's equation (2.1.1) is

$$J[v] = \frac{1}{2} \int_\Omega |\nabla v|^2 dx - \int_\Omega f v dx.$$

If $u \in H_0^1(\Omega)$ is an extremal of $J[v]$ in $H_0^1(\Omega)$, then, for any $\varphi \in H_0^1(\Omega)$, as a function of ε,

$$F(\varepsilon) = J[u + \varepsilon\varphi] = \frac{1}{2} \int_\Omega |\nabla(u + \varepsilon\varphi)|^2 dx - \int_\Omega f(u + \varepsilon\varphi)dx$$

achieves its extremum at $\varepsilon = 0$ and hence $F'(0) = 0$. Since

$$F'(\varepsilon) = \int_\Omega (\nabla u + \varepsilon\nabla\varphi)\nabla\varphi dx - \int_\Omega f\varphi dx,$$

$F'(0) = 0$ implies (2.1.5) for any $\varphi \in H_0^1(\Omega)$. Thus we arrive at

Proposition 2.1.1   *If $u \in H_0^1(\Omega)$ is an extremal of the functional $J[v]$ in $H_0^1(\Omega)$, then u is a weak solution of the Dirichlet problem (2.1.1), (2.1.2).*

The solvability of the Dirichlet problem (2.1.1), (2.1.2) is then transformed to the existence of extremals of its corresponding variational problem.

2.1.4 *Existence of minimizers of the corresponding functional*

Lemma 2.1.1   *For any $f \in L^2(\Omega)$, the functional $J[v]$ is bounded from below in $H_0^1(\Omega)$.*

Proof.   By Poincaré's inequality (§1.3.6) and Cauchy's inequality with ε (§1.1.1), we have, for any $v \in H_0^1(\Omega)$,

$$J[v] \geq \frac{1}{2\mu} \int_\Omega v^2 dx - \int_\Omega \left(\frac{\varepsilon}{2}v^2 + \frac{1}{2\varepsilon}f^2\right) dx$$

$$= \frac{1}{2}\left(\frac{1}{\mu} - \varepsilon\right) \int_\Omega v^2 dx - \frac{1}{2\varepsilon} \int_\Omega f^2 dx,$$

where $\mu > 0$ is the constant in Poincaré's inequality, $\varepsilon > 0$ is a constant to be chosen such that $\varepsilon \leq \dfrac{1}{\mu}$. Then from the above inequality, we obtain

$$J[v] \geq -\frac{1}{2\varepsilon} \int_\Omega f^2 dx$$

and the boundedness from below of $J[v]$ in $H_0^1(\Omega)$ is proved. $\square$

Lemma 2.1.2   *For any $v \in H_0^1(\Omega)$ and $f \in L^2(\Omega)$,*

$$\int_\Omega |\nabla v|^2 dx \leq 4\mu \int_\Omega f^2 dx + 4J[v], \qquad (2.1.7)$$

$$\int_\Omega v^2 dx \leq 4\mu^2 \int_\Omega f^2 dx + 4\mu J[v], \tag{2.1.8}$$

where $\mu > 0$ *is the constant in Poincaré's inequality.*

Proof. Using Cauchy's inequality with ε and Poincaré's inequality, we have

$$\int_\Omega |\nabla v|^2 dx = 2 \int_\Omega fv dx + 2J[v]$$
$$\leq \varepsilon \int_\Omega v^2 dx + \frac{1}{\varepsilon} \int_\Omega f^2 dx + 2J[v]$$
$$\leq \varepsilon\mu \int_\Omega |\nabla v|^2 dx + \frac{1}{\varepsilon} \int_\Omega f^2 dx + 2J[v].$$

Choose $\varepsilon = \dfrac{1}{2\mu}$. Then

$$\int_\Omega |\nabla v|^2 dx \leq \frac{1}{2} \int_\Omega |\nabla v|^2 dx + 2\mu \int_\Omega f^2 dx + 2J[v]$$

and (2.1.7) follows.

Using Poincaré's inequality to the left side of (2.1.7) we further obtain (2.1.8). $\qquad\square$

By Lemma 2.1.1, $J[v]$ is bounded from below in $H_0^1(\Omega)$ and hence $\inf\limits_{H_0^1(\Omega)} J[v]$ is a finite number. The definition of infimum then implies that there exists $u_k \in H_0^1(\Omega)$, such that

$$\lim_{k \to \infty} J[u_k] = \inf_{H_0^1(\Omega)} J[v].$$

$\{u_k\}$ is called a minimizing sequence of $J[v]$ in $H_0^1(\Omega)$.

Existence of the limit $\lim\limits_{k \to \infty} J[u_k]$ implies the boundedness of $J[u_k]$, i.e. for some constant M,

$$|J[u_k]| \leq M, \quad k = 1, 2, \cdots.$$

From this and Lemma 2.1.2, we see that $\{u_k\}$ is bounded in $H_0^1(\Omega)$, i.e. $\{u_k\}$ and $\{\nabla u_k\}$ are bounded in $L^2(\Omega)$, which implies the existence of a subsequence $\{u_{k_i}\}$ of $\{u_k\}$ and a function $u \in H_0^1(\Omega)$, such that $u_{k_i} \rightharpoonup u$, $\nabla u_{k_i} \rightharpoonup \nabla u$ $(i \to \infty)$ in $L^2(\Omega)$. In particular, we have

$$\lim_{i \to \infty} \int_\Omega f u_{k_i} dx = \int_\Omega f u dx.$$

Moreover, from

$$\int_\Omega |\nabla(u_{k_i} - u)|^2 dx \geq 0$$

i.e.

$$\int_\Omega |\nabla u_{k_i}|^2 dx \geq 2 \int_\Omega \nabla u_{k_i} \cdot \nabla u dx - \int_\Omega |\nabla u|^2 dx,$$

it follows

$$\varliminf_{i\to\infty} \int_\Omega |\nabla u_{k_i}|^2 dx \geq 2 \lim_{i\to\infty} \int_\Omega \nabla u_{k_i} \cdot \nabla u dx - \int_\Omega |\nabla u|^2 dx$$

$$= 2 \int_\Omega |\nabla u|^2 dx - \int_\Omega |\nabla u|^2 dx$$

$$= \int_\Omega |\nabla u|^2 dx.$$

So

$$\varliminf_{i\to\infty} J[u_{k_i}] = \frac{1}{2} \varliminf_{i\to\infty} \int_\Omega |\nabla u_{k_i}|^2 dx - \lim_{i\to\infty} \int_\Omega f u_{k_i} dx$$

$$\geq \frac{1}{2} \int_\Omega |\nabla u|^2 dx - \int_\Omega f u dx$$

$$= J[u]$$

and hence

$$\inf_{H_0^1(\Omega)} J[v] \leq J[u] \leq \varliminf_{i\to\infty} J[u_{k_i}] = \lim_{i\to\infty} J[u_{k_i}] = \inf_{H_0^1(\Omega)} J[v].$$

Thus $J[u] = \inf\limits_{H_0^1(\Omega)} J[v]$, i.e. u is a minimizer of $J[v]$ in $H_0^1(\Omega)$.

Proposition 2.1.2 *For any $f \in L^2(\Omega)$, the functional $J[v]$ admits a minimizer in $H_0^1(\Omega)$.*

Combining Proposition 2.1.1 with Proposition 2.1.2, we obtain again the existence of weak solutions of (2.1.1), (2.1.2).

The uniqueness of weak solutions can also be proved in the following way. Let $u_1, u_2 \in H_0^1(\Omega)$ be weak solutions of (2.1.1), (2.1.2). Then by the definition of weak solutions, we have

$$\int_\Omega \nabla u_i \cdot \nabla \varphi dx = \int_\Omega f \varphi dx, \quad \varphi \in C_0^\infty(\Omega) \quad (i = 1, 2),$$

and hence

$$\int_\Omega \nabla u_i \cdot \nabla \varphi dx = \int_\Omega f\varphi dx, \quad \varphi \in H_0^1(\Omega) \quad (i = 1, 2).$$

Denote $u = u_1 - u_2$. Then

$$\int_\Omega \nabla u \cdot \nabla \varphi dx = 0, \quad \varphi \in H_0^1(\Omega).$$

In particular, choosing $\varphi = u$ gives

$$\int_\Omega |\nabla u|^2 dx = 0.$$

Thus $\nabla u = 0$ a.e. in Ω and using the homogeneous boundary value condition yields $u = 0$ a.e. in Ω, which can also be derived from Poincaré's inequality

$$\int_\Omega u^2 dx \le C(n, \Omega) \int_\Omega |\nabla u|^2 dx = 0.$$

2.2 Regularity of Weak Solutions of Poisson's Equation

The weak solution obtained in §2.1 is a function in $H_0^1(\Omega)$. In this section we investigate the regularity of weak solutions in $H_0^1(\Omega)$. The finite difference method is one of the important methods in studying the regularity of solutions. The basic idea of this method is to obtain the differentiability of the solution by investigating its difference quotient. Although the method can be used to the general elliptic equations, we confine ourselves to Poisson's equation in order to make the exposition simple and concise.

2.2.1 *Difference operators*

Definition 2.2.1 For a function $u(x)$ in $\mathbb{R}^n$, denote

$$\Delta_h^i u(x) = \frac{u(x + he_i) - u(x)}{h}$$

and call Δ_h^i the difference operator in x_i, where e_i is the unit vector in the direction x_i.

Proposition 2.2.1 *Difference operators possess the following properties:*

i) The conjugate operator of Δ_h^i, denoted by $\Delta_h^i{}^*$ is just the operator $-\Delta_{-h}^i$, i.e. for any $f(x)$, $g(x) \in L^2(\mathbb{R}^n)$ with compact support, there holds

$$\int_{\mathbb{R}^n} f(x)\Delta_h^i g(x)dx = -\int_{\mathbb{R}^n} g(x)\Delta_{-h}^i f(x)dx;$$

ii) Δ_h^i is commutative with any differential operator, i.e. for any weakly differentiable function u, there hold

$$D_j\Delta_h^i u = \Delta_h^i D_j u \quad (j = 1, 2, \cdots, n);$$

iii) The difference of the product can be expressed as

$$\Delta_h^i(f(x)g(x)) = \Delta_h^i f(x)T_h^i g(x) + f(x)\Delta_h^i g(x),$$

where T_h^i is the translation operator in the direction x_i, defined by

$$T_h^i u(x) = u(x + he_i).$$

We leave the proof to the reader.

The following important properties on the difference of functions in Sobolev spaces are needed for our argument.

Proposition 2.2.2 *Let $\Omega \subset \mathbb{R}^n$ be a domain and $i = 1, \cdots, n$.*

i) If $u \in H^1(\Omega)$ and $\Omega' \subset\subset \Omega$, then for sufficiently small $|h| > 0$, $\Delta_h^i u \in L^2(\Omega')$ and

$$\|\Delta_h^i u\|_{L^2(\Omega')} \le \|D_i u\|_{L^2(\Omega)}.$$

ii) If $u \in L^2(\Omega)$ and for $\Omega' \subset\subset \Omega$ and sufficiently small $|h| > 0$,

$$\|\Delta_h^i u\|_{L^2(\Omega')} \le K$$

with constant K independent of h, then $D_i u \in L^2(\Omega')$ and

$$\|D_i u\|_{L^2(\Omega')} \le K.$$

Proof. i) Suppose for the moment $u \in C^1(\Omega) \cap H^1(\Omega)$. Then

$$\Delta_h^i u(x) = \frac{u(x + he_i) - u(x)}{h}$$

$$= \frac{1}{h}\int_0^h D_i u(x_1, \cdots, x_{i-1}, x_i + \theta, x_{i+1}, \cdots, x_n)d\theta.$$

Using Schwartz's inequality gives

$$|\Delta_h^i u(x)|^2 \le \frac{1}{|h|} \left| \int_0^h |D_i u(x_1, \cdots, x_{i-1}, x_i + \theta, x_{i+1}, \cdots, x_n)|^2 d\theta \right|.$$

Integrating over Ω' leads to

$$\int_{\Omega'} |\Delta_h^i u(x)|^2 dx$$

$$\le \frac{1}{|h|} \left| \int_0^h \int_{\Omega'} |D_i u(x_1, \cdots, x_{i-1}, x_i + \theta, x_{i+1}, \cdots, x_n)|^2 dx d\theta \right|$$

$$\le \frac{1}{|h|} \left| \int_0^h \int_{\Omega'_{|h|}} |D_i u(x_1, \cdots, x_{i-1}, x_i, x_{i+1}, \cdots, x_n)|^2 dx d\theta \right|$$

$$= \int_{\Omega'_{|h|}} |D_i u(x_1, \cdots, x_{i-1}, x_i, x_{i+1}, \cdots, x_n)|^2 dx$$

$$\le \int_{\Omega} |D_i u(x)|^2 dx,$$

where $\Omega'_{|h|} = \{x \in \mathbb{R}^n; \text{dist}(x, \Omega') \le |h|\} \subset \Omega$ for small $|h|$. Thus the desired conclusion is proved for $u \in C^1(\Omega) \cap H^1(\Omega)$. By approximation it can be carried over to $u \in H^1(\Omega)$.

ii) By the weak compactness of any bounded subset in $L^2(\Omega')$, there exist a sequence $\{h_k\}$ and a function $v \in L^2(\Omega')$ with $\|v\|_{L^2(\Omega')} \le K$, such that for any $\varphi \in C_0^\infty(\Omega')$,

$$\int_{\Omega'} \varphi \Delta_{h_k}^i u \, dx \to \int_{\Omega'} \varphi v \, dx.$$

However

$$\int_{\Omega'} \varphi \Delta_{h_k}^i u \, dx = - \int_{\Omega'} u \Delta_{-h_k}^i \varphi \, dx \to - \int_{\Omega'} u D_i \varphi \, dx.$$

Thus

$$\int_{\Omega'} \varphi v \, dx = - \int_{\Omega'} u D_i \varphi \, dx,$$

i.e. $v = D_i u$. $\qquad\square$

Similarly we can prove

Proposition 2.2.3 *Let* $B_1^+(0) = \{x \in B_1(0); x_n > 0\}, 0 < \rho < 1, i = 1, 2 \cdots, n-1$.

i) If $u \in H^1(B_1^+(0))$, then for sufficiently small $|h| > 0$, $\Delta_h^i u \in H^1(B_\rho^+(0))$ and

$$\|\Delta_h^i u\|_{L^2(B_\rho^+(0))} \leq \|D_i u\|_{L^2(B_1^+(0))}.$$

ii) If $u \in L^2(B_1^+(0))$ and for sufficiently small $|h| > 0$,

$$\|\Delta_h^i u\|_{L^2(B_\rho^+(0))} \leq K$$

with constant K independent of h, then $D_i u \in L^2(B_\rho^+(0))$ and

$$\|D_i u\|_{L^2(B_\rho^+(0))} \leq K.$$

2.2.2 Interior regularity

Now we proceed to discuss the regularity of weak solutions of Poisson's equation. First we discuss the interior regularity.

Theorem 2.2.1   *Let $f \in L^2(\Omega)$, and $u \in H^1(\Omega)$ be a weak solution of equation (2.1.1). Then for any subdomain $\Omega' \subset\subset \Omega$, $u \in H^2(\Omega')$ and*

$$\|u\|_{H^2(\Omega')} \leq C \left(\|u\|_{H^1(\Omega)} + \|f\|_{L^2(\Omega)} \right),$$

where C is a constant depending only on n and $\mathrm{dist}\{\Omega', \partial\Omega\}$.

Proof.   For fixed $\Omega' \subset\subset \Omega$, denote $d = \dfrac{1}{4}\mathrm{dist}(\Omega', \partial\Omega)$. Choose a cut-off function on Ω relative to Ω', i.e. a function $\eta(x) \in C_0^\infty(\Omega)$, such that

$$0 \leq \eta(x) \leq 1, \quad \eta(x) \equiv 1 \text{ in } \Omega', \quad \mathrm{dist}\{\mathrm{supp}\eta, \partial\Omega\} \geq 2d.$$

To prove the conclusion of Theorem 2.2.1, by Proposition 2.2.2, it suffices to derive the estimate

$$\int_\Omega \eta^2 |\Delta_h^i \nabla u|^2 dx \leq C(\|u\|_{H^1(\Omega)}^2 + \|f\|_{L^2(\Omega)}^2) \tag{2.2.1}$$

for some constant C independent of h.

To establish any estimate on weak solutions, the original starting point is the definition of weak solutions, i.e. the integral identity

$$\int_\Omega \nabla u \cdot \nabla \varphi dx = \int_\Omega f\varphi dx, \quad \forall \varphi \in H_0^1(\Omega). \tag{2.2.2}$$

The crucial step is to choose a suitable test function φ. In the present case, since using Proposition 2.2.1 i), ii), we have

$$\int_\Omega \eta^2 |\Delta_h^i \nabla u|^2 dx$$

$$= \int_\Omega \Delta_h^i \nabla u \cdot \nabla(\eta^2 \Delta_h^i u) dx - 2 \int_\Omega \Delta_h^i \nabla u \cdot \eta \Delta_h^i u \nabla \eta dx$$

$$= \int_\Omega \nabla u \cdot \nabla \Delta_h^{i^*}(\eta^2 \Delta_h^i u) dx - 2 \int_\Omega \eta \Delta_h^i u \nabla \eta \cdot \Delta_h^i \nabla u dx, \qquad (2.2.3)$$

it is natural to choose $\varphi = \Delta_h^{i^*}(\eta^2 \Delta_h^i u)$ (clearly it belongs to $H_0^1(\Omega)$) in (2.2.2). Thus we obtain

$$\int_\Omega \nabla u \cdot \nabla \Delta_h^{i^*}(\eta^2 \Delta_h^i u) dx = \int_\Omega f \Delta_h^{i^*}(\eta^2 \Delta_h^i u) dx,$$

which combined with (2.2.3) gives

$$\int_\Omega \eta^2 |\Delta_h^i \nabla u|^2 dx = \int_\Omega f \Delta_h^{i^*}(\eta^2 \Delta_h^i u) dx - 2 \int_\Omega \eta \Delta_h^i u \nabla \eta \cdot \Delta_h^i \nabla u dx.$$

Now we use Cauchy's inequality with ε to the integrals on the right side of the above formula to obtain

$$\int_\Omega \eta^2 |\Delta_h^i \nabla u|^2 dx$$

$$\leq \frac{1}{2\varepsilon} \int_\Omega f^2 dx + 2\varepsilon \int_\Omega |\Delta_h^{i^*}(\eta^2 \Delta_h^i u)|^2 dx$$

$$+ \frac{1}{\varepsilon} \int_\Omega |\nabla \eta|^2 |\Delta_h^i u|^2 dx + \varepsilon \int_\Omega \eta^2 |\Delta_h^i \nabla u|^2 dx. \qquad (2.2.4)$$

Using Proposition 2.2.2 i), we have

$$\int_\Omega |\Delta_h^{i^*}(\eta^2 \Delta_h^i u)|^2 dx = \int_{\text{supp}\eta} |\Delta_{-h}^i(\eta^2 \Delta_h^i u)|^2 dx$$

$$\leq \int_\Omega |D_i(\eta^2 \Delta_h^i u)|^2 dx$$

$$\leq \int_\Omega |\nabla(\eta^2 \Delta_h^i u)|^2 dx$$

$$= \int_\Omega |\eta^2 \nabla \Delta_h^i u + 2\eta \Delta_h^i u \nabla \eta|^2 dx$$

$$\leq 2 \int_\Omega |\eta^2 \nabla \Delta_h^i u|^2 dx + 2 \int_\Omega |2\eta \Delta_h^i u \nabla \eta|^2 dx$$

$$\leq 2 \int_\Omega \eta^2 |\nabla \Delta_h^i u|^2 dx + 8 \int_\Omega |\nabla \eta|^2 |\Delta_h^i u|^2 dx$$

and

$$\int_\Omega |\nabla \eta|^2 |\Delta_h^i u|^2 dx \leq C \int_{\mathrm{supp}\eta} |\Delta_h^i u|^2 dx \leq C \int_\Omega |D_i u|^2 dx \leq C \int_\Omega |\nabla u|^2 dx.$$

Combining these with (2.2.4), we finally obtain

$$(1 - 5\varepsilon) \int_\Omega \eta^2 |\Delta_h^i \nabla u|^2 dx \leq C \left(\frac{1}{\varepsilon} + 16\varepsilon \right) \int_\Omega |\nabla u|^2 dx + \frac{1}{2\varepsilon} \int_\Omega f^2 dx$$

and the desired conclusion (2.2.1) follows by choosing ε suitably small. $\square$

Corollary 2.2.1   *Let u be a weak solution of equation (2.1.1). If $f \in H^k(\Omega)$ for some nonnegative integer k, then, for any subdomain $\Omega' \subset\subset \Omega$, $u \in H^{k+2}(\Omega')$ and*

$$\|u\|_{H^{k+2}(\Omega')} \leq C \left(\|u\|_{H^1(\Omega)} + \|f\|_{H^k(\Omega)} \right),$$

where C is a constant depending only on k, n and $\mathrm{dist}\{\Omega', \partial\Omega\}$.

Proof.   First consider the case $k = 1$. By the definition of weak solutions,

$$\int_\Omega \nabla u \cdot \nabla D_i \varphi dx = \int_\Omega f D_i \varphi dx, \quad \forall \varphi \in C_0^\infty(\Omega) \quad (i = 1, \cdots, n).$$

Since, by Theorem 2.2.1, for any subdomain $\Omega' \subset\subset \Omega$, $u \in H^2(\Omega')$ and $f \in H^1(\Omega)$ is assumed, we can integrate by parts in the above formula to derive

$$\int_\Omega \nabla D_i u \cdot \nabla \varphi dx = \int_\Omega D_i f \varphi dx, \quad \forall \varphi \in C_0^\infty(\Omega) \quad (i = 1, \cdots, n).$$

This shows that $D_i u$ is a weak solution of the equation

$$-\Delta v = D_i f, \quad x \in \Omega \quad (i = 1, \cdots, n)$$

and hence we can use Theorem 2.2.1 to assert $D_i u \in H^2(\Omega')$ for any subdomain $\Omega' \subset\subset \Omega$ and obtain the estimate

$$\|u\|_{H^3(\Omega')} \leq C \left(\|u\|_{H^1(\Omega)} + \|f\|_{H^1(\Omega)} \right).$$

By induction, we can prove the conclusion of Corollary 2.2.1 for any positive integer k. $\square$

Corollary 2.2.2   *If $f \in H^k(\Omega)$ with $k > \dfrac{n}{2}$, then the weak solution u of equation (2.1.1) satisfies the equation $-\Delta u = f(x)$ in Ω in the classical sense.*

Proof.   By Corollary 2.2.1, for any subdomain $\Omega' \subset\subset \Omega$, $u \in H^{k+2}(\Omega')$. Since $k > \dfrac{n}{2}$, by the embedding theorem, $H^{k+2}(\Omega') \hookrightarrow C^{2,\alpha}(\Omega')$ with $0 < \alpha \leq 1 - \dfrac{n}{2k}$. $\square$

2.2.3   Regularity near the boundary

Proposition 2.2.4   *Let $\Omega \subset \mathbb{R}^n$ be a bounded domain with $\partial\Omega \in C^2$ and $y = \Psi(x)$ be a local flatting mapping in a neighborhood of the given point $x^0 \in \partial\Omega$ (see §1.1.6). Then for any weak solution of Poisson's equation (2.1.1), $\hat{u}(y) = u(\Psi^{-1}(y))$ is a weak solution of the equation*

$$-\frac{\partial}{\partial y_j}\left(\hat{a}_{ij}(y)\frac{\partial \hat{u}}{\partial y_i}\right) = \hat{f}(y), \quad y \in B_1^+ = B_1^+(0),$$

namely, for any $\varphi \in C_0^\infty(B_1^+)$, there holds

$$\int_{B_1^+} \hat{a}_{ij}(y)\frac{\partial \hat{u}}{\partial y_i} \cdot \frac{\partial \varphi}{\partial y_j}dy = \int_{B_1^+} \hat{f}(y)\varphi(y)dy, \tag{2.2.5}$$

where $\hat{a}_{ij}(y)$ is the (i,j) element of the matrix

$$(\hat{a}_{ij}(y))_{n\times n} = |J(y)|\Psi'(\Psi^{-1}(y))\Psi'(\Psi^{-1}(y))^T,$$

$$\hat{f}(y) = |J(y)|f(\Psi^{-1}(y)),$$

and $x = \Psi^{-1}(y)$ is the inverse mapping of $y = \Psi(x)$, $J(y)$ the Jacobi determinant of the mapping $x = \Psi^{-1}(y)$, $\Psi'(x)$ the derivative matrix of the mapping $y = \Psi(x)$ and $\Psi'(x)^T$ the transposed matrix of $\Psi'(x)$.

In this book, repeated indices denote summation from 1 to n if there is no other indication. The proof of the above proposition is left to the reader.

Remark 2.2.1   *In the formula (2.2.5) and the formula before it, repeated indices imply a summation from 1 to n. Such summation convention will be adopted frequently in the sequel.*

Theorem 2.2.2   *Let $f \in L^2(\Omega)$ and $u \in H_0^1(\Omega)$ be a weak solution of the Dirichlet problem (2.1.1), (2.1.2). If $\partial\Omega \in C^2$, then for any $x^0 \in \partial\Omega$, there* *exists a neighborhood U of x^0, such that $u \in H^2(U \cap \Omega)$ and*

$$\|u\|_{H^2(U \cap \Omega)} \le C \left(\|u\|_{H^1(\Omega)} + \|f\|_{L^2(\Omega)} \right),$$

where C is a constant depending only on n and $\Omega \cap U$.

Proof. By Proposition 2.2.4, there exists a neighborhood U_1 of x^0 and a C^2 invertible mapping $\Psi : U_1 \to B_1(0)$, such that

$$\Psi(U_1 \cap \Omega) = B_1^+ = B_1^+(0), \quad \Psi(U_1 \cap \partial\Omega) = \partial B_1^+ \cap \{y \in \mathbb{R}^n; y_n = 0\}$$

and $\hat{u}(y) = u(\Psi^{-1}(y))$ satisfies (2.2.5) for any $\varphi \in C_0^\infty(B_1^+)$ and hence for any $\varphi \in H_0^1(B_1^+)$.

Choose a cut-off function $\eta(y)$ in B_1 relative to $B_{1/2}$. We first estimate the integral

$$\int_{B_1^+} \eta^2 |\Delta_h^k \nabla \hat{u}|^2 dy,$$

where $\Delta_h^k (k = 1, 2, \cdots, n-1)$ is the tangential difference operator.

Similar to the derivation of (2.2.1), we choose $\varphi = \Delta_h^{k^*}(\eta^2 \Delta_h^k \hat{u})$ in (2.2.5). Here and below, the repeated indices for k do not mean a summation. Since $u \in H_0^1(\Omega)$ and Δ_h^k is a tangential difference operator, it is easy to see that for sufficiently small $|h|$, $\varphi = \Delta_h^{k^*}(\eta^2 \Delta_h^k \hat{u}) \in H_0^1(B_1^+)$. Thus we can take φ as a test function in (2.2.5) to obtain

$$\int_{B_1^+} \hat{a}_{ij} \frac{\partial \hat{u}}{\partial y_i} \cdot \frac{\partial}{\partial y_j} \left[\Delta_h^{k^*}(\eta^2 \Delta_h^k \hat{u}) \right] dy = \int_{B_1^+} \hat{f}(y) \Delta_h^{k^*}(\eta^2 \Delta_h^k \hat{u}) dy.$$

Using Proposition 2.2.1 i), ii) further gives

$$\int_{B_1^+} \Delta_h^k \left(\hat{a}_{ij} \frac{\partial \hat{u}}{\partial y_i} \right) \frac{\partial}{\partial y_j} (\eta^2 \Delta_h^k \hat{u}) dy = \int_{B_1^+} \hat{f}(y) \Delta_h^{k^*}(\eta^2 \Delta_h^k \hat{u}) dy.$$

Since, using Proposition 2.2.1 iii), ii), we have

$$\Delta_h^k \left(\hat{a}_{ij} \frac{\partial \hat{u}}{\partial y_i} \right) = T_h^k \hat{a}_{ij} \frac{\partial \Delta_h^k \hat{u}}{\partial y_i} + \Delta_h^k \hat{a}_{ij} \frac{\partial \hat{u}}{\partial y_j},$$

we are led to

$$\int_{B_1^+} \eta^2 T_h^k \hat{a}_{ij} \frac{\partial \Delta_h^k \hat{u}}{\partial y_i} \cdot \frac{\partial \Delta_h^k \hat{u}}{\partial y_j} dy$$

$$= -\int_{B_1^+} \eta^2 \Delta_h^k \hat{a}_{ij} \frac{\partial \Delta_h^k \hat{u}}{\partial y_j} \cdot \frac{\partial \hat{u}}{\partial y_i} dy - 2 \int_{B_1^+} \eta \frac{\partial \eta}{\partial y_j} \Delta_h^k \hat{a}_{ij} \Delta_h^k \hat{u} \frac{\partial \hat{u}}{\partial y_i} dy$$

$$-2\int_{B_1^+}\eta\frac{\partial\eta}{\partial y_j}T_h^k\hat{a}_{ij}\Delta_h^k\hat{u}\frac{\partial\Delta_h^k\hat{u}}{\partial y_i}dy+\int_{B_1^+}\hat{f}(y)\Delta_h^{k*}(\eta^2\Delta_h^k\hat{u})dy.$$

From this we can proceed similar to the proof of the interior estimate in Theorem 2.2.1 to derive

$$\int_{B_{1/2}^+}|\Delta_h^k\nabla\hat{u}|^2dy\leq C\int_{B_1^+}|\nabla\hat{u}|^2dy+C\int_{B_1^+}\hat{f}^2dy. \qquad (2.2.6)$$

To do this, it is to be noted that, since $\partial\Omega\in C^2$, we have $\hat{a}_{ij}\in C^1$ and hence $|\Delta_h^k\hat{a}_{ij}|\leq M$ for some constant M. Another more important fact to be noted is that there exist constants $\lambda>0$, $h_0>0$, such that $T_h^k\hat{a}_{ij}\xi_i\xi_j\geq\lambda|\xi|^2$, provided $0<|h|\leq h_0$.

Using Proposition 2.2.3 ii), from (2.2.6) we see that for any $k=1,2,\cdots,n-1$ and $j=1,2,\cdots,n$, $\dfrac{\partial^2\hat{u}}{\partial y_k\partial y_j}\in L^2(B_{1/2}^+)$ and

$$\int_{B_{1/2}^+}\left|\frac{\partial^2\hat{u}}{\partial y_k\partial y_j}\right|^2dy\leq C\left(\int_{B_1^+}|\nabla\hat{u}|^2dy+\int_{B_1^+}\hat{f}^2dy\right). \qquad (2.2.7)$$

Now we rewrite (2.2.5) as

$$\int_{B_1^+}\hat{a}_{nn}\frac{\partial\hat{u}}{\partial y_n}\cdot\frac{\partial\varphi}{\partial y_n}dy=\int_{B_1^+}\hat{f}(y)\varphi(y)dy-\sum_{i+j<2n}\int_{B_1^+}\hat{a}_{ij}\frac{\partial\hat{u}}{\partial y_i}\cdot\frac{\partial\varphi}{\partial y_j}dy,$$

or, after integrating by parts in the second term of the right side,

$$\int_{B_1^+}\hat{a}_{nn}\frac{\partial\hat{u}}{\partial y_n}\cdot\frac{\partial\varphi}{\partial y_n}dy=\int_{B_1^+}\left(\hat{f}(y)+\sum_{i+j<2n}\frac{\partial}{\partial y_j}\left(\hat{a}_{ij}\frac{\partial\hat{u}}{\partial y_i}\right)\right)\varphi(y)dy.$$

From this and (2.2.7) it follows that $\dfrac{\partial}{\partial y_n}\left(\hat{a}_{nn}\dfrac{\partial\hat{u}}{\partial y_n}\right)\in L^2(B_{1/2}^+)$. Since it is easy to verify that $\hat{a}_{nn}\neq 0$ in $B_{1/2}^+$, we further have $\dfrac{\partial^2\hat{u}}{\partial y_n^2}\in L^2(B_{1/2}^+)$ and

$$\int_{B_{1/2}^+}\left|\frac{\partial^2\hat{u}}{\partial y_n^2}\right|^2dy\leq C\int_{B_1^+}|\nabla\hat{u}|^2dy+C\int_{B_1^+}\hat{f}^2dy.$$

This combined with (2.2.7) implies $\hat{u}(y)\in H^2(B_{1/2}^+)$. Changing the variable y to the original one shows that $u\in H^2(U\cap\Omega)$ with $U=\Psi^{-1}(B_{1/2}^+)$ and u satisfies the estimate in Theorem 2.2.2. $\qquad\square$

Similar to the interior regularity, we also have

Corollary 2.2.3   *Let $u \in H_0^1(\Omega)$ be a weak solution of the Dirichlet problem (2.1.1), (2.1.2). If $\partial\Omega \in C^{k+2}$ and $f \in H^k(\Omega)$ for some nonnegative integer k, then for any $x^0 \in \partial\Omega$, there exists a neighborhood U of x^0, such that $u \in H^{k+2}(U \cap \Omega)$ and*

$$\|u\|_{H^{k+2}(U\cap\Omega)} \le C \left(\|u\|_{H^1(\Omega)} + \|f\|_{H^k(\Omega)} \right),$$

where C is a constant depending only on k, n and $\Omega \cap U$.

Corollary 2.2.4   *If $\partial\Omega \in C^{k+2}$, $f \in H^k(\Omega)$ and $k > \dfrac{n}{2}$, then for any $x^0 \in \partial\Omega$, there exists a neighborhood U of x^0, such that any weak solution u of the Dirichlet problem (2.1.1), (2.1.2) belongs to $C^{2,\alpha}(\overline{U \cap \Omega})$ with $0 < \alpha \le 1 - \dfrac{n}{2k}$.*

2.2.4   Global regularity

To prove the global regularity of weak solutions, we choose a finite open covering of $\overline{\Omega}$ and decompose the solutions by means of the partition of unity (§1.1.5).

Theorem 2.2.3   *Let $f \in L^2(\Omega)$ and $u \in H_0^1(\Omega)$ be a weak solution of the Dirichlet problem (2.1.1), (2.1.2). If $\partial\Omega \in C^2$, then $u \in H^2(\Omega)$ and*

$$\|u\|_{H^2(\Omega)} \le C \left(\|u\|_{H^1(\Omega)} + \|f\|_{L^2(\Omega)} \right), \tag{2.2.8}$$

where C is a constant depending only on n and Ω.

Proof.   By Theorem 2.2.2, for every $x^0 \in \partial\Omega$, there exists a neighborhood $U(x^0)$ such that $u \in H^2(U(x^0) \cap \Omega)$ and

$$\|u\|_{H^2(U(x^0)\cap\Omega)} \le C \left(\|u\|_{H^1(\Omega)} + \|f\|_{L^2(\Omega)} \right).$$

Using the finite covering theorem we can choose such neighborhoods of finite number $U_1, \cdots, U_N$ to cover $\partial\Omega$. Denote $K = \Omega \setminus \bigcup_{i=1}^{N} U_i$. Then K is a closed subset of Ω and there exists a subdomain $U_0 \subset\subset \Omega$, such that $U_0 \supset K$. Theorem 2.2.1 shows that $u \in H^2(U_0)$, and

$$\|u\|_{H^2(U_0)} \le C \left(\|u\|_{H^1(\Omega)} + \|f\|_{L^2(\Omega)} \right).$$

Using the theorem on the partition of unity, we can choose functions $\eta_0, \eta_1, \cdots, \eta_N$, such that

$$0 \le \eta_i(x) \le 1, \quad \forall x \in U_i \quad (i = 0, 1, \cdots, N),$$

$$\sum_{i=1}^{N} \eta_i(x) = 1, \quad x \in \overline{\Omega}.$$

Thus

$$\|u\|_{H^2(\Omega)} = \left\| \sum_{i=0}^{N} \eta_i u \right\|_{H^2(\Omega)} \le \sum_{i=0}^{N} \|\eta_i u\|_{H^2(\Omega)}$$

$$\le C \left(\|u\|_{H^1(\Omega)} + \|f\|_{L^2(\Omega)} \right).$$

$\square$

Remark 2.2.2 *Under the assumptions of Theorem 2.2.3, we have*

$$\|u\|_{H^2(\Omega)} \le C \|f\|_{L^2(\Omega)}, \tag{2.2.9}$$

where C is a constant depending only on n and Ω.

Proof. We first set $\varphi = u$ in

$$\int_{\Omega} \nabla u \cdot \nabla \varphi \, dx = \int_{\Omega} f \varphi \, dx, \quad \forall \varphi \in H_0^1(\Omega)$$

and use Cauchy's inequality with ε and Poincaré's inequality to obtain

$$\int_{\Omega} |\nabla u|^2 dx = \int_{\Omega} f u \, dx$$

$$\le \frac{\varepsilon}{2} \int_{\Omega} u^2 dx + \frac{1}{2\varepsilon} \int_{\Omega} f^2 dx$$

$$\le \frac{\varepsilon \mu}{2} \int_{\Omega} |\nabla u|^2 dx + \frac{1}{2\varepsilon} \int_{\Omega} f^2 dx,$$

where $\mu > 0$ is the constant in Poincaré's inequality. Choosing $\varepsilon = \dfrac{1}{\mu}$ then gives

$$\int_{\Omega} |\nabla u|^2 dx \le \mu \int_{\Omega} f^2 dx.$$

and further by Poincaré's inequality,

$$\int_{\Omega} u^2 dx \le \mu^2 \int_{\Omega} f^2 dx.$$

Thus the desired inequality (2.2.9) follows by combining the above two estimates and substituting into (2.2.8). □

In the proof of Theorem 2.2.2, the normal derivative of second order is estimated via the equation and the estimates for the tangential derivatives. Repeating this procedure we can estimate higher order derivatives in normal direction.

Theorem 2.2.4 *Let $u \in H_0^1(\Omega)$ be a weak solution of the Dirichlet problem (2.1.1), (2.1.2). If $\partial\Omega \in C^{k+2}$ and $f \in H^k(\Omega)$ for some nonnegative integer k, then $u \in H^{k+2}(\Omega)$ and*

$$\|u\|_{H^{k+2}(\Omega)} \leq C \left(\|u\|_{H^1(\Omega)} + \|f\|_{H^k(\Omega)} \right),$$

where C is a constant depending only on k, n and Ω.

As an immediate corollary of this theorem, we have

Theorem 2.2.5 *Let $u \in H_0^1(\Omega)$ be a weak solution of the Dirichlet problem (2.1.1), (2.1.2). If $\partial\Omega \in C^\infty$ and $f \in C^\infty(\overline{\Omega})$, then $u \in C^\infty(\overline{\Omega})$.*

Remark 2.2.3 *Theorem 2.2.4 shows that $\partial\Omega \in C^{k+2}$, $f \in H^k(\Omega)$ imply $u \in H^{k+2}(\Omega)$ and Theorem 2.2.5 shows that $\partial\Omega \in C^\infty$, $f \in C^\infty(\overline{\Omega})$ imply $u \in C^\infty(\overline{\Omega})$. This means that the conclusions on the regularity of weak solutions are complete both in $H^k(\Omega)$ and $C^\infty(\overline{\Omega})$. In Chapter 8 it will be proved that $f \in C^\alpha(\overline{\Omega})$ ($\alpha \in (0,1)$) implies $u \in C^{2,\alpha}(\overline{\Omega})$, which means that the conclusion on the regularity of solutions is also complete in Hölder spaces. However it is impossible to assert $u \in C^2(\overline{\Omega})$ from $f \in C(\overline{\Omega})$.*

2.2.5 *Study of regularity by means of smoothing operators*

Smoothing operators

$$u_\varepsilon = J_\varepsilon u = \int_\Omega j_\varepsilon(x-y)u(y)dy, \qquad (2.2.10)$$

instead of difference operators can also be applied to the study of regularity of weak solutions, where $j_\varepsilon(x)$ is an arbitrary mollifier. In (2.2.10), u is regarded as zero outside of Ω.

It is easy to check the following facts which are an analog of Proposition 2.2.1 i), ii).

Proposition 2.2.5 *Let $\Omega \subset \mathbb{R}^n$ be a domain.*

i) For any $u, v \in L^1(\Omega)$ vanishing outside of Ω,

$$\int_\Omega u_\varepsilon v dx = \int_\Omega u v_\varepsilon^- dx,$$

where

$$v_\varepsilon^- = J_\varepsilon^- v = \int_\Omega j_\varepsilon^-(x-y)u(y)dy, \quad j_\varepsilon^-(x) = j_\varepsilon(-x).$$

ii) For any $\Omega' \subset\subset \Omega$ and sufficiently small $\varepsilon > 0$,

$$D_i u_\varepsilon = (D_i u)_\varepsilon \quad in \ \Omega' \quad (i = 1, 2, \cdots, n).$$

However we do not have the analog of Proposition 2.2.1 iii) for smoothing operators. This will restrict the application of the present method in the study of regularity.

We also have the following proposition which corresponds to Proposition 2.2.2 and can be proved similarly.

Proposition 2.2.6 *Let $\Omega \subset \mathbb{R}^n$ be a domain and $i = 1, 2, \cdots, n$.*
i) If $u \in H^1(\Omega)$ and $\Omega' \subset\subset \Omega$, then for sufficiently small $\varepsilon > 0$,

$$\|D_i u_\varepsilon\|_{L^2(\Omega')} \leq \|D_i u\|_{L^2(\Omega)}.$$

ii) If $u \in H^1(\Omega)$, $\Omega' \subset\subset \Omega$ and for sufficiently small $\varepsilon > 0$,

$$\|D_i u_\varepsilon\|_{L^2(\Omega')} \leq K$$

with constant K independent of ε, then $D_i u \in L^2(\Omega')$ and

$$\|D_i u\|_{L^2(\Omega')} \leq K.$$

The proof is left to the reader.

We do not state the analog of Proposition 2.2.3, although it does hold.

Now we proceed to use smoothing operators to establish the interior regularity, i.e. to prove Theorem 2.2.1.

Choose a cut-off function $\eta(x)$ as in the proof of Theorem 2.2.1. By Proposition 2.2.6 ii), it suffices to establish the estimate

$$\int_\Omega \eta^2 |D_i \nabla u_\varepsilon|^2 dx \leq C \left(\|u\|_{H^1(\Omega)}^2 + \|f\|_{L^2(\Omega)}^2 \right) \tag{2.2.11}$$

for some constant C independent of ε.

Using Proposition 2.2.5 and integrating by parts, we have

$$\int_\Omega \eta^2 |D_i \nabla u_\varepsilon|^2 dx$$

$$= \int_\Omega D_i \nabla u_\varepsilon \cdot \nabla(\eta^2 D_i u_\varepsilon) dx - 2 \int_\Omega \eta D_i u_\varepsilon \nabla \eta \cdot D_i \nabla u_\varepsilon dx$$

$$= -\int_\Omega \nabla u_\varepsilon \cdot \nabla(D_i(\eta^2 D_i u_\varepsilon)) dx - 2 \int_\Omega \eta D_i u_\varepsilon \nabla \eta \cdot D_i \nabla u_\varepsilon dx$$

$$= -\int_\Omega \nabla u \cdot \nabla(D_i(\eta^2 D_i u_\varepsilon))_\varepsilon^- dx - 2 \int_\Omega \eta D_i u_\varepsilon \nabla \eta \cdot D_i \nabla u_\varepsilon dx.$$

Choosing $\varphi = (D_i(\eta^2 D_i u_\varepsilon))_\varepsilon^-$ in (2.2.2) and combining the resulting equality with the above formula lead to

$$\int_\Omega \eta^2 |D_i \nabla u_\varepsilon|^2 dx = -\int_\Omega f(D_i(\eta^2 D_i u_\varepsilon))_\varepsilon^- dx - 2 \int_\Omega \eta D_i u_\varepsilon \nabla \eta \cdot D_i \nabla u_\varepsilon dx.$$

From this we may deduce (2.2.11) similar to the proof of (2.2.1).

2.3 L^2 Theory of General Elliptic Equations

2.3.1   *Weak solutions*

Now we turn to the following general elliptic equations in divergence form

$$Lu = -D_j(a_{ij} D_i u) + b_i D_i u + cu = f + D_i f^i, \qquad (2.3.1)$$

where a_{ij}, b_i, $c \in L^\infty(\Omega)$, $f \in L^2(\Omega)$, $f^i \in L^2(\Omega)$, and $a_{ij} = a_{ji}$ satisfy the uniform ellipticity condition, i.e. for some constants $0 < \lambda \le \Lambda$,

$$\lambda |\xi|^2 \le a_{ij}(x)\xi_i \xi_j \le \Lambda |\xi|^2, \quad \forall \xi \in \mathbb{R}^n, \ x \in \Omega.$$

In this case, we call (2.3.1) uniformly elliptic equations. Here, repeated indices imply a summation from 1 up to n. As in the preceding sections, only the Dirichlet problem with the homogeneous boundary value condition

$$u\Big|_{\partial\Omega} = 0 \qquad (2.3.2)$$

is discussed.

Remark 2.3.1 *If a nonhomogeneous boundary value condition*

$$u\Big|_{\partial\Omega} = g$$

is prescribed with $g \in H^1(\Omega)$, then, setting $w = u - g$, we can change (2.3.1) to an equation for the new unknown function w,

$$Lw = \tilde{f} + D_i \tilde{f}^i,$$

where

$$\tilde{f} = f - b_i D_i g - cg, \quad \tilde{f}^i = f^i + a_{ij} D_j g.$$

The boundary value condition which w satisfies is then a homogeneous one.

If $u \in C^2(\Omega)$ is a solution of equation (2.3.1), then multiplying (2.3.1) with any $\varphi \in C_0^\infty(\Omega)$ and integrating over Ω lead to, after integrating by parts,

$$\int_\Omega (a_{ij} D_i u D_j \varphi + b_i D_i u \varphi + cu\varphi)\, dx = \int_\Omega (f\varphi - f^i D_i \varphi) dx. \qquad (2.3.3)$$

Conversely, if $u \in C^2(\Omega)$ and for any $\varphi \in C_0^\infty(\Omega)$, (2.3.3) holds, then u satisfies (2.3.1) in the classical sense.

Definition 2.3.1 A function $u \in H^1(\Omega)$ is said to be a weak solution of (2.3.1), if for any $\varphi \in C_0^\infty(\Omega)$, (2.3.3) holds. If, in addition, $u \in H_0^1(\Omega)$, then u is said to be a weak solution of (2.3.1), (2.3.2).

2.3.2 *Riesz's representation theorem and its application*

Riesz's representation theorem applied to Poisson's equation can be carried over to equation (2.3.1) with $b_i = 0 \, (i = 1, \cdots, n)$, i.e. the equation

$$-D_j(a_{ij} D_i u) + cu = f + D_i f^i. \qquad (2.3.4)$$

To this purpose, we define a new inner product in $H_0^1(\Omega)$ as follows:

$$\langle u, v \rangle = \int_\Omega (a_{ij} D_i u D_j v + cuv)\, dx,$$

whose corresponding norm is denoted by $||| \cdot |||$. It is easy to verify that if $c \geq c_0$ for a certain constant c_0, then $\langle \cdot, \cdot \rangle$ possesses all properties of inner product. For example, using the ellipticity condition and Poincaré's inequality, we have, for $u \in H_0^1(\Omega)$,

$$\begin{aligned}
|||u|||^2 = \langle u, u \rangle &= \int_\Omega \left(a_{ij} D_i u D_j u + cu^2 \right) dx \\
&\geq \lambda \int_\Omega |\nabla u|^2 dx + c_0 \int_\Omega u^2 dx \\
&\geq \frac{\lambda}{2} \int_\Omega |\nabla u|^2 dx + \left(\frac{\lambda}{2\mu} + c_0 \right) \int_\Omega u^2 dx \\
&\geq \alpha \|u\|^2,
\end{aligned}$$

provided $\dfrac{\lambda}{2\mu} + c_0 > 0$, where $\|u\|$ is the norm in $H_0^1(\Omega)$, $\mu > 0$ is the constant in Poincaré's inequality and $\alpha = \min\left\{\dfrac{\lambda}{2}, \dfrac{\lambda}{2\mu} + c_0\right\}$. From this it follows that $\langle u, u \rangle = 0$ implies $u = 0$.

Clearly, for a certain constant $\beta > 0$,

$$\|\|u\|\|^2 \le \beta \|u\|^2.$$

Thus

$$\alpha \|u\|^2 \le \|\|u\|\|^2 \le \beta \|u\|^2, \tag{2.3.5}$$

which implies, in particular, that

$$F[v] = \int_\Omega \left(fv - f^i D_i v\right) dx$$

is a bounded linear functional in $\hat{H}_0^1(\Omega)$, the same space as $H_0^1(\Omega)$ endowed with the inner product $\langle \cdot, \cdot \rangle$. Hence, by Riesz's representation theorem (see §2.1.2), there exists a unique $u \in \hat{H}_0^1(\Omega)$ such that

$$\langle u, v \rangle = \int_\Omega \left(a_{ij} D_i u D_j v + cuv\right) dx$$

$$= F[v] = \int_\Omega \left(fv - f^i D_i v\right) dx, \quad \forall v \in \hat{H}_0^1(\Omega).$$

From (2.3.5) we have $u \in H_0^1(\Omega)$ and the above formula implies

$$\int_\Omega \left(a_{ij} D_i u D_j \varphi + cu\varphi\right) dx = \int_\Omega \left(f\varphi - f^i D_i \varphi\right) dx \quad \forall \varphi \in C_0^\infty(\Omega),$$

which means that u is a unique weak solution of (2.3.4), (2.3.2).

Theorem 2.3.1 *There exists a constant c_0 such that for any $f \in L^2(\Omega)$, $f^i \in L^2(\Omega)$ $(i = 1, \cdots, n)$, the Dirichlet problem (2.3.4), (2.3.2) admits a unique weak solution provided $c \ge c_0$.*

2.3.3 *Variational method*

Theorem 2.3.1 can also be proved by means of variational method. The functional corresponding to (2.3.4) is

$$J[v] = \frac{1}{2} \int_\Omega \left(a_{ij} D_i v D_j v + cv^2\right) dx - \int_\Omega \left(fv - f^i D_i v\right) dx.$$

In fact, arguing as in §2.1.3, we may prove that if $u \in H_0^1(\Omega)$ is an extremal of the functional $J[v]$ in $H_0^1(\Omega)$, then u is a weak solution of the Dirichlet problem (2.3.4), (2.3.2).

To prove the existence of a minimizer of $J[v]$ in $H_0^1(\Omega)$, we first establish the boundedness from below of $J[v]$ in $H_0^1(\Omega)$. Using the ellipticity condition and Cauchy's inequality with ε, we have

$$J[v] \geq \frac{\lambda}{2} \int_\Omega |\nabla v|^2 dx + \frac{c_0}{2} \int_\Omega v^2 dx - \frac{\varepsilon}{2} \int_\Omega v^2 dx$$
$$- \frac{1}{2\varepsilon} \int_\Omega f^2 dx - \frac{\varepsilon}{2} \int_\Omega |\nabla v|^2 dx - \frac{1}{2\varepsilon} \int_\Omega f^i f^i dx. \qquad (2.3.6)$$

This and Poincaré's inequality further give

$$J[v] \geq \frac{1}{2} \left(\frac{\lambda - \varepsilon}{\mu} + c_0 - \varepsilon \right) \int_\Omega v^2 dx - \frac{1}{2\varepsilon} \int_\Omega f^2 dx - \frac{1}{2\varepsilon} \int_\Omega f^i f^i dx, \qquad (2.3.7)$$

where $\mu > 0$ is the constant in Poincaré's inequality. If c_0 satisfies $\dfrac{\lambda}{2\mu} + c_0 > 0$, then we may choose $\varepsilon > 0$ so small that $\dfrac{1}{2} \left(\dfrac{\lambda - \varepsilon}{\mu} + c_0 - \varepsilon \right) > 0$. The boundedness from bellow of $J[v]$ in $H_0^1(\Omega)$ is thus proved.

Combining (2.3.6) with (2.3.7), we obtain

$$\int_\Omega v^2 dx + \int_\Omega |\nabla v|^2 dx \leq C_1 + C_2 J[v], \quad \forall v \in H_0^1(\Omega), \qquad (2.3.8)$$

where C_1, C_2 are constants independent of $v \in H_0^1(\Omega)$.

Let $\{u_k\}$ be a minimizing sequence of $J[v]$ in $H_0^1(\Omega)$. From (2.3.8) it follows that both $\{u_k\}$ and $\{\nabla u_k\}$ are bounded in $L^2(\Omega)$ and hence there exist a subsequence $\{u_{k_l}\}$ of $\{u_k\}$ and a function $u \in H_0^1(\Omega)$ such that

$$u_{k_l} \rightharpoonup u, \quad \nabla u_{k_l} \rightharpoonup \nabla u \quad \text{in } L^2(\Omega) \text{ as } l \to \infty.$$

In particular, we have

$$\lim_{l \to \infty} \int_\Omega f u_{k_l} dx = \int_\Omega f u dx, \qquad (2.3.9)$$

$$\lim_{l \to \infty} \int_\Omega f^i D_i u_{k_l} dx = \int_\Omega f^i D_i u dx \quad (i = 1, \cdots, n). \qquad (2.3.10)$$

Moreover, from

$$\int_\Omega \left[a_{ij} D_i(u_{k_l} - u) D_j(u_{k_l} - u) + c(u_{k_l} - u)^2 \right] dx \geq 0$$

i.e.

$$\int_\Omega (a_{ij} D_i u_{k_l} D_j u_{k_l} + c u_{k_l}^2) dx$$

$$\geq 2 \int_\Omega (a_{ij} D_i u_{k_l} D_j u + c u_{k_l} u) dx - \int_\Omega (a_{ij} D_i u D_j u + c u^2) dx,$$

it follows that

$$\lim_{l \to \infty} \int_\Omega (a_{ij} D_i u_{k_l} D_j u_{k_l} + c u_{k_l}^2) dx \geq \int_\Omega (a_{ij} D_i u D_j u + c u^2) dx. \quad (2.3.11)$$

Combining (2.3.9), (2.3.10) and (2.3.11),

$$\lim_{k \to \infty} J[u_k] = \lim_{l \to \infty} J[u_{k_l}] \geq J[u].$$

Thus u is a minimizer of $J[v]$ in $H_0^1(\Omega)$ and the existence of weak solutions of (2.3.4), (2.3.2) is proved.

2.3.4   *Lax-Milgram's theorem and its application*

It is to be noted that not any elliptic equation of the form (2.3.1) has its corresponding functional so that the variational method can be applied to the Dirichlet problem. If b_i $(i = 1, \cdots, n)$ are not all equal to zero, then Riesz's representation theorem also can not be applied. In this case, we need to slightly extend the representation theorem. Lax-Milgram's theorem is one of the very useful results obtained in this direction.

Definition 2.3.2   Let $a(u, v)$ be a bilinear form in the Hilbert space H, i.e. $a(u, v)$ is linear in u and in v respectively.

   i) $a(u, v)$ is said to be bounded, if for some constant $M \geq 0$,

$$|a(u, v)| \leq M \|u\| \|v\|, \quad \forall u, v \in H;$$

   ii) $a(u, v)$ is said to be coercive, if for some constant $\delta > 0$,

$$a(u, u) \geq \delta \|u\|^2, \quad \forall u \in H.$$

Lax-Milgram's Theorem   *If $a(u, v)$ is a bounded and coercive bilinear form in the Hilbert space H, then for any bounded linear functional $F(v)$ in H, there exists a unique $u \in H$, such that*

$$F(v) = a(u, v), \quad \forall v \in H \quad (2.3.12)$$

and

$$\|u\| \leq \frac{1}{\delta}\|F\|.$$

Proof. Since $a(u, v)$ is bilinear and bounded, for any fixed $u \in H$, $a(u, \cdot)$ is a bounded linear functional in H. By Riesz's representation theorem, there exists a unique $Au \in H$, such that

$$a(u, v) = (Au, v), \quad \forall v \in H. \tag{2.3.13}$$

It is easy to verify from the bilinearity of $a(u, v)$ that the operator A is linear. The boundedness of $a(u, v)$ implies the same property of A:

$$\|Au\| \leq M\|u\|, \quad \forall u \in H.$$

In addition, since $a(u, v)$ is coercive, we have

$$\delta\|u\|^2 \leq a(u, u) = (Au, u) \leq \|Au\|\|u\|, \quad \forall u \in H.$$

Hence

$$\delta\|u\| \leq \|Au\|, \quad \forall u \in H,$$

which shows the existence of A^{-1}.

Now we prove that the range of A, denoted by $R(A)$, is the whole space H. First of all, $R(A)$ is a closed subset. In fact, if $\{Au_k\} \subset R(A)$ is a convergent sequence:

$$\lim_{k \to \infty} Au_k = v,$$

then, from

$$\delta\|u_j - u_k\| \leq \|Au_j - Au_k\|$$

we see that $\{u_k\}$ is a Cauchy sequence and hence $\{u_k\}$ is also a convergent sequence:

$$\lim_{k \to \infty} u_k = u.$$

Hence, by the continuity of the operator A,

$$\lim_{k \to \infty} Au_k = Au.$$

Therefore $Au = v$, i.e. $v \in R(A)$.

Suppose $R(A) \neq H$. Then there exists a nonzero element $w \in H$, such that

$$(Au, w) = 0, \quad \forall u \in H.$$

In particular, if we choose $u = w$, then

$$(Aw, w) = a(w, w) = 0,$$

which and the coercivity of $a(u, v)$ imply $w = 0$, a contradiction. Thus $R(A) = H$.

For any bounded linear functional $F(v)$ in H, by Riesz's representation theorem, there exists a unique $w \in H$, such that $\|w\| = \|F\|$ and

$$F(v) = (w, v), \quad v \in H.$$

Choose $u = A^{-1}w$. Then

$$\|u\| \leq \|A^{-1}\| \|w\| \leq \frac{1}{\delta} \|F\|$$

and

$$F(v) = (Au, v), \quad v \in H$$

which combined with (2.3.13) leads to (2.3.12). $\qquad \square$

As an application of Lax-Milgram's theorem, we have

Theorem 2.3.2 *There exists a constant c_0, such that for any $f \in L^2(\Omega)$ and $f^i \in L^2(\Omega)$ $(i = 1, \cdots, n)$, the Dirichlet problem (2.3.1), (2.3.2) admits a unique weak solution provided $c \geq c_0$.*

Proof. Denote

$$a(u, v) = \int_\Omega (a_{ij} D_i u D_j v + b_i D_i u v dx + c u v) \, dx, \quad \forall u, v \in H_0^1(\Omega).$$

Obviously, $a(u, v)$ is bilinear and the boundedness of a_{ij}, b_i, c implies the same property of $a(u, v)$:

$$\begin{aligned}
|a(u, v)| \leq & C \int_\Omega (|\nabla u||\nabla v| + |\nabla u||v| + |u||v|) dx \\
\leq & C \big(\|\nabla u\|_{L^2(\Omega)} \|\nabla v\|_{L^2(\Omega)} + \|\nabla u\|_{L^2(\Omega)} \|v\|_{L^2(\Omega)} \\
& + \|u\|_{L^2(\Omega)} \|v\|_{L^2(\Omega)} \big) \\
\leq & C \|u\|_{H_0^1(\Omega)} \|v\|_{H_0^1(\Omega)}.
\end{aligned}$$

Moreover, using the ellipticity condition and Cauchy's inequality with ε, we derive, for $u \in H_0^1(\Omega)$,

$$a(u, u) \geq \lambda \|\nabla u\|_{L^2(\Omega)}^2 - C \int_\Omega |\nabla u||u| dx + c_0 \|u\|_{L^2(\Omega)}^2$$

$$\geq \lambda \|\nabla u\|_{L^2(\Omega)}^2 - \frac{\varepsilon}{2} \|\nabla u\|_{L^2(\Omega)}^2 - \frac{C^2}{2\varepsilon} \|u\|_{L^2(\Omega)}^2 + c_0 \|u\|_{L^2(\Omega)}^2$$

$$= \left(\lambda - \frac{\varepsilon}{2}\right) \|\nabla u\|_{L^2(\Omega)}^2 + \left(c_0 - \frac{C^2}{2\varepsilon}\right) \|u\|_{L^2(\Omega)}^2.$$

Choosing ε such that $0 < \varepsilon < 2\lambda$ and then taking $c_0 \geq \dfrac{C^2}{2\varepsilon}$, it follows that for some constant $\delta > 0$,

$$a(u, u) \geq \delta \|u\|_{H_0^1(\Omega)}^2, \quad \forall u \in H_0^1(\Omega),$$

i.e. $a(u, v)$ is coercive.

Now we can apply Lax-Milgram's theorem to the bounded linear functional

$$F(v) = \int_\Omega \left(fv - f^i D_i v\right) dx, \quad v \in H_0^1(\Omega)$$

to conclude that there exists a unique $u \in H_0^1(\Omega)$, such that

$$\|u\|_{H_0^1(\Omega)} \leq \frac{1}{\delta} \|F\|$$

and

$$a(u, v) = F(v), \quad \forall u \in H_0^1(\Omega),$$

which means that $u \in H_0^1(\Omega)$ is the unique weak solution of the Dirichlet problem (2.3.1), (2.3.2). $\qquad\square$

2.3.5 *Fredholm's alternative theorem and its application*

Theorem 2.3.2 merely affirms the weak solvability of the Dirichlet problem (2.3.1), (2.3.2) for the case $c \geq c_0$ (some constant). To investigate the general case we need the following result (see [Zhong, Fan and Chen (1998)]).

Fredholm's Alternative Theorem *Let V be a linear space endowed with a norm, $A : V \to V$ a compact linear operator and I the identity operator. Then there is exact one of the following alternatives:*

i) Either the equation

$$x - Ax = 0 \qquad (2.3.14)$$

has a nontrivial solution $x \in V$;
   ii) Or the equation

$$x - Ax = y \qquad (2.3.15)$$

admits a unique solution $x \in V$ for any $y \in V$. In other words, if the homogeneous equation (2.3.14) merely has a trivial solution, then the nonhomogeneous equation (2.3.15) admits a unique solution for any $y \in V$, or if for some $y \in V$, the solution of the nonhomogeneous equation (2.3.15) is unique, then for any $y \in V$, the nonhomogeneous equation (2.3.15) has a unique solution.

As an immediate application of this theorem, we have

Theorem 2.3.3   *There is exact one of the following alternatives:*
   i) Either the boundary value problem

$$Lu = 0, \quad u\Big|_{\partial\Omega} = 0$$

has a nontrivial weak solution;
   ii) Or the boundary value problem

$$Lu = f + D_i f^i, \quad u\Big|_{\partial\Omega} = 0$$

has a unique weak solution for any $f \in L^2(\Omega)$ and $f^i \in L^2(\Omega)$ ($i = 1, \cdots, n$).

Proof.   According to Theorem 2.3.2, there exists a constant ν_0, such that the equation

$$Lu + \nu u = f + D_i f^i$$

has a unique weak solution $u \in H_0^1(\Omega)$ for any $f \in L^2(\Omega)$ and $f^i \in L^2(\Omega)$ ($i = 1, \cdots, n$) provided $\nu \geq \nu_0$, i.e. the operator $L + \nu I$ has its inverse $(L + \nu I)^{-1}$. Thus

$$Lu = h = f + D_i f^i$$

is equivalent to

$$u = (L + \nu I)^{-1} h + \nu (L + \nu I)^{-1} u$$

or

$$u - \nu(L + \nu I)^{-1}u = (L + \nu I)^{-1}h$$

i.e.

$$u - Au = w,$$

where $A = \nu(L + \nu I)^{-1}$, $w = (L + \nu I)^{-1}h \in H_0^1(\Omega)$.

To apply Fredholm's alternative theorem, it suffices to prove the compactness of the operator $A : H_0^1(\Omega) \rightarrow H_0^1(\Omega)$. In fact, A can be regarded as a linear operator from $L^2(\Omega)$ to $H_0^1(\Omega)$. If we use E to denote the embedding operator from $H_0^1(\Omega)$ to $L^2(\Omega)$, then we have $A = AE : H_0^1(\Omega) \rightarrow H_0^1(\Omega)$. Since the embedding operator from $H_0^1(\Omega)$ to $L^2(\Omega)$ is compact and A is a bounded linear operator as shown in the proof of Theorem 2.3.2, we can assert that the operator $A = AE : H_0^1(\Omega) \rightarrow H_0^1(\Omega)$ is also compact. Thus the conclusion of the theorem follows from Fredholm's alternative theorem.

□

Exercises

1. Introduce the definition of weak solutions of the boundary value problem

$$\begin{cases} \Delta^2 u = f, & x \in \Omega, \\ u = \dfrac{\partial u}{\partial \nu} = 0, & x \in \partial\Omega \end{cases}$$

and prove the existence and uniqueness, where $\Omega \subset \mathbb{R}^n$ is a bounded domain, $f \in L^2(\Omega)$ and ν is the unit normal vector outward to $\partial\Omega$.

2. Define weak solutions of the Neumann problem for Poisson's equation

$$\begin{cases} -\Delta u = f, & x \in \Omega, \\ \dfrac{\partial u}{\partial \nu} = 0, & x \in \partial\Omega, \end{cases}$$

where $\Omega \subset \mathbb{R}^n$ is a bounded domain, $f \in L^2(\Omega)$, ν is the unit normal vector outward to $\partial\Omega$. And prove that the problem has a weak solution if and only if

$$\int_\Omega f(x)dx = 0.$$

3. Assume $\lambda > 0$. Define weak solutions of the equation

$$-\Delta u + \lambda u = f, \quad x \in \mathbb{R}^n$$

and prove the existence and uniqueness, where $f \in L^2(\mathbb{R}^n)$.

4. Prove Proposition 2.2.1.

5. Prove Proposition 2.2.4.

6. Let B be the unit ball in $\mathbb{R}^n$ and $u \in H^1(B)$ be a weak solution of Laplace's equation

$$-\Delta u = 0, \quad x \in B.$$

i) Prove $u \in C^\infty(B)$;

ii) Prove that if there exists a function $v \in C^\infty(\mathbb{R}^n)$ such that $u - v \in H_0^1(B)$, then $u \in C^\infty(\overline{B})$.

7. Let $u \in C_0^1(\mathbb{R}^n)$ be a weak solution of the semi-linear equation

$$-\Delta u + u^p = f, \quad x \in \mathbb{R}^n$$

where $f \in L^2(\mathbb{R}^n)$, $p > 0$. Prove $u \in H^2(\mathbb{R}^n)$.

8. Establish the theory of regularity for weak solutions of general elliptic equations.

Chapter 3

L^2 Theory of Linear Parabolic Equations

This chapter is a description of the L^2 theory of linear parabolic equations parallel to the previous chapter. As in treating elliptic equations, we first discuss a typical equation, i.e. the heat equation in greater detail and then discuss the general linear parabolic equations in divergence form in a brief fashion.

3.1   Energy Method

In this section we introduce the energy method, one of the basic methods available to parabolic equations. Let $\Omega \subset \mathbb{R}^n$ be a bounded domain with smooth boundary $\partial \Omega$ and $T > 0$ be a constant. Consider the heat equation

$$\frac{\partial u}{\partial t} - \Delta u = f(x,t), \quad (x,t) \in Q_T = \Omega \times (0,T). \tag{3.1.1}$$

Different from elliptic equations, we are not permitted to prescribe the condition on the whole boundary of Q_T. One of the typical conditions to determine the solution is

$$u\Big|_{\partial p Q_T} = g(x,t),$$

where $\partial p Q_T$ is the parabolic boundary of Q_T, i.e.

$$\partial p Q_T = \partial Q_T \setminus (\Omega \times \{t = T\}).$$

The problem of finding solutions of equation (3.1.1) satisfying this condition is called the first initial-boundary value problem. For simplicity, we merely consider the case $g(x,t)\Big|_{\partial_l Q_T} \equiv 0$, i.e. the condition

$$u(x,t) = 0, \qquad\qquad (x,t) \in \partial \Omega \times (0,T), \tag{3.1.2}$$

$$u(x,0) = u_0(x), \qquad\qquad x \in \Omega. \qquad\qquad (3.1.3)$$

Sometimes we even assume $g(x,t) \equiv 0$, i.e. consider the zero initial-boundary value condition

$$u\Big|_{\partial_p Q_T} = 0. \qquad\qquad (3.1.4)$$

If $g(x,t)$ is appropriately smooth in Q_T, then we may introduce a new unknown function $w = u - g$ to transform the original initial-boundary value condition into the latter.

3.1.1   Definition of weak solutions

Definition 3.1.1   A function $u \in \overset{\circ}{W}{}_2^{1,1}(Q_T)$ is said to be a weak solution of the first initial-boundary value problem (3.1.1), (3.1.2), (3.1.3), if for any $\varphi \in \overset{\circ}{C}{}^\infty(\overline{Q}_T)$, there holds

$$\iint_{Q_T} (u_t \varphi + \nabla u \cdot \nabla \varphi) dx dt = \iint_{Q_T} f \varphi dx dt \qquad\qquad (3.1.5)$$

and $\gamma u(x,0) = u_0(x)$ a.e. on Ω.

Remark 3.1.1   *Since $\overset{\circ}{C}{}^\infty(\overline{Q}_T)$ is dense in $\overset{\circ}{W}{}_2^{1,0}(Q_T)$, the test function φ can be chosen as any function in $\overset{\circ}{W}{}_2^{1,0}(Q_T)$.*

Sometimes, we merely discuss the equation itself and no initial-boundary value condition is concerned. In this case the following definition is needed.

Definition 3.1.2   A function $u \in W_2^{1,1}(Q_T)$ is said to be a weak solution of equation (3.1.1), if for any $\varphi \in C_0^\infty(Q_T)$, the integral identity (3.1.5) holds.

Remark 3.1.2   *It is not difficult to prove that if $u \in W_2^{1,1}(Q_T)$ is a weak solution of equation (3.1.1), then for any $\varphi \in \overset{\circ}{C}{}^\infty(\overline{Q}_T)$ and hence for any $\varphi \in \overset{\circ}{W}{}_2^{1,0}(Q_T)$, the integral identity (3.1.5) holds.*

The following propositions provide some equivalent descriptions of Definition 3.1.1, which are frequently used in the sequel.

Proposition 3.1.1   *A function $u \in \overset{\circ}{W}{}_2^{1,1}(Q_T)$ satisfies (3.1.5) for any $\varphi \in \overset{\circ}{C}{}^\infty(\overline{Q}_T)$ if and only if u satisfies*

$$\iint_{Q_T} (u_t \varphi_t + \nabla u \cdot \nabla \varphi_t) dx dt = \iint_{Q_T} f \varphi_t dx dt \qquad\qquad (3.1.6)$$

for any $\varphi \in \overset{\circ}{C}{}^\infty(\overline{Q}_T)$.

Proof. Suppose $u \in \overset{\circ}{W}{}_2^{1,1}(Q_T)$ satisfies (3.1.5) for any $\varphi \in \overset{\circ}{C}{}^\infty(\overline{Q}_T)$. Then, for any $\varphi \in \overset{\circ}{C}{}^\infty(\overline{Q}_T)$, since $\varphi_t \in \overset{\circ}{C}{}^\infty(\overline{Q}_T)$, (3.1.6) holds.

Inversely, if $u \in \overset{\circ}{W}{}_2^{1,1}(Q_T)$ satisfies (3.1.6) for any $\varphi \in \overset{\circ}{C}{}^\infty(\overline{Q}_T)$, then, since $\psi \in \overset{\circ}{C}{}^\infty(\overline{Q}_T)$ implies $\int_0^t \psi(x, s)ds \in \overset{\circ}{C}{}^\infty(\overline{Q}_T)$, we may choose $\int_0^t \psi(x, s)ds$ as a test function in (3.1.6) to derive (3.1.5). □

Proposition 3.1.2 *A function $u \in \overset{\circ}{W}{}_2^{1,1}(Q_T)$ satisfies (3.1.5) for any $\varphi \in \overset{\circ}{C}{}^\infty(\overline{Q}_T)$ if and only if u satisfies*

$$\iint_{Q_T} (u_t\varphi_t + \nabla u \cdot \nabla\varphi_t)e^{-\theta t}dxdt = \iint_{Q_T} f\varphi_t e^{-\theta t}dxdt \qquad (3.1.7)$$

for any $\varphi \in \overset{\circ}{C}{}^\infty(\overline{Q}_T)$, where θ is an arbitrary constant.

Proof. Suppose $u \in \overset{\circ}{W}{}_2^{1,1}(Q_T)$ satisfies (3.1.5) for any $\varphi \in \overset{\circ}{C}{}^\infty(\overline{Q}_T)$. Then, since $\varphi_t e^{-\theta t} \in \overset{\circ}{C}{}^\infty(\overline{Q}_T)$, we have

$$\iint_{Q_T} (u_t\varphi_t e^{-\theta t} + \nabla u \cdot \nabla\varphi_t e^{-\theta t})dxdt = \iint_{Q_T} f\varphi_t e^{-\theta t}dxdt,$$

namely, (3.1.7) holds.

Inversely, suppose $u \in \overset{\circ}{W}{}_2^{1,1}(Q_T)$ satisfies (3.1.7) for any $\varphi \in \overset{\circ}{C}{}^\infty(\overline{Q}_T)$. Then, since $\psi \in \overset{\circ}{C}{}^\infty(\overline{Q}_T)$ implies $\psi(x, t)e^{\theta t} - \theta \int_0^t \psi(x, s)e^{\theta s}ds \in \overset{\circ}{C}{}^\infty(\overline{Q}_T)$, we can choose the latter as a test function in (3.1.7) to derive (3.1.6) and also (3.1.5) by Proposition 3.1.1. □

Remark 3.1.3 *Since $\overset{\circ}{C}{}^\infty(\overline{Q}_T) \subset V(Q_T) \subset \overset{\circ}{W}{}_2^{1,1}(Q_T)$ and $\overset{\circ}{C}{}^\infty(\overline{Q}_T)$ is dense in $\overset{\circ}{W}{}_2^{1,1}(Q_T)$, we may choose any $\varphi \in V(Q_T)$ as the test function in (3.1.6) and (3.1.7).*

3.1.2 A modified Lax-Milgram's theorem

To prove the existence of weak solutions of the problem considered, we will apply a modified Lax-Milgram's theorem. First we prove

Lemma 3.1.1 *Let H be a Hilbert space, $V \subset H$ a dense subspace of H and $T : V \to H$ a bounded linear operator. If T^{-1} exists and is bounded,*

then the range of the conjugate operator T^* of T is the whole space H, i.e.
$R(T^*) = H$.

Proof.   We want to prove that for any $h \in H$, there exists $u \in H$, such
that $T^*u = h$. To this purpose, we consider the linear functional

$$F(z) = (h, T^{-1}z), \quad \forall z \in R(T)$$

defined in $R(T) = D(T^{-1})$ (domain of T^{-1}). Since

$$\|F\| = \sup_{\|z\|=1} |F(z)| \le \|h\| \|T^{-1}\|,$$

$F(z)$ is bounded. Now we extend $F(z)$ to be a bounded linear functional
in $\overline{R(T)}$ which is a Hilbert space and apply Riesz's representation theorem
(§2.1.2) to assert the existence of $u \in \overline{R(T)}$ satisfying

$$(u, z) = F(z) = (h, T^{-1}z), \quad \forall z \in R(T),$$

i.e.

$$(u, Ty) = (h, y), \quad \forall y \in V$$

or

$$(T^*u, y) = (h, y), \quad \forall y \in V.$$

This and the density of V in H lead to

$$(T^*u, y) = (h, y), \quad \forall y \in H.$$

Thus $T^*u = h$.   □

Modified Lax-Milgram's Theorem   *Let H be a Hilbert space, $V \subset H$
a dense subspace and $a(u, v)$ a bilinear form in $H \times V$ satisfying the following
conditions:*
    i) For some constant $M \ge 0$,

$$|a(u, v)| \le M \|u\|_H \|v\|_V, \quad \forall u \in H, \forall v \in V;$$

    ii) For some constant $\delta > 0$,

$$a(v, v) \ge \delta \|v\|_H^2, \quad \forall v \in V.$$

*Then for any bounded linear functional $F(v)$ in H, there exists $u \in H$, such
that*

$$F(v) = a(u, v), \quad \forall v \in V. \tag{3.1.8}$$

Proof. Since for any fixed $v \in V$, $a(\cdot, v)$ is a bounded linear functional in H, whose boundedness follows from the condition i), Riesz's representation theorem can be applied to assert the unique existence of $Av \in H$, such that

$$a(u, v) = (u, Av)_H, \quad \forall u \in H. \tag{3.1.9}$$

From the bilinearity of $a(u, v)$ and the condition i), it follows immediately that the operator $A : V \to H$ thus defined is bounded and linear. Condition ii) and (3.1.9) imply

$$(v, Av)_H \geq \delta \|v\|_H^2, \quad \forall v \in V$$

and

$$\|Av\|_H \geq \delta \|v\|_H, \quad \forall v \in V.$$

Thus A^{-1} exists and is bounded. Therefore we can apply Lemma 3.1.1 to assert that the range of A^*, the conjugate of A, is the whole space $H : R(A^*) = H$.

Now Riesz's representation theorem is applied, from which it follows that there exists a unique $h \in H$, such that

$$F(v) = (h, v)_H, \quad \forall v \in H. \tag{3.1.10}$$

Since $R(A^*) = H$, there exists $u \in H$, such that $A^*u = h$, and hence

$$(u, Av)_H = (A^*u, v)_H = (h, v)_H, \quad \forall v \in V,$$

which combined with (3.1.9), (3.1.10) leads to (3.1.8). □

3.1.3 *Existence and uniqueness of the weak solution*

Lemma 3.1.2 *Let $u \in \overset{\circ}{W}{}_2^{1,1}(Q_T)$. Then for almost all $t \in (0, T)$,*

$$h(t) - \int_\Omega (\gamma u(x, 0))^2 dx = 2 \int_0^t \int_\Omega u \frac{\partial u}{\partial t} dx ds,$$

where

$$h(t) = \int_\Omega u^2(x, t) dx, \quad t \in (0, T).$$

Proof. According to the definition of $\overset{\circ}{W}{}_2^{1,1}(Q_T)$, there exists a sequence $\{u_m\} \subset \overset{\circ}{C}{}^\infty(\overline{Q}_T)$, such that

$$\lim_{m \to \infty} \|u_m - u\|_{W_2^{1,1}(Q_T)} = 0.$$

From this and Fubini's theorem, it follows that for almost all $t \in (0, T)$,

$$\lim_{m \to \infty} h_m(t) = h(t),$$

where $h_m(t) = \int_\Omega u_m^2(x, t)dx$. Letting $m \to \infty$ in

$$h_m(t) - h_m(0) = \int_0^t h_m'(s)ds = 2\int_0^t \int_\Omega u_m \frac{\partial u_m}{\partial t} dx ds$$

and using

$$\lim_{m \to \infty} \int_0^t \int_\Omega u_m \frac{\partial u_m}{\partial t} dx ds = \int_0^t \int_\Omega u \frac{\partial u}{\partial t} dx ds$$

and (see Remark 1.5.1)

$$\lim_{m \to \infty} h_m(0) = \lim_{m \to \infty} \int_\Omega u_m^2(x, 0)dx = \int_\Omega (\gamma u(x, 0))^2 dx,$$

we are led to the conclusion of the lemma. $\qquad\square$

Theorem 3.1.1 *For any $f \in L^2(Q_T)$, the first initial-boundary value problem (3.1.1), (3.1.2), (3.1.3) admits at most one weak solution.*

Proof. Let u_1, u_2 be weak solutions of problem (3.1.1), (3.1.2), (3.1.3). Then by the definition of weak solutions (Definition 3.1.1) and Remark 3.1.1, $u = u_1 - u_2 \in \overset{\circ}{W}{}_2^{1,1}(Q_T)$, $\gamma u(x, 0) = 0$ and u satisfies

$$\iint_{Q_T} (u_t \varphi + \nabla u \cdot \nabla \varphi)dx dt = 0, \quad \forall \varphi \in \overset{\circ}{W}{}_2^{1,0}(Q_T).$$

Choosing $\varphi = u\chi_{[0,s]}$ leads to

$$\iint_{Q_s} (uu_t + |\nabla u|^2)dx dt = 0,$$

where $\chi_{[0,s]}(t)$ is the characteristic function of the segment $[0, s]$ $(0 < s \le T)$. Hence

$$\iint_{Q_s} uu_t dx dt = -\iint_{Q_s} |\nabla u|^2 dx dt \le 0.$$

From this, using Lemma 3.1.2 and noticing $\gamma u(x, 0) = 0$, we deduce

$$\int_{\Omega} u^2(x, s)dx \leq 0, \quad \text{a.e. } s \in (0, T).$$

Therefore $u = 0$ a.e. in Q_T, i.e. $u_1 = u_2$ a.e. in Q_T. $\qquad\square$

Now we are ready to prove the existence of weak solutions. In this section, we consider the problem with zero initial-boundary value condition (3.1.4) and apply the modified Lax-Milgram's theorem stated above to the existence for this problem. Other methods will be introduced in §3.2 and §3.3 to the existence of weak solutions for problem (3.1.1), (3.1.2), (3.1.3) with general initial values.

Theorem 3.1.2 *For any $f \in L^2(Q_T)$, the first initial-boundary value problem (3.1.1), (3.1.4) admits a weak solution $u \in \overset{\bullet}{W}_2^{1,1}(Q_T)$.*

Proof. Denote

$$a(u, v) = \iint_{Q_T} (u_t v_t + \nabla u \cdot \nabla v_t)e^{-\theta t}dxdt, \quad u \in \overset{\bullet}{W}_2^{1,1}(Q_T), \, v \in V(Q_T),$$

where $\theta > 0$ is a constant. Obviously

$$|a(u, v)| \leq \|u\|_{W_2^{1,1}(Q_T)}\|v\|_{V(Q_T)}, \quad u \in \overset{\bullet}{W}_2^{1,1}(Q_T), \, v \in V(Q_T). \quad (3.1.11)$$

On the other hand, for $v \in V(Q_T)$, we have

$$\iint_{Q_T} \nabla v \cdot \nabla v_t e^{-\theta t}dxdt$$

$$= \frac{1}{2} \iint_{Q_T} e^{-\theta t}\frac{\partial}{\partial t}|\nabla v|^2 dxdt$$

$$= \frac{1}{2} \iint_{Q_T} \frac{\partial}{\partial t}\left(|\nabla v|^2 e^{-\theta t}\right) dxdt + \frac{\theta}{2} \iint_{Q_T} |\nabla v|^2 e^{-\theta t}dxdt$$

$$= \frac{e^{-\theta T}}{2} \int_{\Omega} |\nabla v|^2 \Big|_{t=T} dx - \frac{1}{2} \int_{\Omega} \gamma|\nabla v|^2 \Big|_{t=0} dx + \frac{\theta}{2} \iint_{Q_T} |\nabla v|^2 e^{-\theta t}dxdt,$$

where $\gamma|\nabla v|^2 \Big|_{t=0}$ denotes the trace of $|\nabla v|^2$ when $t = 0$. From this, noticing that $v \in V(Q_T)$ implies $\gamma\nabla v \Big|_{t=0} = 0$ and hence

$$\int_{\Omega} \gamma|\nabla v|^2 \Big|_{t=0} dx = 0,$$

we obtain

$$\iint_{Q_T} \nabla v \cdot \nabla v_t e^{-\theta t} dx dt \geq \frac{\theta e^{-\theta T}}{2} \iint_{Q_T} |\nabla v|^2 dx dt. \qquad (3.1.12)$$

Since $\overset{\bullet}{C}{}^\infty(\overline{Q}_T)$ is dense in $V(Q_T)$, Poincaré's inequality of the form

$$\iint_{Q_T} v^2 dx dt \leq \mu \iint_{Q_T} |\nabla v|^2 dx dt$$

still holds. From (3.1.12) we are led to

$$\iint_{Q_T} \nabla v \cdot \nabla v_t e^{-\theta t} dx dt$$
$$\geq \frac{\theta e^{-\theta T}}{4} \iint_{Q_T} |\nabla v|^2 dx dt + \frac{\theta e^{-\theta T}}{4\mu} \iint_{Q_T} v^2 dx dt.$$

Therefore

$$a(v,v) \geq \delta \|v\|^2_{W_2^{1,1}(Q_T)}, \quad \forall v \in V(Q_T), \qquad (3.1.13)$$

where

$$\delta = \min\left\{ e^{-\theta T}, \frac{\theta e^{-\theta T}}{4}, \frac{\theta e^{-\theta T}}{4\mu} \right\}.$$

Choose $H = \overset{\bullet}{W}{}_2^{1,1}(Q_T)$, $V = V(Q_T)$. Then, by Proposition 1.4.1 of Chapter 1, $V \subset H$ is a dense subspace of H and (3.1.11), (3.1.13) show that the conditions i), ii) in the modified Lax-Milgram's theorem are satisfied. Obviously, $\iint_{Q_T} f v_t e^{-\theta t} dx dt$ is a bounded linear functional of v in H. Therefore there exists a $u \in H = \overset{\bullet}{W}{}_2^{1,1}(Q_T)$, such that

$$a(u,v) = \iint_{Q_T} f v_t e^{-\theta t} dx dt, \quad \forall v \in V(Q_T),$$

namely,

$$\iint_{Q_T} (u_t v_t + \nabla u \cdot \nabla v_t) e^{-\theta t} dx dt = \iint_{Q_T} f v_t e^{-\theta t} dx dt, \quad \forall v \in V(Q_T).$$

This means, by Proposition 3.1.2, u is a weak solution of problem (3.1.1), (3.1.4). $\qquad \square$

3.2   Rothe's Method

In this section we present another important method, called Rothe's method or semi-difference method, which is available to the study of existence for parabolic equations. The basic idea is to difference the equation with respect to the time variable, solve the obtained elliptic equations to construct approximating solutions and use the estimates for approximating solutions to complete the limiting process to arrive at the desired solution.

Let $\Omega \subset \mathbb{R}^n$ be a bounded domain with piecewise smooth boundary. Consider the first initial-boundary value problem (3.1.1), (3.1.2), (3.1.3).

Theorem 3.2.1   *Let $f \in L^2(Q_T)$ and $u_0 \in H_0^1(\Omega)$. Then problem (3.1.1), (3.1.2), (3.1.3) admits a weak solution $u \in \overset{\circ}{W}_2^{1,1}(Q_T)$.*

Proof.   Suppose for the moment, $f \in C(\overline{Q}_T)$. We proceed to prove the existence of weak solutions in several steps.

Step 1   Difference the equation with respect to the time variable t to construct the approximating solutions.

For any positive integer m and function $w(x,t)$, denote

$$w^{m,j}(x) = w(x, jh) \quad (j = 0, 1, \cdots, m),$$

where $h = T/m$. Consider the approximating equation of (3.1.1)

$$\frac{u^{m,j} - u^{m,j-1}}{h} - \Delta u^{m,j} = f^{m,j} \quad (j = 1, 2, \cdots, m). \tag{3.2.1}$$

According to the condition of the theorem, $u^{m,0} = u_0 \in H_0^1(\Omega)$. Suppose that $u^{m,j-1} \in H_0^1(\Omega)$ is known. We want to prove that equation (3.2.1) admits a weak solution $u^{m,j} \in H_0^1(\Omega)$. Denote $v = u^{m,j}$. Then (3.2.1) can be written as

$$-\Delta v + \frac{1}{h}v = f^{m,j} + \frac{u^{m,j-1}}{h} \tag{3.2.2}$$

which is an elliptic equation. It follows from Theorem 2.3.1 of Chapter 2, (3.2.2) admits a unique weak solution $v = u^{m,j} \in H_0^1(\Omega)$. Thus, by induction, we obtain

$$u^{m,1}, u^{m,2}, \cdots, u^{m,m}$$

in $H_0^1(\Omega)$, which is the weak solution of (3.2.1) for $j = 1, 2, \cdots, m$, successively, namely, for $j = 1, 2, \cdots, m$ and for any $\varphi \in H_0^1(\Omega)$, there holds

$$\int_\Omega \left(\frac{1}{h}(u^{m,j} - u^{m,j-1})\varphi + \nabla u^{m,j} \cdot \nabla \varphi \right) dx = \int_\Omega f^{m,j} \varphi dx. \qquad (3.2.3)$$

So far we merely obtain an approximation of the required solution on the line $t = jh = jT/m$ $(j = 1, 2, \cdots, m)$. In order to obtain an approximating solution in the whole domain Q_T, we define

$$w^m(x, t) = \sum_{j=1}^m \chi^{m,j}(t)u^{m,j}(x) \qquad (3.2.4)$$

and

$$u^m(x,t) = \sum_{j=1}^m \chi^{m,j}(t)[\lambda^{m,j}(t)u^{m,j}(x) + (1 - \lambda^{m,j}(t))u^{m,j-1}(x)]$$

$$= \sum_{j=1}^m \chi^{m,j}(t)u^{m,j-1}(x)$$

$$+ \sum_{j=1}^m \chi^{m,j}(t)\lambda^{m,j}(t)(u^{m,j}(x) - u^{m,j-1}(x)), \qquad (3.2.5)$$

where $\chi^{m,j}$ is the characteristic function of the segment $[(j-1)h, jh)$ and

$$\lambda^{m,j}(t) = \begin{cases} \dfrac{t}{h} - (j-1), & t \in [(j-1)h, jh), \\ 0, & \text{otherwise.} \end{cases}$$

For fixed $x \in \Omega$, (3.2.4) is a step function of t, which equals $u^{m,j}(x)$ on $[(j-1)h, jh)$, and (3.2.5) is a broken line function of t, which equals $u^{m,j-1}(x)(u^{m,j}(x))$ at $t = (j-1)h(t = jh)$.

Denote

$$f^m(x,t) = \sum_{j=1}^m \chi^{m,j}(t)f^{m,j}(x).$$

Then from (3.2.3) we see that for $\varphi \in H_0^1(\Omega)$, $t \in (0, T)$,

$$\int_\Omega \left(\frac{\partial u^m}{\partial t}\varphi + \nabla w^m \cdot \nabla \varphi \right) dx = \int_\Omega f^m \varphi dx. \qquad (3.2.6)$$

Step 2   Estimate the approximating solutions.

We need to prove the following estimates

$$\left\|\frac{\partial u^m}{\partial t}\right\|^2_{L^2(Q_T)} \leq M_m, \qquad (3.2.7)$$

$$\|\nabla u^m\|^2_{L^2(Q_T)} \leq 4TM_m, \qquad (3.2.8)$$

$$\|w^m - u^m\|^2_{L^2(Q_T)} \leq h^2 M_m, \qquad (3.2.9)$$

where $M_m = \|\nabla u_0\|^2_{L^2(\Omega)} + \|f^m\|^2_{L^2(Q_T)}$.

To this purpose, we choose $\varphi = u^{m,j} - u^{m,j-1}$ in (3.2.3) to obtain

$$\int_\Omega \left(\frac{1}{h}(u^{m,j} - u^{m,j-1})^2 + \nabla u^{m,j} \cdot (\nabla u^{m,j} - \nabla u^{m,j-1})\right) dx$$
$$= \int_\Omega f^{m,j}(u^{m,j} - u^{m,j-1}) dx.$$

From this we can deduce

$$\frac{1}{h}\|u^{m,j} - u^{m,j-1}\|^2_{L^2(\Omega)} + \|\nabla u^{m,j}\|^2_{L^2(\Omega)}$$
$$\leq \|\nabla u^{m,j} \cdot \nabla u^{m,j-1}\|_{L^1(\Omega)} + \|f^{m,j}(u^{m,j} - u^{m,j-1})\|_{L^1(\Omega)}$$
$$\leq \frac{1}{2}\|\nabla u^{m,j}\|^2_{L^2(\Omega)} + \frac{1}{2}\|\nabla u^{m,j-1}\|^2_{L^2(\Omega)}$$
$$+ \frac{1}{2h}\|u^{m,j} - u^{m,j-1}\|^2_{L^2(\Omega)} + \frac{h}{2}\|f^{m,j}\|^2_{L^2(\Omega)}.$$

Hence

$$\frac{1}{h}\|u^{m,j} - u^{m,j-1}\|^2_{L^2(\Omega)} + \|\nabla u^{m,j}\|^2_{L^2(\Omega)}$$
$$\leq \|\nabla u^{m,j-1}\|^2_{L^2(\Omega)} + h\|f^{m,j}\|^2_{L^2(\Omega)}, \quad (j-1,2,\cdots,m). \qquad (3.2.10)$$

In particular,

$$\|\nabla u^{m,j}\|^2_{L^2(\Omega)} \leq \|\nabla u^{m,j-1}\|^2_{L^2(\Omega)} + h\|f^{m,j}\|^2_{L^2(\Omega)}. \qquad (3.2.11)$$

Iterating (3.2.11) j times yields

$$\|\nabla u^{m,j}\|^2_{L^2(\Omega)} \leq \|\nabla u_0\|^2_{L^2(\Omega)} + h\sum_{i=1}^{j}\|f^{m,i}\|^2_{L^2(\Omega)} \leq M_m \qquad (3.2.12)$$

and summing (3.2.10) on j from 1 up to m yields

$$\frac{1}{h}\sum_{j=1}^{m}\|u^{m,j} - u^{m,j-1}\|^2_{L^2(\Omega)}$$

$$\leq \|\nabla u_0\|^2_{L^2(\Omega)} + h \sum_{j=1}^{m} \|f^{m,j}\|^2_{L^2(\Omega)} = M_m. \tag{3.2.13}$$

Now by the definition of u^m, we have

$$\frac{\partial u^m}{\partial t} = \frac{1}{h} \sum_{j=1}^{m} \chi^{m,j}(u^{m,j} - u^{m,j-1}),$$

$$\nabla u^m = \sum_{j=1}^{m} \chi^{m,j}\left(\nabla u^{m,j-1} + \lambda^{m,j}(\nabla u^{m,j} - \nabla u^{m,j-1})\right).$$

Thus, using (3.2.13) leads to

$$\left\|\frac{\partial u^m}{\partial t}\right\|^2_{L^2(Q_T)} = \frac{1}{h^2} \sum_{j=1}^{m} h\|u^{m,j} - u^{m,j-1}\|^2_{L^2(\Omega)} \leq M_m$$

and using (3.2.12) leads to

$$\|\nabla u^m\|^2_{L^2(Q_T)}$$

$$= \sum_{j=1}^{m} \int_0^T \chi^{m,j} \int_\Omega |(1 - \lambda^{m,j})\nabla u^{m,j-1} + \lambda^{m,j}\nabla u^{m,j}|^2 dx dt$$

$$\leq 2 \sum_{j=1}^{m} \int_0^T \chi^{m,j} \left(\|\nabla u^{m,j-1}\|^2_{L^2(\Omega)} + \|\nabla u^{m,j}\|^2_{L^2(\Omega)}\right) dt$$

$$\leq 2 \sum_{j=1}^{m} h \left(\|\nabla u^{m,j-1}\|^2_{L^2(\Omega)} + \|\nabla u^{m,j}\|^2_{L^2(\Omega)}\right)$$

$$\leq 4T M_m.$$

(3.2.7) and (3.2.8) are then proved.

By the definition of w^m and u^m, we have

$$w^m - u^m = \sum_{j=1}^{m} \chi^{m,j}(1 - \lambda^{m,j})(u^{m,j} - u^{m,j-1})$$

which and (3.2.13) imply

$$\|w^m - u^m\|^2_{L^2(Q_T)} \leq \sum_{j=1}^{m} h\|u^{m,j} - u^{m,j-1}\|^2_{L^2(\Omega)} \leq h^2 M_m$$

and prove (3.2.9).

Step 3 Complete the limiting process.

Since $f \in C(\overline{Q}_T)$, f^m converges to f in $L^2(Q_T)$. Hence

$$\lim_{m \to \infty} M_m = \|\nabla u_0\|_{L^2(\Omega)}^2 + \|f\|_{L^2(Q_T)}^2,$$

which implies, in particular, that $\{M_m\}_{m=1}^{\infty}$ is bounded. Therefore, it follows from (3.2.7), (3.2.8) and Poincaré's inequality

$$\|u^m\|_{L^2(Q_T)}^2 \leq \mu \|\nabla u^m\|_{L^2(Q_T)}^2$$

that $\{u^m\}_{m=1}^{\infty}$ is bounded in $W_2^{1,1}(Q_T)$, which implies the existence of a subsequence of $\{u^m\}_{m=1}^{\infty}$, supposed to be $\{u^m\}_{m=1}^{\infty}$ itself, and a function $u \in W_2^{1,1}(Q_T)$, such that u^m converges to u, and $\dfrac{\partial u^m}{\partial t}$ and ∇u^m converge weakly to $\dfrac{\partial u}{\partial t}$ and ∇u respectively, in $L^2(Q_T)$.

Now we proceed to prove that u is a weak solution of problem (3.1.1), (3.1.2), (3.1.3). First prove $u \in \overset{\circ}{W}_2^{1,1}(Q_T)$ and $\gamma u(x,0) = u_0(x)$ a.e. in Ω. To this end, it suffices to check $u^m \in \overset{\circ}{W}_2^{1,1}(Q_T)$, $\gamma u^m(x,0) = u_0(x)$ a.e. in Ω. For every positive integer m, choose $\{u_k^m\}_{k=1}^{\infty} \subset \overset{\circ}{C}^{\infty}(Q_T)$, such that

$$\lim_{k \to \infty} \|u_k^m - u^m\|_{W_2^{1,1}(Q_T)} = 0, \tag{3.2.14}$$

$$\lim_{k \to \infty} \int_{\Omega} |u_k^m(x,0) - u_0(x)| dx = 0. \tag{3.2.15}$$

For example, we can construct u_k^m as follows: first choose $\{u_k^{m,j}\}_{k=1}^{\infty} \subset C_0^{\infty}(\Omega)$, such that

$$\lim_{k \to \infty} \|u_k^{m,j} - u^{m,j}\|_{H^1(\Omega)} = 0 \quad (j = 1, 2, \cdots, m),$$

and then replace $u^{m,j}$ by $u_k^{m,j}$ in the expression (3.2.5) of u^m, followed by a mollification with respect to t. $u^m \in \overset{\circ}{W}_2^{1,1}(Q_T)$ then follows from (3.2.14), and $\gamma u^m(x,0) = u_0(x)$ a.e. in Ω follows from (3.2.15), Remark 1.5.1 and Remark 1.5.2 of Chapter 1.

In order to verify that u satisfies the integral identity in the definition of weak solutions, we integrate (3.2.6), which holds for any $\varphi \in \overset{\circ}{C}^{\infty}(\overline{Q}_T)$, with respect to t over $(0, T)$ and integrate by parts in x to obtain

$$\iint_{Q_T} \left(\frac{\partial u^m}{\partial t} \varphi - w^m \Delta \varphi \right) dx dt = \iint_{Q_T} f^m \varphi dx dt. \tag{3.2.16}$$

Since from (3.2.9) we see that the fact that u^m converges to u in $L^2(Q_T)$ implies the convergence of w^m to u in $L^2(Q_T)$, we can let $m \to \infty$ in

(3.2.16) to deduce

$$\iint_{Q_T} \left(\frac{\partial u}{\partial t} \varphi - u \Delta \varphi \right) dx dt = \iint_{Q_T} f \varphi dx dt,$$

which is equivalent to

$$\iint_{Q_T} \left(\frac{\partial u}{\partial t} \varphi + \nabla u \cdot \nabla \varphi \right) dx dt = \iint_{Q_T} f \varphi dx dt, \quad \varphi \in \overset{\circ}{C}{}^\infty(\overline{Q}_T)$$

due to $u \in \overset{\circ}{W}{}_2^{1,1}(Q_T)$. Summing up, we have proved that u is a weak solution of problem (3.1.1), (3.1.2), (3.1.3). Moreover, letting $m \to \infty$ in (3.2.7), (3.2.8), we obtain

$$\left\| \frac{\partial u}{\partial t} \right\|_{L^2(Q_T)}^2 \leq \|\nabla u_0\|_{L^2(\Omega)}^2 + \|f\|_{L^2(Q_T)}^2, \tag{3.2.17}$$

$$\|\nabla u\|_{L^2(Q_T)}^2 \leq 4T \left(\|\nabla u_0\|_{L^2(\Omega)}^2 + \|f\|_{L^2(Q_T)}^2 \right). \tag{3.2.18}$$

Now we turn to the general case $f \in L^2(Q_T)$. Choose $\{f_k\}_{k=1}^\infty \subset C(\overline{Q}_T)$ such that

$$\lim_{k \to \infty} \|f_k - f\|_{L^2(Q_T)} = 0.$$

Let $u_k \in \overset{\circ}{W}{}_2^{1,1}(Q_T)$ be the weak solution of the problem

$$\begin{cases} \dfrac{\partial u_k}{\partial t} - \Delta u_k = f_k, & (x,t) \in Q_T, \\[2mm] u_k(x,t) = 0, & (x,t) \in \partial\Omega \times (0,T), \\[2mm] u_k(x,0) = u_0(x), & x \in \Omega \end{cases}$$

as constructed above. From (3.2.17), (3.2.18),

$$\left\| \frac{\partial u_k}{\partial t} \right\|_{L^2(Q_T)}^2 \leq \|\nabla u_0\|_{L^2(\Omega)}^2 + \|f_k\|_{L^2(Q_T)}^2,$$

$$\|\nabla u_k\|_{L^2(Q_T)}^2 \leq 4T \left(\|\nabla u_0\|_{L^2(\Omega)}^2 + \|f_k\|_{L^2(Q_T)}^2 \right).$$

These show that $\{u_k\}_{k=1}^\infty$ is bounded in $\overset{\circ}{W}{}_2^{1,1}(Q_T)$. Hence we can choose a subsequence of $\{u_k\}_{k=1}^\infty$, supposed to be $\{u_k\}_{k=1}^\infty$ itself, and a function $u \in \overset{\circ}{W}{}_2^{1,1}(Q_T)$, such that u_k converges to u, and $\dfrac{\partial u_k}{\partial t}$ and ∇u_k converge weakly to $\dfrac{\partial u}{\partial t}$ and ∇u respectively, in $L^2(Q_T)$. Letting $k \to \infty$ in

$$\iint_{Q_T} \left(\frac{\partial u_k}{\partial t} \varphi + \nabla u_k \cdot \nabla \varphi \right) dxdt = \iint_{Q_T} f_k \varphi dxdt,$$

we see that u satisfies the integral identity in the definition of weak solutions. Since $\gamma u_k(x, 0) = u_0(x)$, from

$$\int_\Omega |u(x, t) - u_0(x)|^2 dx$$
$$\leq \int_\Omega |u(x, t) - u_k(x, t)|^2 dx + \int_\Omega |u_k(x, t) - \gamma u_k(x, 0)|^2 dx,$$

it follows that $\gamma u(x, 0) = u_0(x)$. $\qquad\qquad\square$

3.3 Galerkin's Method

In this section another important method available to parabolic equations, called Galerkin's method, is introduced. This method is efficient in both theory and practical computation. The basic idea of the method is to choose a suitable basic space X and a standard orthogonal basis $\{\omega_i(x)\}$ and then to find a solution of the form $\displaystyle\sum_{i=1}^{\infty} c_i(t)\omega_i(x)$.

As a typical example, we still consider problem (3.1.1), (3.1.2), (3.1.3). In order to apply Galerkin's method to prove the existence of weak solutions, we need the following

Hilbert-Schmidt's Theorem*(see [Jiang and Sun (1994)]) Let H be a separable Hilbert space, A be a bounded and self-adjoint compact operator and $\{\lambda_i\}$ be all eigenvalues of A. Then there exists an orthonormal basis $\{\omega_i\}$, such that $A\omega_i = \lambda_i\omega_i$.*

The existence of weak solutions is proved in four steps:
Step 1 Construct basis.
Define operator

$$A = (-\Delta)^{-1} : L^2(\Omega) \to L^2(\Omega), \quad f \mapsto Af,$$

where Af is the unique solution of the problem

$$\begin{cases} -\Delta u = f, & x \in \Omega, \\ u\big|_{\partial\Omega} = 0. \end{cases}$$

From the L^2 theory of elliptic equations (see Theorem 2.1.1 and Remark 2.2.2 of Chapter 2) we see that $Af \in H^2(\Omega) \cap H_0^1(\Omega)$ and

$$\|Af\|_{H^2(\Omega)} \le C\|f\|_{L^2(\Omega)},$$

i.e. A is a bounded operator. Integrating by parts leads to

$$\begin{aligned} \int_\Omega g(Af)dx &= -\int_\Omega \Delta(Ag)(Af)dx \\ &= -\int_\Omega \Delta(Af)(Ag)dx = \int_\Omega f(Ag)dx, \quad \forall f, g \in L^2(\Omega), \end{aligned}$$

which means that the operator A is self-adjoint. Since $H_0^1(\Omega)$ can be compactly embedded into $L^2(\Omega)$, the operator A is compact. Therefore, by Hilbert-Schmidt's theorem, there exists a standard orthogonal basis $\{\omega_i\}_{i=1}^\infty$, such that $A\omega_i = \lambda_i \omega_i$. Since

$$\begin{cases} -\Delta u = 0, & x \in \Omega, \\ u\big|_{\partial\Omega} = 0 \end{cases}$$

admits only a trivial solution, it is certainly $\lambda_i \ne 0$ and hence

$$-\Delta\omega_i = \frac{1}{\lambda_i}\omega_i.$$

Using Theorem 2.2.4 of Chapter 2 and the embedding theorem, we conclude $\omega_i \in C^2(\overline{\Omega})$ provided $\partial\Omega$ is appropriately smooth.

Step 2 Construct approximating solutions.

Set $u_0 = \sum_{i=1}^\infty c_i\omega_i$ and let

$$u_m(x,t) = \sum_{i=1}^m c_i^m(t)\omega_i(x)$$

satisfy

$$\left(\frac{\partial u_m}{\partial t}, \omega_k\right) = (\Delta u_m, \omega_k) + (f, \omega_k), \quad k = 1, 2, \cdots, m, \tag{3.3.1}$$

where $(\cdot, \cdot)$ is the inner product in $L^2(\Omega)$. Since

$$\left(\frac{\partial u_m}{\partial t}, \omega_k\right) = \sum_{i=1}^{m} \frac{d}{dt} c_i^m(t)(\omega_i, \omega_k) = \frac{d}{dt} c_k^m(t),$$

$$(\Delta u_m, \omega_k) = \sum_{i=1}^{m} c_i^m(t)(\Delta \omega_i, \omega_k) = -\frac{1}{\lambda_k} c_k^m(t),$$

(3.3.1) implies

$$\frac{d}{dt} c_k^m(t) = -\frac{1}{\lambda_k} c_k^m(t) + f_k(t), \qquad (3.3.2)$$

where $f_k(t) = (f, \omega_k)$. Hence

$$c_k^m(t) = e^{-t/\lambda_k} \left(c_k + \int_0^t e^{\tau/\lambda_k} f_k(\tau) d\tau \right).$$

Step 3 Estimate approximating solutions.

Multiplying (3.3.1) by $c_k^m(t)$ and then summing on k from 1 up to m yield

$$\left(\frac{\partial u_m}{\partial t}, u_m\right) = (\Delta u_m, u_m) + (f, u_m),$$

namely,

$$\frac{1}{2} \frac{d}{dt} \|u_m(\cdot, t)\|_{L^2(\Omega)}^2 = -\|\nabla u_m(\cdot, t)\|_{L^2(\Omega)}^2 + (f(\cdot, t), u_m(\cdot, t)).$$

Integrating over $(0, t)$, we further obtain

$$\frac{1}{2} \|u_m(\cdot, t)\|_{L^2(\Omega)}^2 - \frac{1}{2} \|u_m(\cdot, 0)\|_{L^2(\Omega)}^2$$

$$= -\iint_{Q_t} |\nabla u_m|^2 dx dt + \iint_{Q_t} f u_m dx dt.$$

Using Poincaré's inequality and Cauchy's inequality with ε leads to

$$\sup_{t \in [0,T]} \|u_m(\cdot, t)\|_{L^2(\Omega)}^2 + \iint_{Q_T} |\nabla u_m|^2 dx dt$$

$$\leq \|u_m(\cdot, 0)\|_{L^2(\Omega)}^2 + \mu \iint_{Q_T} f^2 dx dt, \qquad (3.3.3)$$

where $\mu > 0$ is the constant in Poincaré's inequality.

Next we multiply both sides of (3.3.1) by $\dfrac{d}{dt}c_k^m(t)$ and sum on k to obtain

$$\left(\frac{\partial u_m}{\partial t}, \frac{\partial u_m}{\partial t}\right) = \left(\Delta u_m, \frac{\partial u_m}{\partial t}\right) + \left(f, \frac{\partial u_m}{\partial t}\right).$$

Integrating over $(0, T)$, integrating by parts with respect to x and using Cauchy's inequality we are led to

$$\iint_{Q_T} \left|\frac{\partial u_m}{\partial t}\right|^2 dxdt + \|\nabla u_m(\cdot, T)\|_{L^2(\Omega)}^2$$
$$\leq \|\nabla u_m(\cdot, 0)\|_{L^2(\Omega)}^2 + \iint_{Q_T} f^2 dxdt. \tag{3.3.4}$$

Combining (3.3.4) with (3.3.3) yields

$$\iint_{Q_T} \left(|u_m|^2 + |\nabla u_m|^2 + \left|\frac{\partial u_m}{\partial t}\right|^2\right) dxdt \leq C, \tag{3.3.5}$$

where C is a constant independent of m.

Step 4 Complete the limiting process.

The estimate (3.3.5) implies the existence of a subsequence of $\{u_m\}$, supposed to be $\{u_m\}$ itself, and a function $u \in W_2^{1,1}(Q_T)$, such that u_m converges to u, and $\dfrac{\partial u_m}{\partial t}$ and ∇u_m converge weakly to $\dfrac{\partial u}{\partial t}$ and ∇u respectively, in $L^2(Q_T)$.

The function u is expected to be a weak solution of problem (3.1.1), (3.1.2), (3.1.3).

First we have $u \in \overset{\circ}{W}_2^{1,1}(Q_T)$. In fact, by Theorem 1.5.4 of Chapter 1, $u_m \in \overset{\circ}{W}_2^{1,1}(Q_T)$ and u is the weak limit of u_m in $\overset{\circ}{W}_2^{1,1}(Q_T)$.

Let $h \in C^2(\overline{\Omega})$ and $\psi \in C^2[0, T]$ be arbitrarily given functions such that $h\big|_{\partial \Omega} = 0$, $\psi(0) = \psi(T) = 0$. Choose a sequence

$$h_j(x) = \sum_{k=1}^{j} \alpha_{jk} \omega_k(x),$$

converging to h in $H^1(\Omega)$. Multiply (3.3.1) by $\psi(t)$, integrate over $(0, T)$ and integrate by parts with respect to x. Then let $m \to \infty$ to obtain

$$\iint_{Q_T} \frac{\partial u}{\partial t} \omega_k \psi dxdt = -\iint_{Q_T} \nabla u \cdot \nabla \omega_k \psi dxdt + \iint_{Q_T} f \omega_k \psi dxdt.$$

From this it follows by multiplying by α_{jk} and summing on k from 1 up to j, that

$$\iint_{Q_T} \frac{\partial u}{\partial t} h_j \psi \, dx dt = - \iint_{Q_T} \nabla u \cdot \nabla h_j \psi \, dx dt + \iint_{Q_T} f h_j \psi \, dx dt.$$

Letting $j \to \infty$ then leads to

$$\iint_{Q_T} \frac{\partial u}{\partial t} h \psi \, dx dt = - \iint_{Q_T} \nabla u \cdot \nabla h \psi \, dx dt + \iint_{Q_T} f h \psi \, dx dt.$$

Because of the arbitrariness of h and ψ, we may assert that for any $\varphi \in C_0^\infty(Q_T)$ (for instance, use Lemma 3.5.1 of §3.5.3)

$$\iint_{Q_T} \frac{\partial u}{\partial t} \varphi \, dx dt = - \iint_{Q_T} \nabla u \cdot \nabla \varphi \, dx dt + \iint_{Q_T} f \varphi \, dx dt.$$

This formula can be further proved without difficulty to hold for any $\varphi \in \overset{\circ}{C}{}^\infty(\overline{Q}_T)$, namely, u is a weak solution of (3.1.1).

It remains to verify $\gamma u(x, 0) = u_0(x)$. We have

$$\int_\Omega |u(x, t) - u_0(x)|^2 dx$$

$$\leq \int_\Omega |u(x, t) - u_m(x, t)|^2 dx + \int_\Omega \left| \sum_{i=1}^m (c_i^m(t) - c_i) \omega_i(x) \right|^2 dx$$

$$+ \int_\Omega \left| \sum_{i=m+1}^\infty c_i \omega_i(x) \right|^2 dx$$

$$= I_1 + I_2 + I_3.$$

Evidently, I_1 and I_3 can be made arbitrarily small by choosing m large enough. Once m is fixed, I_2 can be made arbitrarily small if $t > 0$ is small enough. Thus we have

$$\lim_{t \to \infty} \int_\Omega |u(x, t) - u_0(x)|^2 dx = 0.$$

3.4 Regularity of Weak Solutions

We first discuss the interior regularity.

Theorem 3.4.1   *Let $f \in L^2(Q_T)$ and $u \in \overset{\bullet}{W}{}_2^{1,1}(Q_T)$ be a weak solution of problem (3.1.1), (3.1.4). Denote*

$$\Omega_\delta = \{x \in \Omega; \operatorname{dist}(x, \partial\Omega) > \delta\}, \quad Q_T^\delta = \Omega_\delta \times (0, T).$$

Then $u \in W_2^{2,1}(Q_T^\delta)$ and

$$\|u\|_{W_2^{2,1}(Q_T^\delta)} \le C(\|u\|_{W_2^{1,0}(Q_T)} + \|f\|_{L^2(Q_T)}) \tag{3.4.1}$$

with constant C depending only on n and δ.

Proof.   By the definition of weak solutions,

$$\iint_{Q_T} (u_t\varphi + \nabla u \cdot \nabla\varphi)dxdt = \iint_{Q_T} f\varphi dxdt, \quad \forall\varphi \in \overset{\circ}{W}{}_2^{1,0}(Q_T). \tag{3.4.2}$$

Choose a cut-off function $\eta(x) \in C_0^\infty(\Omega)$, such that $\eta \equiv 1$ on Ω_δ, $0 \le \eta(x) \le 1$ and $|\nabla\eta(x)| \le \dfrac{C}{\delta}$. Since for small h, $\varphi = \Delta_h^{i}{}^*(\eta^2\Delta_h^i u)\chi_{[0,s]} \in \overset{\circ}{W}{}_2^{1,0}(Q_T)$, we can substitute it into (3.4.2) to obtain

$$\iint_{Q_s} \left[u_t\Delta_h^{i}{}^*(\eta^2\Delta_h^i u) + \nabla u \cdot \nabla(\Delta_h^{i}{}^*(\eta^2\Delta_h^i u))\right]dxdt$$

$$= \iint_{Q_s} f\Delta_h^{i}{}^*(\eta^2\Delta_h^i u)dxdt, \tag{3.4.3}$$

where $i = 1, 2, \cdots, n$, $\chi_{[0,s]}(t)$ is the characteristic function of the segment $[0, s]$ $(0 < s \le T)$. Using the properties of difference operators and Lemma 3.1.2, yields

$$\iint_{Q_s} u_t\Delta_h^{i}{}^*(\eta^2\Delta_h^i u)dxdt = \iint_{Q_s} \frac{\partial\Delta_h^i u}{\partial t}\eta^2\Delta_h^i udxdt$$

$$= \frac{1}{2}\iint_{Q_s} \frac{\partial}{\partial t}\left(\eta^2(\Delta_h^i u)^2\right)dxdt$$

$$= \frac{1}{2}\int_\Omega \eta^2(\Delta_h^i u(x,t))^2 dx\Big|_{t=0}^{t=s}$$

$$= \frac{1}{2}\int_\Omega \eta^2(\Delta_h^i u(x,s))^2 dx.$$

Thus, from (3.4.3), we obtain

$$\frac{1}{2}\int_\Omega \eta^2(\Delta_h^i u(x,s))^2 dx + \iint_{Q_s} \nabla u \cdot \nabla(\Delta_h^{i}{}^*(\eta^2\Delta_h^i u))dxdt$$

$$\le \iint_{Q_s} f\Delta_h^{i}{}^*(\eta^2\Delta_h^i u)dxdt, \quad 0 < s \le T, \quad (i = 1, 2, \cdots, n)$$

which implies

$$\frac{1}{2} \sup_{0 < s < T} \int_\Omega \eta^2 (\Delta_h^i u(x,s))^2 dx + \iint_{Q_T} \nabla u \cdot \nabla(\Delta_h^{i*}(\eta^2 \Delta_h^i u)) dx dt$$

$$\leq \iint_{Q_T} f \Delta_h^{i*}(\eta^2 \Delta_h^i u) dx dt, \qquad (i = 1, 2, \cdots, n). \qquad (3.4.4)$$

Starting from this, we can proceed as we did in Chapter 2 for Poisson's equation to derive

$$\sup_{0 < s < T} \int_\Omega \eta^2 (\Delta_h^i u(x,s))^2 dx + \iint_{Q_T^\delta} |\Delta_h^i \nabla u|^2 dx dt$$

$$\leq C \left(\|\nabla u\|_{L^2(Q_T)}^2 + \|f\|_{L^2(Q_T)}^2 \right), \qquad (i = 1, 2, \cdots, n) \qquad (3.4.5)$$

which implies, by virtue of Proposition 2.2.2 of Chapter 2, that $\eta |\nabla u| \in L^\infty((0,T); L^2(\Omega))$, $D^2 u \in L^2(Q_T^\delta)$ and

$$\sup_{0 < s < T} \int_\Omega \eta^2 |\nabla u(x,s)|^2 dx \leq C \left(\|\nabla u\|_{L^2(Q_T)}^2 + \|f\|_{L^2(Q_T)}^2 \right), \qquad (3.4.6)$$

$$\iint_{Q_T^\delta} |D^2 u|^2 dx dt \leq C \left(\|\nabla u\|_{L^2(Q_T)}^2 + \|f\|_{L^2(Q_T)}^2 \right). \qquad (3.4.7)$$

The difference is that here there is an additional nonnegative term on the left side of the inequality and the integrals are with respect to both the space variables and the time variable. In addition, as the starting point of our derivation, (3.4.4) is an inequality rather than an equality as we got for Poisson's equation. However this does not prevent us from deriving the desired estimate.

To derive the estimate on u_t, we use the difference operator in t, which is denoted by Δ_h^0 for simplicity. Extend u to $\Omega \times (-\infty, +\infty)$ by setting $u = 0$ outside Q_T. Take $\varphi = \eta^2 \Delta_h^0 u \in \overset{\circ}{W}_2^{1,0}(Q_T)$ to obtain

$$\iint_{Q_T} \left(\frac{\partial u}{\partial t} \eta^2 \Delta_h^0 u + \nabla u \cdot \nabla(\eta^2 \Delta_h^0 u) \right) dx dt = \iint_{Q_T} f \eta^2 \Delta_h^0 u dx dt$$

or

$$\iint_{Q_T} \eta^2 \frac{\partial u}{\partial t} \Delta_h^0 u dx dt = \iint_{Q_T} \eta^2 f \Delta_h^0 u dx dt - \iint_{Q_T} \Delta_h^0 (\eta^2 |\nabla u|^2) dx dt$$

$$- 2 \iint_{Q_T} \eta \Delta_h^0 u \nabla u \cdot \nabla \eta dx dt.$$

Using Hölder inequality, Cauchy's inequality with ε and (3.4.6) gives

$$\iint_{Q_T} \eta^2 \frac{\partial u}{\partial t} \Delta_h^0 u \, dx dt$$

$$\leq \frac{1}{4} \iint_{Q_T} \eta^2 (\Delta_h^0 u)^2 dx dt + \iint_{Q_T} \eta^2 f^2 dx dt + \left| \iint_{Q_T} \Delta_h^0 (\eta^2 |\nabla u|^2) dx dt \right|$$

$$+ \frac{1}{4} \iint_{Q_T} \eta^2 (\Delta_h^0 u)^2 dx dt + 4 \iint_{Q_T} |\nabla \eta|^2 |\nabla u|^2 dx dt$$

$$\leq \frac{1}{2} \iint_{Q_T} \eta^2 (\Delta_h^0 u)^2 dx dt + \iint_{Q_T} f^2 dx dt + 4 \sup_{0 < s < T} \int_\Omega \eta^2 |\nabla u(x,s)|^2 dx$$

$$+ \left(\frac{C}{\delta} \right)^2 \iint_{Q_T} |\nabla u|^2 dx dt$$

$$\leq \frac{1}{2} \iint_{Q_T} \eta^2 (\Delta_h^0 u)^2 dx dt + C \left(\|\nabla u\|_{L^2(Q_T)}^2 + \|f\|_{L^2(Q_T)}^2 \right).$$

By Proposition 2.2.2, we get

$$\iint_{Q_T} \eta^2 \left(\frac{\partial u}{\partial t} \right)^2 dx dt \leq \frac{1}{2} \iint_{Q_T} \eta^2 \left(\frac{\partial u}{\partial t} \right)^2 dx dt$$

$$+ C \left(\|\nabla u\|_{L^2(Q_T)}^2 + \|f\|_{L^2(Q_T)}^2 \right),$$

which implies

$$\iint_{Q_T^\delta} \left(\frac{\partial u}{\partial t} \right)^2 dx dt \leq C \left(\|\nabla u\|_{L^2(Q_T)}^2 + \|f\|_{L^2(Q_T)}^2 \right). \tag{3.4.8}$$

Combining (3.4.8) and (3.4.7), we obtain the desired estimate. $\square$

Since we can treat the integrals containing u_t in deriving the estimate near the lateral boundary similar to the interior estimate, we can obtain the estimate near the boundary for equation (3.1.1), and hence combine it with the interior estimate to obtain the following result on the global regularity.

Theorem 3.4.2   *Let $f \in L^2(Q_T)$ and $u \in \overset{\bullet}{W}_2^{1,1}(Q_T)$ be a weak solution of problem (3.1.1), (3.1.4). If $\partial\Omega \in C^2$, then $u \in W_2^{2,1}(Q_T)$ and*

$$\|u\|_{W_2^{2,1}(Q_T)} \leq C(\|u\|_{W_2^{1,0}(Q_T)} + \|f\|_{L^2(Q_T)}), \tag{3.4.9}$$

with constant C depending only on n and Ω.

Remark 3.4.1   *Under the assumptions of Theorem 3.4.2, there holds*

$$\|u\|_{W_2^{2,1}(Q_T)} \leq C\|f\|_{L^2(Q_T)} \tag{3.4.10}$$

with constant C depending only on n and Ω.

Proof.   By the definition of weak solutions,

$$\iint_{Q_T} (u_t\varphi + \nabla u \cdot \nabla\varphi)dxdt = \iint_{Q_T} f\varphi dxdt, \quad \forall\varphi \in \overset{\circ}{W}_2^{1,0}(Q_T).$$

Choose $\varphi = u$ and use Cauchy's inequality with ε and Poincaré's inequality on Ω to obtain

$$\frac{1}{2}\int_\Omega u^2(x,t)dx\Big|_{t=0}^{t=T} + \iint_{Q_T} |\nabla u|^2 dxdt$$

$$= \iint_{Q_T} fu dxdt$$

$$\leq \frac{1}{2\mu}\iint_{Q_T} u^2 dxdt + \frac{\mu}{2}\iint_{Q_T} f^2 dxdt$$

$$\leq \frac{1}{2}\iint_{Q_T} |\nabla u|^2 dxdt + \frac{\mu}{2}\iint_{Q_T} f^2 dxdt,$$

where $\mu > 0$ is the constant in Poincaré's inequality. Since u satisfies the zero initial value condition, this leads to

$$\iint_{Q_T} |\nabla u|^2 dxdt \leq \mu \iint_{Q_T} f^2 dxdt, \tag{3.4.11}$$

from which we obtain by using Poincaré's inequality on Ω

$$\iint_{Q_T} u^2 dxdt \leq \mu^2 \iint_{Q_T} f^2 dxdt. \tag{3.4.12}$$

Combining (3.4.11) with (3.4.12) gives

$$\|u\|_{W_2^{1,0}(Q_T)} \leq C\|f\|_{L^2(Q_T)},$$

Finally we substitute it into (3.4.9) to obtain (3.4.10).   □

Furthermore, we have

Theorem 3.4.3   *Let $u \in \overset{\bullet}{W}_2^{1,1}(Q_T)$ be a weak solution of problem (3.1.1), (3.1.4). If $\partial\Omega \in C^{2k+2}$ and $f \in W_2^{2k,k}(Q_T)$ for some nonnegative integer k, then $u \in W_2^{2k+2,k+1}(Q_T)$ and*

$$\|u\|_{W_2^{2k+2,k+1}(Q_T)} \leq C\left(\|u\|_{W_2^{1,0}(Q_T)} + \|f\|_{W_2^{2k,k}(Q_T)}\right),$$

where C is a constant depending only on k, n and Ω.

Corollary 3.4.1   *Let* $u \in \overset{\circ}{W}_2^{1,1}(Q_T)$ *be a weak solution of problem (3.1.1), (3.1.4). If* $\partial\Omega \in C^\infty$ *and* $f \in C^\infty(\overline{Q}_T)$, *then* $u \in C^\infty(\overline{Q}_T)$.

3.5 L^2 Theory of General Parabolic Equations

Now we turn to the general parabolic equations

$$Lu = \frac{\partial u}{\partial t} - D_j(a_{ij}D_i u) + b_i D_i u + cu = f, \qquad (3.5.1)$$

where a_{ij}, b_i, $c \in L^\infty(Q_T)$, $f \in L^2(Q_T)$ and $a_{ij} = a_{ji}$, satisfy the uniform parabolicity condition

$$\lambda|\xi|^2 \le a_{ij}(x,t)\xi_i\xi_j \le \Lambda|\xi|^2, \quad \forall\xi \in \mathbb{R}^n, \ (x,t) \in Q_T,$$

where λ, Λ are constants with $0 < \lambda \le \Lambda$. Here, as before, repeated indices imply a summation from 1 up to n. As in the preceding sections, we consider the problem for (3.5.1) with the initial-boundary value conditions

$$\begin{aligned} u(x,t) &= 0, & (x,t) &\in \partial\Omega \times (0,T), & (3.5.2) \\ u(x,0) &= u_0(x), & x &\in \Omega. & (3.5.3) \end{aligned}$$

Existence of weak solutions of the problem will be treated by means of the methods which we have applied to the heat equation in the preceding sections. The weak solution $u \in \overset{\circ}{W}_2^{1,1}(Q_T)$ of problem (3.5.1), (3.5.2), (3.5.3) is defined by

$$\iint_{Q_T} (u_t\varphi + a_{ij}D_i u D_j\varphi + b_i D_i u\varphi + cu\varphi)\, dxdt$$

$$= \iint_{Q_T} f\varphi dxdt, \quad \forall\varphi \in \overset{\circ}{C}^\infty(\overline{Q}_T). \qquad (3.5.4)$$

All of the methods will be described in a brief fashion.

3.5.1   *Energy method*

As shown in §3.1, the application of the method to the existence of weak solutions is based on a modified Lax-Milgram's theorem. This theorem can also be used to equations of the form (3.5.1) whose coefficients a_{ij} depend only on the space variable x. To demonstrate this fact, as we did in §3.1, we need the following identity which is equivalent to (3.5.4):

$$\iint_{Q_T} (u_t\varphi_t + a_{ij}D_iuD_j\varphi_t + b_iD_iu\varphi_t + cu\varphi_t)\,e^{-\theta t}dxdt$$

$$= \iint_{Q_T} f\varphi_t e^{-\theta t}dxdt, \quad \forall\varphi \in \overset{\circ}{C}{}^{\infty}(\overline{Q}_T),$$

where $\theta > 0$ is a constant which can be chosen arbitrarily.

As in §3.1, we consider only the case $u_0 = 0$. Denote the bilinear form $a(u, v)$ as

$$a(u, v) = \iint_{Q_T} (u_t v_t + a_{ij}D_iuD_jv_t + b_iD_iuv_t + cuv_t)e^{-\theta t}dxdt,$$

$$u \in \overset{\bullet}{W}{}^{1,1}_2(Q_T), v \in V(Q_T),$$

whose boundedness is obvious. To prove the coercivity of $a(u, v)$, i.e. for some constant $\delta > 0$,

$$a(v, v) \geq \delta\|v\|^2_{W^{1,1}_2(Q_T)}, \quad \forall v \in V(Q_T), \tag{3.5.5}$$

we need to estimate all terms in the expression of $a(v, v)$.

Since a_{ij} are independent of t, we may use the parabolicity condition to derive, similar to the proof of Theorem 3.1.2, for $v \in V(Q_T)$,

$$\iint_{Q_T} a_{ij}D_ivD_jv_t e^{-\theta t}dxdt$$

$$= \frac{1}{2}\iint_{Q_T} \frac{\partial}{\partial t}(a_{ij}D_ivD_jv)\,e^{-\theta t}dxdt$$

$$= \frac{1}{2}\iint_{Q_T} \frac{\partial}{\partial t}\left(a_{ij}D_ivD_jve^{-\theta t}\right)dxdt + \frac{\theta}{2}\iint_{Q_T} a_{ij}D_ivD_jve^{-\theta t}dxdt$$

$$= \frac{e^{-\theta T}}{2}\int_{\Omega} a_{ij}D_ivD_jv\Big|_{t=T}dx + \frac{\theta}{2}\iint_{Q_T} a_{ij}D_ivD_jve^{-\theta t}dxdt$$

$$\geq \frac{\lambda e^{-\theta T}}{2}\int_{\Omega} |\nabla v|^2\Big|_{t=T}dx + \frac{\theta\lambda}{2}\iint_{Q_T} |\nabla v|^2 e^{-\theta t}dxdt$$

$$\geq \frac{\theta\lambda}{2}\iint_{Q_T} |\nabla v|^2 e^{-\theta t}dxdt. \tag{3.5.6}$$

In addition, using Cauchy's inequality with ε, we get

$$\left|\iint_{Q_T} b_iD_ivv_t e^{-\theta t}dxdt\right|$$

$$\leq \varepsilon \iint_{Q_T} v_t^2 e^{-\theta t} dx dt + \frac{C}{\varepsilon} \iint_{Q_T} |\nabla v|^2 e^{-\theta t} dx dt, \qquad (3.5.7)$$

and

$$\left| \iint_{Q_T} cvv_t e^{-\theta t} dx dt \right|$$

$$\leq \varepsilon \iint_{Q_T} v_t^2 e^{-\theta t} dx dt + \frac{C}{\varepsilon} \iint_{Q_T} v^2 e^{-\theta t} dx dt, \qquad (3.5.8)$$

where the constant C depends only on the bound of $|b_i|$ and $|c|$.

Combining (3.5.6), (3.5.7), (3.5.8) and using Poincaré's inequality, we are led to

$$a(v, v) \geq (1 - 2\varepsilon) \iint_{Q_T} v_t^2 e^{-\theta t} dx dt$$

$$+ \left(\frac{\theta \lambda}{4} - \frac{C}{\varepsilon} \right) \iint_{Q_T} |\nabla v|^2 e^{-\theta t} dx dt$$

$$+ \left(\frac{\theta \lambda}{4\mu} - \frac{C}{\varepsilon} \right) \iint_{Q_T} v^2 e^{-\theta t} dx dt,$$

where $\mu > 0$ is the constant in Poincaré's inequality. From this, (3.5.5) follows immediately by choosing $\varepsilon > 0$ small enough and $\theta > 0$ large enough.

Remark 3.5.1 *Recalling Theorem 2.3.1 of Chapter 2, we observe that to prove the existence of weak solutions for elliptic equations, the coefficient c is required to be less than some constant c_0. However, as stated above, for parabolic equations, no other conditions in addition to the boundedness of c are required. This is an essential difference between parabolic equations and elliptic equations.*

3.5.2 *Rothe's method*

In applying Rothe's method to the heat equation in §3.2, we transformed the problem into the one of solving elliptic equations and establishing some necessary estimates; the resulting elliptic equations are treated by means of variational method. All of these are available to more general parabolic equations in divergence form. However, since the variational method can not be applied to elliptic equations involving terms of first order derivatives, here we merely discuss equation (3.5.1) with $b_i = 0 \, (i = 1, \cdots, n)$. We stress that, different from the case of elliptic equations, in treating parabolic equations, no other conditions in addition to the boundedness of the coefficient c, are required.

In the present case, the elliptic equation obtained by discreticizing (3.5.1) with respect to t is

$$-D_j(a_{ij}D_iv) + \left(\frac{1}{h} + c\right)v = f^{m,k} + \frac{u^{m,k-1}}{h},$$

whose corresponding functional is

$$J(v) = \frac{1}{2}\int_\Omega \left(a_{ij}D_ivD_jv + \left(\frac{1}{h} + c\right)v^2 \right.$$
$$\left. -2\left(f^{m,k} + \frac{u^{m,k-1}}{h}\right)v\right)dx, \quad \forall v \in H_0^1(\Omega).$$

Since in the expression of $J(v)$, the coefficient of the term v^2 is $\frac{1}{h} + c$ which can be made nonnegative if $h > 0$ is small enough, because of the boundedness of c. Thus the boundedness of $J(v)$ from below is obvious.

3.5.3 Galerkin's method

In applying Galerkin's method, the key step is to choose a suitable basic space and a standard orthogonal basis in it. For general parabolic equations (3.5.1) in divergence form, we choose $L^2(\Omega)$ as the basic space. The existence of the needed basis is proved in the following

Lemma 3.5.1 *Assume $\partial\Omega \subset C^2$. Then there exists an orthonormal basis $\{\omega_i\}_{i=1}^\infty$ in $L^2(\Omega)$ satisfying the following conditions:*

i) $\omega_i \in C^2(\overline{\Omega})$, $\omega_i\big|_{\partial\Omega} = 0$, $(\omega_i, \omega_j)_{L^2(\Omega)} = \delta_{ij}$;

ii) For any $\varphi \in H_0^1(\Omega)$ and $\varepsilon > 0$, there exists a function of the form

$$\varphi_N(x) = \sum_{i=1}^N c_i\omega_i(x), \quad c_i \in \mathbb{R},$$

such that

$$\|\varphi - \varphi_N\|_{H^1(\Omega)} < \varepsilon;$$

iii) For any $v \in C^2(\overline{Q}_T)$ vanishing near the lateral boundary $\partial_l Q_T$ and

$\varepsilon > 0$, *there exists a function of the form*

$$v_N(x, t) = \sum_{i=1}^{N} c_i(t)\omega_i(x), \quad c_i(t) \in C^2([0, T]),$$

such that

$$\iint_{Q_T} (|v - v_N|^2 + |\nabla v - \nabla v_N|^2)dxdt < \varepsilon.$$

Proof.   Since $\partial\Omega \in C^2$, by means of local flatting of the boundary, finite covering and partition of unity, we may assert the existence of a function $\zeta(x) \in C^2(\overline{\Omega})$, such that $\zeta\big|_{\partial\Omega} = 0$ and $\zeta(x) > 0$ in Ω. Let $\Omega \subset \{x \in \mathbb{R}^n; 0 < x_i < l, i = 1, 2, \cdots, n\}$ and denote

$$\bar{\omega}_k(x) = \left(\frac{1}{l}\right)^{n/2} \prod_{j=1}^{n} \sin(\frac{\pi k_j x_j}{l}),$$

where $k = (k_1, k_2, ..., k_n)$ and k_j $(j = 1, \cdots, k_n)$ are positive integers.

Given $\varphi \in C_0^\infty(\Omega)$. By the definition of ζ, $\dfrac{\varphi}{\zeta} \in C_0^2(\Omega)$. Hence for $h > 0$ small enough, $\left(\dfrac{\varphi}{\zeta}\right)_h \in C_0^\infty(\Omega)$ and

$$\left\|\frac{\varphi}{\zeta} - \left(\frac{\varphi}{\zeta}\right)_h\right\|_{H^1(\Omega)} < \frac{\varepsilon}{2},$$

where f_h denotes the mollification of f with radius h. Now we expand $\left(\dfrac{\varphi}{\zeta}\right)_h$ in a Fourier series with respect to $\{\bar{\omega}_k\}$. Since this series and the series obtained by differentiating each term formally converge uniformly in $\overline{\Omega}$, for any $\varepsilon > 0$, there exists a function of the form $\displaystyle\sum_{1 \leq k_j \leq N} c_k\bar{\omega}_k(x)$, such that

$$\left\|\left(\frac{\varphi}{\zeta}\right)_h - \sum_{1 \leq k_j \leq N} c_k\bar{\omega}_k\right\|_{H^1(\Omega)} < \frac{\varepsilon}{2},$$

i.e.

$$\left\|\left(\frac{\varphi}{\zeta}\right)_h - \sum_{1 \leq k_j \leq N} c_k\bar{\omega}_k\right\|_{L^2(\Omega)}$$

$$+ \sum_{i=1}^{n} \left\| \frac{\partial}{\partial x_i} \left(\frac{\varphi}{\zeta} \right)_h - \sum_{1 \le k_j \le N} c_k \frac{\partial \bar{\omega}_k}{\partial x_i} \right\|_{L^2(\Omega)} < \frac{\varepsilon}{2}.$$

Thus

$$\left\| \frac{\varphi}{\zeta} - \sum_{1 \le k_j \le N} c_k \bar{\omega}_k \right\|_{L^2(\Omega)} < \varepsilon,$$

$$\sum_{i=1}^{n} \left\| \frac{\partial}{\partial x_i} \left(\frac{\varphi}{\zeta} \right) - \sum_{1 \le k_j \le N} c_k \frac{\partial \bar{\omega}_k}{\partial x_i} \right\|_{L^2(\Omega)} < \varepsilon.$$

Setting $\omega_k^*(x) = \zeta(x) \bar{\omega}_k(x)$ leads to

$$\left\| \varphi - \sum_{1 \le k_j \le N} c_k \omega_k^* \right\|_{L^2(\Omega)} = \left\| \zeta \left(\frac{\varphi}{\zeta} - \sum_{1 \le k_j \le N} c_k \bar{\omega}_k \right) \right\|_{L^2(\Omega)} < C\varepsilon$$

and

$$\sum_{i=1}^{n} \left\| \frac{\partial \varphi}{\partial x_i} - \sum_{1 \le k_j \le N} c_k \frac{\partial \omega_k^*}{\partial x_i} \right\|_{L^2(\Omega)}$$

$$\le \sum_{i=1}^{n} \left\| \frac{\partial \zeta}{\partial x_i} \left(\frac{\varphi}{\zeta} - \sum_{1 \le k_j \le N} c_k \bar{\omega}_k \right) \right\|_{L^2(\Omega)}$$

$$+ \sum_{i=1}^{n} \left\| \zeta \left(\frac{\partial}{\partial x_i} \left(\frac{\varphi}{\zeta} \right) - \sum_{1 \le k_j \le N} c_k \frac{\partial \bar{\omega}_k}{\partial x_i} \right) \right\|_{L^2(\Omega)}$$

$$\le C\varepsilon.$$

Hence

$$\left\| \varphi - \sum_{1 \le k_j \le N} c_k \omega_k^* \right\|_{H^1(\Omega)} \le C\varepsilon,$$

where the constant C depends only on ζ and is independent of φ. Since $C_0^\infty(\Omega)$ is dense in $H_0^1(\Omega)$, we may obtain a basis $\{\omega_k\}$ satisfying i), ii), by orthonormalizing $\{\omega_k^*\}$ in $L^2(\Omega)$.

It remains to prove that $\{\omega_k\}$ satisfies iii). To this end, we first expand $v(x,t)$ in $\overline{Q}_T$ with respect to $\left\{\sin\dfrac{2m\pi}{T}t\right\}_{m=1}^{\infty}$ and $\left\{\cos\dfrac{2m\pi}{T}t\right\}_{m=0}^{\infty}$:

$$v(x,t) = \sum_{m=1}^{\infty} \alpha_m(x) \sin\frac{2m\pi}{T}t + \sum_{m=0}^{\infty} \beta_m(x) \cos\frac{2m\pi}{T}t,$$

where $\alpha_m, \beta_m \in C^2(\overline{\Omega})$ and $\alpha_m\big|_{\partial\Omega} = \beta_m\big|_{\partial\Omega} = 0$. Since this series and the series obtained by differentiating each term formally converge uniformly in $\overline{Q}_T$, for any $\varepsilon > 0$, there exists a integer N_1, such that

$$\iint_{Q_T} \left(|v - \bar{v}_{N_1}|^2 + |\nabla v - \nabla \bar{v}_{N_1}|^2\right) dxdt < \varepsilon,$$

where

$$\bar{v}_{N_1}(x,t) = \sum_{m=1}^{N_1} \alpha_m(x)\sin\frac{2m\pi}{T}t + \sum_{m=0}^{N_1} \beta_m(x)\cos\frac{2m\pi}{T}t.$$

From ii), it follows that there exists a positive integer N, such that

$$\left\|\alpha_m - \sum_{1\leq k_j\leq N} c'_{k,m}\omega_k\right\|_{H_1(\Omega)}^2 < \frac{\varepsilon}{N_1}, \quad m = 1, 2, \cdots, N_1,$$

$$\left\|\beta_m - \sum_{1\leq k_j\leq N} c''_{k,m}\omega_k\right\|_{H_1(\Omega)}^2 < \frac{\varepsilon}{N_1}, \quad m = 0, 1, \cdots, N_1.$$

Hence

$$\iint_{Q_T} \left(|v - v_N|^2 + |\nabla v - \nabla v_N|^2\right) dxdt \leq C\varepsilon,$$

where

$$v_N(x,t) = \sum_{m=1}^{N_1} \sin\frac{2m\pi}{T}t \sum_{1\leq k_j\leq N} c'_{k,m}\omega_k(x)$$

$$+ \sum_{m=0}^{N_1} \cos\frac{2m\pi}{T}t \sum_{1\leq k_j\leq N} c''_{k,m}\omega_k(x)$$

$$\equiv \sum_{i=1}^{nN} c_i(t)\omega_i(x).$$

The proof is complete. □

Now we use Galerkin's method to prove the following

Theorem 3.5.1 *Assume that* $\partial\Omega \in C^2$, a_{ij}, b_i, c, $\dfrac{\partial a_{ij}}{\partial t} \in L^\infty(\Omega)$, $f \in L^2(\Omega)$, $a_{ij} = a_{ji}$ *satisfy the parabolicity condition and* $u_0 \in H_0^1(\Omega)$. *Then problem (3.5.1), (3.5.2), (3.5.3) admits a weak solution in* $\overset{\circ}{W}_2^{1,1}(Q_T)$.

Proof. We merely prove the conclusion of the theorem for the case $a_{ij} \in C^1(\overline{Q}_T)$, $c, f \in C(\overline{Q}_T)$. The conclusion under the assumptions of the theorem can be carried over by approximation.

We proceed first to construct the approximating solutions. Since $u_0 \in H_0^1(\Omega)$, by Lemma 3.5.1 ii), there exists $u_0^m = \displaystyle\sum_{i=1}^m c_i^m \omega_i$ converging to u_0 in $H_0^1(\Omega)$. Approximating solutions to be found are of the form

$$u_m(x,t) = \sum_{i=1}^m g_i^m(t)\omega_i(x) \quad (m = 1, 2, \cdots),$$

satisfying

$$\int_\Omega \left(\frac{\partial u_m}{\partial t}\omega_k + a_{ij}\frac{\partial u_m}{\partial x_i} \cdot \frac{\partial \omega_k}{\partial x_j} + b_i\frac{\partial u_m}{\partial x_i}\omega_k + cu_m\omega_k - f\omega_k \right) dx = 0,$$

$$k = 1, 2, \cdots, m, \quad (m = 1, 2, \cdots)$$

or $g_i^m(t)$ $(i = 1, 2, \cdots, m)$ satisfying equations of the form

$$\frac{d}{dt}g_i^m(t) = f_i(t, g_1^m(t), \cdots, g_m^m(t)) \quad (i = 1, 2, \cdots, m), \tag{3.5.9}$$

where f_i are some linear functions of $g_1^m, g_2^m, \cdots, g_m^m$. In addition, the initial value conditions

$$g_i^m(0) = c_i^m \quad (i = 1, 2, \cdots, m) \tag{3.5.10}$$

should be satisfied. By the theory of ordinary differential equations, problem (3.5.9), (3.5.10) admit solutions $g_i^m(t) \in C^1[0, T]$.

For approximating solutions thus constructed, we can prove

$$\|u_m\|_{W_2^{1,1}(Q_T)} \le C, \quad (m = 1, 2, \cdots)$$

without difficulty and then take limit of a subsequence of $\{u_m\}$ to obtain the desired weak solution. □

Exercises

1. Prove Theorem 3.4.2.

2. Establish the theory of regularity of weak solutions for general parabolic equations.

3. Define weak solutions of the Cauchy problem

$$
\begin{cases}
\dfrac{\partial u}{\partial t} - \Delta u + \lambda u = 0, & (x,t) \in \mathbb{R}^n \times (0,T), \\
u(x,0) = u_0(x), & x \in \mathbb{R}^n
\end{cases}
$$

and prove the existence and uniqueness, where $\lambda \in \mathbb{R}$ and $u_0 \in H^1(\mathbb{R}^n)$.

4. Consider the first initial-boundary value problem

$$
\begin{cases}
\dfrac{\partial u}{\partial t} - \Delta u = 0, & (x,t) \in Q_T = \Omega \times (0,T), \\
u(x,t) = 0, & (x,t) \in \partial\Omega \times (0,T), \\
u(x,0) = u_0(x), & x \in \Omega,
\end{cases}
$$

where $\Omega \subset \mathbb{R}^n$ is a bounded domain and $u_0 \in L^2(\Omega)$.

i) Define weak solutions of the above problem and prove the existence and uniqueness;

ii) Prove that if u is a weak solution of the problem, then $u \in C^\infty(Q_T)$ and if, in addition, $\partial\Omega \in C^\infty$, then $u \in C^\infty(\overline{\Omega} \times (0,T))$.

5. Let $u \in C_0^1(Q_T)$ be a weak solution of the equation

$$
\frac{\partial u}{\partial t} - \Delta u + |\nabla u| + \left|\frac{\partial u}{\partial t}\right| + u^p = f, \quad (x,t) \in Q_T = \Omega \times (0,T),
$$

where $\Omega \subset \mathbb{R}^n$ is a bounded domain, $f \in L^2(Q_T)$, $p > 0$. Prove $u \in H^2(Q_T)$.

6. Define weak solutions of the second initial-boundary value problem

$$
\begin{cases}
\dfrac{\partial u}{\partial t} - \Delta u = f, & (x,t) \in Q_T = \Omega \times (0,T), \\
\dfrac{\partial u}{\partial \nu} = g(x,t), & (x,t) \in \partial\Omega \times (0,T), \\
u(x,0) = u_0(x), & x \in \Omega
\end{cases}
$$

and prove the uniqueness, where $\Omega \subset \mathbb{R}^n$ is a bounded domain, $f \in L^2(Q_T)$, $g \in L^2(\partial\Omega \times (0,T))$ and $u_0 \in L^2(\Omega)$.

7. Define weak solutions of the initial-boundary value problem

$$\begin{cases} \dfrac{\partial u}{\partial t} + \Delta^2 u = f, & (x,t) \in Q_T = \Omega \times (0,T), \\[2mm] u = \dfrac{\partial u}{\partial \nu} = 0, & (x,t) \in \partial\Omega \times (0,T), \\[2mm] u(x,0) = 0, & x \in \Omega \end{cases}$$

and prove the existence and uniqueness, where $\Omega \subset \mathbb{R}^n$ is a bounded domain, $f \in L^2(Q_T)$ and ν is the unit normal vector outward to $\partial\Omega$.

Chapter 4

De Giorgi Iteration and Moser Iteration

This chapter is devoted to a discussion of properties of weak solutions. Two powerful techniques, the De Giorgi iteration and the Moser iteration, are introduced, which can be applied not only to linear elliptic and parabolic equations in divergence form, but also to quasilinear equations, not only to the estimate of maximum norm, but also to the study of other properties, such as regularity of weak solutions. In order to expose the basic idea and main points of these techniques in a limited space, we confine ourselves basically to Poisson's equation and the heat equation, and apply the techniques merely to the estimate of maximum norm of weak solutions.

4.1 Global Boundedness Estimates of Weak Solutions of Poisson's Equation

In this section we illustrate the De Giorgi iteration by applying it to the estimate of maximum norm of weak solutions for Poisson's equation.

4.1.1 *Weak maximum principle for solutions of Laplace's equation*

Definition 4.1.1 Let $u \in H^1(\Omega)$. The least upper bound and the greatest lower bound of u on Ω and $\partial\Omega$ are defined as

$$\sup_{\Omega} u = \inf\{l; (u-l)_+ = 0, \text{ a.e. in } \Omega\},$$

$$\sup_{\partial\Omega} u = \inf\{l; (u-l)_+ \in H_0^1(\Omega)\},$$

$$\inf_{\Omega} u = -\sup_{\Omega}(-u), \quad \inf_{\partial\Omega} u = -\sup_{\partial\Omega}(-u),$$

where $s_+ = \max\{s, 0\}$.

In case u is continuous on $\overline{\Omega}$, the definition of $\sup_{\Omega} u$, $\inf_{\Omega} u$ and $\sup_{\partial\Omega} u$, $\inf_{\partial\Omega} u$ coincides with the usual one. For $\sup_{\Omega} u$ and $\inf_{\Omega} u$, this is obvious and for $\sup_{\partial\Omega} u$ and $\inf_{\partial\Omega} u$, this follows from the discussion of the trace of functions in $H^1(\Omega)$ (see §1.5).

Let $\Omega \subset \mathbb{R}^n$ be a bounded domain. Consider Laplace's equation

$$-\Delta u = 0, \quad x \in \Omega. \tag{4.1.1}$$

Proposition 4.1.1   *Let $u \in H^1(\Omega)$ be a weak solution of Laplace's equation (4.1.1). Then*

$$\sup_{\Omega} u \le \sup_{\partial\Omega} u.$$

Proof.   By the definition of weak solutions, u satisfies

$$\int_{\Omega} \nabla u \cdot \nabla \varphi \, dx = 0 \tag{4.1.2}$$

for any $\varphi \in C_0^\infty(\Omega)$ and hence for any $\varphi \in H_0^1(\Omega)$. Set $l = \sup_{\partial\Omega} u$. Then for any $k > l$, $(u - k)_+ \in H_0^1(\Omega)$. From Proposition 1.3.10 of Chapter 1,

$$\frac{\partial(u-k)_+}{\partial x_i} = \begin{cases} \dfrac{\partial u}{\partial x_i}, & \text{if } u > k, \\ 0, & \text{if } u \le k. \end{cases}$$

Choosing $\varphi = (u - k)_+$ in (4.1.2) gives

$$\int_{\Omega} |\nabla(u - k)_+|^2 dx = 0.$$

Thus from Poincaré's inequality (Theorem 1.3.4 of Chapter 1), we obtain

$$\int_{\Omega} |(u - k)_+|^2 dx \le \mu \int_{\Omega} |\nabla(u - k)_+|^2 dx = 0,$$

where $\mu > 0$ is the constant in Poincaré's inequality, which implies $(u - k)_+ = 0$, or $u \le k$ a.e. in Ω. Thus the conclusion of the proposition follows from the arbitrariness of $k > l$. $\square$

Corollary 4.1.1   *Let $u \in H^1(\Omega)$ be a weak solution of Laplace's equation (4.1.1). Then*

$$\inf_{\Omega} u \ge \inf_{\partial\Omega} u.$$

Choosing functions of the form $(u - k)_+$ as test functions to derive the estimate of maximum norm is an important technique in establishing a priori estimates. However, the argument as simple as above cannot be used to the same estimate for general equations, even Poisson's equation. For such equations, instead, one has to proceed by means of some iteration techniques, among them is the De Giorgi iteration introduced in the following.

4.1.2 Weak maximum principle for solutions of Poisson's equation

Lemma 4.1.1 *Let $\varphi(t)$ be a nonnegative and nonincreasing function on $[k_0, +\infty)$, satisfying*

$$\varphi(h) \leq \left(\frac{M}{h-k}\right)^{\alpha} [\varphi(k)]^{\beta}, \quad \forall h > k \geq k_0 \qquad (4.1.3)$$

for some constants $M > 0$, $\alpha > 0$, $\beta > 1$. Then there exists $d > 0$ such that

$$\varphi(h) = 0, \quad \forall h \geq k_0 + d.$$

Proof. Set

$$k_s = k_0 + d - \frac{d}{2^s}, \quad s = 0, 1, 2, \cdots$$

with constant $d > 0$ to be determined. Then from (4.1.3) we obtain the recursive formula

$$\varphi(k_{s+1}) \leq \frac{M^{\alpha} 2^{(s+1)\alpha}}{d^{\alpha}} [\varphi(k_s)]^{\beta} \quad (s = 0, 1, 2, \cdots). \qquad (4.1.4)$$

From this we can prove, by induction,

$$\varphi(k_s) \leq \frac{\varphi(k_0)}{r^s} \quad (s = 0, 1, 2, \cdots) \qquad (4.1.5)$$

with constant $r > 1$ to be chosen. Once this is proved, letting $s \to \infty$ then derives $\varphi(k_0 + d) = 0$ and the conclusion of the lemma by the nonincreasingness of $\varphi(t)$. Suppose that (4.1.5) is valid for s, then using (4.1.4) gives

$$\varphi(k_{s+1}) \leq \frac{M^{\alpha} 2^{(s+1)\alpha}}{d^{\alpha}} [\varphi(k_s)]^{\beta} \leq \frac{\varphi(k_0)}{r^{s+1}} \cdot \frac{M^{\alpha} 2^{(s+1)\alpha}}{d^{\alpha} r^{s(\beta-1)-1}} [\varphi(k_0)]^{\beta-1}.$$

Now we choose $r = 2^{\alpha/(\beta-1)}$. Then

$$\varphi(k_{s+1}) \leq \frac{\varphi(k_0)}{r^{s+1}} \cdot \frac{M^\alpha 2^{\alpha\beta/(\beta-1)}}{d^\alpha} [\varphi(k_0)]^{\beta-1}.$$

From this, we see that if $d > 0$ satisfies

$$\frac{M^\alpha 2^{\alpha\beta/(\beta-1)}}{d^\alpha} [\varphi(k_0)]^{\beta-1} \leq 1,$$

i.e.

$$d \geq M 2^{\beta/(\beta-1)} [\varphi(k_0)]^{(\beta-1)/\alpha},$$

then (4.1.5) is also valid for s replaced by $s + 1$. □

Now we turn to Poisson's equation

$$-\Delta u = f(x), \quad x \in \Omega. \tag{4.1.6}$$

Theorem 4.1.1　*Let $f \in L^\infty(\Omega)$ and $u \in H^1(\Omega)$ be a weak solution of Poisson's equation (4.1.6). Then*

$$\sup_\Omega u \leq \sup_{\partial\Omega} u + C\|f\|_{L^\infty(\Omega)},$$

where C is a constant depending only on n and Ω.

Proof.　By the definition of weak solutions, u satisfies

$$\int_\Omega \nabla u \cdot \nabla \varphi dx = \int_\Omega f\varphi dx$$

for any $\varphi \in C_0^\infty(\Omega)$ and hence for any $\varphi \in H_0^1(\Omega)$. As did in the proof of Proposition 4.1.1, set $l = \sup_{\partial\Omega} u$ and choose $\varphi = (u - k)_+$ with $k > l$ in the above identity. Then we obtain

$$\int_\Omega |\nabla\varphi|^2 dx = \int_\Omega f\varphi dx$$

and hence

$$\int_\Omega |\nabla\varphi|^2 dx \leq \int_\Omega |f\varphi| dx. \tag{4.1.7}$$

Using the embedding theorem gives

$$\left(\int_\Omega |\varphi|^p dx \right)^{2/p} \leq C \int_\Omega |f\varphi| dx,$$

where the constant C depends only on n and Ω and

$$2 < p < \begin{cases} +\infty, & n = 1, 2, \\ \dfrac{2n}{n-2}, & n \geq 3. \end{cases}$$

In other words

$$\left(\int_{A(k)} |\varphi|^p dx \right)^{2/p} \leq C \int_{A(k)} |f\varphi| dx,$$

where

$$A(k) = \{ x \in \Omega; u(x) > k \}.$$

From this, using Hölder's inequality

$$\int_{A(k)} |f\varphi| dx \leq \left(\int_{A(k)} |\varphi|^p dx \right)^{1/p} \left(\int_{A(k)} |f|^q dx \right)^{1/q},$$

where q is the conjugate exponent of p, i.e.

$$\frac{1}{p} + \frac{1}{q} = 1,$$

we obtain

$$\left(\int_{A(k)} |\varphi|^p dx \right)^{1/p} \leq C \left(\int_{A(k)} |f|^q dx \right)^{1/q}. \tag{4.1.8}$$

Since $h > k$ implies $A(h) \subset A(k)$, and $\varphi \geq h - k$ on $A(h)$, we have

$$\int_{A(k)} |\varphi|^p dx \geq \int_{A(h)} |\varphi|^p dx \geq (h-k)^p |A(h)|,$$

where, as before, $|E|$ denotes the measure of E. This combined with (4.1.8) gives

$$(h-k)|A(h)|^{1/p} \leq C\|f\|_{L^\infty(\Omega)} |A(k)|^{1/q},$$

i.e.

$$|A(h)| \leq \left(\frac{C\|f\|_{L^\infty(\Omega)}}{h-k} \right)^p |A(k)|^{p/q}.$$

Since $p > 2$ implies $p > q$, from Lemma 4.1.1 we obtain

$$|A(l+d)| = 0,$$

where

$$d = C\|f\|_{L^\infty(\Omega)}|A(l)|^{(p-q)/(pq)}2^{p/(p-q)}$$
$$\leq C|\Omega|^{(p-q)/(pq)}2^{p/(p-q)}\|f\|_{L^\infty(\Omega)}.$$

By the definition of $A(k)$, this means that for almost all $x \in \Omega$,

$$u \leq l + C|\Omega|^{(p-q)/(pq)}2^{p/(p-q)}\|f\|_{L^\infty(\Omega)}.$$

$\square$

Corollary 4.1.2 *Let $f \in L^\infty(\Omega)$ and $u \in H^1(\Omega)$ be a weak solution of (4.1.6). Then*

$$\inf_\Omega u \geq \inf_{\partial\Omega} u - C\|f\|_{L^\infty(\Omega)},$$

where C is a constant depending only on n and Ω.

The De Giorgi iteration technique can also be applied to more general elliptic equations in divergence form. For instance, for the slightly general equation

$$-\Delta u + c(x)u = f(x) + \operatorname{div}\vec{f}(x), \quad x \in \Omega, \tag{4.1.9}$$

we have

Theorem 4.1.2 *Assume that $p > n \geq 3$, $0 \leq c(x) \leq M$, $f \in L^{p_*}(\Omega)$, $\vec{f} \in L^p(\Omega; \mathbb{R}^n)$ and $u \in H^1(\Omega)$ is a weak solution of equation (4.1.9). Then*

$$\sup_\Omega u \leq \sup_{\partial\Omega} u_+ + C\left(\|f\|_{L^{p_*}(\Omega)} + \|\vec{f}\|_{L^p(\Omega)}\right)|\Omega|^{1/n-1/p},$$
$$\inf_\Omega u \geq \inf_{\partial\Omega} u_- - C\left(\|f\|_{L^{p_*}(\Omega)} + \|\vec{f}\|_{L^p(\Omega)}\right)|\Omega|^{1/n-1/p},$$

where $p_* = np/(n+p)$, $s_- = \min\{s, 0\}$ and C is a constant depending only on n, p, M and Ω, but is independent of the lower bound of $|\Omega|$.

For the proof we leave to the reader.

4.2 Global Boundedness Estimates for Weak Solutions of the Heat Equation

In this section we apply the De Giorgi iteration technique to the estimates of maximum norm for weak solutions of the heat equation.

4.2.1 Weak maximum principle for solutions of the homogeneous heat equation

Definition 4.2.1 Let $u \in W_2^{1,1}(Q_T)$, where $Q_T = \Omega \times (0, T)$ with $\Omega \subset \mathbb{R}^n$ being a bounded domain. Define the least upper bound and the greatest lower bound of u on Q_T and $\partial_p Q_T$ as

$$\sup_{Q_T} u = \inf\{l; (u - l)_+ = 0, \text{ a.e. in } Q_T\},$$

$$\sup_{\partial_p Q_T} u = \inf\{l; (u - l)_+ \in \overset{\bullet}{W}_2^{1,1}(Q_T)\},$$

$$\inf_{Q_T} u = -\sup_{Q_T}(-u), \quad \inf_{\partial_p Q_T} u = -\sup_{\partial_p Q_T}(-u).$$

First consider the homogeneous heat equation

$$\frac{\partial u}{\partial t} - \Delta u = 0, \quad (x, t) \in Q_T. \tag{4.2.1}$$

Proposition 4.2.1 Let $u \in W_2^{1,1}(Q_T)$ be a weak solution of the homogeneous heat equation (4.2.1). Then

$$\sup_{Q_T} u \leq \sup_{\partial_p Q_T} u.$$

Proof. By the definition of weak solutions, u satisfies

$$\iint_{Q_T} (u_t \varphi + \nabla u \cdot \nabla \varphi) \, dx dt = 0 \tag{4.2.2}$$

for any $\varphi \in \overset{\circ}{C}^\infty(\overline{Q}_T)$ and hence for any $\varphi \in \overset{\circ}{W}_2^{1,1}(Q_T)$. Set $l = \sup_{\partial_p Q_T} u$ and choose $\varphi = (u - k)_+$ with $k > l$. Then

$$\varphi \in \overset{\bullet}{W}_2^{1,1}(Q_T) \subset \overset{\circ}{W}_2^{1,1}(Q_T).$$

Substituting φ into (4.2.2) and using the operation rules of weak derivatives (see Proposition 1.3.10 of Chapter 1), give

$$\iint_{Q_T} (u-k)_t (u-k)_+ dx dt + \iint_{Q_T} \nabla(u-k) \cdot \nabla(u-k)_+ dx dt = 0,$$

i.e.

$$\frac{1}{2} \iint_{Q_T} \frac{\partial}{\partial t}(u-k)_+^2 dx dt + \iint_{Q_T} |\nabla(u-k)_+|^2 dx dt = 0.$$

Thus, we obtain from Lemma 3.1.2 of Chapter 3,

$$\frac{1}{2} \int_\Omega (u(x,T)-k)_+^2 dx - \frac{1}{2} \int_\Omega (\gamma((u(x,0)-k)_+))^2 dx$$
$$+ \iint_{Q_T} |\nabla(u-k)_+|^2 dx dt = 0.$$

Since by Corollary 1.5.6 of Chapter 1,

$$\int_\Omega (\gamma((u(x,0)-k)_+))^2 dx = 0,$$

we have

$$\iint_{Q_T} |\nabla(u-k)_+|^2 dx dt \le 0.$$

Combining this with Poincaré's inequality, we further obtain

$$\iint_{Q_T} (u-k)_+^2 dx dt \le \mu \iint_{Q_T} |\nabla(u-k)_+|^2 dx dt \le 0,$$

where $\mu > 0$ is the constant in Poincaré's inequality. Hence $u(x,t) \le k$ a.e. in Q_T and the conclusion of the proposition follows from the arbitrariness of $k > l$. $\qquad\square$

Corollary 4.2.1   *Let $u \in W_2^{1,1}(Q_T)$ be a weak solution of (4.2.1). Then*

$$\inf_{Q_T} u \ge \inf_{\partial_p Q_T} u.$$

4.2.2   Weak maximum principle for solutions of the nonhomogeneous heat equation

Now we turn to the nonhomogeneous heat equation

$$\frac{\partial u}{\partial t} - \Delta u = f(x,t), \quad (x,t) \in Q_T. \tag{4.2.3}$$

Theorem 4.2.1   *Let $f \in L^\infty(Q_T)$ and $u \in W_2^{1,1}(Q_T)$ be a weak solution of the nonhomogeneous heat equation (4.2.3). Then*

$$\sup_{Q_T} u \le \sup_{\partial_p Q_T} u + C\|f\|_{L^\infty(Q_T)},$$

where C is a constant depending only on n and Ω.

Proof.   Denote $\sup\limits_{\partial_p Q_T} u = l$. For $k > l$ and $0 \le t_1 < t_2 \le T$, we have $\varphi = (u - k)_+ \chi_{[t_1,t_2]} \in \overset{\circ}{W}_2^{1,0}(Q_T)$, where $\chi_{[t_1,t_2]}(t)$ is the characteristic function of the interval $[t_1, t_2]$. Thus we may choose φ as a test function in the definition of weak solutions (see Remark 3.1.2 of Chapter 3) to obtain

$$\iint_{Q_T} (u - k)_t (u - k)_+ \chi_{[t_1,t_2]} dxdt + \iint_{Q_T} \chi_{[t_1,t_2]} |\nabla(u - k)_+|^2 dxdt$$
$$= \iint_{Q_T} f(u - k)_+ \chi_{[t_1,t_2]} dxdt.$$

Hence

$$\frac{1}{2} \int_{t_1}^{t_2} \frac{d}{dt} \int_\Omega (u - k)_+^2 dxdt + \int_{t_1}^{t_2} \int_\Omega |\nabla(u - k)_+|^2 dxdt$$
$$\le \int_{t_1}^{t_2} \int_\Omega |f|(u - k)_+ dxdt,$$

i.e.

$$\frac{1}{2}(I_k(t_2) - I_k(t_1)) + \int_{t_1}^{t_2} \int_\Omega |\nabla(u - k)_+|^2 dxdt$$
$$\le \int_{t_1}^{t_2} \int_\Omega |f|(u - k)_+ dxdt,$$

where

$$I_k(t) = \int_\Omega (u - k)_+^2 dx.$$

Assume that the absolutely continuous function $I_k(t)$ attains its maximum at $\sigma \in [0, T]$. Since $I_k(0) = 0$, $I_k(t) \ge 0$, we may suppose $\sigma > 0$. Taking $t_1 = \sigma - \varepsilon$, $t_2 = \sigma$ with $\varepsilon > 0$ small enough so that $\sigma - \varepsilon > 0$ and noticing that

$$I_k(\sigma) - I_k(\sigma - \varepsilon) \ge 0,$$

we obtain

$$\int_{\sigma-\varepsilon}^{\sigma}\int_{\Omega}|\nabla(u-k)_+|^2dxdt \leq \int_{\sigma-\varepsilon}^{\sigma}\int_{\Omega}|f|(u-k)_+dxdt.$$

Thus it follows from

$$\frac{1}{\varepsilon}\int_{\sigma-\varepsilon}^{\sigma}\int_{\Omega}|\nabla(u-k)_+|^2dxdt \leq \frac{1}{\varepsilon}\int_{\sigma-\varepsilon}^{\sigma}\int_{\Omega}|f|(u-k)_+dxdt$$

by letting $\varepsilon \to 0^+$ that

$$\int_{\Omega}|\nabla(u(x,\sigma)-k)_+|^2dx \leq \int_{\Omega}|f(x,\sigma)|(u(x,\sigma)-k)_+dx.$$

This is an analog of (4.1.7) with $\varphi = (u-k)_+$ obtained for Poisson's equation. Having this in hand, we may process as in the proof of Theorem 4.1.1 to establish the desired estimate. To this end, denote

$$A_k(t) = \{x; u(x,t) > k\}, \quad \mu_k = \sup_{0<t<T}|A_k(t)|.$$

Then, similar to the derivation of (4.1.8), we may deduce

$$\left(\int_{A_k(\sigma)}(u-k)_+^pdx\right)^{1/p} \leq C\left(\int_{A_k(\sigma)}|f|^qdx\right)^{1/q}$$
$$\leq C\|f\|_{L^\infty(Q_T)}|A_k(\sigma)|^{1/q} \leq C\|f\|_{L^\infty(Q_T)}\mu_k^{1/q},$$

where

$$2 < p < \begin{cases} +\infty, & n=1,2, \\ \dfrac{2n}{n-2}, & n \geq 3, \end{cases} \quad q = \frac{p}{p-1}.$$

Applying Hölder's inequality to $I_k(\sigma)$ and combining the result with the above estimate, we are led to

$$I_k(\sigma) \leq \left(\int_{A_k(\sigma)}(u-k)_+^pdx\right)^{2/p}|A_k(\sigma)|^{(p-2)/p}$$
$$\leq (C\|f\|_{L^\infty(Q_T)})^2\mu_k^{(3p-4)/p}.$$

Hence, for any $t \in [0,T]$,

$$I_k(t) \leq I_k(\sigma) \leq (C\|f\|_{L^\infty(Q_T)})^2\mu_k^{(3p-4)/p}. \tag{4.2.4}$$

Since for any $h > k$ and $t \in [0, T]$,

$$I_k(t) \geq \int_{A_h(t)} (u - k)_+^2 \, dx \geq (h - k)^2 |A_h(t)|,$$

from (4.2.4) we obtain

$$(h - k)^2 \mu_h \leq (C\|f\|_{L^\infty(Q_T)})^2 \mu_k^{(3p-4)/p},$$

i.e.

$$\mu_h \leq \left(\frac{C\|f\|_{L^\infty(Q_T)}}{h - k} \right)^2 \mu_k^{(3p-4)/p}.$$

Using Lemma 4.1.1 and noticing that $p > 2$ implies

$$\frac{3p - 4}{p} = 1 + \frac{2p - 4}{p} > 1,$$

we finally arrive at

$$\mu_{l+d} = \sup_{0 < t < T} |A_{l+d}(t)| = 0,$$

where

$$d = C\|f\|_{L^\infty(Q_T)} \mu_l^{1-2/p} 2^{(3p-4)/(2p-4)}$$
$$\leq C|\Omega|^{1-2/p} 2^{(3p-4)/(2p-4)} \|f\|_{L^\infty(Q_T)}.$$

This means, by the definition of $A(k)$,

$$u \leq l + C|\Omega|^{1-2/p} 2^{(3p-4)/(2p-4)} \|f\|_{L^\infty(Q_T)}, \qquad \text{a.e. in } Q_T.$$

$\square$

Corollary 4.2.2 *Let $f \in L^\infty(Q_T)$ and $u \in W_2^{1,1}(Q_T)$ be a weak solution of (4.2.3). Then*

$$\inf_{Q_T} u \geq \inf_{\partial_p Q_T} u - C\|f\|_{L^\infty(Q_T)},$$

where C is a constant depending only on n and Ω.

The De Giorgi iteration technique can also be applied to more general parabolic equations in divergence form to establish the weak maximum principle. For instance, for the equation

$$\frac{\partial u}{\partial t} - \Delta u + c(x, t)u = f(x, t) + \operatorname{div} \vec{f}(x, t), \quad (x, t) \in Q_T, \qquad (4.2.5)$$

we have

Theorem 4.2.2 *Assume $p > n \geq 3$, $0 \leq c(x,t) \leq M$, $f \in L^\infty(0,T; L^{p_*}(\Omega))$, $\vec{f} \in L^\infty(0,T; L^p(\Omega, \mathbb{R}^n))$, and $u \in W_2^{1,1}(Q_T)$ is a weak solution of equation (4.2.5). Then*

$$\sup_{Q_T} u \leq \sup_{\partial_p Q_T} u_+ + C \left(\sup_{0<t<T} \|f\|_{L^{p_*}(\Omega)} + \sup_{0<t<T} \|\vec{f}\|_{L^p(\Omega)} \right) |\Omega|^{1/n - 1/p},$$

$$\inf_{Q_T} u \geq \inf_{\partial_p Q_T} u_- - C \left(\sup_{0<t<T} \|f\|_{L^{p_*}(\Omega)} + \sup_{0<t<T} \|\vec{f}\|_{L^p(\Omega)} \right) |\Omega|^{1/n - 1/p},$$

where $p_* = np/(n+p)$ and C is a constant depending only on n, p, M and Ω, but is independent of the lower bound of $|\Omega|$.

The proof is left to the reader.

Remark 4.2.1 *Among the assumptions of Theorem 4.2.2, $c(x,t) \geq 0$ is not necessary. If we replace this condition by $|c(x,t)| \leq M$ for some constant M, then the same estimates hold, but the constant C depends on T in addition to n, p, M and Ω. In fact, under such condition, we may introduce a new unknown function $w = e^{-\lambda t} u$ with $\lambda > 0$ to be determined and change equation (4.2.5) into*

$$\frac{\partial w}{\partial t} - \Delta w + (\lambda + c)w = e^{-\lambda t}(f + \operatorname{div} \vec{f}).$$

If we choose $\lambda = \|c\|_{L^\infty(Q_T)}$, then $\lambda + c(x,t) \geq 0$ and thus the conclusion of Theorem 4.2.2 can be applied to the new equation to obtain the desired estimate.

4.3 Local Boundedness Estimates for Weak Solutions of Poisson's Equation

Another important technique in estimating the maximum norm of solutions is the Moser iteration. In this section, the main points of this technique will be illustrated by applying it to Poisson's equation in establishing the local boundedness estimate of weak solutions.

4.3.1 *Weak subsolutions (supersolutions)*

In order to further discuss the local boundedness of weak solutions of Poisson's equation, we introduce the concept of weak subsolutions (supersolutions) of equation (4.1.9).

Definition 4.3.1 A function $u \in H^1(\Omega)$ is said to be a weak subsolution (supersolution), if for any nonnegative function $\varphi \in C_0^\infty(\Omega)$,

$$\int_\Omega (\nabla u \cdot \nabla \varphi + cu\varphi)\, dx \leq (\geq) \int_\Omega \left(f\varphi dx - \vec{f} \cdot \nabla \varphi \right) dx.$$

Sometimes the weak subsolution (supersolution) is said to be a function in $H^1(\Omega)$ satisfying

$$-\Delta u + c(x)u \leq (\geq) f(x) + \operatorname{div} \vec{f}(x)$$

in the weak sense.

Proposition 4.3.1 *Assume that $f \equiv 0$, $\vec{f} \equiv 0$, $c(x) \geq 0$ and $u \in H^1(\Omega) \cap L^\infty(\Omega)$ is a weak subsolution of equation (4.1.9). If $g''(s) \geq 0$, $g'(s) \geq 0$, $g(0) = 0$, then $w = g(u)$ is also a subsolution of equation (4.1.9).*

Proof. By the definition of weak subsolutions, for any nonnegative function $\varphi \in C_0^\infty(\Omega)$,

$$\int_\Omega (\nabla u \cdot \nabla \varphi + cu\varphi) dx \leq 0.$$

Since $g'(s) \geq 0$ and $u \in H^1(\Omega) \cap L^\infty(\Omega)$, we have $0 \leq g'(u)\varphi \in H_0^1(\Omega)$ and hence $g'(u)\varphi$ can be chosen as a test function. Substituting it into the above inequality we obtain

$$\int_\Omega \left(g'(u)\nabla u \cdot \nabla \varphi + g''(u)|\nabla u|^2 \varphi + cug'(u)\varphi \right) dx \leq 0.$$

Hence

$$\int_\Omega (\nabla w \cdot \nabla \varphi + cw\varphi) dx$$

$$\leq -\int_\Omega g''(u)|\nabla u|^2 \varphi dx + \int_\Omega c(g(u) - ug'(u))\varphi dx$$

$$\leq \int_\Omega c(g(u) - ug'(u))\varphi dx.$$

Here we have used the condition $g''(s) \geq 0$. To prove that $w = g(u)$ is a subsolution of (4.1.9), besides $c \geq 0$, $\varphi \geq 0$, it suffices to note that $g(0) = 0$ and $g''(s) \geq 0$ imply

$$g(s) - sg'(s) \leq 0, \quad \forall s \in \mathbb{R}.$$

$\square$

Remark 4.3.1   *Conditions $g''(s) \geq 0$ and $g'(s) \geq 0$ in Proposition 4.3.1 can be replaced by that $g(s)$ is a nondecreasing convex Lipschitz function; a typical example is*

$$g(s) = s_+^p, \quad s \in \mathbb{R} \quad (p \geq 1).$$

We leave the proof to the reader.

4.3.2   Local boundedness estimate for weak solutions of Laplace's equation

Theorem 4.3.1   *Let $x^0 \in \Omega$, $B_R = B_R(x^0) \subset \Omega$ and $u \in H^1(\Omega) \cap L^\infty(\Omega)$ be a weak subsolution of Laplace's equation (4.1.1). Then*

$$\sup_{B_{R/2}} u \leq C \left(\frac{1}{R^n} \int_{B_R} u_+^2 \, dx \right)^{1/2},$$

where C is a constant depending only on n.

Proof.   From Proposition 4.3.1, we see that u_+ is also a weak subsolution of Laplace's equation (4.1.1). So we may assume $u \geq 0$. The proof will be proceeded in three steps.

   Step 1   Prove the inverse Poincaré's inequality, namely, for any $p \geq 2$,

$$\int_{B_R} \eta^2 |\nabla u^{p/2}|^2 dx \leq C \int_{B_R} u^p |\nabla \eta|^2 dx, \tag{4.3.1}$$

where $\eta(x)$ is the zero extension of a cut-off function on $B_{\rho'}$, relative to B_ρ $(0 < \rho < \rho' \leq R)$, namely, $\eta \in C_0^\infty(B_{\rho'})$, $0 \leq \eta(x) \leq 1$, $\eta(x) = 1$ on B_ρ, $\eta(x) = 0$ on $\Omega \backslash B_{\rho'}$ and $|\nabla \eta(x)| \leq \dfrac{C}{\rho' - \rho}$.

   Choose $\eta^2 u^{p-1}$ as a test function in the definition of weak subsolutions. Then

$$\int_\Omega \nabla u \cdot \nabla(\eta^2 u^{p-1}) dx \leq 0,$$

or

$$(p-1) \int_{B_R} \eta^2 u^{p-2} |\nabla u|^2 dx + 2 \int_{B_R} \eta u^{p-1} \nabla u \cdot \nabla \eta \, dx \leq 0.$$

Hence

$$\frac{p-1}{p} \int_{B_R} \eta^2 |\nabla u^{p/2}|^2 dx + \int_{B_R} \eta u^{p/2} \nabla u^{p/2} \cdot \nabla \eta \, dx \leq 0.$$

Since using Cauchy's inequality with ε gives

$$\int_{B_R} \eta u^{p/2} \nabla u^{p/2} \cdot \nabla \eta \, dx \leq \frac{\varepsilon}{2} \int_{B_R} \eta^2 |\nabla u^{p/2}|^2 dx + \frac{1}{2\varepsilon} \int_{B_R} u^p |\nabla \eta|^2 dx,$$

(4.3.1) can be obtained by choosing $\varepsilon > 0$ small enough.

Step 2 Prove the inverse Hölder's inequality, namely, for any $p \geq 2$,

$$\left(\frac{1}{R^n} \int_{B_\rho} u^{pq} dx \right)^{1/q} \leq C \left(\frac{1}{R^{n-2}(\rho' - \rho)^2} \int_{B_{\rho'}} u^p dx \right), \qquad (4.3.2)$$

where

$$1 < q < \begin{cases} +\infty, & n = 1, 2, \\ \dfrac{n}{n-2}, & n \geq 3. \end{cases}$$

From Remark 1.3.6 of Chapter 1, we have

$$\left(\frac{1}{R^n} \int_{B_\rho} \eta^{2q} u^{pq} dx \right)^{1/(2q)} \leq R^{-n/(2q)} \left(\int_{B_R} (\eta u^{p/2})^{2q} dx \right)^{1/(2q)}$$

$$\leq CR^{1-n/2} \left(\int_{B_R} |\nabla(\eta u^{p/2})|^2 dx \right)^{1/2},$$

or

$$\left(\frac{1}{R^n} \int_{D_\rho} \eta^{2q} u^{pq} dx \right)^{1/q} \leq \frac{C}{R^{n-2}} \int_{B_R} |\nabla(\eta u^{p/2})|^2 dx,$$

where C is a constant depending only on n. On the other hand, using the inverse Poincaré's inequality gives

$$\int_{B_R} |\nabla(\eta u^{p/2})|^2 dx = \int_{B_R} |\eta \nabla u^{p/2} + u^{p/2} \nabla \eta|^2 dx$$

$$\leq 2 \int_{B_R} \eta^2 |\nabla u^{p/2}|^2 dx + 2 \int_{B_R} u^p |\nabla \eta|^2 dx$$

$$\leq C \int_{B_R} u^p |\nabla \eta|^2 dx.$$

Therefore (4.3.2) holds.

Step 3 Iterate.

Denote

$$\rho_k = \frac{R}{2}\left(1 + \frac{1}{2^k}\right), \quad k = 0, 1, \cdots$$

and choose $p = 2q^k$, $\rho = \rho_{k+1}$, $\rho' = \rho_k$ in (4.3.2). Then

$$\left(\frac{1}{R^n}\int_{B_{\rho_{k+1}}}|u|^{2q^{k+1}}dx\right)^{1/(2q^{k+1})}$$

$$\leq C^{1/(2q^k)}4^{(k+2)/(2q^k)}\left(\frac{1}{R^n}\int_{B_{\rho_k}}|u|^{2q^k}dx\right)^{1/(2q^k)}.$$

Iterating repeatedly leads to

$$\left(\frac{1}{R^n}\int_{B_{\rho_{k+1}}}|u|^{2q^{k+1}}dx\right)^{1/(2q^{k+1})} \leq C^\alpha 4^\beta\left(\frac{1}{R^n}\int_{B_R}u^2dx\right)^{1/2},$$

where

$$\alpha = \sum_{k=0}^{\infty}\frac{1}{2q^k}, \quad \beta = \sum_{k=0}^{\infty}\frac{k+2}{2q^k}.$$

Since $q > 1$ implies the convergence of these series, α, β are both finite numbers. Thus for some constant C depending only on n,

$$\left(\frac{1}{R^n}\int_{B_{R/2}}|u|^{2q^{k+1}}dx\right)^{1/(2q^{k+1})} \leq C\left(\frac{1}{R^n}\int_{B_R}u^2dx\right)^{1/2}.$$

From this the desired conclusion follows by letting $k \to \infty$.   □

4.3.3   Local boundedness estimate for solutions of Poisson's equation

Now we turn to equation (4.1.9) with $\vec{f} \equiv 0$, namely,

$$-\Delta u + c(x)u = f(x), \quad x \in \Omega. \tag{4.3.3}$$

Theorem 4.3.2   Let $0 \leq c(x) \leq M$, $f \in L^\infty(\Omega)$ and $u \in H^1(\Omega) \cap L^\infty(\Omega)$ be a weak subsolution of equation (4.3.3). Then there exists a constant $R_0 > 0$ depending only on M, such that for any $x^0 \in \Omega$, there holds

$$\sup_{B_{R/2}} u \leq C\left(\frac{1}{R^n}\int_{B_R}u^2dx\right)^{1/2} + C\|f\|_{L^\infty(\Omega)},$$

provided $0 < R \leq R_0$ *and* $B_R = B_R(x^0) \subset \Omega$, *where* C *is a constant depending only on* n, R_0 *and* M.

Proof.　For any $x^0 \in \Omega$, denote

$$\overline{u}(x) = u(x) + |x - x^0|^2 \|f\|_{L^\infty(\Omega)}, \quad x \in \Omega.$$

Then, in the weak sense

$$
\begin{aligned}
& -\Delta \overline{u} + c\overline{u} \\
= & -\Delta u - 2n\|f\|_{L^\infty(\Omega)} + cu + c|x - x^0|^2 \|f\|_{L^\infty(\Omega)} \\
\leq & f + c|x - x^0|^2 \|f\|_{L^\infty(\Omega)} - 2n\|f\|_{L^\infty(\Omega)} \\
\leq & f + M|x - x^0|^2 \|f\|_{L^\infty(\Omega)} - 2n\|f\|_{L^\infty(\Omega)}, \quad \text{in } \Omega.
\end{aligned}
$$

Choose $R_0 > 0$ such that $R_0^2 \leq \dfrac{1}{M+1}$. Then, in the weak sense,

$$-\Delta \overline{u} + c\overline{u} \leq 0, \quad \text{in } \Omega \cap B_{R_0}(x^0),$$

namely, $\overline{u}$ is a weak subsolution of equation $-\Delta v + cv = 0$ in $\Omega \cap B_{R_0}(x^0)$. By Proposition 4.3.1, $\overline{u}_+$ also satisfies

$$-\Delta \overline{u}_+ + c\overline{u}_+ \leq 0, \quad \text{in } \Omega \cap B_{R_0}(x^0)$$

and hence

$$-\Delta \overline{u}_+ \leq 0, \quad \text{in } \Omega \cap B_{R_0}(x^0)$$

in the weak sense, namely, $\overline{u}_+$ is a weak subsolution of Laplace's equation in $\Omega \cap B_{R_0}(x^0)$. Thus, from Theorem 4.3.1, we obtain

$$\sup_{B_{R/2}} \overline{u}_+ \leq C \left(\frac{1}{R^n} \int_{B_R} |\overline{u}_+|^2 dx \right)^{1/2}, \quad 0 < R \leq R_0$$

and hence the conclusion of the theorem follows. $\qquad\square$

Remark 4.3.2　*What we have adopted to establish the local boundedness estimates in Theorem 4.3.1 and Theorem 4.3.2 is the so-called Moser iteration technique, a technique of extreme importance. The method is based on the fact*

$$\|u\|_{L^\infty} = \lim_{p \to \infty} \|u\|_{L^p}.$$

The basic idea in establishing the boundedness estimates is to choose suitably ρ_k and p_k such that $\rho_0 = R$, $\lim_{k \to \infty} \rho_k = R/2$ and $\lim_{k \to \infty} p_k = +\infty$, and then try to prove that

$$A_k = \|u\|_{L^{p_k}(B_{\rho_k})}$$

satisfies the recursive formula

$$A_{k+1} \leq C^{\alpha_k} A_k$$

with $\alpha_k \geq 0$ such that the series $\displaystyle\sum_{k=0}^{\infty} \alpha_k$ is convergent.

4.3.4   Estimate near the boundary for weak solutions of Poisson's equation

All estimates presented above are established in a neighborhood of the interior points of Ω. In addition to these interior estimates, we need to establish estimates near the boundary points. As an example, we will do this for a domain of the form $Q^+ = \{x \in \mathbb{R}^n; |x_i| < 1 \, (1 \leq i \leq n), x_n > 0\}$. For general domains, we may transform the neighborhood of the boundary point to a domain of this special form by means of local flatting of the boundary. Of course, after a local flatting transformation, the shape of the equation will be changed.

Theorem 4.3.3   *Assume that $0 \leq c(x) \leq M$, $f \in L^\infty(Q^+)$ and $u \in H^1(Q^+) \cap L^\infty(Q^+)$ is a weak subsolution of equation (4.3.3) with trace vanishing on the bottom of Q^+, i.e. $\gamma u(x_1, \cdots, x_{n-1}, 0) = 0$. Then there exists a constant $R_0 \in (0, 1]$ depending only on M, such that for any point x^0 on the bottom of Q^+,*

$$\sup_{B_{R/2}^+} u \leq C \left(\frac{1}{R^n} \int_{B_R^+} u^2 dx \right)^{1/2} + C\|f\|_{L^\infty(Q^+)},$$

provided $0 < R \leq R_0$ and $B_R^+ = B_R^+(x^0) \subset Q^+$, where $B_R^+(x^0) = B_R(x^0) \cap \{x \in \mathbb{R}^n; x_n > 0\}$ and C is a constant depending only on n, R_0 and M.

The proof is similar as the one of Theorem 4.3.2. We leave it to the reader.

4.4 Local Boundedness Estimates for Weak Solutions of the Heat Equation

In this section, we use the Moser iteration technique to the local boundedness estimates for weak solutions of the nonhomogeneous heat equation.

4.4.1 *Weak subsolutions (supersolutions)*

We first introduce weak subsolutions (supersolutions) of equation (4.2.5).

Definition 4.4.1 A function $u \in W_2^{1,1}(Q_T)$ is said to be a weak subsolution (supersolution) of (4.2.5), if for any nonnegative function $\varphi \in C_0^\infty(Q_T)$, there holds

$$\iint_{Q_T} (u_t\varphi + \nabla u \cdot \nabla\varphi + cu\varphi)dxdt \leq (\geq) \iint_{Q_T} \left(f\varphi - \vec{f} \cdot \nabla\varphi\right)dxdt.$$

Sometimes, a weak subsolution (supersolution) u of (4.2.5) is said to be a function satisfying

$$\frac{\partial u}{\partial t} - \Delta u + c(x,t)u \leq (\geq)f(x,t) + \mathrm{div}\vec{f}(x,t)$$

in the weak sense.

Proposition 4.4.1 *Assume that $f \equiv 0$, $\vec{f} \equiv 0$, $c(x,t) \geq 0$ and $u \in W_2^{1,1}(Q_T) \cap L^\infty(Q_T)$ is a weak subsolution of (4.2.5). If $g''(s) \geq 0$, $g'(s) \geq 0$ and $g(0) = 0$, then $g(u)$ is also a weak subsolution of (4.2.5).*

The proof is similar to the elliptic case (see Proposition 4.3.1).

Remark 4.4.1 *Conditions $g''(s) \geq 0$ and $g'(s) \geq 0$ of Proposition 4.4.1 can be replaced by that $g(s)$ is a nondecreasing convex Lipschitz function; one of the typical cases is*

$$g(s) = s_+^p, \quad s \in \mathbb{R} \quad (p \geq 1).$$

4.4.2 *Local boundedness estimate for weak solutions of the homogeneous heat equation*

Theorem 4.4.1 *Let $(x^0, t_0) \in Q_T$, $Q_R = Q_R(x^0, t_0) = B_R(x^0) \times (t_0 - R^2, t_0 + R^2) \subset Q_T$ and $u \in W_2^{1,1}(Q_T) \cap L^\infty(Q_T)$ be a weak subsolution of*

(4.2.1). Then

$$\sup_{Q_{R/2}} u \le C \left(\frac{1}{R^{n+2}} \iint_{Q_R} u^2 dx dt \right)^{1/2}$$

with C depending only on n.

Proof. Similar to the discussion for Poisson's equation, we proceed in three steps.

    Step 1 Derive an estimate similar to the inverse Poincaré's inequality.

    For any ρ, ρ' such that $R/2 \le \rho < \rho' \le R$, choose a cut-off function $\eta(x)$ on $B_{\rho'}$, relative to B_ρ, namely, $\eta \in C_0^\infty(B_{\rho'})$, $0 \le \eta(x) \le 1$, $\eta(x) = 1$ on B_ρ, and $|\nabla \eta(x)| \le \dfrac{C}{\rho' - \rho}$ and extend η to be zero for $x \in \Omega \backslash B_{\rho'}$. For any $s \in (t_0 - \rho^2, t_0 + R^2)$, choose $\xi \in C^\infty(-\infty, s]$, such that $\xi(t) = 1$ on $[t_0 - \rho^2, s]$, $\xi(t) = 0$ on $(-\infty, t_0 - \rho'^2]$ and $0 \le \xi'(t) \le \dfrac{C}{(\rho' - \rho)^2}$ for $t \le s$, and extend it to be zero for $t > s$.

    We may assume that $u \ge 0$; otherwise we use u^+ instead of u. Choose $\varphi = \xi^2 \eta^2 u$ as a test function in the definition of weak subsolutions to obtain

$$\iint_{Q_T} \left(u_t \xi^2 \eta^2 u + \nabla u \cdot \nabla(\xi^2 \eta^2 u) \right) dx dt \le 0,$$

namely,

$$\frac{1}{2} \iint_{Q_{\rho'}^s} \frac{\partial}{\partial t}(\xi^2 \eta^2 u^2) dx dt - \iint_{Q_{\rho'}^s} \xi \xi' \eta^2 u^2 dx dt$$
$$+ \iint_{Q_{\rho'}^s} \xi^2 \eta^2 |\nabla u|^2 dx dt + 2 \iint_{Q_{\rho'}^s} u \xi^2 \eta \nabla u \cdot \nabla \eta \, dx dt \le 0,$$

where $Q_\rho^t = B_\rho \times (t_0 - \rho^2, t)$. Since $\xi(t) = 0$ at $t = t_0 - \rho'^2$, we have

$$\iint_{Q_{\rho'}^s} \frac{\partial}{\partial t}(\xi^2 \eta^2 u^2) dx dt = \int_{B_{\rho'}} \xi^2 \eta^2 u^2 \Big|_{t=s} dx.$$

Using Cauchy's inequality with ε gives

$$2 \left| \iint_{Q_{\rho'}^s} u \xi^2 \eta \nabla u \cdot \nabla \eta \, dx dt \right|$$
$$\le \frac{1}{2} \iint_{Q_{\rho'}^s} \xi^2 \eta^2 |\nabla u|^2 dx dt + 2 \iint_{Q_{\rho'}^s} \xi^2 u^2 |\nabla \eta|^2 dx dt.$$

Hence

$$
\frac{1}{2} \int_{B_{\rho'}} \xi^2 \eta^2 u^2 \Big|_{t=s} dx + \iint_{Q^s_{\rho'}} \xi^2 \eta^2 |\nabla u|^2 dx dt
$$

$$
\leq \frac{1}{2} \iint_{Q^s_{\rho'}} \xi^2 \eta^2 |\nabla u|^2 dx dt + 2 \iint_{Q^s_{\rho'}} \xi^2 u^2 |\nabla \eta|^2 dx dt + \iint_{Q^s_{\rho'}} \xi \xi' \eta^2 u^2 dx dt
$$

$$
\leq \frac{1}{2} \iint_{Q^s_{\rho'}} \xi^2 \eta^2 |\nabla u|^2 dx dt + \frac{C}{(\rho' - \rho)^2} \iint_{Q^s_{\rho'}} u^2 dx dt.
$$

Therefore

$$
\sup_{t_0 - \rho^2 \leq t \leq t_0 + \rho^2} \int_{B_{\rho'}} \eta^2 u^2(x, t) dx + \int_{t_0 - \rho^2}^{t_0 + \rho^2} \int_{B_{\rho'}} \eta^2 |\nabla u|^2 dx dt
$$

$$
\leq \frac{C}{(\rho' - \rho)^2} \iint_{Q_{\rho'}} u^2 dx dt.
$$

Step 2   Derive an estimate similar to the inverse Hölder's inequality.

Let $\chi(t)$ be the characteristic function of the segment $[t_0 - \rho^2, t_0 + \rho^2]$. Then $\chi(t)\eta u \in V_2(Q_R)$ and by the t-anisotropic embedding theorem (Theorem 1.4.2 of Chapter 1),

$$
\left(\frac{1}{R^{n+2}} \iint_{Q_R} (\chi(t)\eta u)^{2q} dx dt \right)^{1/q}
$$

$$
\leq C(n) R^{-n} \Big(\sup_{t_0 - R^2 \leq t \leq t_0 + R^2} \int_{B_R} (\chi(t)\eta u(x,t))^2 dx
$$

$$
+ \iint_{Q_R} |\chi(t)\nabla(\eta u)|^2 dx dt \Big).
$$

$$
= C(n) R^{-n} \Big(\sup_{t_0 - \rho^2 \leq t \leq t_0 + \rho^2} \int_{B_{\rho'}} \eta^2 u^2(x,t) dx
$$

$$
+ \int_{t_0 - \rho^2}^{t_0 + \rho^2} \int_{B_{\rho'}} |\eta \nabla u + u \nabla \eta|^2 dx dt \Big),
$$

where

$$
q = \begin{cases} 5/3, & n = 1, 2, \\ 1 + 2/n, & n \geq 3. \end{cases}
$$

Using the inequality obtained in Step 1, we further deduce

$$\left(\frac{1}{R^{n+2}}\iint_{Q_\rho} u^{2q}\,dxdt\right)^{1/q}$$

$$\leq \left(\frac{1}{R^{n+2}}\iint_{Q_R}(\chi(t)\eta u)^{2q}\,dxdt\right)^{1/q}$$

$$\leq C(n)R^{-n}\left(\sup_{t_0-\rho^2\leq t\leq t_0+\rho^2}\int_{B_{\rho'}}\eta^2 u^2(x,t)\,dx\right.$$

$$\left.+\int_{t_0-\rho^2}^{t_0+\rho^2}\int_{B_{\rho'}}|\eta\nabla u + u\nabla\eta|^2\,dxdt\right)$$

$$\leq C(n)R^{-n}\left(\sup_{t_0-\rho^2\leq t\leq t_0+\rho^2}\int_{B_{\rho'}}\eta^2 u^2(x,t)\,dx\right.$$

$$\left.+2\int_{t_0-\rho^2}^{t_0+\rho^2}\int_{B_{\rho'}}\eta^2|\nabla u|^2\,dxdt+2\int_{t_0-\rho^2}^{t_0+\rho^2}\int_{B_{\rho'}}u^2|\nabla\eta|^2\,dxdt\right)$$

$$\leq\frac{C}{R^n(\rho'-\rho)^2}\iint_{Q_{\rho'}}u^2\,dxdt.$$

Proposition 4.4.1 shows that u^{q^k} is also a weak subsolution of (4.2.1). So with u^{q^k} in place of u, we obtain from the above inequality,

$$\left(\frac{1}{R^{n+2}}\iint_{Q_\rho}u^{2q^{k+1}}\,dxdt\right)^{1/2q^{k+1}}\leq\left(\frac{C}{R^n(\rho'-\rho)^2}\iint_{Q_{\rho'}}u^{2q^k}\,dxdt\right)^{1/2q^k}.$$

Step 3 Iterate.

Similar to the case of Poisson's equation. □

4.4.3 *Local boundedness estimate for weak solutions of the nonhomogeneous heat equation*

Now we turn to equation (4.2.5) with $\vec{f}\equiv 0$, namely,

$$\frac{\partial u}{\partial t}-\Delta u + c(x,t)u = f(x,t),\quad (x,t)\in Q_T. \tag{4.4.1}$$

Theorem 4.4.2 Let $|c(x,t)|\leq M$, $f\in L^\infty(Q_T)$ and $u\in W_2^{1,1}(Q_T)\cap L^\infty(Q_T)$ be a weak subsolution of (4.4.1). Then for any $(x^0,t_0)\in Q_T$ and

$R > 0$ *such that* $Q_R = Q_R(x^0, t_0) \subset Q_T$,

$$\sup_{Q_{R/2}} u \leq C \left(\frac{1}{R^{n+2}} \iint_{Q_R} u^2 dx dt \right)^{1/2} + C\|f\|_{L^\infty(Q_T)}$$

with C depending only on n, M and T.

Proof. Let

$$w(x,t) = e^{-Mt} u(x,t), \quad (x,t) \in Q_T.$$

Then, in the weak sense, w satisfies

$$\frac{\partial w}{\partial t} - \Delta w + (M+c)w = e^{-Mt} \left(\frac{\partial u}{\partial t} - \Delta u + cu \right) \leq e^{-Mt} f, \quad \text{in } Q_T.$$

So, without loss of generality, we may assume that $c(x,t)$ in (4.4.1) satisfies $0 \leq c(x,t) \leq 2M$ in Q_T. Set

$$\bar{u}(x,t) = u(x,t) - t\|f\|_{L^\infty(Q_T)}, \quad \text{in } Q_T.$$

Then, in the weak sense, $\bar{u}$ satisfies

$$\frac{\partial \bar{u}}{\partial t} - \Delta \bar{u} + c\bar{u}$$
$$= \frac{\partial u}{\partial t} - \Delta u + cu - \|f\|_{L^\infty(Q_T)} - ct\|f\|_{L^\infty(Q_T)}$$
$$= f - \|f\|_{L^\infty(Q_T)} - ct\|f\|_{L^\infty(Q_T)}$$
$$\leq 0, \quad \text{in } Q_T,$$

namely, $\bar{u}$ is a weak subsolution of (4.4.1) with $f \equiv 0$. So

$$\frac{\partial \bar{u}_+}{\partial t} - \Delta \bar{u}_+ + c\bar{u}_+ \leq 0, \quad \text{in } Q_T$$

and hence

$$\frac{\partial \bar{u}_+}{\partial t} - \Delta \bar{u}_+ \leq 0, \quad \text{in } Q_T,$$

namely, $\bar{u}_+$ is a weak subsolution of (4.2.1) in Q_T. Thus, from Theorem 4.4.1, we have

$$\sup_{Q_{R/2}} \bar{u}_+ \leq C \left(\frac{1}{R^{n+2}} \int_{Q_R} |\bar{u}_+|^2 dx \right)^{1/2}.$$

Hence

$$\sup_{Q_{R/2}} u \le C \left(\frac{1}{R^{n+2}} \int_{Q_R} u^2 dx \right)^{1/2} + C\|f\|_{L^\infty(Q_T)}.$$

$\square$

Similar to the case of Poisson's equation, we can further establish the estimate near the boundary for the heat equation.

Exercises

1. Prove Theorem 4.1.2.

2. Let $\lambda > 0$, $\Omega \subset \mathbb{R}^n$ be a bounded domain, $f \in L^\infty(\Omega)$ and $u \in H_0^1(\Omega) \cap L^\infty(\Omega)$ be a weak solution of

$$-\text{div}\Big((u^2 + \lambda)\nabla u\Big) + u = f, \quad x \in \Omega.$$

i) Use the De Giorgi iteration technique to establish the maximum norm estimate for u;

ii) Whether the maximum norm estimate holds in case $\lambda = 0$?

3. Prove Theorem 4.2.2.

4. Let $\Omega \subset \mathbb{R}^n$ be a bounded domain, $p > 2$ and $f \in L^\infty(Q_T)$. Consider the first initial-boundary value problem for p-Laplace's equation

$$\begin{cases} \dfrac{\partial u}{\partial t} - \text{div}\big(|\nabla u|^{p-2}\nabla u\big) = f, & (x,t) \in Q_T = \Omega \times (0,T), \\ u\Big|_{\partial_p Q_T} = 0. \end{cases}$$

i) Define weak solutions of this problem and prove the existence and uniqueness;

ii) Use the De Giorgi iteration technique to establish the maximum norm estimate for weak solutions of the problem.

5. Prove Remark 4.3.1.

6. Prove Theorem 4.3.3.

7. Let $\Omega \subset \mathbb{R}^n$ be a bounded domain and $u \in L^\infty(\Omega)$. Show that

$$\|u\|_{L^\infty(\Omega)} = \lim_{p \to \infty} \|u\|_{L^p(\Omega)}.$$

8. Let u_1 and $u_2 \in H_0^1(\Omega)$ be a weak solution and a weak subsolution of Poisson's equation

$$-\Delta u = f, \quad x \in \Omega,$$

where $\Omega \subset \mathbb{R}^n$ is a bounded domain and $f \in L^2(\Omega)$. Show that

$$u_1(x) \geq u_2(x), \quad \text{a.e. } x \in \Omega.$$

9. Prove that $u \in W_2^{1,1}(Q_T)$ is a weak solution of the equation

$$\frac{\partial u}{\partial t} - \Delta u + c(x,t)u = f, \quad (x,t) \in Q_T = \Omega \times (0,T)$$

if and only if u is both the weak supersolution and the weak subsolution of the equation, where $\Omega \subset \mathbb{R}^n$ is a bounded domain, $c \in L^\infty(Q_T)$ and $f \in L^2(Q_T)$.

10. Prove Proposition 4.4.1 and Remark 4.4.1.

11. Let $u \in W_2^{1,1}(Q_T) \cap L^\infty(Q_T)$ be a weak solution of the equation

$$\frac{\partial u}{\partial t} - \operatorname{div}\left(\left(\frac{u^2+1}{u^2+3}\right)\nabla u\right) + c(x,t)u = f, \quad (x,t) \in Q_T = \Omega \times (0,T)$$

where $\Omega \subset \mathbb{R}^n$ is a bounded domain, $c \in L^\infty(Q_T)$ and $f \in L^2(Q_T)$. Use the Moser iteration technique to derive the local boundedness estimate for u.

12. Use the Moser iteration technique to derive the local boundedness estimate near the boundary for weak solutions of the heat equation.

Chapter 5

Harnack's Inequalities

In this chapter, we continue our study of properties of weak solutions. we will concentrate our attention on Harnack's inequalities which reveal deeply the properties of solutions of elliptic and parabolic equations. Such kind of inequalities hold not only for general linear elliptic and parabolic equations in divergence form, but also for quasilinear equations (see Chapter 10). However, we merely illustrate the argument for the simplest equations such as Laplace's equation and the homogeneous heat equation, although the basic idea is available to general linear and quasilinear equations.

5.1 Harnack's Inequalities for Solutions of Laplace's Equation

In this section, we are concerned with Laplace's equation

$$-\Delta u = 0, \quad x \in \mathbb{R}^n. \tag{5.1.1}$$

5.1.1 Mean value formula

Theorem 5.1.1 Let $u \in C^2(\mathbb{R}^n)$ be a solution of (5.1.1). Then for any ball $B_R = B_R(y) \subset \mathbb{R}^n$,

$$u(y) = \frac{1}{n\omega_n R^{n-1}} \int_{\partial B_R} u(x) ds, \tag{5.1.2}$$

$$u(y) = \frac{1}{\omega_n R^n} \int_{B_R} u(x) dx, \tag{5.1.3}$$

where ω_n is the measure of the unit ball in $\mathbb{R}^n$.

Proof. Integrating (5.1.1) over the ball $B_\rho = B_\rho(y)$ with $\rho \in (0, R)$ gives

$$\int_{B_\rho} \Delta u \, dx = \int_{\partial B_\rho} \frac{\partial u}{\partial \vec{\nu}} \, ds = 0,$$

where $\vec{\nu}$ denotes the unit normal vector outward to ∂B_ρ. Introduce the polar coordinate $\rho = |x - y|$, $z = \dfrac{x - y}{\rho}$. Then from the above formula, we have, with $u = u(x) = u(y + \rho z)$,

$$
\begin{aligned}
0 &= \int_{\partial B_\rho} \frac{\partial u}{\partial \vec{\nu}} \, ds \\
&= \int_{\partial B_\rho(0)} \frac{\partial}{\partial \rho} u(y + \rho z) \, ds \\
&= \rho^{n-1} \int_{\partial B_1(0)} \frac{\partial}{\partial \rho} u(y + \omega) \, ds \\
&= \rho^{n-1} \frac{\partial}{\partial \rho} \int_{\partial B_1(0)} u(y + \omega) \, ds \\
&= \rho^{n-1} \frac{\partial}{\partial \rho} \left[\rho^{1-n} \int_{\partial B_\rho(0)} u(y + \rho z) \, ds \right].
\end{aligned}
$$

Hence

$$
\begin{aligned}
\rho^{1-n} \int_{\partial B_\rho(0)} u(y + \rho z) \, ds &= R^{1-n} \int_{\partial B_R(0)} u(y + Rz) \, ds \\
&= R^{1-n} \int_{\partial B_R} u(x) \, ds.
\end{aligned}
$$

Sending $\rho \to 0^+$ and noting

$$\lim_{\rho \to 0^+} \rho^{1-n} \int_{\partial B_\rho(0)} u(y + \rho z) \, ds = n \omega_n u(y),$$

we then derive (5.1.2).

Since (5.1.2) holds for any $R > 0$, we have

$$n \omega_n \rho^{n-1} u(y) = \int_{\partial B_\rho} u(x) \, ds,$$

Integrating over $(0, R)$ with respect to ρ then gives (5.1.3). $\qquad\square$

5.1.2 Classical Harnack's inequality

Theorem 5.1.2 *Let $u \in C^2(\mathbb{R}^n)$ be a nonnegative solution of (5.1.1). Then for any ball $B_R = B_R(y) \subset \mathbb{R}^n$,*

$$\sup_{B_R} u \leq C \inf_{B_R} u,$$

where the constant C depends only on n. (In fact, we may take $C = 3^n$).

Proof. For any x^1, $x^2 \in B_R(y)$,

$$B_R(x^1) \subset B_{2R}(y) \subset B_{3R}(x^2).$$

Using the mean value formula, we have

$$
\begin{aligned}
u(x^1) &= \frac{1}{\omega_n R^n} \int_{B_R(x^1)} u(x)dx \\
&\leq \frac{1}{\omega_n R^n} \int_{B_{2R}(y)} u(x)dx \\
&\leq \frac{1}{\omega_n R^n} \int_{B_{3R}(x^2)} u(x)dx \\
&= 3^n u(x^2).
\end{aligned}
$$

The desired conclusion then follows from the arbitrariness of x^1 and x^2. $\square$

We have proved Harnack's inequality for classical solutions of Laplace's equation. In fact, such kind of inequality also holds for weak solutions. To prove, we proceed in several steps.

5.1.3 Estimate of $\sup_{B_{\theta R}} u$

Lemma 5.1.1 *Let $0 \leq T_0 < T_1$ and $\varphi(t)$ be a bounded and nonnegative function on $[T_0, T_1]$, satisfying*

$$\varphi(t) \leq \theta \varphi(s) + \frac{A}{(s-t)^\alpha} + B$$

for any t, s such that $T_0 \leq t < s \leq T_1$, where $\theta < 1$, A, B and α are nonnegative constants. Then

$$\varphi(\rho) \leq C \left(\frac{A}{(R-\rho)^\alpha} + B \right), \quad T_0 \leq \rho < R \leq T_1,$$

where C is a constant depending only on α and θ.

Proof. Let $T_0 \leq \rho < R \leq T_1$. Denote

$$t_0 = \rho, \quad t_{i+1} = t_i + (1 - \tau)\tau^i (R - \rho) \quad (i = 0, 1, \cdots)$$

with $\tau \in (0, 1)$ to be specified. By assumptions of the lemma, we have

$$\varphi(t_i) \leq \theta\varphi(t_{i+1}) + \frac{A}{((1 - \tau)\tau^i (R - \rho))^\alpha} + B, \quad i = 0, 1, \cdots.$$

For any integer $k \geq 1$, iterating gives

$$\varphi(t_0) \leq \theta^k \varphi(t_k) + \left(\frac{A}{(1 - \tau)^\alpha (R - \rho)^\alpha} + B\right) \sum_{i=0}^{k-1} (\theta\tau^{-\alpha})^i.$$

The desired conclusion follows by choosing τ such that $\theta\tau^{-\alpha} < 1$ in the above inequality and sending $k \to \infty$. $\qquad \square$

Theorem 5.1.3 *Let* $u \in H^1_{\text{loc}}(\mathbb{R}^n)$ *be a bounded weak subsolution of (5.1.1). Then for any $p > 0$ and $0 < \theta < 1$,*

$$\sup_{B_{\theta R}} u \leq C \left(\frac{1}{|B_R|} \int_{B_R} (u_+)^p dx\right)^{1/p},$$

where C is a constant depending only on n, p and $(1 - \theta)^{-1}$.

Proof. From Proposition 4.3.1 of Chapter 4, u_+ is also a weak subsolution of Laplace's equation (4.1.1). Hence, we may assume that $u \geq 0$. Using Theorem 4.3.1 of Chapter 4 (There we have treated the special case $\theta = 1/2$; the general case $0 < \theta < 1$ can be treated similarly) we can derive the conclusion for $p = 2$. We can prove the conclusion for $p > 2$ in a similar way or by using Hölder's inequality directly on the conclusion for $p = 2$.

Now we consider the case $0 < p < 2$. We first use the result for $p = 2$ to obtain

$$\sup_{B_{\theta R}} u \leq C((1 - \theta)R)^{-n/2} \left(\int_{B_R} u^2 dx\right)^{1/2}$$

$$\leq C((1 - \theta)R)^{-n/2} \left(\sup_{B_R} u\right)^{1-p/2} \left(\int_{B_R} u^p dx\right)^{1/2},$$

and then use Young's inequality with ε to derive

$$\sup_{B_{\theta R}} u \leq \frac{1}{2} \sup_{B_R} u + C((1 - \theta)R)^{-n/p} \|u\|_{L^p(B_R)}.$$

Denote $\varphi(s) = \sup\limits_{B_s} u$ and set $s = \theta R$, $t = R$ in the above inequality. Then

$$\varphi(s) \leq \frac{1}{2}\varphi(t) + \frac{C}{(t-s)^{n/p}}\|u\|_{L^p(B_R)}, \quad 0 < s < t \leq R.$$

Thus from Lemma 5.1.1, we deduce the desired conclusion

$$\varphi(\theta R) \leq \frac{C}{((1-\theta)R)^{n/p}}\|u\|_{L^p(B_R)}.$$

$\square$

Checking this proof, we may get that

Remark 5.1.1 *Theorem 5.1.3 still holds if $u \in H^1_{\text{loc}}(\mathbb{R}^n)$ is replaced by $u \in H^1(B_R)$.*

5.1.4 *Estimate of $\inf\limits_{B_{\theta R}} u$*

Lemma 5.1.2 *Let $\Phi(s)$ be a smooth function on $\mathbb{R}$ with $\Phi''(s) \geq 0$ and $u \in H^1_{\text{loc}}(\mathbb{R}^n)$ be a bounded weak solution of (5.1.1). Then $v = \Phi(u) \in H^1_{\text{loc}}(\mathbb{R}^n)$ is a weak subsolution of (5.1.1), namely,*

$$\int_{\mathbb{R}^n} \nabla v \cdot \nabla \varphi \leq 0, \quad \forall 0 \leq \varphi \in C_0^\infty(\mathbb{R}^n).$$

Proof. In fact, for any nonnegative function $\varphi \in C_0^\infty(\mathbb{R}^n)$, we have

$$\int_{R^n} \nabla v \cdot \nabla \varphi dx$$
$$= \int_{R^n} \Phi'(u)\nabla u \cdot \nabla \varphi dx$$
$$= \int_{R^n} \nabla u \cdot \nabla(\Phi'(u)\varphi)dx - \int_{R^n} \Phi''(u)|\nabla u|^2 \varphi dx$$
$$\leq \int_{R^n} \nabla u \cdot \nabla(\Phi'(u)\varphi)dx$$
$$= 0.$$

$\square$

Remark 5.1.2 *The smoothness condition for $\Phi(s)$ can be weakened to local Lipschitz continuity.*

Lemma 5.1.3 *Assume that $w \in L^2(B_2)$ and satisfies*

$$\left(\int_{B_2}(\eta^2|w|^{2h})^q dx\right)^{1/q} \leq C(2h)^{2h} + Ch^2\int_{B_2}(|\nabla \eta| + \eta)^2|w|^{2h}dx$$

for any $h \geq 1$ and any cut-off function η on B_2, where $q > 1$ and the constant C is independent of h and η. Then there exists a constant $C' > 0$, such that for any integer $m \geq 2$,

$$\left(\int_{B_1} |w|^m dx \right)^{1/m} \leq C'm.$$

Proof.   Set

$$h_i = q^{i-1}, \quad \delta_0 = 2, \quad \delta_i = \delta_{i-1} - \frac{1}{2^i}, \quad i = 1, 2, \cdots.$$

Choose η to be a cut-off function on $B_{\delta_{i-1}}$, relative to B_{δ_i}, namely, $\eta \in C_0^\infty(B_{\delta_{i-1}})$, $0 \leq \eta(x) \leq 1$, $\eta(x) = 1$ in B_{δ_i} and $|\nabla \eta(x)| \leq 2^i C$. Then from the assumption of the lemma, we see that

$$\left(\int_{B_{\delta_i}} |w|^{2q^i} dx \right)^{1/q} \leq Cq^{2(i-1)q^{i-1}} + C(2q)^{2(i-1)} \int_{B_{\delta_{i-1}}} |w|^{2q^{i-1}} dx.$$

With $I_i = \left(\int_{B_{\delta_i}} |w|^{2q^i} dx \right)^{1/(2q^i)}$, we further have

$$I_i \leq C^{1/(2q^{i-1})} q^{i-1} + C^{1/(2q^{i-1})} (2q)^{(i-1)/q^{i-1}} I_{i-1}, \quad i = 1, 2, \cdots.$$

Iterating then gives

$$I_j \leq C \sum_{i=1}^{j} q^{i-1} + C^{\alpha_j} (2q)^{\beta_j} I_0, \quad j = 1, 2, \cdots,$$

where $\alpha_j = \sum_{i=1}^{j} \frac{1}{2q^{i-1}}$, $\beta_j = \sum_{i=1}^{j} \frac{i-1}{q^{i-1}}$. Since $\sum_{i=1}^{j} q^{i-1} \leq Cq^j$, the above inequality implies

$$I_j \leq Cq^j + CI_0, \quad j = 1, 2, \cdots.$$

From this, noting that for any fixed integer $m \geq 2$, there exists an integer j, such that $2q^{j-1} \leq m \leq 2q^j$, and using Hölder's inequality, we have

$$\left(\int_{B_1} |w|^m dx \right)^{1/m} \leq CI_j \leq Cm + CI_0.$$

Thus, with another constant C' independent of m, we finally obtain

$$\left(\int_{B_1} |w|^m dx\right)^{1/m} \le C'm.$$

$\square$

Lemma 5.1.4 *Assume that* $w \in H^1_{\text{loc}}(\mathbb{R}^n)$, $\int_{B_2} w(x)dx = 0$ *and*

$$\Delta w + |\nabla w|^2 \le 0, \quad x \in \mathbb{R}^n$$

in the sense of distributions, i.e.

$$-\int_{\mathbb{R}^n} \nabla w \cdot \nabla \varphi dx + \int_{\mathbb{R}^n} |\nabla w|^2 \varphi dx \le 0, \quad \forall 0 \le \varphi \in C_0^\infty(\mathbb{R}^n).$$

Then there exists a constant $p > 0$ depending only on n, such that

$$\int_{B_1} \frac{(p|w|)^m}{m!} \le 2^{-m}, \quad m = 2, 3, \cdots. \tag{5.1.4}$$

Proof. We proceed in two steps: first prove the conclusion for $m = 2$, and then use the standard Moser iteration to reach the conclusion for $m > 2$.

Choose $\eta \in C_0^\infty(B_3)$ such that $0 \le \eta \le 1$, $\eta = 1$ in B_2 and $|\nabla \eta| \le C$ and set $\varphi = \eta^2$. Then

$$-\int_{\mathbb{R}^n} \nabla(\eta^2) \cdot \nabla w dx + \int_{\mathbb{R}^n} |\nabla w|^2 \eta^2 dx \le 0.$$

Integrating by parts and using Cauchy's inequality with ε, we deduce

$$\int_{B_3} |\nabla w|^2 \eta^2 dx \le \int_{B_3} \nabla(\eta^2) \cdot \nabla w dx$$

$$= 2\int_{B_3} \eta \nabla \eta \cdot \nabla w dx$$

$$\le \frac{1}{2}\int_{B_3} |\nabla w|^2 \eta^2 dx + 2\int_{B_3} |\nabla \eta|^2 dx.$$

Thus

$$\int_{B_2} |\nabla w|^2 dx \le C. \tag{5.1.5}$$

From this it follows by noting $\int_{B_2} w(x)dx = 0$ and using Poincaré's inequality,

$$\int_{B_2} w^2(x)dx \le \mu \int_{B_2} |\nabla w|^2 dx \le C,$$

where $\mu > 0$ is the constant in Poincaré's inequality. Choosing $p \leq (2C)^{-1/2}$ then gives (5.1.4) for $m = 2$.

To prove the conclusion in the case $m > 2$, we take $\varphi = \eta^2 |w|^{2h}$ with $h \geq 1$ and η being a cut-off function on B_2. Then

$$\int_{B_2} \eta^2 |w|^{2h} |\nabla w|^2 dx$$

$$\leq \int_{B_2} \nabla(\eta^2 |w|^{2h}) \cdot \nabla w \, dx$$

$$= 2 \int_{B_2} \eta |w|^{2h} \nabla \eta \cdot \nabla w \, dx + 2h \int_{B_2} \eta^2 |w|^{2h-1} \mathrm{sgn} w |\nabla w|^2 dx.$$

Using Cauchy's inequality and Young's inequality with ε,

$$2\eta |\nabla \eta \cdot \nabla w| \leq \frac{1}{4h} \eta^2 |\nabla w|^2 + 4h |\nabla \eta|^2,$$

$$2h |w|^{2h-1} \leq \frac{2h-1}{2h} |w|^{2h} + (2h)^{2h-1},$$

we then derive

$$\int_{B_2} \eta^2 |w|^{2h} |\nabla w|^2 dx$$

$$\leq \frac{1}{4h} \int_{B_2} \eta^2 |w|^{2h} |\nabla w|^2 dx + 4h \int_{B_2} |\nabla \eta|^2 |w|^{2h} dx$$

$$+ \frac{2h-1}{2h} \int_{B_2} \eta^2 |w|^{2h} |\nabla w|^2 dx + (2h)^{2h-1} \int_{B_2} \eta^2 |\nabla w|^2 dx$$

$$= \left(1 - \frac{1}{4h}\right) \int_{B_2} \eta^2 |w|^{2h} |\nabla w|^2 dx + (2h)^{2h-1} \int_{B_2} \eta^2 |\nabla w|^2 dx$$

$$+ 4h \int_{B_2} |\nabla \eta|^2 |w|^{2h} dx,$$

which combined with (5.1.5) leads to

$$\int_{B_2} \eta^2 |w|^{2h} |\nabla w|^2 dx \leq C(2h)^{2h} + 16h^2 \int_{B_2} |\nabla \eta|^2 |w|^{2h} dx.$$

Thus

$$\int_{B_2} |\nabla(\eta^2 |w|^{2h})| dx$$

$$\leq 2h \int_{B_2} \eta^2 |w|^{2h-1} |\nabla w| dx + 2 \int_{B_2} \eta |\nabla \eta| |w|^{2h} dx$$

$$\leq \int_{B_2} \eta^2 |w|^{2h} |\nabla w|^2 dx + h^2 \int_{B_2} \eta^2 |w|^{2(h-1)} dx$$

$$+ 2 \int_{B_2} \eta |\nabla \eta| |w|^{2h} dx$$

$$\leq \int_{B_2} \eta^2 |w|^{2h} |\nabla w|^2 dx + h^2 \int_{B_2} \eta^2 \left(\frac{h-1}{h} |w|^{2h} + \frac{1}{h} \right) dx$$

$$+ 2 \int_{B_2} \eta |\nabla \eta| |w|^{2h} dx$$

$$\leq C(2h)^{2h} + 16 h^2 \int_{B_2} (|\nabla \eta| + \eta)^2 |w|^{2h} dx.$$

From the embedding theorem, we further obtain

$$\left(\int_{B_2} (\eta^2 |w|^{2h})^q dx \right)^{1/q} \leq C \int_{B_2} |\nabla (\eta^2 |w|^{2h})| dx$$

$$\leq C(2h)^{2h} + C h^2 \int_{B_2} (|\nabla \eta| + \eta)^2 |w|^{2h} dx,$$

where

$$1 < q < \begin{cases} +\infty, & n = 1, 2, \\ \dfrac{n}{n-2}, & n \geq 3. \end{cases}$$

Therefore, using Lemma 5.1.3 we infer

$$\int_{B_1} |w|^m dx \leq (Cm)^m, \quad m = 2, 3, \cdots.$$

Since $m^m \leq e^m m!$, we finally arrive at

$$\int_{B_1} |w|^m dx \leq (Ce)^m m!,$$

which is just (5.1.4), if we take $p = (2Ce)^{-1}$. □

Theorem 5.1.4 *Let $u \in H^1_{\text{loc}}(\mathbb{R}^n)$ be a nonnegative and bounded weak solution of (5.1.1). Then for any $0 < \theta < 1$, there exist constants $p_0 > 0$ and $C > 0$ depending only on n and $(1 - \theta)^{-1}$, such that*

$$\inf_{B_{\theta R}} u \geq \frac{1}{C} \left(\frac{1}{|B_R|} \int_{B_R} u^{p_0} dx \right)^{1/p_0}. \tag{5.1.6}$$

Proof. Without loss of generality, we may assume that $\inf_{\mathbb{R}^n} u > 0$; otherwise we may replace u by $u + \varepsilon$ ($\varepsilon > 0$) and then let $\varepsilon \to 0$ in the inequality for $u + \varepsilon$. For simplicity, we assume $R = 1$; for the general case, we may use the rescaling technique.

For any $p > 0$, set $\Phi(s) = \dfrac{1}{s^p}$. Then $\Phi''(s) \geq 0$ and Lemma 5.1.2 shows that $\Phi(u)$ is a weak subsolution of $-\Delta u = 0$. Thus from Theorem 5.1.3, we have

$$\sup_{B_\theta} \frac{1}{u^p} \leq C \int_{B_1} \frac{1}{u^p} dx,$$

and hence

$$\left(\inf_{B_\theta} u \right)^p \geq \frac{1}{C} \left(\int_{B_1} u^{-p} dx \right)^{-1},$$

namely,

$$\inf_{B_\theta} u \geq \frac{1}{C^{1/p}} \left(\int_{B_1} u^{-p} dx \right)^{-1/p}$$

$$= \frac{1}{C^{1/p}} \left(\int_{B_1} u^{-p} dx \int_{B_1} u^p dx \right)^{-1/p} \left(\int_{B_1} u^p dx \right)^{1/p}.$$

It remains to prove that there exists a constant $p_0 > 0$ such that

$$\int_{B_1} u^{-p_0} dx \int_{B_1} u^{p_0} dx \leq C. \tag{5.1.7}$$

Set $w = \ln u - \beta$ with $\beta = \dfrac{1}{|B_2|} \displaystyle\int_{B_2} \ln u \, dx$. Then $\displaystyle\int_{B_2} w(x) dx = 0$. It is easy to see that (5.1.7) holds if

$$\int_{B_1} e^{p_0 |w|} dx \leq C. \tag{5.1.8}$$

In fact, (5.1.8) implies that

$$\int_{B_1} e^{p_0 (\beta - \ln u)} dx \leq C, \quad \int_{B_1} e^{p_0 (\ln u - \beta)} dx \leq C,$$

and hence (5.1.7) holds.

We now use Lemma 5.1.4 to prove the existence of $p_0 > 0$ such that (5.1.8) holds. To this end, we need to check the condition of Lemma 5.1.4, namely, to prove that, in the sense of distributions,

$$\Delta w + |\nabla w|^2 \le 0, \quad x \in \mathbb{R}^n. \tag{5.1.9}$$

Since u is a bounded weak solution of (5.1.1) and $\inf_{\mathbb{R}^n} u > 0$, we have, for any nonnegative function $\varphi \in C_0^\infty(\mathbb{R}^n)$,

$$\int_{\mathbb{R}^n} \nabla u \cdot \nabla \left(\frac{\varphi}{u} \right) dx = 0,$$

namely,

$$\int_{\mathbb{R}^n} \frac{1}{u} \nabla u \cdot \nabla \varphi dx - \int_{\mathbb{R}^n} \frac{1}{u^2} \varphi |\nabla u|^2 dx = 0$$

or

$$\int_{\mathbb{R}^n} \nabla w \cdot \nabla \varphi dx - \int_{\mathbb{R}^n} |\nabla w|^2 \varphi dx = 0, \quad \forall 0 \le \varphi \in C_0^\infty(\mathbb{R}^n),$$

which shows that w satisfies (5.1.9) in the sense of distributions. □

Checking this proof and the proofs of Lemmas 5.1.2–5.1.4, we may get that

Remark 5.1.3 *Theorem 5.1.4 still holds if $u \in H^1_{\text{loc}}(\mathbb{R}^n)$ is replaced by $u \in H^1(B_{3R})$.*

If we replace B_1 and B_2 by $B_{(\theta+1)/2}$ and $B_{(\theta+3)/4}$, respectively, in the proof of Theorem 5.1.4, we may further get by a similar process of proof with some modifications that

Remark 5.1.4 *Theorem 5.1.4 still holds if $u \in H^1_{\text{loc}}(\mathbb{R}^n)$ is replaced by $u \in H^1(B_R)$.*

5.1.5 Harnack's inequality

Theorem 5.1.5 *Let $u \in H^1(B_{3R})$ be a nonnegative and bounded weak solution of (5.1.1). Then for any $0 < \theta < 1$ and $R > 0$,*

$$\sup_{B_{\theta R}} u \le C \inf_{B_{\theta R}} u$$

with the constant C depending only on n and $(1 - \theta)^{-1}$.

Proof. From Theorem 5.1.3 and Remark 5.1.1, for any $p > 0$,

$$\sup_{B_{\theta R}} u \leq C \left(\frac{1}{|B_R|} \int_{B_R} u^p dx \right)^{1/p}.$$

On the other hand, from Theorem 5.1.4 and Remark 5.1.3, there exists a constant $p_0 > 0$, such that

$$\inf_{B_{\theta R}} u \geq \frac{1}{C} \left(\frac{1}{|B_R|} \int_{B_R} u^{p_0} dx \right)^{1/p_0}.$$

Combining these two inequalities derives the desired conclusion. $\qquad\square$

Furthermore, the condition $u \in H^1(B_{3R})$ in Theorem 5.1.5 may be replaced by $u \in H^1(B_R)$, according to Remark 5.1.4 or by the finite covering theorem. More general, we have

Theorem 5.1.6   *Let $\Omega \subset \mathbb{R}^n$ be a domain and $u \in H^1(\Omega)$ be a nonnegative and bounded weak solution of (5.1.1). Then for any bounded subdomain $\Omega' \subset\subset \Omega$,*

$$\sup_{\Omega'} u \leq C \inf_{\Omega'} u \qquad\qquad (5.1.10)$$

with the constant C depending only on n, Ω' and Ω.

Proof. Fix

$$0 < R \leq \frac{1}{4} \mathrm{dist}(\Omega', \partial\Omega).$$

For any $x \in \overline{\Omega}'$, from Theorem 5.1.5, there exists a constant C depending only on n such that

$$\sup_{B_R(x)} u \leq C \inf_{B_R(x)} u. \qquad\qquad (5.1.11)$$

Choose $x^1, x^2 \in \overline{\Omega}'$ so that

$$\sup_{B_R(x^1)} u = \sup_{\Omega'} u, \qquad \inf_{B_R(x^2)} u = \inf_{\Omega'} u. \qquad\qquad (5.1.12)$$

Let $\Gamma \subset \overline{\Omega}'$ be a closed arc joining x^1 and x^2. By virtue of the finite covering theorem, Γ can be covered by a finite number N (depending only on Ω' and Ω) of balls of radius R. Applying the estimate (5.1.11) in each ball and

$B_R(x^1)$, $B_R(x^2)$, respectively, and combining the resulting inequalities, we obtain

$$\sup_{B_R(x^1)} u \le C^{N+2} \inf_{B_R(x^2)} u.$$

Then, (5.1.10) follows from this and (5.1.12). $\qquad\square$

5.1.6 *Hölder's estimate*

The following auxiliary lemma is useful in proving the Hölder continuity of solutions.

Lemma 5.1.5 *Let $\omega(R)$ be a nonnegative and nondecreasing function on $[0, R_0]$. If there exist $\theta, \eta \in (0,1)$, $\gamma \in (0,1]$ and $K \ge 0$, such that*

$$\omega(\theta R) \le \eta \omega(R) + K R^\gamma, \quad 0 < R \le R_0, \qquad (5.1.13)$$

then there exist constants $\alpha \in (0, \gamma)$ and $C > 0$ depending only on θ, η, γ, such that

$$\omega(R) \le C \left(\frac{R}{R_0} \right)^\alpha [\omega(R_0) + K R_0^\gamma], \quad 0 < R \le R_0. \qquad (5.1.14)$$

Proof. We may take η so close to 1 that $\theta^{-\gamma}\eta > 1$ and (5.1.13) still holds. Let $\tilde{R}_0 \in (\theta R_0, R_0]$ and denote

$$\tilde{R}_s = \theta^s \tilde{R}_0, \quad s = 0, 1, 2, \cdots.$$

Then, from (5.1.13) we have

$$\omega(\tilde{R}_{s+1}) \le \eta \omega(\tilde{R}_s) + K \tilde{R}_s^\gamma, \quad s - 0, 1, 2, \cdots.$$

Iterating gives, for $s = 0, 1, 2, \cdots$,

$$\omega(\tilde{R}_s) \le \eta^s \omega(\tilde{R}_0) + \sum_{m=0}^{s-1} K \eta^m \tilde{R}_{s-m-1}^\gamma$$

$$\le \eta^s \omega(\tilde{R}_0) + K \tilde{R}_0^\gamma \theta^{\gamma(s-1)} \sum_{m=0}^{s-1} (\theta^{-\gamma}\eta)^m$$

$$= \eta^s \omega(\tilde{R}_0) + K \tilde{R}_0^\gamma \theta^{\gamma(s-1)} \frac{(\theta^{-\gamma}\eta)^s - 1}{\theta^{-\gamma}\eta - 1}$$

$$\le \eta^s \omega(\tilde{R}_0) + K \tilde{R}_0^\gamma \theta^{\gamma(s-1)} \frac{(\theta^{-\gamma}\eta)^s}{\theta^{-\gamma}\eta - 1}$$

$$= \eta^s [\omega(\tilde{R}_0) + C K \tilde{R}_0^\gamma],$$

where $C = \dfrac{\theta^{-\gamma}}{\theta^{-\gamma}\eta - 1}$. Since $s = \log_\theta \dfrac{\tilde{R}_s}{\tilde{R}_0}$, we have

$$\omega(\tilde{R}_s) \leq \left(\frac{\tilde{R}_s}{\tilde{R}_0}\right)^\alpha [\omega(\tilde{R}_0) + CK\tilde{R}_0^\gamma]$$

$$\leq C\left(\frac{\tilde{R}_s}{R_0}\right)^\alpha [\omega(R_0) + KR_0^\gamma], \quad s = 0, 1, 2, \cdots,$$

where $\alpha = \dfrac{\ln \eta}{\ln \theta} \in (0, \gamma)$. Let $\tilde{R}_0$ vary over $(\theta R_0, R_0]$. Then $\tilde{R}_s$ ($s = 0, 1, 2, \cdots$) varies over $(0, R_0]$. Thus we can obtain (5.1.14) from the above inequality. $\qquad\square$

Theorem 5.1.7 *Let $\Omega \subset \mathbb{R}^n$ be a domain and $u \in H^1(\Omega)$ be a bounded weak solution of (5.1.1). Then for any bounded subdomain $\Omega' \subset\subset \Omega$, there exists a constant $\alpha \in (0, 1)$ such that*

$$[u]_{\alpha;\Omega'} \leq C,$$

where C is a constant depending only on n, Ω' and Ω.

Proof. For any fixed $x^0 \in \Omega'$ and $0 < R < \dfrac{1}{3}\mathrm{dist}(\Omega', \partial\Omega)$, denote

$$m(R) = \inf_{B_R} u, \quad M(R) = \sup_{B_R} u,$$

where $B_R = B_R(x^0)$. Let

$$v(x) = u(x) - m(R), \quad w(x) = M(R) - u(x).$$

Then $v, w \in H^1(B_{3R})$ are nonnegative and bounded and $-\Delta v = -\Delta w = 0$ in B_{3R} in the sense of distributions. Using Harnack's inequality to v and w gives

$$\sup_{B_{R/2}} v \leq C \inf_{B_{R/2}} v, \quad \sup_{B_{R/2}} w \leq C \inf_{B_{R/2}} w,$$

namely,

$$M(R/2) - m(R) \leq C[m(R/2) - m(R)], \quad M(R) - m(R/2) \leq C[M(R) - M(R/2)].$$

We may assume $C > 1$; otherwise we replace C by $C + 1$. From the above two inequalities, we see that

$$M(R/2) - m(R/2) \leq \frac{C-1}{C+1}[M(R) - m(R)].$$

Denote $f(R) = M(R) - m(R)$, $\eta = \dfrac{C-1}{C+1}$. Then $f(R)$ is nonnegative and nondecreasing and satisfies

$$f(R/2) \leq \eta f(R).$$

By the iteration lemma (Lemma 5.1.5), there exists $\alpha \in (0,1)$, such that

$$f(R) \leq CR^\alpha,$$

namely,

$$[u]_{\alpha; B_R(x^0)} \leq C.$$

The conclusion of the theorem can then be completed by an easy covering argument. $\square$

5.2 Harnack's Inequalities for Solutions of the Homogeneous Heat Equation

In §5.1, we have proved that for nonnegative and bounded weak solution $u \in H^1_{\mathrm{loc}}(\mathbb{R}^n)$ of (5.1.1), there holds

$$\sup_{B_{\theta R}} u \leq C(n, \theta) \inf_{B_{\theta R}} u, \quad \forall \theta \in (0,1).$$

A natural question is that for the homogeneous heat equation

$$\frac{\partial u}{\partial t} - \Delta u = 0, \quad (x,t) \in \mathbb{R}^n \times \mathbb{R}_+ \tag{5.2.1}$$

whether the analogous inequality

$$\sup_{Q_{\theta R}} u \leq C(n, \theta) \inf_{Q_{\theta R}} u, \quad \forall \theta \in (0,1) \tag{5.2.2}$$

holds? The following example shows that the answer is negative.

Example 5.2.1 The equation $u_t - u_{xx} = 0$ has a nonnegative and bounded solution on $(-R, R) \times [0, R^2]$,

$$u(x,t) = (t + R^2)^{-1/2} \exp\left\{ -\frac{(x+\xi)^2}{4(t+R^2)} \right\},$$

where ξ is a constant. Let $\theta \in (0,1)$. Then for fixed $x \in (-\theta R, 0) \cup (0, \theta R)$ and $t \in [0, R^2]$, we have

$$\frac{u(0,t)}{u(x,t)} = \exp\left\{\frac{2x\xi + x^2}{4(t+R^2)}\right\} \to 0, \quad \text{as } \xi \text{sgn} x \to -\infty,$$

which shows that (5.2.2) does not hold.

However, for equation (5.2.1), there holds another version of Harnack's inequality. We proceed to establish such kind of Harnack's inequality in several steps.

Let $(x^0, t_0) \in \mathbb{R}^n \times \mathbb{R}_+$, and $R^2 < t_0$. Denote

$$B_R = B_R(x^0) = \{x \in \mathbb{R}^n; |x - x^0| < R\},$$
$$Q_R = Q_R(x^0, t_0) = B_R(x^0) \times (t_0 - R^2, t_0),$$
$$\Theta_{\theta R} = B_{\theta R}(x^0) \times (t_0 - R^2 - \theta R^2, t_0 - R^2).$$

5.2.1 Weak Harnack's inequality

Lemma 5.2.1 *Let $\Omega \subset \mathbb{R}^n$ be a bounded convex domain, $\mathcal{N}$ be a measurable subset of Ω and $u \in W^{1,p}(\Omega)$ with $1 \leq p < +\infty$. Then*

$$\|u - u_{\mathcal{N}}\|_{L^p(\Omega)} \leq C \frac{1}{|\mathcal{N}|} (\text{diam}\Omega)^{n+1} \|\nabla u\|_{L^p(\Omega)},$$

where $u_{\mathcal{N}} = \frac{1}{|\mathcal{N}|} \int_{\mathcal{N}} u(x)dx$ and C is a constant depending only on n.

Proof. Since $C^\infty(\overline{\Omega})$ is dense in $W^{1,p}(\Omega)$, it suffices to prove the conclusion for $u \in C^\infty(\overline{\Omega})$.

Let $u \in C^\infty(\overline{\Omega})$. Then for $x, y \in \Omega$,

$$u(x) - u(y) = -\int_0^{|x-y|} \frac{\partial u(x + r\omega)}{\partial r} dr, \quad \omega = \frac{y - x}{|x - y|}.$$

Integrating over $\mathcal{N}$ with respect to y, gives

$$|\mathcal{N}|(u(x) - u_{\mathcal{N}}) = -\int_{\mathcal{N}} \int_0^{|x-y|} \frac{\partial u(x + r\omega)}{\partial r} dr dy.$$

Denote $d = \text{diam}\Omega$ and

$$V(x + r\omega) = \begin{cases} \left|\dfrac{\partial u(x + r\omega)}{\partial r}\right|, & \text{when } x + r\omega \in \Omega, \\ 0, & \text{when } x + r\omega \in \mathbb{R}^n \backslash \Omega. \end{cases}$$

Then

$$|u(x) - u_{\mathcal{N}}| \leq \frac{1}{|\mathcal{N}|} \int_{|x-y|<d} \int_0^d V(x + r\omega) dr dy$$

$$= \frac{1}{|\mathcal{N}|} \int_0^d \int_{|\omega|=1} \int_0^d \rho^{n-1} V(x + r\omega) d\rho d\omega dr$$

$$= \frac{d^n}{n|\mathcal{N}|} \int_0^d \int_{|\omega|=1} V(x + r\omega) d\omega dr$$

$$\leq \frac{d^n}{n|\mathcal{N}|} \int_\Omega |x - y|^{1-n} |\nabla u(y)| dy.$$

Hence

$$\int_\Omega |u(x) - u_{\mathcal{N}}|^p dx$$

$$\leq \left(\frac{d^n}{n|\mathcal{N}|}\right)^p \int_\Omega \left(\int_\Omega |x - y|^{1-n} |\nabla u(y)| dy\right)^p dx. \qquad (5.2.3)$$

If $p = 1$, then the conclusion of the lemma follows by exchanging the order of the integral in (5.2.3); if $p > 1$, $n = 1$, then the conclusion follows immediately by using Hölder's inequality to the integral on the right side of (5.2.3).

Now we discuss the case $p > 1$, $n \geq 2$. Choose $\mu \in (0, 1)$, such that

$$1 - \frac{n}{p(n-1)} < \mu < \frac{n(p-1)}{p(n-1)}$$

or

$$\mu(1 - n)\frac{p}{p-1} > -n, \quad (1 - \mu)(1 - n)p > -n;$$

the existence of μ is obvious. Using Hölder's inequality, we have

$$\left(\int_\Omega |x - y|^{1-n} |\nabla u(y)| dy\right)^p$$

$$= \left(\int_\Omega \left(|x - y|^{\mu(1-n)}\right)\left(|x - y|^{(1-\mu)(1-n)} |\nabla u(y)|\right) dy\right)^p$$

$$\leq \left(\int_\Omega |x - y|^{\mu(1-n)p/(p-1)} dy\right)^{p-1} \int_\Omega |x - y|^{(1-\mu)(1-n)p} |\nabla u(y)|^p dy$$

$$\leq C d^{\mu(1-n)p+n(p-1)} \int_\Omega |x - y|^{(1-\mu)(1-n)p} |\nabla u(y)|^p dy.$$

Integrating over Ω with respect to x and exchanging the order of the integral, we further obtain

$$\int_\Omega \left(\int_\Omega |x-y|^{1-n} |\nabla u(y)| dy \right)^p dx$$

$$\leq C d^{\mu(1-n)p+n(p-1)} \int_\Omega \int_\Omega |x-y|^{(1-\mu)(1-n)p} |\nabla u(y)|^p dy dx$$

$$= C d^{\mu(1-n)p+n(p-1)} \int_\Omega \left(\int_\Omega |x-y|^{(1-\mu)(1-n)p} dx \right) |\nabla u(y)|^p dy$$

$$\leq C d^{\mu(1-n)p+n(p-1)} d^{(1-\mu)(1-n)p+n} \int_\Omega |\nabla u(y)|^p dy$$

$$= C d^p \int_\Omega |\nabla u(y)|^p dy.$$

Substituting this into the right side of (5.2.3) then derives the desired conclusion. □

Lemma 5.2.2 *For any constant $\gamma > 0$, there exists a nonnegative function $g(s) \in C^2(0, +\infty)$ with the following properties:*
  i) For any $s > 0$, $g''(s) \geq [g'(s)]^2 - \gamma g'(s)$, $g'(s) \leq 0$;
  ii) $g(s) \sim -\ln s$ as $s \to 0^+$;
  iii) $g(s) = 0$ for $s \geq 1$.

Proof. We first observe that if $g(s)$ satisfies i), then

$$f(s) = -e^{-g(s)}, \quad s > 0$$

satisfies $f''(s) + \gamma f'(s) = e^{-g(s)} [g''(s) - (g'(s))^2 + \gamma g'(s)] \geq 0$, namely, $h(s) = f'(s) + \gamma f(s)$ is nondecreasing on $(0, +\infty)$. We try to consider the function

$$g_0(s) = \left(-\ln \frac{1-e^{-\gamma s}}{1-e^{-\gamma}} \right)^+, \quad s > 0.$$

It is easy to see that $g_0(s)$ is nonincreasing and satisfies conditions ii) and iii). In addition, the function

$$f_0(s) = -e^{-g_0(s)} = \max\left\{ -\frac{1-e^{-\gamma s}}{1-e^{-\gamma}}, -1 \right\}, \quad s > 0$$

satisfies $f_0''(s) + \gamma f_0'(s) = e^{-g_0(s)} [g_0''(s) - (g_0'(s))^2 + \gamma g_0'(s)] = 0$ for $s \neq 1$. Roughly speaking, $g_0(s)$ satisfies all conditions i), ii), iii) except at $s = 1$, where $g_0(s)$ loses the smoothness.

The above analysis leads us to find a suitable smooth approximation of $g_0(s)$. A simple calculation shows that

$$h_0(s) = f_0'(s) + \gamma f_0(s) = \begin{cases} -\dfrac{\gamma}{1 - e^{-\gamma}}, & \text{for } s \in [0, 1), \\ -\gamma, & \text{for } s \in (1, +\infty) \end{cases}$$

and $\lim\limits_{s \to 1^-} h_0(s) < \lim\limits_{s \to 1^+} h_0(s)$. Moreover

$$\int_0^2 e^{\gamma s} h_0(s)\, ds = -e^{2\gamma}.$$

Now we construct a smooth approximation of $h_0(s)$, denoted by $h(s)$, which satisfies the following conditions: $h(s) \in C^\infty[0, +\infty)$, $h(s) < 0$, $h'(s) \geq 0$ and

$$h(s) = h_0(s) \text{ for } s \in [0, 1/2] \cup [2, +\infty), \quad \int_0^2 e^{\gamma s} h(s)\, ds = -e^{2\gamma}.$$

Then we determine $\tilde{f}(s)$ by solving the equation $\tilde{f}'(s) + \gamma \tilde{f}(s) = h(s)$ with $\tilde{f}(0) = 0$. $\tilde{f}(s)$ can be expressed as

$$\tilde{f}(s) = e^{-\gamma s} \int_0^s e^{\gamma s} h(s)\, ds, \quad s > 0.$$

A simple calculation shows that

$$\tilde{f}(s) = -1, \quad s \geq 2.$$

Finally, we define

$$g(s) = -\ln(-\tilde{f}(s)), \quad s > 0.$$

It is easy to verify that $g(s)$ is the required function. In fact, since

$$(\tilde{f}'(s) e^{\gamma s})' = (\tilde{f}''(s) + \gamma \tilde{f}'(s)) e^{\gamma s} = h'(s) e^{\gamma s} \geq 0, \quad s > 0,$$

$\tilde{f}'(s) e^{\gamma s}$ is nondecreasing. For $s \geq 2$, we have $\tilde{f}'(s) e^{\gamma s} = 0$, so for $s \geq 0$, $\tilde{f}'(s) e^{\gamma s} \leq 0$, namely, $\tilde{f}'(s) \leq 0$, and hence $\tilde{g}'(s) = -(\tilde{f}'(s)/\tilde{f}(s)) \leq 0$. This together with $h'(s) \geq 0$ shows that $g(s)$ satisfies i). $\square$

Remark 5.2.1 *If $g(s)$ satisfies i), then $G(s) = g(\alpha s + \beta)$ possesses the same property, where $\alpha \geq 1$, $\beta > 0$ are constants.*

Lemma 5.2.3   *Let $u \in H^1_{loc}(\mathbb{R}^n \times \mathbb{R}_+)$ be a nonnegative weak solution of (5.2.1). If*

$$\text{mes}\{(x,t) \in Q_R; u(x,t) \geq 1\} \geq \mu \text{mes} Q_R, \quad 0 < \mu < 1, \qquad (5.2.4)$$

then, for any $\sigma \in (0, \mu)$ and $\beta \in (\mu, 1)$ satisfying $\dfrac{1-\mu}{1-\sigma}\beta^{-n} = \dfrac{2}{3}$, there exists a constant $h \in (0, 1)$ depending only on n and μ, such that

$$\text{mes}\{x \in B_{\beta R}; u(x,t) \geq h\} \geq \frac{1}{4}\text{mes} B_{\beta R}, \quad t_0 - \sigma R^2 \leq t \leq t_0.$$

Proof.   Let ζ be a cut-off function on B_R relative to $B_{\beta R}$, namely, $\zeta \in C_0^\infty(B_R)$, $0 \leq \zeta(x) \leq 1$, $\zeta(x) = 1$ on $B_{\beta R}$ and $|\nabla \zeta(x)| \leq \dfrac{C}{(1-\beta)R}$. Take $\varphi = \zeta^2 \chi_{[t_1,t_2]} G'(u)$ as a test function in the definition of weak solutions of (5.2.1), where $\chi_{[t_1,t_2]}$ is the characteristic function of the segment $[t_1, t_2]$, $t_0 - R^2 \leq t_1 < t_2 \leq t_0$, and $G(s) \in C^2(\mathbb{R})$ is a function satisfying $G'(s) \leq 0$, $G''(s) - (G'(s))^2 \geq 0$. Then

$$\int_{t_1}^{t_2} \int_{B_R} (\zeta^2 G'(u)u_t + \nabla(\zeta^2 G'(u)) \cdot \nabla u) dx dt = 0$$

or

$$\int_{t_1}^{t_2} \int_{B_R} (\zeta^2 G'(u)u_t + \zeta^2 G''(u)|\nabla u|^2 + G'(u)\nabla u \cdot \nabla(\zeta^2)) dx dt = 0.$$

With $w = G(u)$, we have

$$\int_{t_1}^{t_2} \int_{B_R} (\zeta^2 w_t + \zeta^2 |\nabla w|^2 + \nabla w \cdot \nabla(\zeta^2)) dx dt$$

$$= \int_{t_1}^{t_2} \int_{B_R} \zeta^2[(G'(u))^2 - G''(u)]|\nabla u|^2 dx dt \leq 0.$$

Using Cauchy's inequality with ε to obtain

$$|\nabla w \cdot \nabla(\zeta^2)| = 2\zeta|\nabla w \cdot \nabla \zeta| \leq \frac{1}{2}\zeta^2|\nabla w|^2 + 2|\nabla \zeta|^2,$$

we further derive

$$\int_{t_1}^{t_2} \int_{B_R} \zeta^2 w_t dx dt + \frac{1}{2}\int_{t_1}^{t_2} \int_{B_R} \zeta^2|\nabla w|^2 dx dt$$

$$\leq 2 \int_{t_1}^{t_2} \int_{B_R} |\nabla \zeta|^2 dx dt \leq CR^n \leq C\text{mes} B_R. \qquad (5.2.5)$$

Now we take $w = G(u) = g(u + h)$ with g being the function constructed in Lemma 5.2.2 and h to be specified later. Denote

$$\bar{\mu}(t) = \text{mes}\{x \in B_R; u(x,t) \geq 1\}, \quad N_t = \{x \in B_{\beta R}; u(x,t) \geq h\}.$$

Then by the assumption of the lemma, we have

$$\int_{t_0 - R^2}^{t_0} \bar{\mu}(t) dt \geq \mu \text{mes} Q_R = R^2 \mu \text{mes} B_R.$$

On the other hand, obviously

$$\int_{t_0 - \sigma R^2}^{t_0} \bar{\mu}(t) dt \leq \sigma R^2 \text{mes} B_R.$$

So

$$\int_{t_0 - R^2}^{t_0 - \sigma R^2} \bar{\mu}(t) dt \geq (\mu - \sigma) R^2 \text{mes} B_R.$$

Hence, by the mean value theorem, we see that there exists $\tau \in [t_0 - R^2, t_0 - \sigma R^2]$, such that

$$\bar{\mu}(\tau) \geq \frac{\mu - \sigma}{1 - \sigma} \text{mes} B_R.$$

Take $t_1 = \tau$, $t_2 \in [t_0 - \sigma R^2, t_0]$ in (5.2.5) and note that $\beta \in (\mu, 1)$. Then

$$\int_{\tau}^{t_2} \int_{B_R} \zeta^2 w_t dx dt \leq C \text{mes} B_R \leq C(\mu) \text{mes} B_{\mu R} \leq C(\mu) \text{mes} B_{\beta R}.$$

Thus

$$\int_{B_R} \zeta^2(x) w(x, t_2) dx$$

$$= \int_{\tau}^{t_2} \int_{B_R} \zeta^2 w_t dx dt + \int_{B_R} \zeta^2(x) w(x, \tau) dx$$

$$\leq C(\mu) \text{mes} B_{\beta R} + \int_{B_R} \zeta^2(x) w(x, \tau) dx. \tag{5.2.6}$$

Since $w = g(u + h)$ and $g'(s) \leq 0$, we have

$$\int_{B_R} \zeta^2(x) w(x, t_2) dx \geq \int_{B_{\beta R} \backslash N_{t_2}} w(x, t_2) dx$$

$$\geq \text{mes}(B_{\beta R} \backslash N_{t_2}) g(2h). \tag{5.2.7}$$

Again note that $g(s) = 0$ for $s \geq 1$. We also have

$$\int_{B_R} \zeta^2(x) w(x,\tau) dx \leq \int_{B_R} w(x,\tau) dx$$

$$= \int_{\{x \in B_R; u(x,\tau) < 1\}} w(x,\tau) dx \leq (\mathrm{mes} B_R - \overline{\mu}(\tau)) g(h)$$

$$\leq \left(1 - \frac{\mu - \sigma}{1 - \sigma}\right) g(h) \mathrm{mes} B_R = \frac{1 - \mu}{(1 - \sigma)\beta^n} g(h) \mathrm{mes} B_{\beta R}. \qquad (5.2.8)$$

Combining (5.2.7), (5.2.8) with (5.2.6) and noting that $\dfrac{1 - \mu}{1 - \sigma}\beta^{-n} = \dfrac{2}{3}$, we arrive at

$$\mathrm{mes}(B_{\beta R} \backslash N_{t_2}) \leq \frac{3C(\mu) + 2g(h)}{3g(2h)} \mathrm{mes} B_{\beta R}.$$

Since $g(s) \sim -\ln s \ (s \to 0^+)$, we may choose h so small that

$$\mathrm{mes}(B_{\beta R} \backslash N_{t_2}) \leq \frac{3}{4} \mathrm{mes} B_{\beta R}.$$

Therefore

$$\mathrm{mes} N_{t_2} \geq \frac{1}{4} \mathrm{mes} B_{\beta R}, \quad \forall t_2 \in [t_0 - \sigma R^2, t_0],$$

namely,

$$\mathrm{mes}\{x \in B_{\beta R}; u(x,t) \geq h\} \geq \frac{1}{4} \mathrm{mes} B_{\beta R}, \quad t_0 - \sigma R^2 \leq t \leq t_0. \qquad \square$$

Remark 5.2.2 *If condition (5.2.4) is replaced by*

$$\mathrm{mes}\{(x,t) \in Q_R; u(x,t) \geq \varepsilon\} \geq \mu \mathrm{mes} Q_R, \quad 0 < \mu < 1,$$

then, since $\dfrac{u}{\varepsilon}$ is still a nonnegative weak solution of (5.2.1), we can use Lemma 5.2.3 to derive

$$\mathrm{mes}\{x \in B_{\beta R}; u(x,t) \geq \varepsilon h\} \geq \frac{1}{4} \mathrm{mes} B_{\beta R}, \quad t_0 - \sigma R^2 \leq t \leq t_0.$$

Lemma 5.2.4 *Let $u \in H^1_{\mathrm{loc}}(\mathbb{R}^n \times \mathbb{R}_+)$ be a nonnegative weak solution of (5.2.1) satisfying*

$$\mathrm{mes}\{x \in B_{\beta R}; u(x,t) \geq h\} \geq \nu \mathrm{mes} B_{\beta R}, \quad t_0 - \sigma R^2 \leq t \leq t_0,$$

where $0 < \nu < 1$. *Then for* $\theta = 1/2 \min(\beta, \sqrt{\sigma})$, *there exists a constant* $\gamma > 0$ *depending only on* n, ν, h *and* θ, *such that*

$$u(x,t) > \gamma, \quad (x,t) \in Q_{\theta R}.$$

Proof. Denote $w = G(u)$, where $G(s) \in C^2(\mathbb{R})$ satisfies $G'(s) \leq 0$, $G''(s) - (G'(s))^2 \geq 0$. Then, in the sense of distributions, w satisfies

$$w_t - \Delta w = G'(u)u_t - G'(u)\Delta u - G''(u)|\nabla u|^2 = -G''(u)|\nabla u|^2 \leq 0.$$

This means that w is a weak subsolution of (5.2.1). Thus we can use the local boundedness estimate for solutions of the homogeneous heat equation (see §4.4) to obtain

$$\sup_{Q_{\theta R}} w^2 \leq \frac{C}{R^{n+2}} \|w\|^2_{L^2(Q_{2\theta R})}, \tag{5.2.9}$$

where the constant C depends only on n and θ. Let ζ be the function used in Lemma 5.2.3. Take $t_1 = t_0 - \sigma R^2$, $t_2 = t_0$ in (5.2.5). Then we obtain

$$\int_{t_0-\sigma R^2}^{t_0} \int_{B_R} (\zeta^2 w)_t dx dt + \frac{1}{2} \int_{t_0-\sigma R^2}^{t_0} \int_{B_R} \zeta^2 |\nabla w|^2 dx dt \leq CR^n. \tag{5.2.10}$$

If we take, in particular, $w = G(u) = g\left(\dfrac{u+k}{h}\right) (0 < k < h)$ with g being the function constructed in Lemma 5.2.2, then by the monotonicity of $g(s)$, $w \leq g\left(\dfrac{k}{h}\right)$ and hence

$$\int_{t_0-\sigma R^2}^{t_0} \int_{B_R} (\zeta^2 w)_t dx dt$$

$$= \int_{B_R} \zeta^2(x)w(x,t)dx \bigg|_{t=t_0-\sigma R^2}^{t=t_0}$$

$$\geq -\int_{B_R} \zeta^2(x)w(x,t_0-\sigma R^2)dx \geq -CR^n g\left(\frac{k}{h}\right). \tag{5.2.11}$$

Combining (5.2.10) with (5.2.11) gives

$$\int_{t_0-\sigma R^2}^{t_0} \int_{B_{\beta R}} |\nabla w|^2 dx dt \leq CR^n \left(1 + g\left(\frac{k}{h}\right)\right). \tag{5.2.12}$$

By the assumption of the lemma and the fact that $g(s) = 0$ for $s \geq 1$, we have

$$\text{mes}\{x \in B_{\beta R}; w(x,t) = 0\} \geq \nu \text{mes} B_{\beta R}, \quad t_0 - \sigma R^2 \leq t \leq t_0.$$

Using Lemma 5.2.1 with $\Omega = B_{\beta R}$ and $\mathcal{N} = \{x \in B_{\beta R}; w(x,t) = 0\}$, we see that, for any $t_0 - \sigma R^2 \leq t \leq t_0$,

$$
\begin{aligned}
\int_{B_{\beta R}} w^2(x,t) dx &\leq C \frac{(\beta R)^{2n+2}}{|\mathcal{N}|^2} \int_{B_{\beta R}} |\nabla w(x,t)|^2 dx \\
&\leq C \frac{(\beta R)^{2n+2}}{|B_{\beta R}|^2} \int_{B_{\beta R}} |\nabla w(x,t)|^2 dx \\
&\leq C R^2 \int_{B_{\beta R}} |\nabla w(x,t)|^2 dx.
\end{aligned}
$$

From (5.2.12), we further obtain

$$
\begin{aligned}
\int_{t_0 - \sigma R^2}^{t_0} \int_{B_{\beta R}} w^2(x,t) dx dt &\leq C R^2 \int_{t_0 - \sigma R^2}^{t_0} \int_{B_{\beta R}} |\nabla w|^2 dx dt \\
&\leq C R^{n+2} \left(1 + g\left(\frac{k}{h}\right) \right).
\end{aligned}
$$

Combining this with (5.2.9) yields

$$
\sup_{Q_{\theta R}} w^2 \leq \frac{C}{R^{n+2}} \|w\|^2_{L^2(Q_{2\theta R})} \leq C \left(1 + g\left(\frac{k}{h}\right) \right). \tag{5.2.13}
$$

Choosing γ so small that $2\gamma < h$ and

$$
\left(g\left(\frac{2\gamma}{h}\right) \right)^2 > C \left(1 + g\left(\frac{\gamma}{h}\right) \right), \tag{5.2.14}
$$

we must have $u \geq \gamma$ on $Q_{\theta R}$. Suppose, to the contrary, there exists $(\tilde{x}, \tilde{t}) \in Q_{\theta R}$, such that $u(\tilde{x}, \tilde{t}) < \gamma$. Then from (5.2.13), we would have

$$
\left(g\left(\frac{2\gamma}{h}\right) \right)^2 \leq \left(g\left(\frac{u(\tilde{x}, \tilde{t}) + k}{h}\right) \right)^2 = (w(\tilde{x}, \tilde{t}))^2 \leq C \left(1 + g\left(\frac{k}{h}\right) \right),
$$

which contradicts (5.2.14). $\qquad\square$

From Remark 5.2.2 and Lemma 5.2.4, we obtain

Theorem 5.2.1 *(Weak Harnack's Inequality) Assume that $u \in H^1_{\text{loc}}(\mathbb{R}^n \times \mathbb{R}_+)$ is a nonnegative weak solution of (5.2.1) and for some constants $\varepsilon > 0$ and $\mu \in (0,1)$,*

$$
\text{mes}\{(x,t) \in Q_R; u(x,t) \geq \varepsilon\} \geq \mu \text{mes} Q_R.
$$

Then there exist a constant $\theta \in (0, 1/2)$ depending only on n and μ and a constant $\gamma > 0$ depending only on n, μ, ε and θ, such that

$$u(x, t) \geq \gamma, \quad \forall (x, t) \in Q_{\theta R}.$$

5.2.2 Hölder's estimate

Applying the weak Harnack's inequality, we can establish interior Hölder's estimate for solutions of the homogeneous heat equation.

Lemma 5.2.5 *Let $u \in H^1(Q_{R_0})$ be a bounded weak solution of (5.2.1) in Q_{R_0}. Then there exist constants $\theta \in (0, 1/2)$ and $\sigma \in (0, 1)$ depending only on n, such that for any $0 < R \leq R_0/2$, either*

i) $\underset{Q_{\theta R}}{\operatorname{osc}} u \leq CR$

or

ii) $\underset{Q_{\theta R}}{\operatorname{osc}} u \leq \sigma \underset{Q_R}{\operatorname{osc}} u.$

Proof. Denote $M = \underset{Q_R}{\sup} u$. Without loss of generality, we may assume that $\omega(R) = \underset{Q_R}{\operatorname{osc}} u = 2M$; otherwise we consider $v = u - \dfrac{1}{2} \left(\underset{Q_R}{\sup} u + \underset{Q_R}{\inf} u \right)$ instead of u, which is a bounded weak solution of (5.2.1) with

$$\underset{Q_R}{\operatorname{osc}} v = \underset{Q_R}{\operatorname{osc}} u = 2 \underset{Q_R}{\sup} v.$$

If $M < R$, then for $\theta \in (0, 1)$, we have

$$\underset{Q_{\theta R}}{\operatorname{osc}} u \leq \underset{Q_R}{\operatorname{osc}} u = 2M < 2R,$$

which implies i).

To prove ii) in case $M \geq R$, note that one of the following two cases must be valid:

$$\operatorname{mes}\{(x, t) \in Q_R, u \geq 0\} \geq \frac{1}{2} \operatorname{mes} Q_R$$

and

$$\operatorname{mes}\{(x, t) \in Q_R, -u \geq 0\} \geq \frac{1}{2} \operatorname{mes} Q_R.$$

For definiteness, we assume that the first one is valid. Let $\bar{u} = 1 + \dfrac{u}{M}$. Then $\bar{u} \geq 0$ and

$$\text{mes}\{(x,t) \in Q_R, \bar{u} \geq 1\} \geq \frac{1}{2}\text{mes}Q_R.$$

From the weak Harnack's inequality (Theorem 5.2.1), it follows that there exist $\theta \in (0, 1/2)$ and $0 < \gamma < 1$, such that

$$\bar{u}(x,t) \geq \gamma, \quad (x,t) \in Q_{\theta R}.$$

Thus

$$-M(1 - \gamma) \leq u(x,t) \leq M, \quad (x,t) \in Q_{\theta R}.$$

Hence

$$\omega(\theta R) = \sup_{Q_{\theta R}} u - \inf_{Q_{\theta R}} u \leq 2M\left(1 - \frac{\gamma}{2}\right) = \sigma\omega(R)$$

with $\sigma = 1 - \dfrac{\gamma}{2}$.   □

Using the iteration lemma (Lemma 5.1.5), we can obtain

Corollary 5.2.1 *Let $u \in H^1(Q_{R_0})$ be a bounded weak solution of (5.2.1). Then there exist constants $\alpha \in (0,1)$ and $C > 0$ depending only on n, such that*

$$\underset{Q_R}{\text{osc}}\, u \leq C\left(\frac{R}{R_0}\right)^\alpha \left[\underset{Q_{R_0}}{\text{osc}}\, u + R_0\right], \quad 0 < R \leq R_0.$$

Furthermore, we have

Theorem 5.2.2 *Let $u \in H^1_{\text{loc}}(\mathbb{R}^n \times \mathbb{R}_+)$ be a bounded weak solution of (5.2.1). Then there exists a constant $\alpha \in (0,1)$, such that for any $Q_R \subset \overline{Q}_R \subset \mathbb{R}^n \times \mathbb{R}_+$,*

$$[u]_{\alpha, Q_R} \leq C,$$

where the constant C depends only on n and Q_R.

5.2.3   Harnack's inequality

Using the weak Harnack's inequality (Theorem 5.2.1) and Hölder's estimate (Theorem 5.2.2). We can derive the following Harnack's inequality.

Theorem 5.2.3   *Let $u \in H^1_{\text{loc}}(\mathbb{R}^n \times \mathbb{R}_+)$ be a nonnegative and bounded weak solution of (5.2.1). If $4R^2 < t_0$, then there exists a constant $\theta \in (0,1)$ depending only on n, such that*

$$\sup_{\Theta_R} u \leq C \inf_{Q_{\theta R}} u,$$

where the constant C depends only on n and θ.

Proof.   We may assume that $R = 1$, since in the general case, we can transform the problem to the case $R = 1$ by rescaling.

Suppose first $\sup_{\Theta_1} u = 1$. From Theorem 5.2.2, there exist constants $\alpha \in (0,1)$ and $C > 0$ depending only on n, such that $[u]_{\alpha, Q_2} \leq C$. Since $\Theta_1 \subset Q_2$, there must be constants $\varepsilon > 0$ and $\mu \in (0,1)$ depending only on n but independent of u, such that

$$\text{mes}\{(x,t) \in Q_2; u(x,t) \geq \varepsilon\} \geq \mu \text{mes} Q_2.$$

From Theorem 5.2.1, it follows that there exist a constant $\theta \in (0,1)$ depending only on n, μ and a constant γ depending only on n, μ, ε and θ, such that

$$u(x,t) \geq \gamma, \quad (x,t) \in Q_\theta.$$

Thus

$$\sup_{\Theta_1} u = 1 \leq \frac{1}{\gamma} \inf_{Q_\theta} u$$

and the conclusion of the theorem follows with $C = \dfrac{1}{\gamma}$.

The case $\sup_{\Theta_1} u = 0$ is trivial. For the general case $\sup_{\Theta_1} u > 0$, we consider $w = \dfrac{1}{\sup_{\Theta_1} u} u$, which is also a nonnegative and bounded weak solution of (5.2.1) with $\sup_{\Theta_1} w = 1$. Thus, from what we have proved,

$$\sup_{\Theta_1} w = 1 \leq C \inf_{Q_\theta} w.$$

Multiplying both sides by $\sup_{Q_1} u$ leads to

$$\sup_{\Theta_1} u \leq C \inf_{Q_\theta} u.$$

$\square$

Exercises

1. Assume that $\Omega \subset \mathbb{R}^n$ is a bounded domain and $u \in C(\Omega)$ satisfies the mean value equality, namely, for any $y \in \Omega$ and $R > 0$,

$$u(y) = \frac{1}{n\omega_n R^{n-1}} \int_{\partial B_R} u(x)ds,$$

provided $B_R(y) \subset \overline{B}_R(y) \subset \Omega$. Prove that $u \in C^2(\Omega)$ and satisfies Laplace's equation

$$-\Delta u = 0, \quad x \in \Omega,$$

where $B_R(y)$ is the ball in $\mathbb{R}^n$ of radius R, centered at y, ω_n is the measure of the unit ball in $\mathbb{R}^n$.

2. Prove Remarks 5.1.1, 5.1.3 and 5.1.4.

3. Assume that B_R is a ball in $\mathbb{R}^n$ of radius R, $f \in L^\infty(B_R)$ and $u \in H^1(B_R)$ is a nonnegative and bounded weak solution of

$$-\Delta u = f, \quad x \in B_R.$$

Set $\tilde{u} = u + R^2 \|f\|_{L^\infty(B_R)}$. Prove that for any $0 < \theta < 1$,

$$\sup_{B_{\theta R}} \tilde{u} \leq C \inf_{B_{\theta R}} \tilde{u},$$

where $C > 0$ is a constant depending only on n, $(1 - \theta)^{-1}$ and $\|u\|_{L^\infty(B_R)}$.

4. Establish the weak Harnack's inequality for solutions of the nonhomogeneous heat equation.

Chapter 6

Schauder's Estimates for Linear Elliptic Equations

In this and next chapters, we introduce Schauder's estimates for linear elliptic equations and linear parabolic equations of second order respectively. These estimates will be applied to the existence theory of classical solutions in Chapter 8. In this chapter, Schauder's estimates will be established first for Poisson's equation. To establish Schauder's estimates for such a typical and simple in form equation, one can easily expound the basic idea of the method and catch the essence of the argument. Based on the results obtained for this equation, we finally complete the estimates for general linear elliptic equations.

To establish Schauder's estimates, we will adopt the theory of Campanato spaces. By means of such approach, the derivation will be more succinct compared with those based on the potential theory or based on mollification of functions given by Trudinger. In addition, this approach is available not only to linear elliptic equations and systems of second order, but also to equations and systems of higher order.

6.1 Campanato Spaces

Schauder's estimates are a priori estimates on the Hölder norms of the derivatives of solutions, which are certain kind of pointwise estimates. It is well-known that in many cases, it is quite difficult to derive pointwise estimates directly from the differential equation considered. However, to derive integral estimates is relatively easy. Thus it is reasonable to ask if there is some approach based on an integral description, instead of the above pointwise estimate. The answer is positive. In this section the Campanato spaces are introduced to describe the integral characteristic of the Hölder continuous functions.

Definition 6.1.1 Let $\Omega \subset \mathbb{R}^n$ be a bounded domain. If there exists a constant A, such that for any $x \in \Omega$ and $0 < \rho < \text{diam}\Omega$,

$$|\Omega_\rho(x)| \geq A\rho^n,$$

where $\Omega_\rho(x) = \Omega \cap B_\rho(x)$, then Ω is called a domain of (A)-type.

Definition 6.1.2 (Campanato Spaces) For $p \geq 1$, $\mu \geq 0$, the subset of all functions u in $L^p(\Omega)$ satisfying

$$[u]_{p,\mu} = [u]_{p,\mu;\Omega} \equiv \sup_{\substack{x \in \Omega \\ 0 < \rho < \text{diam}\Omega}} \left(\rho^{-\mu} \int_{\Omega_\rho(x)} |u(y) - u_{x,\rho}|^p dy \right)^{1/p} < +\infty$$

and endowed with the norm

$$\|u\|_{\mathcal{L}^{p,\mu}} = \|u\|_{\mathcal{L}^{p,\mu}(\Omega)} = [u]_{p,\mu;\Omega} + \|u\|_{L^p(\Omega)}$$

is called a Campanato space, denoted by $\mathcal{L}^{p,\mu}(\Omega)$, where

$$u_{x,\rho} = \frac{1}{|\Omega_\rho(x)|} \int_{\Omega_\rho(x)} u(y)dy.$$

Note that $[u]_{p,\mu;\Omega}$ is a semi-norm rather than a norm, since $[u]_{p,\mu;\Omega} = 0$ dose not imply $u = 0$.

It is easy to verify

Proposition 6.1.1 $\mathcal{L}^{p,\mu}(\Omega)$ *is a Banach space.*

Proposition 6.1.2 *(Property of the Mean Value) Let $\Omega \subset \mathbb{R}^n$ be a domain of (A)-type and $u \in \mathcal{L}^{p,\mu}(\Omega)$. Then for any $x \in \overline{\Omega}$ and ρ such that $0 < \rho < R < \text{diam}\Omega$, there holds*

$$|u_{x,R} - u_{x,\rho}| \leq C[u]_{p,\mu;\Omega}\rho^{-n/p}R^{\mu/p},$$

where the constant C depends only on A and p.

Proof. For any $y \in \Omega_\rho(x)$, we have

$$|u_{x,R} - u_{x,\rho}|^p \leq 2^{p-1}(|u_{x,R} - u(y)|^p + |u_{x,\rho} - u(y)|^p).$$

Integrating over $\Omega_\rho(x) \subset \Omega_R(x)$ with respect to y leads to

$$|\Omega_\rho(x)||u_{x,R} - u_{x,\rho}|^p$$

$$\leq 2^{p-1} \left(\int_{\Omega_R(x)} |u_{x,R} - u(y)|^p dy + \int_{\Omega_\rho(x)} |u_{x,\rho} - u(y)|^p dy \right).$$

Hence

$$A\rho^n |u_{x,R} - u_{x,\rho}|^p \leq C[u]_{p,\mu;\Omega}^p (R^\mu + \rho^\mu) \leq C[u]_{p,\mu;\Omega}^p R^\mu$$

or

$$|u_{x,R} - u_{x,\rho}| \leq C[u]_{p,\mu;\Omega}\rho^{-n/p} R^{\mu/p}$$

with another constant C. $\square$

Theorem 6.1.1 *(Integral Characteristic of Hölder Continuous Functions) If Ω is a domain of (A)-type and $n < \mu \leq n + p$, then $\mathcal{L}^{p,\mu}(\Omega) = C^\alpha(\overline{\Omega})$ and*

$$C_1[u]_{\alpha;\Omega} \leq [u]_{p,\mu;\Omega} \leq C_2[u]_{\alpha;\Omega},$$

where $\alpha = \dfrac{\mu - n}{p}$ and C_1, C_2 are some positive constants depending only on n, A, p, μ.

The precise meaning of $\mathcal{L}^{p,\mu}(\Omega) = C^\alpha(\overline{\Omega})$ is that $C^\alpha(\overline{\Omega}) \subset \mathcal{L}^{p,\mu}(\Omega)$ and for any $u \in \mathcal{L}^{p,\mu}(\Omega)$, there exists a function $\tilde{u} \in C^\alpha(\overline{\Omega})$ such that $\tilde{u} = u$ a.e. in Ω.

Proof. Let $u \in C^\alpha(\overline{\Omega})$. Then for any $x \in \Omega$, $0 < \rho < \text{diam}\Omega$ and $y \in \Omega_\rho(x)$, we have

$$
\begin{aligned}
|u(y) - u_{x,\rho}| &= \frac{1}{|\Omega_\rho(x)|} \left| \int_{\Omega_\rho(x)} (u(y) - u(z))dz \right| \\
&\leq \frac{1}{|\Omega_\rho(x)|} \int_{\Omega_\rho(x)} |u(y) - u(z)|dz \\
&\leq \frac{[u]_{\alpha;\Omega}}{|\Omega_\rho(x)|} \int_{\Omega_\rho(x)} |y - z|^\alpha dz \\
&\leq \frac{[u]_{\alpha;\Omega}}{A\rho^n} \int_{\Omega_{2\rho}(0)} |z|^\alpha dz \\
&= \frac{[u]_{\alpha;\Omega}}{A\rho^n} \int_0^{2\rho} n\omega_n r^{n-1+\alpha} dr \\
&\leq C[u]_{\alpha;\Omega}\rho^\alpha,
\end{aligned}
$$

where ω_n is the measure of the unit ball in $\mathbb{R}^n$. Hence

$$\rho^{-\mu} \int_{\Omega_\rho(x)} |u(y) - u_{x,\rho}|^p dy \leq C^p[u]_{\alpha;\Omega}^p \rho^{p\alpha - \mu} |\Omega_\rho(x)| \leq C^p \omega_n [u]_{\alpha;\Omega}^p$$

or

$$[u]_{p,\mu;\Omega} \leq C[u]_{\alpha;\Omega} \tag{6.1.1}$$

with another constant C depending only on n, A and p. This, together with $\|u\|_{L^p(\Omega)} \leq C|u|_{0;\Omega}$ implies $u \in \mathcal{L}^{p,\mu}(\Omega)$ and

$$\|u\|_{\mathcal{L}^{p,\mu}(\Omega)} \leq C|u|_{\alpha;\Omega}.$$

Conversely, assume $u \in \mathcal{L}^{p,\mu}(\Omega)$. We will prove that there exists a function $\tilde{u} \in C^\alpha(\overline{\Omega})$ such that $\tilde{u} = u$ a.e. in Ω.

Step 1   Construct $\tilde{u}$.

For any fixed $x \in \overline{\Omega}$ and $0 < R < \text{diam}\Omega$, let $R_i = R/2^i$ $(i = 0, 1, 2, \cdots)$. Then by Proposition 6.1.2,

$$|u_{x,R_i} - u_{x,R_{i+1}}| \leq C[u]_{p,\mu;\Omega} R^{(\mu-n)/p} 2^{i(n-\mu)/p+n/p}.$$

Hence for any integer j such that $0 \leq j < i$, we have

$$|u_{x,R_j} - u_{x,R_i}| \leq C[u]_{p,\mu;\Omega} R^{(\mu-n)/p} \sum_{k=j}^{i-1} 2^{k(n-\mu)/p+n/p}$$

$$= C 2^{n/p}[u]_{p,\mu;\Omega} R^{(\mu-n)/p} 2^{j(n-\mu)/p} \frac{1 - \left(2^{(n-\mu)/p}\right)^{i-j}}{1 - 2^{(n-\mu)/p}}$$

or

$$|u_{x,R_j} - u_{x,R_i}| \leq C[u]_{p,\mu;\Omega} R_j^{(\mu-n)/p} \tag{6.1.2}$$

with another constant C depending only on n, A, p and μ. This implies that for any $x \in \overline{\Omega}$ and $0 < R < \text{diam}\Omega$, $\{u_{x,R_i}\}_{i=0}^\infty$ is a Cauchy sequence and hence

$$\tilde{u}_R(x) = \lim_{i \to \infty} u_{x,R_i}, \quad x \in \overline{\Omega}.$$

For any $0 < r < R$, let $r_i = r/2^i$ $(i = 0, 1, 2, \cdots)$. Then by Proposition 6.1.2

$$|u_{x,R_i} - u_{x,r_i}| \leq C[u]_{p,\mu;\Omega} r_i^{-n/p} R_i^{\mu/p}$$

$$= C[u]_{p,\mu;\Omega} \left(\frac{R_i}{r_i}\right)^{n/p} R_i^{(\mu-n)/p}$$

$$= C[u]_{p,\mu;\Omega} \left(\frac{R}{r}\right)^{n/p} R_i^{(\mu-n)/p}.$$

Since $\mu > n$, we have

$$\lim_{i\to\infty} |u_{x,R_i} - u_{x,r_i}| = 0.$$

Hence $\tilde{u}_R(x) = \tilde{u}_r(x)$, which means that $\tilde{u}_R(x)$ is independent of R. Denote

$$\tilde{u}(x) = \tilde{u}_R(x), \quad x \in \overline{\Omega}.$$

Step 2 Prove $\tilde{u} = u$ a.e. in Ω.

Take $j = 0$ in (6.1.2) and let $i \to \infty$. Then we obtain

$$|u_{x,R} - \tilde{u}(x)| \le C[u]_{p,\mu;\Omega} R^{(\mu-n)/p}. \tag{6.1.3}$$

Hence

$$\tilde{u}(x) = \lim_{R\to 0^+} u_{x,R}, \quad x \in \overline{\Omega}.$$

On the other hand, by Lebesgue's theorem,

$$u(x) = \lim_{R\to 0^+} u_{x,R}, \quad \text{a.e. } x \in \Omega.$$

Therefore $\tilde{u} = u$ a.e. in Ω.

Step 3 Prove $\tilde{u} \in C^\alpha(\overline{\Omega})$.

For any $x, y \in \overline{\Omega}$, $x \ne y$, denote $R = |x - y|$. Then

$$|\tilde{u}(x) - \tilde{u}(y)| \le |\tilde{u}(x) - u_{x,2R}| + |u_{x,2R} - u_{y,2R}| + |u_{y,2R} - \tilde{u}(y)|.$$

From (6.1.3), we have

$$|\tilde{u}(x) - u_{x,2R}| + |u_{y,2R} - \tilde{u}(y)| \le C[u]_{p,\mu;\Omega} R^{(\mu-n)/p}.$$

Denote $G = \Omega_{2R}(x) \cap \Omega_{2R}(y)$. Then

$$\int_G |u_{x,2R} - u_{y,2R}| dz$$

$$\le \int_{\Omega_{2R}(x)} |u_{x,2R} - u(z)| dz + \int_{\Omega_{2R}(y)} |u_{y,2R} - u(z)| dz$$

$$\le |\Omega_{2R}(x)|^{1-1/p} \left(\int_{\Omega_{2R}(x)} |u_{x,2R} - u(z)|^p dz \right)^{1/p}$$

$$+ |\Omega_{2R}(y)|^{1-1/p} \left(\int_{\Omega_{2R}(y)} |u_{y,2R} - u(z)|^p dz \right)^{1/p}$$

$$\le (2R)^{\mu/p} |\Omega_{2R}(x)|^{1-1/p} [u]_{p,\mu;\Omega} + (2R)^{\mu/p} |\Omega_{2R}(y)|^{1-1/p} [u]_{p,\mu;\Omega}$$

$$\le C[u]_{p,\mu;\Omega} R^{(\mu-n)/p+n}.$$

Since $\Omega_R(x) \subset G$, we have $AR^n \leq |\Omega_R(x)| \leq |G|$. Thus from the above inequality,

$$|u_{x,2R} - u_{y,2R}| \leq C[u]_{p,\mu;\Omega} R^{(\mu-n)/p}$$

with another constant C. Therefore

$$|\tilde{u}(x) - \tilde{u}(y)| \leq C[u]_{p,\mu;\Omega}|x - y|^{(\mu-n)/p}, \qquad (6.1.4)$$

which implies $\tilde{u} \in C^\alpha(\overline{\Omega})$ and

$$[\tilde{u}]_{\alpha;\Omega} \leq C[u]_{p,\mu;\Omega} \qquad (6.1.5)$$

with constant C depending only on n, A and p.

Summing up, we have shown $\mathcal{L}^{p,\mu}(\Omega) = C^\alpha(\overline{\Omega})$ and completed the proof of the theorem. □

Remark 6.1.1　*For any $u \in \mathcal{L}^{p,\mu}(\Omega) = C^\alpha(\overline{\Omega})$*

$$|u|_{0;\Omega} \leq C\|u\|_{\mathcal{L}^{p,\mu}(\Omega)}.$$

In fact, by the continuity of u, there exists $z \in \overline{\Omega}$, such that

$$\frac{1}{|\Omega|} \int_\Omega u(y)dy = u(z).$$

Thus, for any $x \in \overline{\Omega}$, using (6.1.5), we have

$$\begin{aligned}
|u(x)| &\leq |u(x) - u(z)| + |u(z)| \\
&= |u(x) - u(z)| + \frac{1}{|\Omega|}\left|\int_\Omega u(y)dy\right| \\
&\leq C[u]_{\alpha;\Omega}|x - z|^\alpha + |\Omega|^{-1/p}\|u\|_{L^p(\Omega)} \\
&\leq C(\mathrm{diam}\Omega)^\alpha[u]_{p,\mu;\Omega} + |\Omega|^{-1/p}\|u\|_{L^p(\Omega)} \\
&\leq C\left([u]_{p,\mu;\Omega} + \|u\|_{L^p(\Omega)}\right) \\
&= C\|u\|_{\mathcal{L}^{p,\mu}(\Omega)}.
\end{aligned}$$

Remark 6.1.2　*For any $0 < \lambda < 1$, we may define*

$$[u]_{p,\mu}^{(\lambda)} = [u]_{p,\mu;\Omega}^{(\lambda)} \equiv \sup_{\substack{x \in \Omega \\ 0 < \rho < \lambda\mathrm{diam}\Omega}} \left(\rho^{-\mu}\int_{\Omega_\rho(x)}|u(y) - u_{x,\rho}|^p dy\right)^{1/p},$$

which is also a semi-norm. From the proof of Theorem 6.1.1, we see that, if $\alpha = \dfrac{\mu - n}{p} \in (0,1]$, then this semi-norm is equivalent to the Hölder *semi-norm* $[u]_{\alpha;\Omega}$, *i.e.*

$$C_1[u]_{\alpha;\Omega} \le [u]_{p,\mu;\Omega}^{(\lambda)} \le C_2[u]_{\alpha;\Omega}, \tag{6.1.6}$$

where C_1, C_2 *are positive constants depending only on* n, A, p, μ *and* λ.

In fact, the second part of (6.1.6) follows from (6.1.1) and the obvious inequality

$$[u]_{p,\mu;\Omega}^{(\lambda)} \le [u]_{p,\mu;\Omega}$$

where the constant C_2 can be chosen independent of λ. On the other hand, similar to the proof of (6.1.4), we may obtain

$$[u]_{\alpha;\Omega}^{(\lambda)} \equiv \sup_{\substack{x,y\in\Omega \\ 0<|x-y|<\lambda\mathrm{diam}\Omega}} \frac{|u(x)-u(y)|}{|x-y|^\alpha} \le C[u]_{p,\mu;\Omega}^{(\lambda)},$$

which in combination with

$$[u]_{\alpha;\Omega} \le \left(\frac{1}{\lambda}+1\right)[u]_{\alpha;\Omega}^{(\lambda)},$$

implies that the first part of (6.1.6) also holds, but the constant C_1 depends on λ.

Proposition 6.1.3 *Let* Ω *be a domain of (A)-type. Then for* $\mu > n+p$, *all elements of* $\mathcal{L}^{p,\mu}(\Omega)$ *are constants.*

Proof. From (6.1.4) in the proof of Theorem 6.1.1, for any $x, y \in \overline{\Omega}$,

$$|\tilde{u}(x) - \tilde{u}(y)| \le C[u]_{p,\mu;\Omega}|x-y|^{(\mu-n)/p}.$$

By the assumption, $\mu > n+p$ or $\dfrac{\mu-n}{p} > 1$, thus $\dfrac{\partial\tilde{u}}{\partial x_i}$ $(i = 1, 2, \cdots, n)$ exist and equal zero. $\qquad\square$

6.2 Schauder's Estimates for Poisson's Equation

6.2.1 *Estimates to be established*

From now on, we are devoted to Schauder's estimates for solutions of linear elliptic equations. To obtain the estimate for solutions on a bounded domain Ω, we first establish the local interior estimate, i.e. the estimate on any ball contained in Ω and the local estimate near the boundary $\partial\Omega$, i.e. the estimate on the small neighborhood of any point of $\partial\Omega$, and then use the finite covering technique. Since the smooth boundary can be locally transformed to a superplane by flatting technique, to obtain the local estimate near the boundary, it suffices first to establish the estimate on any small semiball.

We begin our discussion with Poisson's equation

$$-\Delta u(x) = f(x), \quad x \in \mathbb{R}^n_+, \tag{6.2.1}$$

and hope to establish the following estimates:

i) Interior estimate. If $u \in C^{2,\alpha}(\overline{B}_R)\, (0 < \alpha < 1)$ is a solution of equation (6.2.1) in $B_R = B_R(x^0)$, then

$$[D^2 u]_{\alpha;B_{R/2}} \leq C \left(\frac{1}{R^{2+\alpha}} |u|_{0;B_R} + \frac{1}{R^\alpha} |f|_{0;B_R} + [f]_{\alpha;B_R} \right); \tag{6.2.2}$$

ii) Near boundary estimate. If $u \in C^{2,\alpha}(\overline{B}^+_R)\, (0 < \alpha < 1)$ is a solution of equation (6.2.1) in $B^+_R = B^+_R(x^0) = \{x \in B_R(x^0); x_n > 0\}$ satisfying

$$u\Big|_{x_n=0} = 0, \tag{6.2.3}$$

then

$$[D^2 u]_{\alpha;B^+_{R/2}} \leq C \left(\frac{1}{R^{2+\alpha}} |u|_{0;B^+_R} + \frac{1}{R^\alpha} |f|_{0;B^+_R} + [f]_{\alpha;B^+_R} \right). \tag{6.2.4}$$

In (6.2.2) and (6.2.4), C is a constant depending only on n.

Remark 6.2.1 *From the interpolation inequality (Theorem 1.2.2 of Chapter 1), we see that in (6.2.2) and (6.2.4), $[D^2 u]_\alpha$ can be replaced by $|u|_{2,\alpha}$.*

Remark 6.2.2 *If instead of (6.2.3), the boundary value condition is $u\Big|_{x_n=0} = \varphi$ and $\varphi \in C^{2,\alpha}(B^+_R)$, then we may consider the equation for $u - \varphi$.*

Remark 6.2.3 *In the proof of (6.2.2), (6.2.4) (and their preparatory propositions) stated below, we always assume that the solution u considered is sufficiently smooth. This is reasonable, because we have the following proposition.*

Proposition 6.2.1 *If the estimate (6.2.2) ((6.2.4)) holds for any $R > 0$ and any sufficiently smooth solution u of (6.2.1) on $\overline{B}_R = \overline{B}_R(x^0)$ ($\overline{B}^+_R = \overline{B}^+_R(x^0)$ with (6.2.3)), then (6.2.2) ((6.2.4)) also holds for any $R > 0$ and any solution $u \in C^{2,\alpha}(\overline{B}_R)$ ($C^{2,\alpha}(\overline{B}^+_R)$) of (6.2.1).*

Proof. Given $R > 0$. Suppose that $u \in C^{2,\alpha}(\overline{B}_R)$ is a solution of (6.2.1) in B_R. Denote $R_\varepsilon = R - \varepsilon$ with $0 < \varepsilon < R$. Let $\xi \in C_0^\infty(B_R)$ be a cut-off function on B_R relative to B_{R_ε} and $v = \xi u$. Then $v \in C^{2,\alpha}(\overline{B}_R)$ satisfies

$$-\Delta v = g, \quad x \in B_R, \tag{6.2.5}$$

$$v\Big|_{\partial B_R} = 0, \tag{6.2.6}$$

where $g = \xi f - u \Delta \xi - \nabla \xi \cdot \nabla u$. Since $u \in C^{2,\alpha}(\overline{B}_R)$, we have $g \in C^\alpha(\overline{B}_R)$. Now we choose a sequence $\{g_m\} \subset C^\infty(\overline{B}_R)$, converging to g in $C^\alpha(\overline{B}_R)$ as $m \to \infty$ and consider the approximating problem

$$\begin{cases} -\Delta v_m = g_m, & x \in B_R, \\ v_m\Big|_{\partial B_R} = 0. \end{cases}$$

From the L^2 theory (see Theorem 2.2.5 of Chapter 2), this problem has a solution $v_m \in C^\infty(\overline{B}_R)$ and

$$\|v_m - v_l\|_{H^2(B_R)} \leq C \|g_m - g_l\|_{L^2(B_R)}, \quad (m, l = 1, 2, \cdots)$$

where the constant C depends only on n and B_R (see Remark 2.2.2 of Chapter 2). This implies that $\{v_m\}$ converges in $H^2(B_R)$ as $m \to \infty$, whose limit function is obviously a solution of (6.2.5), (6.2.6) and hence it is equal to v almost everywhere in B_R by the uniqueness of the solution.

Since $v_m \in C^\infty(\overline{B}_R)$, by the assumption of the proposition, we have

$$[D^2 v_m]_{\alpha; B_{R/2}} \leq C \left(\frac{1}{R^{2+\alpha}} |v_m|_{0; B_R} + \frac{1}{R^\alpha} |g_m|_{0; B_R} + [g_m]_{\alpha; B_R} \right), \tag{6.2.7}$$

where the constant C depends only on n. According to the maximum principle for Poisson's equation, it is easily seen that $\{v_m\}$ is uniformly bounded and uniformly converges on B_R as $m \to \infty$, whose limit function is just v. Since the right side of (6.2.7) is bounded, using Arzela-Ascoli's theorem, we see that there exist a subsequence $\{v_{m_k}\}$ of $\{v_m\}$ and a function $w \in C^{2,\alpha}(\overline{B}_{R/2})$, such that as $k \to \infty$ we have

$$v_{m_k}(x) \to w(x),$$

$$D_i v_{m_k}(x) \to D_i w(x), \quad 1 \leq i \leq n,$$

$$D_{ij} v_{m_k}(x) \to D_{ij} w(x), \quad 1 \leq i, j \leq n,$$

uniformly on $\overline{B}_{R/2}$.

Now we take $m = m_k$ in

$$[D^2 v_m]_{\alpha; B_{R_\varepsilon/2}} \leq C\left(\frac{1}{R_\varepsilon^{2+\alpha}}|v_m|_{0;B_{R_\varepsilon}} + \frac{1}{R_\varepsilon^\alpha}|g_m|_{0;B_{R_\varepsilon}} + [g_m]_{\alpha;B_{R_\varepsilon}}\right),$$

which holds by the assumption of the proposition, and let $k \to \infty$ to obtain

$$[D^2 w]_{\alpha; B_{R_\varepsilon/2}} \leq C\left(\frac{1}{R_\varepsilon^{2+\alpha}}|v|_{0;B_{R_\varepsilon}} + \frac{1}{R_\varepsilon^\alpha}|g|_{0;B_{R_\varepsilon}} + [g]_{\alpha;B_{R_\varepsilon}}\right),$$

Since clearly $w = v$ on $\overline{B}_{R_\varepsilon/2}$ and $\xi = 1$ on $\overline{B}_{R_\varepsilon}$, the above inequality is just

$$[D^2 u]_{\alpha; B_{R_\varepsilon/2}} \leq C\left(\frac{1}{R_\varepsilon^{2+\alpha}}|u|_{0;B_{R_\varepsilon}} + \frac{1}{R_\varepsilon^\alpha}|f|_{0;B_{R_\varepsilon}} + [f]_{\alpha;B_{R_\varepsilon}}\right),$$

from which (6.2.2) follows by letting $\varepsilon \to 0^+$.

Similarly we can prove the second part of the proposition. $\square$

6.2.2 Caccioppoli's inequalities

we first prove Caccioppoli's inequalities for solution of Poisson's equation.

Theorem 6.2.1 *Let u be a solution of (6.2.1) in B_R. Then for any $0 < \rho < R$ and $\lambda \in \mathbb{R}$, there hold*

$$\int_{B_\rho} |Du|^2 dx \leq C\left[\frac{1}{(R-\rho)^2}\int_{B_R}(u-\lambda)^2 dx + (R-\rho)^2\int_{B_R}f^2 dx\right], \quad (6.2.8)$$

$$\int_{B_\rho} |Dw|^2 dx \leq C\left[\frac{1}{(R-\rho)^2}\int_{B_R}(w-\lambda)^2 dx + \int_{B_R}(f-f_R)^2 dx\right], \quad (6.2.9)$$

where $w = D_i u\,(1 \leq i \leq n)$, $f_R = \dfrac{1}{|B_R|}\displaystyle\int_{B_R}f(x)dx$ and C is a constant depending only on n.

Proof. Let η be a cut-off function on B_R relative to B_ρ, i.e. $\eta \in C_0^\infty(B_R)$ and satisfies

$$0 \leq \eta(x) \leq 1, \quad \eta(x) = 1 \text{ in } B_\rho, \quad |D\eta(x)| \leq \frac{C}{R-\rho}.$$

To prove (6.2.8), we multiply both sides of (6.2.1) by $\eta^2(u - \lambda)$, integrate over B_R and integrate by parts to derive

$$\int_{B_R} \eta^2 |Du|^2 dx = -2\int_{B_R}\eta(u-\lambda)D\eta \cdot Du\,dx + \int_{B_R}\eta^2(u-\lambda)f\,dx.$$

Using Cauchy's inequality with ε to all terms on the right side we are led to

$$\int_{B_R} \eta^2 |Du|^2 dx$$

$$\leq \frac{1}{2} \int_{B_R} \eta^2 |Du|^2 dx + 2 \int_{B_R} (u - \lambda)^2 |D\eta|^2 dx$$

$$+ \frac{1}{2}(R - \rho)^2 \int_{B_R} \eta^2 f^2 dx + \frac{1}{2(R - \rho)^2} \int_{B_R} \eta^2 (u - \lambda)^2 dx$$

$$\leq \frac{1}{2} \int_{B_R} \eta^2 |Du|^2 dx + \frac{C}{(R - \rho)^2} \int_{B_R} (u - \lambda)^2 dx$$

$$+ \frac{1}{2}(R - \rho)^2 \int_{B_R} \eta^2 f^2 dx + \frac{1}{2(R - \rho)^2} \int_{B_R} \eta^2 (u - \lambda)^2 dx$$

which implies (6.2.8).

To prove (6.2.9), we multiply both sides of the equation for w, i.e.

$$-\Delta w(x) = D_i f(x) = D_i(f(x) - f_R), \quad x \in B_R$$

by $\eta^2 (u - \lambda)$, integrate over B_R and integrate by parts to derive

$$\int_{B_R} \eta^2 |Dw|^2 dx$$

$$= -2 \int_{B_R} \eta(w - \lambda) D\eta \cdot Dw dx - \int_{B_R} \eta^2 (f - f_R) D_i w dx$$

$$- 2 \int_{B_R} \eta(w - \lambda)(f - f_R) D_i \eta dx.$$

Then we use Cauchy's inequality with ε to all terms on the right side to obtain

$$\int_{B_R} \eta^2 |Dw|^2 dx$$

$$\leq \frac{1}{2} \int_{B_R} \eta^2 |Dw|^2 dx + C \int_{B_R} (w - \lambda)^2 |D\eta|^2 dx$$

$$+ C \int_{B_R} \eta^2 (f - f_R)^2 dx$$

$$\leq \frac{1}{2} \int_{B_R} \eta^2 |Dw|^2 dx + \frac{C}{(R - \rho)^2} \int_{B_R} (w - \lambda)^2 dx$$

$$+ C \int_{B_R} (f - f_R)^2 dx.$$

and hence (6.2.9) follows.   □

Corollary 6.2.1   *Let u be a solution of (6.2.1) in B_R. Then*

$$\int_{B_{R/2}} |D^2 u|^2 dx \leq C \left(\frac{1}{R^4} \int_{B_R} u^2 dx + R^n |f|_{0;B_R}^2 + R^{n+2\alpha} [f]_{\alpha;B_R}^2 \right),$$

where C is a constant depending only on n.

Proof.   Taking ρ and R to be $\dfrac{R}{2}$ and $\dfrac{3}{4}R$ and $\lambda = 0$ in (6.2.9) gives

$$\int_{B_{R/2}} |Dw|^2 dx \leq \frac{C}{R^2} \int_{B_{3R/4}} w^2 dx + C \int_{B_{3R/4}} (f - f_R)^2 dx;$$

taking ρ and R to be $\dfrac{3}{4}R$ and R respectively and $\lambda = 0$ in (6.2.8) gives

$$\int_{B_{3R/4}} w^2 dx \leq \frac{C}{R^2} \int_{B_R} u^2 dx + CR^2 \int_{B_R} f^2 dx.$$

A combination of these two inequalities leads to

$$\int_{B_{R/2}} |D^2 u|^2 dx$$
$$\leq \frac{C}{R^4} \int_{B_R} u^2 dx + C \int_{B_R} f^2 dx + C \int_{B_{3R/4}} (f - f_R)^2 dx$$
$$\leq \frac{C}{R^4} \int_{B_R} u^2 dx + CR^n |f|_{0;B_R}^2 + CR^{n+2\alpha} [f]_{\alpha;B_R}^2.$$
□

Corollary 6.2.2   *If $f \equiv 0$ in B_1, then for any positive integer k, there holds*

$$\|u\|_{H^k(B_{1/2})} \leq C \|u\|_{L^2(B_1)},$$

where C is a constant depending only on n and k.

Proof.   The conclusion for $k = 1$ follows immediately from Caccioppoli's inequality (6.2.8). Now we consider the case $k = 2$. From (6.2.8), we have

$$\int_{B_\rho} |Du|^2 dx \leq \frac{C}{(R - \rho)^2} \int_{B_R} u^2 dx. \qquad (6.2.10)$$

Applying (6.2.8) to $D_j u$ ($j = 1, 2, \cdots, n$) leads to

$$\int_{B_\rho} |DD_j u|^2 dx \leq \frac{C}{(R - \rho)^2} \int_{B_R} |D_j u|^2 dx. \qquad (6.2.11)$$

Taking $\rho = \dfrac{3}{4}$, $R = 1$ in (6.2.10) and $\rho = \dfrac{1}{2}$, $R = \dfrac{3}{4}$ in (6.2.11), we obtain

$$\int_{B_{3/4}} |Du|^2 dx \leq C \int_{B_1} u^2 dx,$$

$$\int_{B_{1/2}} |DD_j u|^2 dx \leq C \int_{B_{3/4}} |D_j u|^2 dx.$$

Thus

$$\|u\|_{H^2(B_{1/2})} \leq C\|u\|_{L^2(B_1)}.$$

For the case $k > 2$, we may prove by analogy.   □

Corollary 6.2.3   *If $f \equiv 0$ in B_R, then*

$$\sup_{B_{R/2}} |u| \leq C \left(\frac{1}{R^n} \int_{B_R} u^2 dx \right)^{1/2},$$

where C is a constant depending only on n.

Proof.   Assume $R = 1$ for the moment. Choose $k > n/2$ in Corollary 6.2.2 and use the Sobolev embedding theorem. Then we obtain

$$\sup_{B_{1/2}} |u| \leq C\|u\|_{H^k(B_{1/2})} \leq C\|u\|_{L^2(B_1)}.$$

For the general case $R > 0$, the desired conclusion can be obtained by rescaling.   □

Theorem 6.2.2   *Let u be a solution of problem (6.2.1), (6.2.3) in B_R^+. Then for any $0 < \rho < R$, there hold*

$$\int_{B_\rho^+} |Du|^2 dx \leq C\left[\frac{1}{(R-\rho)^2} \int_{B_R^+} u^2 dx + (R-\rho)^2 \int_{B_R^+} f^2 dx \right], \quad (6.2.12)$$

$$\int_{B_\rho^+} |Dw|^2 dx \leq C\left[\frac{1}{(R-\rho)^2} \int_{B_R^+} w^2 dx + \int_{B_R^+} (f - f_R)^2 dx \right], \quad (6.2.13)$$

where $w = D_i u \, (1 \leq i < n)$, $f_R = \dfrac{1}{|B_R^+|} \displaystyle\int_{B_R^+} f(x) dx$ and C is a constant depending only on n.

Proof.   We merely prove (6.2.13). The proof is similar to that of Theorem 6.2.1, the only difference is that here we multiply the equation for w

$$-\Delta w(x) = D_i f(x) = D_i(f(x) - f_R), \quad x \in B_R^+$$

by $\eta^2 w$ and then integrate over B_R^+ to obtain

$$-\int_{B_R^+} \eta^2 w \Delta w \, dx = \int_{B_R^+} \eta^2 w D_i (f - f_R) \, dx.$$

Since $\eta \in C_0^\infty(B_R)$ and $w\big|_{x_n=0} = 0$, the integral over the boundary ∂B_R^+ resulting from integrating by parts is equal to zero. $\square$

Remark 6.2.4 *In Theorem 6.2.2, we use f_R to denote the average of f over the half ball B_R^+, while in Theorem 6.2.1, f_R denotes the average of f over the ball B_R. In order to abbreviate the notations, here we use the same notation to denote slightly different things. But no confusion will be caused.*

Remark 6.2.5 *Using (6.2.13) and the equation*

$$D_{nn}u = -\sum_{k=1}^{n-1} D_{kk}u - f, \quad x \in B_R^+$$

we can derive

$$\int_{B_\rho^+} |D^2 u|^2 dx$$

$$\leq C \left(\sum_{j=1}^{n-1} \int_{B_\rho^+} |DD_j u|^2 dx + \int_{B_\rho^+} f^2 dx \right)$$

$$\leq C \left[\frac{1}{(R-\rho)^2} \int_{B_R^+} |Du|^2 dx + \int_{B_R^+} (f - f_R)^2 dx + \int_{B_\rho^+} f^2 dx \right]$$

$$\leq C \left[\frac{1}{(R-\rho)^2} \int_{B_R^+} |Du|^2 dx + \int_{B_R^+} f^2 dx \right], \tag{6.2.14}$$

where C is a constant depending only on n.

Remark 6.2.6 *We cannot apply the method of the proof of Theorem 6.2.2 to $w = D_n u$, since $u\big|_{x_n=0} = 0$ dose not imply $w\big|_{x_n=0} = 0$.*

Remark 6.2.7 *In the proof of Theorem 6.2.2, we did not use $\eta^2(w-\lambda)$ as a multiplier, because $\eta^2(w-\lambda)\big|_{x_n=0} = -\lambda\eta^2\big|_{x_n=0} = 0$ if and only if $\lambda = 0$.*

Combining (6.2.12) with (6.2.14) gives

Corollary 6.2.4 *Let u be a solution of problem (6.2.1), (6.2.3). Then*

$$\int_{B_{R/2}^+} |D^2u|^2 dx \leq C \left(\frac{1}{R^4} \int_{B_R^+} u^2 dx + R^n |f|_{0;B_R^+}^2 + R^{n+2\alpha}[f]_{\alpha;B_R^+}^2 \right),$$

where C is a constant depending only on n.

Corollary 6.2.5 *If $f \equiv 0$ in B_1^+, then for any positive integer k, there holds*

$$\|u\|_{H^k(B_{1/2}^+)} \leq C\|u\|_{L^2(B_1^+)},$$

where C is a constant depending only on n and k.

Proof. The conclusion for $k = 1$ follows from Caccioppoli's inequality (6.2.12) immediately. Using (6.2.12) for u and $D_i u$ $(i = 1, 2, \cdots, n-1)$ and combining with the equation, one obtains the desired conclusion for $k = 2$. The case $k > 2$ can be discussed by analogy. $\square$

Similar to the proof of Corollary 6.2.3, we can use Corollary 6.2.5 and the embedding theorem to obtain

Corollary 6.2.6 *If $f \equiv 0$ in B_R^+, then*

$$\sup_{B_{R/2}^+} |u| \leq C \left(\frac{1}{R^n} \int_{B_R^+} u^2 dx \right)^{1/2},$$

where the constant C depends only on n.

6.2.3 Interior estimate for Laplace's equation

Theorem 6.2.3 *Let u be a solution of equation (6.2.1) with $f \equiv 0$ in B_R. Then for any $0 < \rho \leq R$, there hold*

$$\int_{B_\rho} u^2 dx \leq C \left(\frac{\rho}{R} \right)^n \int_{B_R} u^2 dx, \tag{6.2.15}$$

$$\int_{B_\rho} (u - u_\rho)^2 dx \leq C \left(\frac{\rho}{R} \right)^{n+2} \int_{B_R} (u - u_R)^2 dx, \tag{6.2.16}$$

where $u_\rho = \dfrac{1}{|B_\rho|} \displaystyle\int_{B_\rho} u(x) dx$ and C is a constant depending only on n.

Proof.   We first prove (6.2.15). By Corollary 6.2.3, we have, for $0 < \rho < R/2$,

$$\int_{B_\rho} u^2 dx \le |B_\rho| \sup_{B_\rho} u^2 \le C\rho^n \sup_{B_{R/2}} u^2 \le C \left(\frac{\rho}{R}\right)^n \int_{B_R} u^2 dx.$$

For $R/2 \le \rho \le R$, obviously

$$\int_{B_\rho} u^2 dx \le \int_{B_R} u^2 dx \le 2^n \left(\frac{\rho}{R}\right)^n \int_{B_R} u^2 dx.$$

A combination of these inequalities leads to (6.2.15) for $0 < \rho \le R$.

Now we prove (6.2.16). Since $D_j u \, (j = 1, 2, \cdots, n)$ satisfy Laplace's equation in B_R, from (6.2.15), we have

$$\int_{B_\rho} (D_j u)^2 dx \le C \left(\frac{\rho}{R}\right)^n \int_{B_R} (D_j u)^2 dx, \quad j = 1, 2, \cdots, n.$$

From this and Poincaré's inequality

$$\frac{1}{\rho^2} \int_{B_\rho} (u - u_\rho)^2 dx \le C \int_{B_\rho} |Du|^2 dx,$$

we obtain, for $0 < \rho < R/2$,

$$\int_{B_\rho} (u - u_\rho)^2 dx \le C\rho^2 \int_{B_\rho} |Du|^2 dx \le C\rho^2 \left(\frac{\rho}{R}\right)^n \int_{B_{R/2}} |Du|^2 dx.$$

On the other hand, if we choose $\rho = R/2$ and $\lambda = u_R$ in (6.2.8), then we have

$$\int_{B_{R/2}} |Du|^2 dx \le \frac{C}{R^2} \int_{B_R} (u - u_R)^2 dx.$$

Hence, for $0 < \rho < R/2$,

$$\int_{B_\rho} (u - u_\rho)^2 dx \le C \left(\frac{\rho}{R}\right)^{n+2} \int_{B_R} (u - u_R)^2 dx.$$

Noticing that $g(\lambda) = \int_{B_\rho} (u - \lambda)^2 dx \, (\lambda \in \mathbb{R})$ attains its minimum at $\lambda = u_\rho$, for $R/2 \le \rho \le R$, we have

$$\int_{B_\rho} (u - u_\rho)^2 dx \le \int_{B_\rho} (u - u_R)^2 dx \le \int_{B_R} (u - u_R)^2 dx$$

$$\le 2^{n+2} \left(\frac{\rho}{R}\right)^{n+2} \int_{B_R} (u - u_R)^2 dx.$$

A combination of these inequalities shows that for any $0 < \rho \leq R$, (6.2.16) holds. $\qquad\qquad\qquad\qquad\qquad\qquad\qquad\qquad\qquad\qquad\qquad\qquad\square$

6.2.4 Near boundary estimate for Laplace's equation

Theorem 6.2.4 *Let u be a solution of problem (6.2.1), (6.2.3) with $f \equiv 0$ in B_R^+. Then for any nonnegative integer i and any $0 < \rho \leq R$, there holds*

$$\int_{B_\rho^+} |D^i u|^2 dx \leq C \left(\frac{\rho}{R}\right)^n \int_{B_R^+} |D^i u|^2 dx,$$

where C is a constant depending only on n.

Proof. We proceed to prove the theorem in five cases.

i) The case $i = 0$. The conclusion can be obtained similar to the interior estimate ((6.2.15) in Theorem 6.2.3).

ii) The case $i = 1$. Choose $k > n/2 + 1$. For $0 < \rho < R/2$, from the embedding theorem and Corollary 6.2.5, we can obtain

$$\int_{B_\rho^+} |Du|^2 dx \leq C\rho^n \sup_{B_{R/2}^+} |Du|^2$$

$$\leq C\rho^n \sum_{j=1}^{k} R^{2(j-1)-n} \int_{B_{R/2}^+} |D^j u|^2 dx$$

$$\leq C \left(\frac{\rho}{R}\right)^n R^{-2} \int_{B_R^+} u^2 dx.$$

Since $u\big|_{x_n=0} = 0$, we have

$$\int_{B_R^+} u^2 dx \leq CR^2 \int_{B_R^+} (D_n u)^2 dx \leq CR^2 \int_{B_R^+} |Du|^2 dx.$$

Thus for $0 < \rho < R/2$,

$$\int_{B_\rho^+} |Du|^2 dx \leq C \left(\frac{\rho}{R}\right)^n \int_{B_R^+} |Du|^2 dx.$$

When $R/2 \leq \rho \leq R$, it suffices to take $C \geq 2^n$.

iii) The case $i = 2$. Since for $j = 1, 2, \ldots, n - 1$, $D_j u\big|_{x_n=0} = 0$, we may use the conclusion for $i = 1$ to assert

$$\int_{B_\rho^+} |DD_j u|^2 dx \leq C \left(\frac{\rho}{R}\right)^n \int_{B_R^+} |DD_j u|^2 dx.$$

By virtue of the equation $D_{nn}u = -\sum_{k=1}^{n-1} D_{kk}u$, we further obtain

$$\int_{B_\rho^+} (D_{nn}u)^2 dx \leq C\left(\frac{\rho}{R}\right)^n \sum_{k=1}^{n-1} \int_{B_R^+} |DD_k u|^2 dx$$

$$\leq C\left(\frac{\rho}{R}\right)^n \int_{B_R^+} |D^2 u|^2 dx.$$

Thus

$$\int_{B_\rho^+} |D^2 u|^2 dx \leq C\left(\frac{\rho}{R}\right)^n \int_{B_R^+} |D^2 u|^2 dx.$$

iv) The case $i = 3$. We first use the conclusion for $i = 2$ to assert that for $j = 1, 2, \ldots, n-1$,

$$\int_{B_\rho^+} |D^2 D_j u|^2 dx \leq C\left(\frac{\rho}{R}\right)^n \int_{B_R^+} |D^2 D_j u|^2 dx.$$

By virtue of the equation $D_{nnn}u = -\sum_{k=1}^{n-1} D_{kkn}u$, we further obtain

$$\int_{B_\rho^+} |D_{nnn}u|^2 dx \leq C\left(\frac{\rho}{R}\right)^n \int_{B_R^+} |D^3 u|^2 dx.$$

Thus

$$\int_{B_\rho^+} |D^3 u|^2 dx \leq C\left(\frac{\rho}{R}\right)^n \int_{B_R^+} |D^3 u|^2 dx.$$

v) The case $i > 3$. We may prove by analogy. □

Theorem 6.2.5   *Let u be a solution of problem (6.2.1), (6.2.3) with $f \equiv 0$ in B_R^+. Then for any $0 < \rho \leq R$, there holds*

$$\int_{B_\rho^+} u^2 dx \leq C\left(\frac{\rho}{R}\right)^{n+2} \int_{B_R^+} u^2 dx, \tag{6.2.17}$$

where C is a constant depending only on n.

Proof.   Since $u\big|_{x_n=0} = 0$, from Theorem 6.2.4 we see that for $0 < \rho < R/2$,

$$\int_{B_\rho^+} u^2 dx \leq C\rho^2 \int_{B_\rho^+} (D_n u)^2 dx \leq C\rho^2 \left(\frac{\rho}{R}\right)^n \int_{B_{R/2}^+} |Du|^2 dx.$$

Using Cacccioppoli's inequality (6.2.12), we have

$$\int_{B_{R/2}^+} |Du|^2 dx \leq \frac{C}{R^2} \int_{B_R^+} u^2 dx.$$

Thus for $0 < \rho < R/2$, (6.2.17) holds. For $R/2 \leq \rho \leq R$, (6.2.17) is obvious; it suffices to take $C \geq 2^{n+2}$.   □

6.2.5   *Iteration lemma*

Lemma 6.2.1   *Assume that $\phi(R)$ is a nonnegative and nondecreasing function on $[0, R_0]$, satisfying*

$$\phi(\rho) \leq A \left(\frac{\rho}{R}\right)^\alpha \phi(R) + BR^\beta, \quad 0 < \rho < R \leq R_0,$$

where α, β are constants with $0 < \beta < \alpha$. Then there exists a constant C depending only on A, α and β, such that

$$\phi(\rho) \leq C \left(\frac{\rho}{R}\right)^\beta [\phi(R) + BR^\beta], \quad 0 < \rho < R \leq R_0.$$

Proof.   Let $\nu = \frac{1}{2}(\alpha + \beta)$ and choose $\tau \in (0,1)$ such that $A\tau^{\alpha-\nu} \leq 1$. Then

$$\begin{aligned}
\phi(\tau R) \leq &A\tau^\alpha \phi(R) + BR^\beta \\
= &A\tau^{\alpha-\nu}\tau^\nu \phi(R) + BR^\beta \leq \tau^\nu \phi(R) + BR^\beta,
\end{aligned}$$

$$\begin{aligned}
\phi(\tau^2 R) \leq &\tau^\nu \phi(\tau R) + B\tau^\beta R^\beta \\
\leq &\tau^{2\nu} \phi(R) + B(\tau^\nu + \tau^\beta)R^\beta,
\end{aligned}$$

$$\begin{aligned}
\phi(\tau^3 R) \leq &\tau^\nu \phi(\tau^2 R) + B\tau^{2\beta} R^\beta \\
\leq &\tau^{3\nu} \phi(R) + B(\tau^{2\nu} + \tau^{\nu+\beta} + \tau^{2\beta})R^\beta,
\end{aligned}$$

$$\cdots \quad \cdots$$

$$\begin{aligned}
\phi(\tau^{k+1} R) \leq &\tau^{(k+1)\nu} \phi(R) + B \left(\tau^{k\nu} + \tau^{(k-1)\nu+\beta} + \cdots + \tau^{k\beta}\right) R^\beta \\
= &\tau^{(k+1)\nu} \phi(R) + B\tau^{k\beta} \left(\tau^{k(\nu-\beta)} + \tau^{(k-1)(\nu-\beta)} + \cdots + 1\right) R^\beta \\
= &\tau^{(k+1)\nu} \phi(R) + B\frac{\tau^{k\beta}(1 - \tau^{(k+1)(\nu-\beta)})}{1 - \tau^{\nu-\beta}} R^\beta \\
\leq &C_1 \tau^{k\beta}[\phi(R) + BR^\beta],
\end{aligned}$$

where $C_1 \geq 1$ is a constant independent of k. Thus

$$\phi(\tau^k R) \leq C_1 \tau^{(k-1)\beta}[\phi(R) + BR^\beta], \quad \forall k \geq 0.$$

For any fixed $0 < \rho < R \leq R_0$, choose a nonnegative integer k, such that $\tau^{k+1}R < \rho \leq \tau^k R$. Then

$$\phi(\rho) \leq \phi(\tau^k R) \leq C_1 \tau^{(k-1)\beta}[\phi(R) + BR^\beta]$$

$$\leq C_1 \tau^{-2\beta} \left(\frac{\rho}{R}\right)^\beta [\phi(R) + BR^\beta]$$

$$= C \left(\frac{\rho}{R}\right)^\beta [\phi(R) + BR^\beta].$$

$\square$

6.2.6 Interior estimate for Poisson's equation

Theorem 6.2.6 *Let u be a solution of equation (6.2.1) in B_{R_0} and $w = D_i u$ $(i = 1, 2, \ldots, n)$. Then for any $0 < \rho \leq R \leq R_0$, there holds*

$$\frac{1}{\rho^{n+2\alpha}} \int_{B_\rho} |Dw - (Dw)_\rho|^2 dx \leq \frac{C}{R^{n+2\alpha}} \int_{B_R} |Dw - (Dw)_R|^2 dx + C[f]_{\alpha;B_R}^2,$$

where C is a constant depending only on n.

Proof. Decompose w as follows: $w = w_1 + w_2$ with w_1 and w_2 satisfying

$$\begin{cases} -\Delta w_1 = 0, & \text{in } B_R, \\ w_1\big|_{\partial B_R} = w, \end{cases}$$

and

$$\begin{cases} -\Delta w_2 = D_i f = D_i(f - f_R), & \text{in } B_R, \\ w_2\big|_{\partial B_R} = 0. \end{cases}$$

Now we apply (6.2.16) to Dw_1 to obtain

$$\int_{B_\rho} |Dw_1 - (Dw_1)_\rho|^2 dx \leq C \left(\frac{\rho}{R}\right)^{n+2} \int_{B_R} |Dw_1 - (Dw_1)_R|^2 dx.$$

Thus for any $0 < \rho \leq R \leq R_0$, we have

$$\int_{B_\rho} |Dw - (Dw)_\rho|^2 dx$$

$$\leq 2 \int_{B_\rho} |Dw_1 - (Dw_1)_\rho|^2 dx + 2 \int_{B_\rho} |Dw_2 - (Dw_2)_\rho|^2 dx$$

$$\leq C \left(\frac{\rho}{R}\right)^{n+2} \int_{B_R} |Dw_1 - (Dw_1)_R|^2 dx + 2 \int_{B_R} |Dw_2 - (Dw_2)_R|^2 dx$$

$$\leq C \left(\frac{\rho}{R}\right)^{n+2} \int_{B_R} |Dw - (Dw)_R|^2 dx + C \int_{B_R} |Dw_2 - (Dw_2)_R|^2 dx$$

$$\leq C \left(\frac{\rho}{R}\right)^{n+2} \int_{B_R} |Dw - (Dw)_R|^2 dx + C \int_{B_R} |Dw_2|^2 dx.$$

Multiply the equation for w_2 by w_2 and integrate over B_R and note that $w_2\big|_{\partial B_R} = 0$. Then we deduce

$$\int_{B_R} |Dw_2|^2 dx = -\int_{B_R} w_2 \Delta w_2 dx$$

$$= \int_{B_R} w_2 D_i (f - f_R) dx$$

$$= -\int_{B_R} (f - f_R) D_i w_2 dx$$

$$\leq \frac{1}{2} \int_{B_R} |Dw_2|^2 dx + \frac{1}{2} \int_{B_R} (f - f_R)^2 dx.$$

Thus

$$\int_{B_R} |Dw_2|^2 dx \leq \int_{B_R} (f - f_R)^2 dx \leq CR^{n+2\alpha} [f]^2_{\alpha;B_R}. \tag{6.2.18}$$

Hence

$$\int_{B_\rho} |Dw - (Dw)_\rho|^2 dx$$

$$\leq C \left(\frac{\rho}{R}\right)^{n+2} \int_{B_R} |Dw - (Dw)_R|^2 dx + CR^{n+2\alpha} [f]^2_{\alpha;B_R}.$$

Using the iteration lemma (Lemma 6.2.1) we finally obtain

$$\int_{B_\rho} |Dw - (Dw)_\rho|^2 dx$$

$$\leq C \left(\frac{\rho}{R}\right)^{n+2\alpha} \left(\int_{B_R} |Dw - (Dw)_R|^2 dx + CR^{n+2\alpha} [f]^2_{\alpha;B_R} \right).$$

$\square$

Theorem 6.2.7   *Let u be a solution of equation (6.2.1) in B_R and $w = D_i u$ $(i = 1, 2, \ldots, n)$. Then for any $0 < \rho \le \dfrac{R}{2}$, there holds*

$$\int_{B_\rho} |Dw - (Dw)_\rho|^2 dx \le C\rho^{n+2\alpha} M_R, \qquad (6.2.19)$$

where C is a constant depending only on n and

$$M_R = \frac{1}{R^{4+2\alpha}} |u|^2_{0;B_R} + \frac{1}{R^{2\alpha}} |f|^2_{0;B_R} + [f]^2_{\alpha;B_R}.$$

Proof.   According to Theorem 6.2.6 and Corollary 6.2.1, we have

$$\int_{B_\rho} |Dw - (Dw)_\rho|^2 dx$$

$$\le C\rho^{n+2\alpha} \left(\frac{1}{R^{n+2\alpha}} \int_{B_{R/2}} |Dw - (Dw)_{R/2}|^2 dx + [f]^2_{\alpha;B_{R/2}} \right)$$

$$\le C\rho^{n+2\alpha} \left(\frac{1}{R^{n+2\alpha}} \int_{B_{R/2}} |Dw|^2 dx + [f]^2_{\alpha;B_{R/2}} \right)$$

$$\le C\rho^{n+2\alpha} \left(\frac{1}{R^{n+4+2\alpha}} \int_{B_R} u^2 dx + \frac{1}{R^{2\alpha}} |f|^2_{0;B_R} + [f]^2_{\alpha;B_R} \right),$$

from which the conclusion of Theorem 6.2.7 follows. $\square$

Theorem 6.2.8   *Let u be a solution of equation (6.2.1) in B_R. Then*

$$[D^2 u]_{\alpha;B_{R/2}} \le C \left(\frac{1}{R^{2+\alpha}} |u|_{0;B_R} + \frac{1}{R^\alpha} |f|_{0;B_R} + [f]_{\alpha;B_R} \right), \qquad (6.2.20)$$

where C is a constant depending only on n.

Proof.   According to Theorem 6.2.7, for $x \in B_{R/2}$, $0 < \rho \le \dfrac{R}{4}$, we have

$$\int_{B_\rho(x) \cap B_{R/2}} |D^2 u(y) - (D^2 u)_{B_\rho(x) \cap B_{R/2}}|^2 dy$$

$$\le \int_{B_\rho(x)} |D^2 u(y) - (D^2 u)_{x,\rho}|^2 dy$$

$$\le C\rho^{n+2\alpha} \left(\frac{1}{R^{4+2\alpha}} |u|^2_{0;B_{R/2}(x)} + \frac{1}{R^{2\alpha}} |f|^2_{0;B_{R/2}(x)} + [f]^2_{\alpha;B_{R/2}(x)} \right)$$

$$\le C\rho^{n+2\alpha} \left(\frac{1}{R^{4+2\alpha}} |u|^2_{0;B_R} + \frac{1}{R^{2\alpha}} |f|^2_{0;B_R} + [f]^2_{\alpha;B_R} \right),$$

where

$$(D^2 u)_{B_\rho(x) \cap B_{R/2}} = \frac{1}{|B_\rho(x) \cap B_{R/2}|} \int_{B_\rho(x) \cap B_{R/2}} D^2 u(y) dy.$$

Hence

$$[D^2 u]_{2,n+2\alpha;B_{R/2}}^{(1/4)} \leq C \left(\frac{1}{R^{4+2\alpha}} |u|_{0;B_R}^2 + \frac{1}{R^{2\alpha}} |f|_{0;B_R}^2 + [f]_{\alpha;B_R}^2 \right)^{1/2}$$

$$\leq C \left(\frac{1}{R^{2+\alpha}} |u|_{0;B_R} + \frac{1}{R^\alpha} |f|_{0;B_R} + [f]_{\alpha;B_R} \right)$$

and (6.2.20) follows by using Remark 6.1.2. □

6.2.7 Near boundary estimate for Poisson's equation

Theorem 6.2.9 *Let u be a solution of problem (6.2.1), (6.2.3) in $B_{R_0}^+$, $w = D_i u$ $(i = 1, 2, \ldots, n-1)$. Then for any $0 < \rho \leq R \leq R_0$, there holds*

$$\frac{1}{\rho^{n+2\alpha}} \int_{B_\rho^+} \left(\sum_{j=1}^{n-1} |D_j w|^2 + |D_n w - (D_n w)_\rho|^2 \right) dx$$

$$\leq \frac{C}{R^{n+2\alpha}} \int_{B_R^+} \left(\sum_{j=1}^{n-1} |D_j w|^2 + |D_n w - (D_n w)_R|^2 \right) dx$$

$$+ C[f]_{\alpha;B_R^+}^2, \tag{6.2.21}$$

where $v_\rho = \frac{1}{|B_\rho^+|} \int_{B_\rho^+} v(x) dx$ and C is a constant depending only on n.

Proof. Decompose w as follows: $w = w_1 + w_2$ with w_1 and w_2 satisfying

$$\begin{cases} -\Delta w_1 = 0, & \text{in } B_R^+, \\ w_1 \big|_{\partial B_R^+} = w \end{cases}$$

and

$$\begin{cases} -\Delta w_2 = D_i f = D_i(f - f_R), & \text{in } B_R^+, \\ w_2 \big|_{\partial B_R^+} = 0. \end{cases}$$

For $j = 1, 2, \ldots, n-1$, from Theorem 6.2.5, we have

$$\int_{B_\rho^+} |D_j w_1|^2 dx \leq C \left(\frac{\rho}{R}\right)^{n+2} \int_{B_R^+} |D_j w_1|^2 dx, \quad 0 < \rho \leq R \leq R_0.$$

Hence, for any $0 < \rho \leq R \leq R_0$,

$$\int_{B_\rho^+} |D_j w|^2 dx$$

$$\leq 2 \int_{B_\rho^+} |D_j w_1|^2 dx + 2 \int_{B_\rho^+} |D_j w_2|^2 dx$$

$$\leq C \left(\frac{\rho}{R}\right)^{n+2} \int_{B_R^+} |D_j w_1|^2 dx + 2 \int_{B_R^+} |D_j w_2|^2 dx$$

$$\leq C \left(\frac{\rho}{R}\right)^{n+2} \int_{B_R^+} |D_j w|^2 dx + C \int_{B_R^+} |D_j w_2|^2 dx$$

$$\leq C \left(\frac{\rho}{R}\right)^{n+2} \int_{B_R^+} |D_j w|^2 dx + C R^{n+2\alpha} [f]_{\alpha; B_R^+}^2, \tag{6.2.22}$$

where we have used the estimate

$$\int_{B_R^+} |D_j w_2|^2 dx \leq \int_{B_R^+} |D w_2|^2 dx \leq C R^{n+2\alpha} [f]_{\alpha; B_R^+}^2,$$

whose proof is similar to (6.2.18).

Thus for $0 < \rho < R/2$, we obtain by using Poincaré's inequality and Theorem 6.2.4,

$$\int_{B_\rho^+} |D_n w - (D_n w)_\rho|^2 dx$$

$$\leq 2 \int_{B_\rho^+} |D_n w_1 - (D_n w_1)_\rho|^2 dx + 2 \int_{B_\rho^+} |D_n w_2 - (D_n w_2)_\rho|^2 dx$$

$$\leq C \rho^2 \int_{B_\rho^+} |DD_n w_1|^2 dx + C \int_{B_\rho^+} |D_n w_2|^2 dx$$

$$\leq C \rho^2 \left(\frac{\rho}{R}\right)^n \int_{B_{R/2}^+} |D^2 w_1|^2 dx + C \int_{B_R^+} |D w_2|^2 dx.$$

Using the equation $D_{nn} w_1 = \sum_{j=1}^{n-1} D_{jj} w_1$ and Caccioppoli's inequality, we further obtain

$$\int_{B_{R/2}^+} |D^2 w_1|^2 dx$$

$$\leq \int_{B_{R/2}^+} |D_{nn} w_1|^2 dx + 2 \sum_{j=1}^{n-1} \int_{B_{R/2}^+} |DD_j w_1|^2 dx$$

$$\leq C \sum_{j=1}^{n-1} \int_{B_{R/2}^+} |DD_j w_1|^2 dx$$

$$\leq C \sum_{j=1}^{n-1} \frac{1}{R^2} \int_{B_R^+} |D_j w_1|^2 dx$$

$$\leq \frac{C}{R^2} \sum_{j=1}^{n-1} \int_{B_R^+} |D_j w|^2 dx + \frac{C}{R^2} \sum_{j=1}^{n-1} \int_{B_R^+} |D_j w_2|^2 dx.$$

Thus, for $0 < \rho < R/2$, we have

$$\int_{B_\rho^+} |D_n w - (D_n w)_\rho|^2 dx$$

$$\leq C \left(\frac{\rho}{R}\right)^{n+2} \sum_{j=1}^{n-1} \int_{B_R^+} |D_j w|^2 dx + C R^{n+2\alpha} [f]_{\alpha;B_R^+}^2. \tag{6.2.23}$$

Combining (6.2.23) with (6.2.22) we see that for $0 < \rho < R/2$,

$$\int_{B_\rho^+} \left(\sum_{j=1}^{n-1} |D_j w|^2 + |D_n w - (D_n w)_\rho|^2 \right) dx$$

$$\leq C \left(\frac{\rho}{R}\right)^{n+2} \int_{B_R^+} \left(\sum_{j=1}^{n-1} |D_j w|^2 + |D_n w - (D_n w)_R|^2 \right) dx$$

$$+ C R^{n+2\alpha} [f]_{\alpha;B_R^+}^2.$$

Therefore (6.2.21) follows for $0 < \rho < R/2$ by using Lemma 6.2.1. For $\frac{R}{2} \leq \rho \leq R$, (6.2.21) obviously holds. The proof is complete. $\qquad\square$

Theorem 6.2.10 *Let u be a solution of problem (6.2.1), (6.2.3) in B_R^+ and $w = D_i u \, (i = 1, 2, \ldots, n)$. Then for any $0 < \rho \leq \frac{R}{2}$, there holds*

$$\int_{B_\rho^+} |Dw - (Dw)_\rho|^2 dx \leq C \rho^{n+2\alpha} M_R, \tag{6.2.24}$$

where C is a constant depending only on n and

$$M_R = \frac{1}{R^{4+2\alpha}} |u|_{0;B_R^+}^2 + \frac{1}{R^{2\alpha}} |f|_{0;B_R^+}^2 + [f]_{\alpha;B_R^+}^2.$$

Proof.   From Theorem 6.2.9 and Corollary 6.2.4, we have

$$\int_{B_\rho^+} \Big(\sum_{j=1}^{n-1} |D_j w|^2 + |D_n w - (D_n w)_\rho|^2 \Big) dx$$

$$\leq C\rho^{n+2\alpha} \Big(\frac{1}{R^{n+2\alpha}} \int_{B_{R/2}^+} \Big(\sum_{j=1}^{n-1} |D_j w|^2 + |D_n w - (D_n w)_{R/2}|^2 \Big) dx$$

$$+ [f]^2_{\alpha; B_{R/2}^+} \Big)$$

$$\leq C\rho^{n+2\alpha} \Big(\frac{1}{R^{n+2\alpha}} \int_{B_{R/2}^+} |Dw|^2 dx + [f]^2_{\alpha; B_{R/2}^+} \Big)$$

$$\leq C\rho^{n+2\alpha} \Big(\frac{1}{R^{n+4+2\alpha}} \int_{B_R^+} u^2 dx + \frac{1}{R^{2\alpha}} |f|^2_{0; B_R^+} + [f]^2_{\alpha; B_R^+} \Big) \qquad (6.2.25)$$

which implies, in particular, that for $j = 1, 2, \cdots, n-1$,

$$\int_{B_\rho^+} |D_j w|^2 dx \leq C\rho^{n+2\alpha} M_R.$$

Hence

$$(D_j w)_\rho^2 = \frac{1}{|B_\rho^+|^2} \Big(\int_{B_\rho^+} |D_j w| dx \Big)^2 \leq \frac{1}{|B_\rho^+|} \int_{B_\rho^+} |D_j w|^2 dx \leq C\rho^{2\alpha} M_R.$$

Thus, for $j = 1, 2, \cdots, n-1$,

$$\int_{B_\rho^+} |D_j w - (D_j w)_\rho|^2 dx$$

$$\leq 2 \int_{B_\rho^+} |D_j w|^2 dx + 2 \int_{B_\rho^+} |(D_j w)_\rho|^2 dx$$

$$\leq C\rho^{n+2\alpha} M_R.$$

Moreover, (6.2.25) implies

$$\int_{B_\rho^+} |D_n w - (D_n w)_\rho|^2 dx \leq C\rho^{n+2\alpha} M_R.$$

Thus

$$\int_{B_\rho^+} |Dw - (Dw)_\rho|^2 dx \leq C\rho^{n+2\alpha} M_R$$

and we have proved (6.2.24) for $w = D_i u$ $(i = 1, 2, \cdots, n-1)$. Again using the equation $D_{nn} u = -\sum_{i=1}^{n-1} D_{ii} u - f$, we derive (6.2.24) for $w = D_n u$. $\qquad\square$

Theorem 6.2.11 *Let u be a solution of problem (6.2.1), (6.2.3) in B_R^+. Then*

$$[D^2 u]_{\alpha; B_{R/2}^+(x^0)}$$

$$\leq C \left(\frac{1}{R^{2+\alpha}} |u|_{0; B_R^+(x^0)} + \frac{1}{R^\alpha} |f|_{0; B_R^+(x^0)} + [f]_{\alpha; B_R^+(x^0)} \right), \qquad (6.2.26)$$

where C is a constant depending only on n.

Proof. According to Theorem 6.2.10, for $x \in \partial B_{R/2}^+(x^0) \cap B_{R/2}(x^0)$, $0 < \rho \leq \dfrac{R}{4}$, we have

$$\int_{B_\rho^+(x)} |D^2 u(y) - (D^2 u)_{B_\rho^+(x)}|^2 dy$$

$$\leq C \rho^{n+2\alpha} \left(\frac{1}{R^{4+2\alpha}} |u|_{0; B_{R/2}^+(x)}^2 + \frac{1}{R^{2\alpha}} |f|_{0; B_{R/2}^+(x)}^2 + [f]_{\alpha; B_{R/2}^+(x)}^2 \right)$$

$$\leq C \rho^{n+2\alpha} M_R, \qquad (6.2.27)$$

where

$$M_R = \frac{1}{R^{4+2\alpha}} |u|_{0; B_R^+(x^0)}^2 + \frac{1}{R^{2\alpha}} |f|_{0; B_R^+(x^0)}^2 + [f]_{\alpha; B_R^+(x^0)}^2.$$

Let $x \in B_{R/2}^+(x^0)$. Denote $\tilde{x} = (x_1, \cdots, x_{n-1}, 0)$. If $0 < x_n < \dfrac{R}{4}$, $x_n \leq \rho \leq \dfrac{R}{4}$, then $B_\rho(x) \cap B_{R/2}^+(x^0) \subset B_{2\rho}^+(\tilde{x})$. Thus from (6.2.27), we have

$$\int_{B_\rho(x) \cap B_{R/2}^+(x^0)} |D^2 u(y) - (D^2 u)_{B_\rho(x) \cap B_{R/2}^+(x^0)}|^2 dy$$

$$\leq \int_{B_{2\rho}^+(\tilde{x})} |D^2 u(y) - (D^2 u)_{B_{2\rho}^+(\tilde{x})}|^2 dy$$

$$\leq C \rho^{n+2\alpha} M_R. \qquad (6.2.28)$$

If $0 < x_n < \dfrac{R}{4}$, $0 < \rho < x_n$, then by Theorem 6.2.6 and (6.2.28) for $\rho = x_n$, we derive

$$\int_{B_\rho(x)} |D^2 u(y) - (D^2 u)_{B_\rho(x)}|^2 dy$$

$$\leq C\rho^{n+2\alpha}\left(\frac{1}{x_n^{n+2\alpha}}\int_{B_{x_n}(x)}|Dw-(Dw)_{B_{x_n}(x)}|^2dx+[f]_{\alpha;B_{x_n}(x)}^2\right)$$

$$\leq C\rho^{n+2\alpha}M_R.$$

Summing up, we have proved that if $0<x_n<\dfrac{R}{4}$, then for $0<\rho\leq\dfrac{R}{4}$,

$$\int_{B_\rho(x)\cap B_{R/2}^+(x^0)}|D^2u(y)-(D^2u)_{B_\rho(x)\cap B_{R/2}^+(x^0)}|^2dy\leq C\rho^{n+2\alpha}M_R.$$

On the other hand, if $\dfrac{R}{4}\leq x_n<\dfrac{R}{2}$, $0<\rho\leq\dfrac{R}{4}$, then $B_\rho(x)\subset B_{R/4}(x)\subset B_{3R/4}^+(x^0)\subset B_R^+(x^0)$ and hence, by Theorem 6.2.7,

$$\int_{B_\rho(x)\cap B_{R/2}^+(x^0)}|D^2u(y)-(D^2u)_{B_\rho(x)\cap B_{R/2}^+(x^0)}|^2dy$$

$$\leq\int_{B_\rho(x)}|D^2u(y)-(D^2u)_{B_\rho(x)}|^2dy$$

$$\leq C\rho^{n+2\alpha}M_R.$$

Thus

$$[D^2u]_{2,n+2\alpha;B_{R/2}^+(x^0)}^{(1/4)}\leq CM_R^{1/2}$$

$$\leq C\left(\frac{1}{R^{2+\alpha}}|u|_{0;B_R^+(x^0)}+\frac{1}{R^\alpha}|f|_{0;B_R^+(x^0)}+[f]_{\alpha;B_R^+(x^0)}\right)$$

and (6.2.26) follows by using Remark 6.1.2.   □

Remark 6.2.8 *If the boundary of the domain considered is not the superplane $x_n=0$, but a superplane of other form, we still have the same near boundary estimate. Of course, in proving we need to replace $w=D_iu\,(i=1,2,\cdots,n-1)$ by the tangential derivatives with respect to the superplane.*

Remark 6.2.9 *Using the interior estimate and the near boundary estimate after local flatting of the boundary, we can obtain the global Schauder's estimate on Ω. However, it is to be noted that, after local flatting of the boundary, Poisson's equation will be changed into another elliptic equation. We will discuss how to establish the global estimates for general linear elliptic equations in the next section.*

6.3 Schauder's Estimates for General Linear Elliptic Equations

Consider the general linear elliptic equation

$$Lu \equiv -a_{ij}(x)D_{ij}u + b_i(x)D_iu + c(x)u = f(x), \quad x \in \Omega, \qquad (6.3.1)$$

where $\Omega \subset \mathbb{R}^n$ is a bounded domain. We merely study the Dirichlet problem, namely, the problem with boundary value condition

$$u\Big|_{\partial\Omega} = \varphi(x). \qquad (6.3.2)$$

The purpose of this section is to establish Schauder's estimates for solutions of problem (6.3.1), (6.3.2) under certain conditions. We have the following theorem.

Theorem 6.3.1 *Assume that $0 < \alpha < 1$, $\partial\Omega \in C^{2,\alpha}$, $a_{ij}, b_i, c \in C^\alpha(\overline{\Omega})$, $a_{ij} = a_{ji}$, and equation (6.3.1) satisfies the uniform ellipticity condition, namely, there exist constants λ, Λ with $0 < \lambda \le \Lambda$, such that*

$$\lambda|\xi|^2 \le a_{ij}(x)\xi_i\xi_j \le \Lambda|\xi|^2, \quad \forall \xi \in \mathbb{R}^n, \ x \in \Omega.$$

In addition, assume $f \in C^\alpha(\overline{\Omega})$, $\varphi \in C^{2,\alpha}(\overline{\Omega})$. If $u \in C^{2,\alpha}(\overline{\Omega})$ is a solution of problem (6.3.1), (6.3.2), then

$$|u|_{2,\alpha;\Omega} \le C(|f|_{\alpha;\Omega} + |\varphi|_{2,\alpha;\Omega} + |u|_{0;\Omega}), \qquad (6.3.3)$$

where C is a constant depending only on n, α, λ, Λ, Ω and the $C^\alpha(\overline{\Omega})$ norms of a_{ij}, b_i, c.

Remark 6.3.1 *It is to be noted that in Theorem 6.3.1, in order (6.3.3) holds, we need not require $u \in C^{2,\alpha}(\overline{\Omega})$. In fact, $u \in C^{2,\alpha}(\Omega) \cap C(\overline{\Omega})$ is enough. Under such condition, we have*

$$|u|_{2,\alpha;\Omega_\varepsilon} \le C(|f|_{\alpha;\Omega} + |\varphi|_{2,\alpha;\Omega} + |u|_{0;\Omega}),$$

where

$$\Omega_\varepsilon = \{x \in \Omega; \operatorname{dist}(x, \partial\Omega) > \varepsilon\}$$

with $\varepsilon > 0$ small enough. From this we finally obtain (6.3.3) by letting $\varepsilon \to 0$.

The proof of Theorem 6.3.1 will be completed by means of simplifying the problem and applying Schauder's estimates for solutions of Poisson's equation and the finite covering argument.

6.3.1 *Simplification of the problem*

First of all, we observe that in establishing the a priori estimate (6.3.3) for equation (6.3.1), without loss of generality, we may assume $\varphi \equiv 0$. In fact, in the general case, we may consider the function $w = u - \varphi$, which satisfies $w\big|_{\partial\Omega} = 0$ and

$$Lw = Lu - L\varphi = f(x) - L\varphi(x) \in C^{\alpha}(\overline{\Omega}).$$

If we have established the estimate (6.3.3) for the special case $\varphi \equiv 0$, then applying it to the above problem gives

$$|w|_{2,\alpha;\Omega} \leq C(|f - L\varphi|_{\alpha;\Omega} + |w|_{0;\Omega}),$$

from which (6.3.3) follows immediately.

Next, we point out that it suffices to prove (6.3.3) for the equation without terms of lower order, namely, the equation of the special form

$$-a_{ij}(x)D_{ij}u = f(x), \quad x \in \Omega. \tag{6.3.4}$$

In fact, if we can prove (6.3.3) for equation (6.3.4) with $\varphi \equiv 0$, then applying it to equation (6.3.1) gives

$$|u|_{2,\alpha;\Omega} \leq C(|f - b_i D_i u - cu|_{\alpha;\Omega} + |u|_{0;\Omega}) \leq C(|f|_{\alpha;\Omega} + |u|_{1,\alpha;\Omega}),$$

and by the interpolation inequality, we obtain

$$|u|_{2,\alpha;\Omega} \leq C(|f|_{\alpha;\Omega} + |u|_{0;\Omega}). \tag{6.3.5}$$

The above discussion shows that we need merely to prove estimate (6.3.3) for the special equation (6.3.4) with the special boundary condition $u\big|_{\partial\Omega} = 0$, namely, to prove the estimate (6.3.5).

6.3.2 *Interior estimate*

We will prove the estimate (6.3.5) by means of the so-called method of solidifying coefficients. The basic idea is to fix a point $x^0 \in \Omega$ and treat (6.3.4) as an equation with constant coefficients

$$-a_{ij}(x^0)D_{ij}u = h(x), \tag{6.3.6}$$

where

$$h(x) = f(x) + g(x), \tag{6.3.7}$$

$$g(x) = (a_{ij}(x) - a_{ij}(x^0))D_{ij}u.$$

In order to estimate the solutions of (6.3.6) for a given smooth function $h(x)$, we will change the variables to further simplify (6.3.6) to the form of Poisson's equation so that we can apply the results obtained in §6.2.

Since $A = (a_{ij}(x^0))$ is a positive definite matrix, there exists a nonsingular matrix P, such that $P^T A P = I_n$, where I_n is an $n \times n$ unit matrix. Let

$$y = P^T x = (P_{ij})x,$$

where we regard the variables x, y as column vectors. Then

$$\frac{\partial u}{\partial x_i} = \frac{\partial \hat{u}}{\partial y_k} \cdot \frac{\partial y_k}{\partial x_i} = P_{ki}\frac{\partial \hat{u}}{\partial y_k} = P_{ki}\hat{D}_k\hat{u},$$

$$\frac{\partial^2 u}{\partial x_i \partial x_j} = P_{ki}\frac{\partial^2 \hat{u}}{\partial y_k \partial y_l} \cdot \frac{\partial y_l}{\partial x_j} = P_{ki}P_{lj}\hat{D}_{kl}\hat{u},$$

where $\hat{D}_k = \dfrac{\partial}{\partial y_k}$, $\hat{D}_{kl} = \dfrac{\partial^2}{\partial y_k \partial y_l}$, $\hat{u}(y) = u((P^T)^{-1}y)$. Hence

$$-a_{ij}(x^0)D_{ij}u = -a_{ij}(x^0)P_{ki}P_{lj}\hat{D}_{kl}\hat{u}.$$

Since $P^T A P = I_n$, we have

$$a_{ij}(x^0)P_{ki}P_{lj} = \delta_{kl} = \begin{cases} 1, & k = l, \\ 0, & k \neq l. \end{cases}$$

Thus equation (6.3.6) becomes

$$-\hat{\Delta}\hat{u} = \hat{h}(y), \tag{6.3.8}$$

where $\hat{\Delta} = \dfrac{\partial^2}{\partial y_1^2} + \dfrac{\partial^2}{\partial y_2^2} + \cdots + \dfrac{\partial^2}{\partial y_n^2}$, $\hat{h}(y) = h((P^T)^{-1}y)$.

We may assert

$$\Lambda^{-1/2}|x^1 - x^2| \leq |P^T x^1 - P^T x^2| \leq \lambda^{-1/2}|x^1 - x^2|, \tag{6.3.9}$$

namely, the distance in the x-space is equivalent to that in y-space. In fact

$$|y| = (x^T P P^T x)^{1/2} = (x^T A^{-1} x)^{1/2}.$$

If we denote by $\underline{\lambda}$ and $\overline{\lambda}$ the minimal and maximal eigenvalues of A then $\overline{\lambda}^{-1}$ and $\underline{\lambda}^{-1}$ are the minimal and maximal eigenvalues of A^{-1}. Hence

$$\overline{\lambda}^{-1/2}|x| \le |y| \le \underline{\lambda}^{-1/2}|x|. \tag{6.3.10}$$

From the ellipticity condition we have $\lambda \le \underline{\lambda} \le \overline{\lambda} \le \Lambda$. Thus (6.3.10) implies (6.3.9).

Now we proceed to apply the interior estimate obtained in §6.2 (Theorem 6.2.8) to equation (6.3.8). Thus, we obtain the following estimate

$$[\hat{D}^2\hat{u}]_{\alpha;\hat{B}_{R/2}} \le C\left(\frac{1}{R^{2+\alpha}}|\hat{u}|_{0;\hat{B}_R} + \frac{1}{R^\alpha}|\hat{h}|_{0;\hat{B}_R} + [\hat{h}]_{\alpha;\hat{B}_R}\right), \tag{6.3.11}$$

where $\hat{B}_R$ denotes the ball in the y-space of radius R centered at $P^T x^0$ such that $\hat{B}_R \subset \hat{\Omega} = \{y = P^T x; x \in \Omega\}$. We assume $0 < R \le 1$.

Now we return to (6.3.6) with $h(x)$ given by (6.3.7). Since

$$[\hat{g}]_{\alpha;\hat{B}_R} \le CR^\alpha[\hat{D}^2\hat{u}]_{\alpha;\hat{B}_R} + [\hat{a}_{ij}]_{\alpha;\hat{B}_R}|\hat{D}_{ij}\hat{u}|_{0;\hat{B}_R}$$
$$\le C(R^\alpha[\hat{D}^2\hat{u}]_{\alpha;\hat{B}_R} + |\hat{u}|_{2;\hat{B}_R}),$$

$$|\hat{g}|_{0;\hat{B}_R} \le CR^\alpha|\hat{D}^2\hat{u}|_{0;\hat{B}_R} \le CR^\alpha|\hat{u}|_{2;\hat{B}_R},$$

by virtue of the interpolation inequality and noticing that $0 < R \le 1$, we obtain

$$[\hat{g}]_{\alpha;\hat{B}_R} \le C\left(R^\alpha[\hat{D}^2\hat{u}]_{\alpha;\hat{B}_R} + \frac{1}{R^2}|\hat{u}|_{0;\hat{B}_R}\right)$$
$$\le C\left(R^\alpha[\hat{D}^2\hat{u}]_{\alpha;\hat{B}_R} + \frac{1}{R^{2+\alpha}}|\hat{u}|_{0;\hat{B}_R}\right),$$

$$|\hat{g}|_{0;\hat{B}_R} \le CR^\alpha\left(R^\alpha[\hat{D}^2\hat{u}]_{\alpha;\hat{B}_R} + \frac{1}{R^2}|\hat{u}|_{0;\hat{B}_R}\right)$$
$$\le CR^\alpha\left(R^\alpha[\hat{D}^2\hat{u}]_{\alpha;\hat{B}_R} + \frac{1}{R^{2+\alpha}}|\hat{u}|_{0;\hat{B}_R}\right).$$

This combined with (6.3.11) leads to

$$[\hat{D}^2\hat{u}]_{\alpha;\hat{B}_{R/2}} \le C\Big(R^\alpha[\hat{D}^2\hat{u}]_{\alpha;\hat{B}_R} + \frac{1}{R^{2+\alpha}}|\hat{u}|_{0;\hat{B}_R}$$
$$+ \frac{1}{R^\alpha}|\hat{f}|_{0;\hat{B}_R} + [\hat{f}]_{\alpha;\hat{B}_R}\Big).$$

Returning to the original variable x, we derive

$$[D^2u]_{\alpha;\hat{B}_{R/2}^{-1}} \leq C\Big(R^\alpha[D^2u]_{\alpha;\hat{B}_R^{-1}} + \frac{1}{R^{2+\alpha}}|u|_{0;\hat{B}_R^{-1}}$$
$$+ \frac{1}{R^\alpha}|f|_{0;\hat{B}_R^{-1}} + [f]_{\alpha;\hat{B}_R^{-1}}\Big),$$

where $\hat{B}_R^{-1} = \{x = (P^T)^{-1}y; y \in \hat{B}_R\}$. Here it should be noted that (6.3.9) ensures

$$[D^2u]_{\alpha;\hat{B}_{R/2}^{-1}} \leq C[\hat{D}^2\hat{u}]_{\alpha;\hat{B}_{R/2}}, \quad [\hat{D}^2\hat{u}]_{\alpha;\hat{B}_R} \leq C[D^2u]_{\alpha;\hat{B}_R^{-1}},$$
$$|\hat{u}|_{0;\hat{B}_R} \leq C|u|_{0;\hat{B}_R^{-1}}, \quad |\hat{f}|_{0;\hat{B}_R} \leq C|f|_{0;\hat{B}_R^{-1}}, \quad [\hat{f}]_{\alpha;\hat{B}_R} \leq C[f]_{\alpha;\hat{B}_R^{-1}}$$

with the constant C independent of $x^0 \in \Omega$. In particular, we have

$$[D^2u]_{\alpha;\hat{B}_{R/2}^{-1}} \leq C\Big(R^\alpha[D^2u]_{\alpha;\Omega} + \frac{1}{R^{2+\alpha}}|u|_{0;\Omega} + \frac{1}{R^\alpha}|f|_{0;\Omega} + [f]_{\alpha;\Omega}\Big).$$

It is easy to check $B_{\lambda^{1/2}R} \subset \hat{B}_R^{-1}$. Thus for $0 < R \leq 1$ small enough, we have

$$[D^2u]_{\alpha;B_{2R}} \leq C\Big(R^\alpha[D^2u]_{\alpha;\Omega} + \frac{1}{R^{2+\alpha}}|u|_{0;\Omega}$$
$$+ \frac{1}{R^\alpha}|f|_{0;\Omega} + [f]_{\alpha;\Omega}\Big) \qquad (6.3.12)$$

with another constant C.

6.3.3 Near boundary estimate

To establish the near boundary estimate we adopt the local platting technique of the boundary. Let $x^0 \in \partial\Omega$. Since $\partial\Omega \in C^{2,\alpha}$, there exist a neighborhood U of x^0 and a $C^{2,\alpha}$ invertible mapping $\Psi : U \to \hat{B}_1(0)$, such that

$$\Psi(U \cap \Omega) = \hat{B}_1^+ = \{y \in \hat{B}_1(0); y_n > 0\},$$
$$\Psi(U \cap \partial\Omega) = \partial\hat{B}_1^+ \cap \{y \in \mathbb{R}^n; y_n = 0\},$$

where $\hat{B}_1(0)$ denotes the unit ball in the y-space. Denote

$$P_{ij} = \frac{\partial\Psi_i}{\partial x_j}, \quad P_{ij}^k = \frac{\partial^2\Psi_k}{\partial x_i\partial x_j},$$

where $\Psi = (\Psi_1, \Psi_2, \cdots, \Psi_n)$. Then

$$\frac{\partial u}{\partial x_i} = \frac{\partial \hat{u}}{\partial y_k} \cdot \frac{\partial y_k}{\partial x_i} = P_{ki}\hat{D}_k\hat{u},$$

$$\frac{\partial^2 u}{\partial x_i \partial x_j} = \frac{\partial^2 \hat{u}}{\partial y_k \partial y_l} \cdot \frac{\partial y_k}{\partial x_i} \cdot \frac{\partial y_l}{\partial x_j} + \frac{\partial \hat{u}}{\partial y_k} \cdot \frac{\partial^2 y_k}{\partial x_i \partial x_j} = P_{ki}P_{lj}\hat{D}_{kl}\hat{u} + P_{ij}^k\hat{D}_k\hat{u},$$

where $\hat{D}_k = \dfrac{\partial}{\partial y_k}$, $\hat{D}_{kl} = \dfrac{\partial^2}{\partial y_k \partial y_l}$, $\hat{u}(y) = u(\Psi^{-1}(y))$. Hence

$$-a_{ij}(x)D_{ij}u = -\hat{a}_{ij}(y)P_{ki}P_{lj}\hat{D}_{kl}\hat{u} - \hat{a}_{ij}(y)P_{ij}^k\hat{D}_k\hat{u},$$

where $\hat{a}_{ij}(y) = a_{ij}(\Psi^{-1}(y))$. Thus, with the transformation $y = \Psi(x)$, equation (6.3.4) turns out to be

$$-\hat{a}_{ij}(y)P_{ki}P_{lj}\hat{D}_{kl}\hat{u} - \hat{a}_{ij}(y)P_{ij}^k\hat{D}_k\hat{u} = \hat{f}(y), \qquad (6.3.13)$$

where $\hat{f}(y) = f(\Psi^{-1}(y))$. Let $G(y) = \Psi'(\Psi^{-1}(y))(\Psi'(\Psi^{-1}(y)))^T$. Since $y = \Psi(x)$ is a $C^{2,\alpha}$ invertible mapping, $\Psi'(\Psi^{-1}(y))$ is nonsingular and hence $G(y)$ is positive definite and continuous. Let $m(y)$ and $M(y)$ be the minimal and maximal eigenvalues of $G(y)$ and denote

$$m = \min_{y \in \hat{B}_1^+} m(y), \quad M = \max_{y \in \hat{B}_1^+} M(y).$$

Then $0 < m \leq M$. Since

$$(\hat{a}_{ij}(y)P_{ki}P_{lj}) = \Psi'(\Psi^{-1}(y))\hat{A}(y)(\Psi'(\Psi^{-1}(y)))^T,$$

where $\hat{A}(y) = (\hat{a}_{ij}(y))$, for any $\xi \in \mathbb{R}^n$, we have

$$\hat{a}_{ij}(y)P_{ki}P_{lj}\xi_k\xi_l = \xi^T\Psi'(\Psi^{-1}(y))\hat{A}(y)(\Psi'(\Psi^{-1}(y)))^T\xi.$$

Let $\eta(y) = (\Psi'(\Psi^{-1}(y)))^T\xi$. Then

$$\lambda|\eta|^2 \leq \hat{a}_{ij}(y)P_{ki}P_{lj}\xi_k\xi_l \leq \Lambda|\eta|^2.$$

Since $|\eta|^2 = \eta^T\eta = \xi^T G(y)\xi$, we have

$$m|\xi|^2 \leq |\eta|^2 \leq M|\xi|^2.$$

Hence

$$\lambda m|\xi|^2 \leq \hat{a}_{ij}(y)P_{ki}P_{lj}\xi_k\xi_l \leq \Lambda M|\xi|^2.$$

This means that equation (6.3.13) is uniformly elliptic. Since Ψ is an invertible $C^{2,\alpha}$ mapping, there exist μ_1, μ_2 with $0 < \mu_1 \le \mu_2$, such that for any $x^1, x^2 \in \overline{\Omega} \cap U$,

$$\mu_1|x^1 - x^2| \le |\Psi(x^1) - \Psi(x^2)| \le \mu_2|x^1 - x^2|,$$

which shows that the distance in the x-space is equivalent to that in the y-space.

To establish the estimate near the boundary for solutions of equation (6.3.13), as we did for the interior estimate, we consider the equation without terms of lower order, treat it by means of the method of solidifying coefficients and change it to Poisson's equation by a transformation of variables followed by using the estimate near the boundary for this special equation. Here it should be noted that after changing variables, the boundary $y_n = 0$ of $\hat{B}_1^+ = \{y \in \hat{B}_1(0); y_n = 0\}$ becomes a superplane of another shape. However as indicated in Remark 6.2.8, in this case, the near boundary estimate for Poisson's equation stated in Theorem 6.2.11 still holds. Using this result, coming back to the variable y, returning to the original equation and coming back to the variable x, we finally obtain

$$[D^2u]_{\alpha;O_R} \le C\Big(R^\alpha[D^2u]_{\alpha;\Omega} + \frac{1}{R^{2+\alpha}}|u|_{0;\Omega} + \frac{1}{R^\alpha}|f|_{0;\Omega} + [f]_{\alpha;\Omega}\Big),$$

where $0 < R \le 1$ and $O_R \subset \Omega$ is such a domain which depends on R and for some constant $\sigma > 0$ independent of R such that $\Omega \cap B_{\sigma R}(x^0) \subset O_R$. Thus for $0 < R \le 1$ small enough, there holds

$$[D^2u]_{\alpha;\Omega_{2R}} \le C\Big(R^\alpha[D^2u]_{\alpha;\Omega} + \frac{1}{R^{2+\alpha}}|u|_{0;\Omega}$$
$$+ \frac{1}{R^\alpha}|f|_{0;\Omega} + [f]_{\alpha;\Omega}\Big), \qquad (6.3.14)$$

with another constant C, where $\Omega_R = \Omega \cap B_R(x^0)$.

6.3.4 *Global estimate*

Now we proceed to combine the interior estimate (6.3.12) and the near boundary estimate (6.3.14) and use the finite covering argument to establish the global Schauder's estimate.

Combining the interior estimate (6.3.12) and the estimate near the boundary (6.3.14), we see that for any $x^0 \in \overline{\Omega}$, there exists $0 < R(x^0) \le 1$, such that for any $0 < R \leq R(x^0)$,

$$[D^2 u]_{\alpha;\Omega_{2R}} \leq C_0 \left(R^\alpha [D^2 u]_{\alpha;\Omega} + \frac{1}{R^{2+\alpha}} |u|_{0;\Omega} + \frac{1}{R^\alpha} |f|_{0;\Omega} + [f]_{\alpha;\Omega} \right),$$

where $\Omega_R = \Omega \cap B_R(x^0)$, C_0 is a constant independent of x^0, R and $R(x^0)$ and either $B_R(x^0)$ is included in Ω or $x^0 \in \partial\Omega$. We will assume that for any $x^0 \in \overline{\Omega}$, $R(x^0) \leq R_0 \leq 1$ with R_0 to be specified later. By the finite covering theorem, there exist a finite number of such open balls $B_{R_1}(x^1)$, $B_{R_2}(x^2), \cdots, B_{R_m}(x^m)$ with $x^j \in \overline{\Omega}$ $(j = 1, 2, \cdots, m)$, covering $\overline{\Omega}$, and for any $0 < R \leq R_j \leq R_0$,

$$[D^2 u]_{\alpha;\Omega_{2R}(x^j)} \leq C_0 \left(R^\alpha [D^2 u]_{\alpha;\Omega} + \frac{1}{R^{2+\alpha}} |u|_{0;\Omega} + \frac{1}{R^\alpha} |f|_{0;\Omega} + [f]_{\alpha;\Omega} \right)$$

$$(j = 1, 2, \cdots, m). \qquad (6.3.15)$$

Let $x', x'' \in \overline{\Omega}$ and assume $x' \in B_{R_{j_0}}(x^{j_0})$. Then one of the following two cases must occur:

   i) $|x' - x''| \geq R_{j_0}$;

   ii) $x'' \in \Omega_{2R_{j_0}}(x^{j_0})$.

If i) occurs, then

$$\frac{|D^2 u(x') - D^2 u(x'')|}{|x' - x''|^\alpha} \leq \frac{2}{R_{j_0}^\alpha} |D^2 u|_{0;\Omega}.$$

If ii) occurs, then from (6.3.15) we obtain

$$\frac{|D^2 u(x') - D^2 u(x'')|}{|x' - x''|^\alpha}$$

$$\leq |D^2 u|_{\alpha;\Omega_{2R_{j_0}}(x^{j_0})}$$

$$\leq C_0 \left(R_{j_0}^\alpha [D^2 u]_{\alpha;\Omega} + \frac{1}{R_{j_0}^{2+\alpha}} |u|_{0;\Omega} + \frac{1}{R_{j_0}^\alpha} |f|_{0;\Omega} + [f]_{\alpha;\Omega} \right).$$

In either case, we have

$$\frac{|D^2 u(x') - D^2 u(x'')|}{|x' - x''|^\alpha}$$

$$\leq C_0 R_{j_0}^\alpha [D^2 u]_{\alpha;\Omega} + \frac{2}{R_{j_0}^\alpha} |D^2 u|_{0;\Omega} + \frac{C_0}{R_{j_0}^{2+\alpha}} |u|_{0;\Omega} + \frac{C_0}{R_{j_0}^\alpha} |f|_{0;\Omega} + C_0 [f]_{\alpha;\Omega}.$$

Hence

$$[D^2 u]_{\alpha;\Omega} \leq C_0 R_0^\alpha [D^2 u]_{\alpha;\Omega} + \frac{2}{\widetilde{R}_0^\alpha} |D^2 u|_{0;\Omega}$$

$$+ \frac{C_0}{\tilde{R}_0^{2+\alpha}} |u|_{0;\Omega} + \frac{C_0}{\tilde{R}_0^{\alpha}} |f|_{0;\Omega} + C_0 [f]_{\alpha;\Omega}. \qquad (6.3.16)$$

where $\tilde{R}_0 = \min\{R_1, R_2, \cdots, R_m\}$. Using the interpolation inequality gives

$$\frac{2}{\tilde{R}_0^{\alpha}} |D^2 u|_{0;\Omega} \le \frac{1}{3} [D^2 u]_{\alpha;\Omega} + C |u|_{0;\Omega}. \qquad (6.3.17)$$

Now we choose $R_0 \le (3C_0)^{-1/\alpha}$. Then

$$C_0 R_0^{\alpha} [D^2 u]_{\alpha;\Omega} \le \frac{1}{3} [D^2 u]_{\alpha;\Omega}, \qquad (6.3.18)$$

Combining (6.3.17), (6.3.18) with (6.3.16) we derive

$$[D^2 u]_{\alpha;\Omega} \le C(|u|_{0;\Omega} + |f|_{\alpha;\Omega})$$

with some constant C. Using the interpolation inequality again then leads to (6.3.5).

Exercises

1. Prove that $\mathcal{L}^{p,\mu}(\Omega)$ is a Banach space, where $\Omega \subset \mathbb{R}^n$ is a bounded domain, $p \ge 1$, $\mu \ge 0$.

2. Let $\Omega \subset \mathbb{R}^n$ be a bounded domain and $u \in L^2(\Omega)$. Prove that the function

$$g(\lambda) = \int_{\Omega} (u(x) - \lambda)^2 dx, \quad \lambda \in \mathbb{R}$$

attains its minimum at

$$\lambda = u_\Omega = \frac{1}{|\Omega|} \int_{\Omega} u(x) dx.$$

3. Let $u \in C^{2,\alpha}(\overline{B})$ be a solution of the boundary value problem

$$\begin{cases} -\Delta u + |\nabla u| = f, & x \in B, \\ u\big|_{\partial B} = 0, \end{cases}$$

where $0 < \alpha < 1$ and B is the unit ball in $\mathbb{R}^n$. Prove that there exists a constant $C > 0$ depending only on n, such that

$$|u|_{2,\alpha;B} \le C(|f|_{\alpha;B} + |u|_{0;B}).$$

4. Establish Schauder's estimates for solutions of the following boundary value problem for the biharmonic equation

$$\begin{cases} \Delta^2 u = f, & x \in \Omega, \\ u = \dfrac{\partial u}{\partial \nu} = 0, & x \in \partial\Omega. \end{cases}$$

Chapter 7

Schauder's Estimates for Linear Parabolic Equations

In this chapter, we introduce Schauder's estimates for solutions of linear parabolic equations of second order. We first consider the heat equation, establishing Schauder's estimates for this equation, and then applying to general linear parabolic equations. To obtain Schauder's estimates for solutions of the heat equation, we also adopt the theory of Campanato spaces.

7.1 t-Anisotropic Campanato Spaces

In Chapter 6, we have introduced the Campanato spaces, and described the integral characteristic of the Hölder continuous functions in spatial domains. In this section, we introduce the t-anisotropic Campanato spaces, and describe the integral characteristic of the Hölder continuous functions in parabolic domains.

Let $\Omega \subset \mathbb{R}^n$ be a bounded domain, $T > 0$. Denote $Q = \Omega \times (0, T)$,

$$I_\rho = I_\rho(t_0) = (t_0 - \rho^2, t_0 + \rho^2), \quad B_\rho = B_\rho(x^0), \quad Q_\rho(x^0, t_0) = B_\rho \times I_\rho.$$

Definition 7.1.1 (Campanato Spaces) Let $p \geq 1$, $\mu \geq 0$. The subspace of all functions u in $L^p(Q)$ satisfying

$$
\begin{aligned}
[u]_{p,\mu} &= [u]_{p,\mu;Q} \\
&\equiv \sup_{\substack{(x,t)\in Q \\ 0<\rho<\mathrm{diam}\Omega}} \left(\rho^{-\mu} \iint_{Q\cap Q_\rho(x,t)} |u(y,s) - u_{x,t,\rho}|^p dy ds \right)^{1/p} \\
&< +\infty
\end{aligned}
$$

endowed with the norm

$$\|u\|_{\mathcal{L}^{p,\mu}} = \|u\|_{\mathcal{L}^{p,\mu}(Q)} = [u]_{p,\mu;Q} + \|u\|_{L^p(Q)}$$

is called a Campanato space, denoted by $\mathcal{L}^{p,\mu}(Q)$, where

$$u_{x,t,\rho} = \frac{1}{|Q \cap Q_\rho(x,t)|} \iint_{Q \cap Q_\rho(x,t)} u(y,s)\,dy\,ds.$$

Comparing with Definition 6.1.2, here instead of $\Omega \cap B_\rho(x)$, we use $Q \cap Q_\rho(x,t)$.

Remark 7.1.1 $[u]_{p,\mu;Q}$ *is a semi-norm, rather than a norm, since* $[u]_{p,\mu;Q} = 0$ *does not imply* $u = 0$.

It is easy to check

Proposition 7.1.1 $\mathcal{L}^{p,\mu}(Q)$ *is a Banach space.*

Theorem 7.1.1 *(Integral Characteristic of Hölder Continuous Functions) Let Ω be an (A)-type domain, $n + 2 < \mu \le n + 2 + p$. Then* $\mathcal{L}^{p,\mu}(Q) = C^{\alpha,\alpha/2}(\overline{Q})$ *and*

$$C_1[u]_{\alpha,\alpha/2,Q} \le [u]_{p,\mu,Q} \le C_2[u]_{\alpha,\alpha/2,Q},$$

where $\alpha = \dfrac{\mu - n - 2}{p}$ *and C_1, C_2 are some positive constants depending only on n, A, p, μ.*

The proof is similar to that of Theorem 6.1.1 in Chapter 6.

Remark 7.1.2 *For $0 < \lambda < 1$, define the new semi-norm*

$$[u]_{p,\mu;Q}^{(\lambda)} \equiv \sup_{\substack{(x,t) \in Q \\ 0 < \rho < \lambda \mathrm{diam}\Omega}} \left(\rho^{-\mu} \iint_{Q \cap Q_\rho(x,t)} |u(y,s) - u_{x,t,\rho}|^p\,dy\,ds \right)^{1/p}.$$

Similar to the case of spatial domains, if $\alpha = \dfrac{\mu - n - 2}{p} \in (0,1]$, then these semi-norms are equivalent to the Hölder semi-norms $[u]_{\alpha,\alpha/2;Q}$, that is

$$C_1[u]_{\alpha,\alpha/2;Q} \le [u]_{p,\mu;Q}^{(\lambda)} \le C_2[u]_{\alpha,\alpha/2;Q},$$

where C_1, C_2 are positive constants depending only on n, A, p, μ and λ.

Proposition 7.1.2 *If $\mu > n + 2 + p$, then all the elements in $\mathcal{L}^{p,\mu}(Q)$ are constants.*

The proof is similar to that of Proposition 6.1.3 in Chapter 6.

7.2 Schauder's Estimates for the Heat Equation

7.2.1 *Estimates to be established*

Now we proceed to establish Schauder's estimates for solutions of the first boundary value problem for linear parabolic equations. Similar to the case of elliptic equations, we first establish the local interior estimates and local estimates near the boundary. Since for parabolic equations, the boundary condition is prescribed on the parabolic boundary, we need to establish the estimates near the bottom, near the lateral and near the lateral-bottom.

We begin our discussion with the heat equation

$$\frac{\partial u}{\partial t} - \Delta u = f(x, t). \tag{7.2.1}$$

i) Interior estimate. If $u \in C^{2+\alpha, 1+\alpha/2}(\overline{Q}_R)$ is a solution of (7.2.1) in $Q_R = Q_R(x^0, t_0)$, then

$$[D^2 u]_{\alpha, \alpha/2; Q_{R/2}} \le C \left(\frac{1}{R^{2+\alpha}} |u|_{0; Q_R} + \frac{1}{R^\alpha} |f|_{0; Q_R} + [f]_{\alpha, \alpha/2; Q_R} \right); \tag{7.2.2}$$

ii) Near bottom estimate. If $u \in C^{2+\alpha, 1+\alpha/2}(\overline{Q}_R^0)$ is a solution of (7.2.1) in $Q_R^0 = Q_R^0(x^0, 0) = B_R(x^0) \times I_R^0 = B_R(x^0) \times (0, R^2)$ satisfying

$$u\Big|_{t=0} = 0, \tag{7.2.3}$$

then

$$[D^2 u]_{\alpha, \alpha/2; Q_{R/2}^0} \le C \left(\frac{1}{R^{2+\alpha}} |u|_{0; Q_R^0} + \frac{1}{R^\alpha} |f|_{0; Q_R^0} + [f]_{\alpha, \alpha/2; Q_R^0} \right); \tag{7.2.4}$$

iii) Near lateral estimate. If $u \in C^{2+\alpha, 1+\alpha/2}(\overline{Q}_R^+)$ is a solution of (7.2.1) in $Q_R^+ = Q_R^+(x^0, t_0) = B_R^+(x^0) \times I_R(t_0) = \{x \in B_R(x^0); x_n > 0\} \times (t_0 - R^2, t_0 + R^2)$ satisfying

$$u\Big|_{x_n=0} = 0, \tag{7.2.5}$$

then

$$[D^2 u]_{\alpha, \alpha/2; Q_{R/2}^+} \le C \left(\frac{1}{R^{2+\alpha}} |u|_{0; Q_R^+} + \frac{1}{R^\alpha} |f|_{0; Q_R^+} + [f]_{\alpha, \alpha/2; Q_R^+} \right); \tag{7.2.6}$$

iv) Near lateral-bottom estimate. If $u \in C^{2+\alpha, 1+\alpha/2}(\overline{Q}_R^{0+})$ is a solution of (7.2.1) in $Q_R^{0+} = Q_R^{0+}(x^0, 0) = B_R^+(x^0) \times I_R^0 = \{x \in B_R(x^0); x_n >$

$0\} \times (0, R^2)$ satisfying (7.2.3), (7.2.5), then

$$[D^2 u]_{\alpha,\alpha/2;Q^{0+}_{R/2}} \leq C \left(\frac{1}{R^{2+\alpha}} |u|_{0;Q^{0+}_R} + \frac{1}{R^{\alpha}} |f|_{0;Q^{0+}_R} + [f]_{\alpha,\alpha/2;Q^{0+}_R} \right). \quad (7.2.7)$$

Remark 7.2.1 *By the interpolation inequality (Theorem 1.2.1 in Chapter 1) and equation (7.2.1), we see that in (7.2.2), (7.2.4), (7.2.6) and (7.2.7), $[D^2 u]_{\alpha,\alpha/2}$ can be replaced by $|u|_{2+\alpha,1+\alpha/2}$.*

Remark 7.2.2 *If the boundary value condition is $u\big|_{\partial_p Q} = \varphi$, and $\varphi \in C^{2+\alpha,1+\alpha/2}(\overline{Q})$, then instead of equation (7.2.1), we may consider the equation for $u - \varphi$.*

Remark 7.2.3 *Similar to the case of elliptic equations, in the following arguments, we may always assume that the solution is sufficiently smooth.*

7.2.2 Interior estimate

Similar to the case of elliptic equations, we need to establish Caccioppoli's inequalities, which are slightly different in form.

Theorem 7.2.1 *Let u be a solution of equation (7.2.1) in Q_R, $w = D_i u \, (1 \leq i \leq n)$. Then for any $0 < \rho < R$ and $\lambda \in \mathbb{R}$, we have*

$$\sup_{I_\rho} \int_{B_\rho} (u - \lambda)^2 dx + \iint_{Q_\rho} |Du|^2 dx dt$$

$$\leq C \left[\frac{1}{(R-\rho)^2} \iint_{Q_R} (u - \lambda)^2 dx dt + (R - \rho)^2 \iint_{Q_R} f^2 dx dt \right], \quad (7.2.8)$$

$$\sup_{I_\rho} \int_{B_\rho} (w - \lambda)^2 dx + \iint_{Q_\rho} |Dw|^2 dx dt$$

$$\leq C \left[\frac{1}{(R-\rho)^2} \iint_{Q_R} (w - \lambda)^2 dx dt + \iint_{Q_R} (f - f_R)^2 dx dt \right], \quad (7.2.9)$$

where $f_R = \frac{1}{|Q_R|} \iint_{Q_R} f(x,t) dx dt$, C is a positive constant depending only on n.

Proof. We only show (7.2.9); the proof of (7.2.8) is similar. Let $\eta(x)$ be a cut-off function defined on B_R related to B_ρ, that is, $\eta \in C_0^\infty(B_R)$, $0 \leq \eta(x) \leq 1$, $\eta \equiv 1$ in B_ρ and $|D\eta(x)| \leq \dfrac{C}{R - \rho}$. Let $\xi \in C^\infty(\mathbb{R})$, $0 \leq \xi(t) \leq 1$, $\xi \equiv 0$ for $t \leq t_0 - R^2$, $\xi \equiv 1$ for $t \geq t_0 - \rho^2$ and $0 \leq \xi'(t) \leq \dfrac{C}{(R - \rho)^2}$.

Multiplying the equation for w

$$\frac{\partial w}{\partial t} - \Delta w = D_i f(x, t) = D_i(f(x, t) - f_R)$$

by $\eta^2 \xi^2(w - \lambda)$ and integrating over $Q_R^s \equiv B_R \times (t_0 - R^2, s)$ $(s \in I_R)$, we have

$$\iint_{Q_R^s} \eta^2 \xi^2 (w - \lambda) \frac{\partial w}{\partial t} dx dt$$

$$= \iint_{Q_R^s} \eta^2 \xi^2 (w - \lambda) \Delta w dx dt + \iint_{Q_R^s} \eta^2 \xi^2 (w - \lambda) D_i(f - f_R) dx dt.$$

Integrating by parts yields

$$\frac{1}{2} \iint_{Q_R^s} \frac{\partial}{\partial t}(\eta^2 \xi^2 (w - \lambda)^2) dx dt - \iint_{Q_R^s} \eta^2 \xi \xi'(w - \lambda)^2 dx dt$$

$$= - \iint_{Q_R^s} \eta^2 \xi^2 |Dw|^2 dx dt - 2 \iint_{Q_R^s} \eta \xi^2 (w - \lambda) D\eta \cdot Dw dx dt$$

$$- \iint_{Q_R^s} \eta^2 \xi^2 (f - f_R) D_i w dx dt$$

$$- 2 \iint_{Q_R^s} \eta \xi^2 (w - \lambda)(f - f_R) D_i \eta dx dt,$$

that is,

$$\frac{1}{2} \int_{B_R} \eta^2 \xi^2 (w - \lambda)^2 dx \bigg|_s + \iint_{Q_R^s} \eta^2 \xi^2 |Dw|^2 dx dt$$

$$= 2 \iint_{Q_R^s} \eta \xi^2 (w - \lambda) D\eta \cdot Dw dx dt - \iint_{Q_R^s} \eta^2 \xi^2 (f - f_R) D_i w dx dt$$

$$- 2 \iint_{Q_R^s} \eta \xi^2 (w - \lambda)(f - f_R) D_i \eta dx dt + \iint_{Q_R^s} \eta^2 \xi \xi'(w - \lambda)^2 dx dt.$$

Applying Cauchy's inequality with ε to the first three terms on the right side of the above formula, we obtain

$$\frac{1}{2} \int_{B_R} \eta^2 \xi^2 (w - \lambda)^2 dx \bigg|_s + \iint_{Q_R^s} \eta^2 \xi^2 |Dw|^2 dx dt$$

$$\leq \frac{1}{2} \iint_{Q_R^s} \eta^2 \xi^2 |Dw|^2 dx dt + C \iint_{Q_R^s} \xi^2 (w - \lambda)^2 |D\eta|^2 dx dt$$

$$+ C \iint_{Q_R^s} \eta^2 \xi^2 (f - f_R)^2 dx dt + \iint_{Q_R^s} \eta^2 \xi |\xi'|(w - \lambda)^2 dx dt$$

$$\le \frac{1}{2} \iint_{Q_R^s} \eta^2 \xi^2 |Dw|^2 \, dx dt + \frac{C}{(R-\rho)^2} \iint_{Q_R^s} (w-\lambda)^2 \, dx dt$$
$$+ C \iint_{Q_R^s} (f - f_R)^2 \, dx dt.$$

Hence

$$\left. \int_{B_R} \eta^2 \xi^2 (w-\lambda)^2 \, dx \right|_s + \iint_{Q_R^s} \eta^2 \xi^2 |Dw|^2 \, dx dt$$
$$\le \frac{C}{(R-\rho)^2} \iint_{Q_R^s} (w-\lambda)^2 \, dx dt + C \iint_{Q_R^s} (f-f_R)^2 \, dx dt.$$

Therefore, for any $s \in I_\rho$, we have

$$\left. \int_{B_\rho} (w-\lambda)^2 \, dx \right|_s \le \left. \int_{B_R} \eta^2 \xi^2 (w-\lambda)^2 \, dx \right|_s$$
$$\le \frac{C}{(R-\rho)^2} \iint_{Q_R^s} (w-\lambda)^2 \, dx dt + C \iint_{Q_R^s} (f-f_R)^2 \, dx dt$$
$$\le \frac{C}{(R-\rho)^2} \iint_{Q_R} (w-\lambda)^2 \, dx dt + C \iint_{Q_R} (f-f_R)^2 \, dx dt$$

and

$$\iint_{Q_\rho} |Dw|^2 \, dx dt \le \iint_{Q_R} \eta^2 \xi^2 |Dw|^2 \, dx dt$$
$$\le \frac{C}{(R-\rho)^2} \iint_{Q_R} (w-\lambda)^2 \, dx dt + C \iint_{Q_R} (f-f_R)^2 \, dx dt,$$

which implies (7.2.9). $\square$

Corollary 7.2.1 *Let u be a solution of equation (7.2.1) in Q_R, $w = D_i u \, (1 \le i \le n)$. Then*

$$\sup_{I_{R/2}} \int_{B_{R/2}} w^2 \, dx + \iint_{Q_{R/2}} |Dw|^2 \, dx dt$$
$$\le \frac{C}{R^4} \iint_{Q_R} u^2 \, dx dt + C R^{n+2} |f|_{0;Q_R}^2 + C R^{n+2+2\alpha} [f]_{\alpha,\alpha/2;Q_R}^2,$$

where C is a constant depending only on n.

Proof. In (7.2.9), choosing ρ and R as $\dfrac{R}{2}$ and $\dfrac{3}{4}R$ respectively and setting $\lambda = 0$, we obtain

$$\sup_{I_{R/2}} \int_{B_{R/2}} w^2 dx + \iint_{Q_{R/2}} |Dw|^2 dxdt$$
$$\leq \frac{C}{R^2} \iint_{Q_{3R/4}} w^2 dxdt + C \iint_{Q_{3R/4}} (f - f_{3R/4})^2 dxdt,$$

and in (7.2.8), choosing ρ and R as $\dfrac{3}{4}R$ and R respectively and setting $\lambda = 0$, we obtain

$$\iint_{Q_{3R/4}} |w|^2 dxdt \leq \frac{C}{R^2} \iint_{Q_R} u^2 dxdt + CR^2 \iint_{Q_R} f^2 dxdt.$$

Combining the above two inequalities, we see that

$$\sup_{I_{R/2}} \int_{B_{R/2}} w^2 dx + \iint_{Q_{R/2}} |Dw|^2 dxdt$$
$$\leq \frac{C}{R^4} \iint_{Q_R} u^2 dxdt + C \iint_{Q_R} f^2 dxdt + C \iint_{Q_{3R/4}} (f - f_{3R/4})^2 dxdt$$
$$\leq \frac{C}{R^4} \iint_{Q_R} u^2 dxdt + CR^{n+2}|f|_{0;Q_R}^2 + CR^{n+2+2\alpha}[f]_{\alpha;Q_R}^2.$$

$\square$

Repeated use of the inequality (7.2.8) leads to

Corollary 7.2.2 *If $f \equiv 0$ in Q_1, then for any nonnegative integer k, we have*

$$\sup_{I_{1/2}} \int_{B_{1/2}} |D^k u|^2 dx + \iint_{Q_{1/2}} |D^{k+1} u|^2 dxdt \leq C \iint_{Q_1} u^2 dxdt, \qquad (7.2.10)$$

where C is a positive constant depending only on n and k.

Corollary 7.2.3 *If $f \equiv 0$ in Q_R, then*

$$\sup_{Q_{R/2}} |u| \leq C \left(\frac{1}{R^{n+2}} \iint_{Q_R} u^2 dxdt \right)^{1/2},$$

where C is a positive constant depending only on n.

Proof. We first assume that $R = 1$. Take a natural number $k > \dfrac{n+1}{2}$. By the equation

$$D_t u = \Delta u$$

and Corollary 7.2.2 we see that, for $i, j = 0, 1, \cdots, k$, there hold

$$\iint_{Q_{1/2}} |D^j D_t^i u|^2 dx dt \leq \iint_{Q_{1/2}} |D^{j+2i} u|^2 dx dt \leq C \iint_{Q_1} u^2 dx dt.$$

Thus

$$\|u\|_{H^k(Q_{1/2})} \leq C \|u\|_{L^2(Q_1)}.$$

Applying the embedding theorem, we obtain

$$\sup_{Q_{1/2}} u^2 \leq C \|u\|_{H^k(Q_{1/2})} \leq C \|u\|_{L^2(Q_1)}.$$

For the general case $R > 0$, we may use the rescaling technique to obtain the desired conclusion.   □

Theorem 7.2.2   *Let u be a solution of equation (7.2.1) in Q_R and $f \equiv 0$ in Q_R. Then for any $0 < \rho \leq R$, we have*

$$\iint_{Q_\rho} u^2 dx dt \leq C \left(\frac{\rho}{R}\right)^{n+2} \iint_{Q_R} u^2 dx dt, \qquad (7.2.11)$$

$$\iint_{Q_\rho} (u - u_\rho)^2 dx dt \leq C \left(\frac{\rho}{R}\right)^{n+4} \iint_{Q_R} (u - u_R)^2 dx dt, \qquad (7.2.12)$$

where $u_R = \dfrac{1}{|Q_R|} \iint_{Q_R} u(x, t) dx dt$, C is a positive constant depending only on n.

Proof.   We first prove (7.2.11). By Corollary 7.2.3 we see that, if $0 < \rho < R/2$, then

$$\iint_{Q_\rho} u^2 dx dt \leq |Q_\rho| \sup_{Q_\rho} u^2 \leq C \rho^{n+2} \sup_{Q_{R/2}} u^2 \leq C \left(\frac{\rho}{R}\right)^{n+2} \iint_{Q_R} u^2 dx dt.$$

So, if $0 < \rho < R/2$, then (7.2.11) is valid, while if $R/2 \leq \rho \leq R$, (7.2.11) is obvious with $C \geq 2^{n+2}$.

Next we prove (7.2.12). By the t-anisotropic Poincaré's inequality (Theorem 1.4.3 of Chapter 1), we see that

$$\iint_{Q_\rho} (u - u_\rho)^2 dx dt \leq C \left(\rho^2 \iint_{Q_\rho} |Du|^2 dx dt + \rho^4 \iint_{Q_\rho} |D_t u|^2 dx dt\right).$$

Using equation (7.2.1) and noticing that $f \equiv 0$ in Q_R, we see that

$$\iint_{Q_\rho} (u - u_\rho)^2 dx dt \leq C \left(\rho^2 \iint_{Q_\rho} |Du|^2 dx dt + \rho^4 \iint_{Q_\rho} |\Delta u|^2 dx dt\right)$$

$$\leq C \left(\rho^2 \iint_{Q_\rho} |Du|^2 dx dt + \rho^4 \iint_{Q_\rho} |D^2 u|^2 dx dt \right).$$

In the above inequality, applying (7.2.11) and Caccioppoli's inequality (7.2.8) to Du and applying (7.2.11), Caccioppoli's inequality (7.2.9) and (7.2.8) to $D^2 u$, we see that if $0 < \rho < R/2$, then

$$\iint_{Q_\rho} (u - u_\rho)^2 dx dt$$

$$\leq C \left(\frac{\rho}{R} \right)^{n+2} \left(\rho^2 \iint_{Q_{R/2}} |Du|^2 dx dt + \rho^4 \iint_{Q_{R/2}} |D^2 u|^2 dx dt \right)$$

$$\leq C \left(\frac{\rho}{R} \right)^{n+2} \left[\left(\frac{\rho}{R} \right)^2 + \left(\frac{\rho}{R} \right)^4 \right] \iint_{Q_R} (u - \lambda)^2 dx dt$$

and (7.2.12) follows by the choice of $\lambda = u_R$. If $R/2 \leq \rho \leq R$, then (7.2.12) is obvious with $C \geq 2^{n+4}$. $\qquad\square$

Similar to the treatment for Poisson's equation, we may apply the interior estimate (7.2.12) for the homogeneous equation and the iteration lemma (Lemma 6.2.1 of Chapter 6) to obtain the interior estimate for solutions of the nonhomogeneous heat equation.

Theorem 7.2.3 *Let u be a solution of equation (7.2.1) in Q_{R_0}, $w = D_i u \, (1 \leq i \leq n)$. Then for any $0 < \rho \leq R \leq R_0$, we have*

$$\frac{1}{\rho^{n+2+2\alpha}} \iint_{Q_\rho} |Dw - (Dw)_\rho|^2 dx dt$$

$$\leq \frac{C}{R^{n+2+2\alpha}} \iint_{Q_R} |Dw - (Dw)_R|^2 dx dt + C[f]^2_{\alpha,\alpha/2;Q_R},$$

where C is a positive constant depending only on n.

Proof. Decompose w as $w = w_1 + w_2$ with w_1 and w_2 satisfying

$$\begin{cases} \dfrac{\partial w_1}{\partial t} - \Delta w_1 = 0, & \text{in } Q_R, \\[2mm] w_1 \Big|_{\partial_p Q_R} = w \end{cases}$$

and

$$\begin{cases} \dfrac{\partial w_2}{\partial t} - \Delta w_2 = D_i f = D_i(f - f_R), & \text{in } Q_R, \\[2mm] w_2\Big|_{\partial_p Q_R} = 0. \end{cases}$$

Applying (7.2.12) to w_1, we obtain

$$\iint_{Q_\rho} |Dw_1 - (Dw_1)_\rho|^2 dxdt$$

$$\leq C \left(\frac{\rho}{R}\right)^{n+4} \iint_{Q_R} |Dw_1 - (Dw_1)_R|^2 dxdt.$$

So, for any $0 < \rho \leq R \leq R_0$,

$$\iint_{Q_\rho} |Dw - (Dw)_\rho|^2 dxdt$$

$$\leq 2 \iint_{Q_\rho} |Dw_1 - (Dw_1)_\rho|^2 dxdt + 2 \iint_{Q_\rho} |Dw_2 - (Dw_2)_\rho|^2 dxdt$$

$$\leq C \left(\frac{\rho}{R}\right)^{n+4} \iint_{Q_R} |Dw_1 - (Dw_1)_R|^2 dxdt + C \iint_{Q_R} |Dw_2|^2 dxdt$$

$$\leq C \left(\frac{\rho}{R}\right)^{n+4} \iint_{Q_R} |Dw - (Dw)_R|^2 dxdt + C \iint_{Q_R} |Dw_2|^2 dxdt.$$

Multiplying the equation for w_2 by w_2, integrating over Q_R, integrating by parts and noticing that $w_2\Big|_{\partial_p Q_R} = 0$, we see that

$$\frac{1}{2} \iint_{Q_R} \frac{\partial}{\partial t}(w_2^2) dxdt + \iint_{Q_R} |Dw_2|^2 dxdt$$

$$= \iint_{Q_R} w_2 D_i(f - f_R) dxdt$$

$$= -\iint_{Q_R} (f - f_R) D_i w_2 dxdt$$

$$\leq \frac{1}{2} \iint_{Q_R} |Dw_2|^2 dxdt + \frac{1}{2} \iint_{Q_R} (f - f_R)^2 dxdt,$$

from which, noticing

$$\iint_{Q_R} \frac{\partial}{\partial t}(w_2^2) dxdt = \int_{B_R} w_2^2(x,t)\Big|_{t=t_0+R^2} dx \geq 0,$$

it follows

$$\iint_{Q_R} |Dw_2|^2 dxdt \leq \iint_{Q_R} (f - f_R)^2 dxdt \leq CR^{n+2+2\alpha}[f]^2_{\alpha,\alpha/2;Q_R}$$

and hence

$$\iint_{Q_\rho} |Dw - (Dw)_\rho|^2 dxdt$$
$$\leq C \left(\frac{\rho}{R}\right)^{n+4} \iint_{Q_R} |Dw - (Dw)_R|^2 dxdt + CR^{n+2+2\alpha}[f]^2_{\alpha,\alpha/2;Q_R}.$$

Using the iteration lemma (Lemma 6.2.1 of Chapter 6), we finally obtain

$$\iint_{Q_\rho} |Dw - (Dw)_\rho|^2 dxdt$$
$$\leq C \left(\frac{\rho}{R}\right)^{n+2+2\alpha} \left(\iint_{Q_R} |Dw - (Dw)_R|^2 dxdt + R^{n+2+2\alpha}[f]^2_{\alpha,\alpha/2;Q_R} \right).$$

$\square$

Theorem 7.2.4 *Let u be a solution of equation (7.2.1) in Q_R, $w = D_i u\,(1 \leq i \leq n)$. Then for any $0 < \rho \leq \dfrac{R}{2}$, we have*

$$\iint_{Q_\rho} |Dw - (Dw)_\rho|^2 dxdt \leq C\rho^{n+2+2\alpha} M_R,$$

where C is a positive constant depending only on n, and

$$M_R = \frac{1}{R^{4+2\alpha}}|u|^2_{0;Q_R} + \frac{1}{R^{2\alpha}}|f|^2_{0;Q_R} + [f]^2_{\alpha,\alpha/2;Q_R}.$$

Proof. According to Theorem 7.2.3 and Corollary 7.2.1, we see that

$$\iint_{Q_\rho} |Dw - (Dw)_\rho|^2 dxdt$$
$$\leq C\rho^{n+2+2\alpha} \left(\frac{1}{R^{n+2+2\alpha}} \int_{Q_{R/2}} |Dw - (Dw)_{R/2}|^2 dxdt + [f]^2_{\alpha,\alpha/2;Q_{R/2}} \right)$$
$$\leq C\rho^{n+2+2\alpha} \left(\frac{1}{R^{n+2+2\alpha}} \iint_{Q_{R/2}} |Dw|^2 dxdt + [f]^2_{\alpha,\alpha/2;Q_{R/2}} \right)$$
$$\leq C\rho^{n+2+2\alpha} \left(\frac{1}{R^{n+6+2\alpha}} \iint_{Q_R} u^2 dxdt + \frac{1}{R^{2\alpha}}|f|^2_{0;Q_R} + [f]^2_{\alpha,\alpha/2;Q_R} \right),$$

from which we get the conclusion of the theorem. $\square$

Theorem 7.2.5   *Let u be a solution of equation (7.2.1) in Q_R. Then*

$$[D^2 u]_{\alpha,\alpha/2;Q_{R/2}}$$

$$\leq C\left(\frac{1}{R^{2+\alpha}}|u|_{0;Q_R} + \frac{1}{R^\alpha}|f|_{0;Q_R} + [f]_{\alpha,\alpha/2;Q_R}\right), \qquad (7.2.13)$$

where C is a positive constant depending only on n.

Proof.   By Theorem 7.2.4, for $(x,t) \in Q_{R/2}$, $0 < \rho \leq \dfrac{R}{4}$, we have

$$\iint_{Q_\rho(x,t)\cap Q_{R/2}} |D^2 u(y,s) - (D^2 u)_{Q_\rho(x,t)\cap Q_{R/2}}|^2 dyds$$

$$\leq \iint_{Q_\rho(x,t)} |D^2 u(y,s) - (D^2 u)_{x,t,\rho}|^2 dyds$$

$$\leq C\rho^{n+2+2\alpha}\left(\frac{1}{R^{4+2\alpha}}|u|^2_{0;Q_{R/2}(x,t)}\right.$$

$$\left.+\frac{1}{R^{2\alpha}}|f|^2_{0;Q_{R/2}(x,t)} + [f]^2_{\alpha,\alpha/2;Q_{R/2}(x,t)}\right)$$

$$\leq C\rho^{n+2+2\alpha}\left(\frac{1}{R^{4+2\alpha}}|u|^2_{0;Q_R} + \frac{1}{R^{2\alpha}}|f|^2_{0;Q_R} + [f]^2_{\alpha,\alpha/2;Q_R}\right),$$

where

$$(D^2 u)_{Q_\rho(x,t)\cap Q_{R/2}} = \frac{1}{|Q_\rho(x,t)\cap Q_{R/2}|}\iint_{Q_\rho(x,t)\cap Q_{R/2}} D^2 u(y,s)dyds.$$

Thus

$$[D^2 u]^{(1/4)}_{2,n+2+2\alpha;Q_{R/2}} \leq C\left(\frac{1}{R^{4+2\alpha}}|u|^2_{0;Q_R} + \frac{1}{R^{2\alpha}}|f|^2_{0;Q_R} + [f]^2_{\alpha,\alpha/2;Q_R}\right)^{1/2}$$

$$\leq C\left(\frac{1}{R^{2+\alpha}}|u|_{0;Q_R} + \frac{1}{R^\alpha}|f|_{0;Q_R} + [f]_{\alpha,\alpha/2;Q_R}\right).$$

From Remark 7.1.2, we obtain (7.2.13) and the proof is complete.   □

7.2.3   Near bottom estimate

Since all the spatial derivatives on the bottom are tangential derivatives, the establishment of near bottom estimate is quite similar to the interior estimate.

Firstly, we establish Caccioppoli's inequalities.

Theorem 7.2.6 *Let u be a solution of equation (7.2.1) in Q_R^0 satisfying (7.2.3), $w = D_i u \, (1 \le i \le n)$. Then for any $0 < \rho < R$, we have*

$$\sup_{I_\rho^0} \int_{B_\rho} u^2 dx + \iint_{Q_\rho^0} |Du|^2 dxdt$$

$$\le C \left[\frac{1}{(R-\rho)^2} \iint_{Q_R^0} u^2 dxdt + (R-\rho)^2 \iint_{Q_R^0} f^2 dxdt \right], \qquad (7.2.14)$$

$$\sup_{I_\rho^0} \int_{B_\rho} w^2 dx + \iint_{Q_\rho^0} |Dw|^2 dxdt$$

$$\le C \left[\frac{1}{(R-\rho)^2} \iint_{Q_R^0} w^2 dxdt + \iint_{Q_R^0} (f - f_R)^2 dxdt \right], \qquad (7.2.15)$$

where $f_R = \dfrac{1}{|Q_R^0|} \iint_{Q_R^0} f(x,t) dxdt$, and C is a positive constant depending only on n.

Proof. We only present the proof of (7.2.15), which is similar to that of Theorem 7.2.1, the only difference is that here we multiply the equation for w

$$\frac{\partial w}{\partial t} - \Delta w = D_i f(x,t) = D_i(f(x,t) - f_R),$$

by $\eta^2 w$ and then integrate over $Q_R^{0,s} \equiv B_R \times (0,s)$ $(s \in I_R^0)$ to obtain

$$\iint_{Q_R^{0,s}} \eta^2 w \frac{\partial w}{\partial t} dxdt = \iint_{Q_R^{0,s}} \eta^2 w \Delta w dxdt + \iint_{Q_R^{0,s}} \eta^2 w D_i(f - f_R) dxdt.$$

Since $\eta \in C_0^\infty(B_R)$, the boundary integral vanishes when integrating by parts with respect to the spatial variables. On the other hand, by $w\big|_{t=0} = 0$, we have

$$\iint_{Q_R^{0,s}} \eta^2 w \frac{\partial w}{\partial t} dxdt = \frac{1}{2} \int_{B_R} \eta^2(x) w^2(x,t) \Big|_{t=0}^{t=s} dx$$

$$= \frac{1}{2} \int_{B_R} \eta^2(x) w^2(x,s) dx.$$

$\square$

Remark 7.2.4 *We use the notation f_R to denote the average of f over Q_R^0 when treating the near bottom estimate, while use f_R to denote the average of f over Q_R when treating the interior estimate. In what follows, we will use the same notation f_R to denote the average of f over Q_R^+ and*

Q_R^{0+} when treating the near lateral estimate and the near lateral-bottom estimate. No confusion will be caused from the context.

Combining the inequality (7.2.14) with (7.2.15) we have

Corollary 7.2.4   *Let u be a solution of equation (7.2.1) in Q_R^0 satisfying (7.2.3), $w = D_i u \, (1 \le i \le n)$, Then*

$$\sup_{I_{R/2}^0} \int_{B_{R/2}} w^2 dx + \iint_{Q_{R/2}^0} |Dw|^2 dxdt$$

$$\le \frac{C}{R^4} \iint_{Q_R^0} u^2 dxdt + CR^{n+2}|f|_{0;Q_R^0}^2 + CR^{n+2+2\alpha}[f]_{\alpha,\alpha/2;Q_R^0}^2,$$

where C is a positive constant depending only on n.

Repeated use of the inequality (7.2.14) leads to

Corollary 7.2.5   *If $f \equiv 0$ in Q_1^0, then for any nonnegative integer k, we have*

$$\sup_{I_{1/2}^0} \int_{B_{1/2}} |D^k u|^2 dx + \iint_{Q_{1/2}^0} |D^{k+1} u|^2 dxdt \le C \iint_{Q_1^0} u^2 dxdt, \qquad (7.2.16)$$

where C is a positive constant depending only on n and k.

By using (7.2.16), similar to the proof of Corollary 7.2.3, it follows that

Corollary 7.2.6   *If $f \equiv 0$ in Q_R^0, then*

$$\sup_{Q_{R/2}^0} |u| \le C \left(\frac{1}{R^{n+2}} \iint_{Q_R^0} u^2 dxdt \right)^{1/2},$$

where C is a positive constant depending only on n.

Applying Corollary 7.2.6, similar to the proof of (7.2.11), we obtain the following

Theorem 7.2.7   *Let u be a solution of equation (7.2.1) in Q_R^0 satisfying (7.2.3) and $f \equiv 0$ in Q_R^0. Then for any $0 < \rho \le R$, we have*

$$\iint_{Q_\rho^0} u^2 dxdt \le C \left(\frac{\rho}{R} \right)^{n+2} \iint_{Q_R^0} u^2 dxdt, \qquad (7.2.17)$$

where C is a positive constant depending only on n.

Furthermore, we have the following

Theorem 7.2.8   *Let u be a solution of equation (7.2.1) in Q_R^0 satisfying (7.2.3) and $f \equiv 0$ in Q_R^0. Then for any $0 < \rho \le R$, we have*

$$\iint_{Q_\rho^0} u^2 dx dt \le C \left(\frac{\rho}{R}\right)^{n+4} \iint_{Q_R^0} u^2 dx dt, \qquad (7.2.18)$$

where C is a positive constant depending only on n.

Proof.   Obviously, Δu satisfies the same equation and initial value condition as u, which implies that (7.2.17) is valid for Δu too, that is

$$\iint_{Q_\rho^0} |\Delta u|^2 dx dt \le C \left(\frac{\rho}{R}\right)^{n+2} \iint_{Q_R^0} |\Delta u|^2 dx dt, \quad 0 < \rho \le R.$$

Noticing $u\big|_{t=0} = 0$, we see that if $0 < \rho < R/2$, then

$$\iint_{Q_\rho^0} u^2 dx dt \le C\rho^4 \iint_{Q_\rho^0} |D_t u|^2 dx dt$$

$$= C\rho^4 \iint_{Q_\rho^0} |\Delta u|^2 dx dt$$

$$\le C\rho^4 \left(\frac{\rho}{R}\right)^{n+2} \iint_{Q_{R/2}^0} |\Delta u|^2 dx dt.$$

Using (7.2.15) and (7.2.14), we obtain

$$\iint_{Q_{R/2}^0} |D^2 u|^2 dx dt \le \frac{C}{R^2} \iint_{Q_{3R/4}^0} |Du|^2 dx dt \le \frac{C}{R^4} \iint_{Q_R^0} u^2 dx dt.$$

Therefore,

$$\iint_{Q_\rho^0} u^2 dx dt \le C \left(\frac{\rho}{R}\right)^{n+6} \iint_{Q_R^0} u^2 dx dt \le C \left(\frac{\rho}{R}\right)^{n+4} \iint_{Q_R^0} u^2 dx dt,$$

that is, for $0 < \rho < R/2$, (7.2.18) is valid. If $R/2 \le \rho \le R$, then (7.2.18) holds obviously with $C \ge 2^{n+4}$. $\square$

Similar to the treatment for the interior estimate, using the near bottom estimate (7.2.18) for the homogeneous equation and the iteration lemma (Lemma 6.2.1), we may obtain the near bottom estimate for the nonhomogeneous heat equation.

Theorem 7.2.9   *Let u be a solution of equation (7.2.1) in $Q^0_{R_0}$ satisfying (7.2.3), $w = D_i u\,(1 \le i \le n)$. Then for any $0 < \rho \le R \le R_0$, we have*

$$\frac{1}{\rho^{n+2+2\alpha}} \iint_{Q^0_\rho} |Dw|^2 dxdt \le \frac{C}{R^{n+2+2\alpha}} \iint_{Q^0_R} |Dw|^2 dxdt + C[f]^2_{\alpha,\alpha/2;Q^0_R},$$

where C is a positive constant depending only on n.

Proof.   Decompose w into $w = w_1 + w_2$ with w_1 and w_2 satisfying

$$\begin{cases} \dfrac{\partial w_1}{\partial t} - \Delta w_1 = 0, & \text{in } Q^0_R, \\[2mm] w_1\big|_{\partial_p Q^0_R} = w \end{cases}$$

and

$$\begin{cases} \dfrac{\partial w_2}{\partial t} - \Delta w_2 = D_i f = D_i(f - f_R), & \text{in } Q^0_R, \\[2mm] w_2\big|_{\partial_p Q^0_R} = 0. \end{cases}$$

Using (7.2.18) for Dw_1, we obtain

$$\iint_{Q^0_\rho} |Dw_1|^2 dxdt \le C \left(\frac{\rho}{R}\right)^{n+4} \iint_{Q^0_R} |Dw_1|^2 dxdt.$$

So, for any $0 < \rho \le R \le R_0$, we have

$$\iint_{Q^0_\rho} |Dw|^2 dxdt$$

$$\le 2 \iint_{Q^0_\rho} |Dw_1|^2 dxdt + 2 \iint_{Q^0_\rho} |Dw_2|^2 dxdt$$

$$\le C \left(\frac{\rho}{R}\right)^{n+4} \iint_{Q^0_R} |Dw|^2 dxdt + C \iint_{Q^0_R} |Dw_2|^2 dxdt.$$

Similar to the proof of Theorem 7.2.3, multiplying the equation for w_2 by w_2, and integrating over Q^0_R, we see that

$$\iint_{Q^0_R} |Dw_2|^2 dxdt \le \iint_{Q^0_R} (f - f_R)^2 dxdt \le C R^{n+2+2\alpha}[f]^2_{\alpha,\alpha/2;Q^0_R}.$$

Therefore

$$\iint_{Q^0_\rho} |Dw|^2 dxdt \le C \left(\frac{\rho}{R}\right)^{n+4} \iint_{Q^0_R} |Dw|^2 dxdt + C R^{n+2+2\alpha}[f]^2_{\alpha,\alpha/2;Q^0_R}.$$

Finally, using the iteration lemma (Lemma 6.2.1) we immediately get the conclusion of the theorem. □

Theorem 7.2.10 *Let u be a solution of equation (7.2.1) in Q_R^0 satisfying (7.2.3), $w = D_i u \, (1 \le i \le n)$. Then for any $0 < \rho \le \dfrac{R}{2}$, we have*

$$\iint_{Q_\rho^0} |Dw - (Dw)_\rho|^2 dxdt \le C\rho^{n+2+2\alpha} M_R,$$

where C is a positive constant depending only on n, and

$$M_R = \frac{1}{R^{4+2\alpha}} |u|_{0;Q_R^0}^2 + \frac{1}{R^{2\alpha}} |f|_{0;Q_R^0}^2 + [f]_{\alpha,\alpha/2;Q_R^0}^2.$$

Proof. By virtue of Theorem 7.2.9 and Corollary 7.2.4, we see that

$$\iint_{Q_\rho^0} |Dw|^2 dxdt$$

$$\le C\rho^{n+2+2\alpha} \Big(\frac{1}{R^{n+2+2\alpha}} \int_{Q_{R/2}^0} |Dw|^2 dxdt + [f]_{\alpha,\alpha/2;Q_{R/2}^0}^2 \Big)$$

$$\le C\rho^{n+2+2\alpha} \Big(\frac{1}{R^{n+6+2\alpha}} \iint_{Q_R^0} u^2 dxdt + \frac{1}{R^{2\alpha}} |f|_{0;Q_R^0}^2 + [f]_{\alpha,\alpha/2;Q_R^0}^2 \Big)$$

$$\le C\rho^{n+2+2\alpha} M_R.$$

Thus

$$(Dw)_\rho^2 = \frac{1}{|Q_\rho^0|^2} \Big(\iint_{Q_\rho^0} |Dw| dxdt \Big)^2 \le \frac{1}{|Q_\rho^0|} \iint_{Q_\rho^0} |Dw|^2 dxdt \le C\rho^{2\alpha} M_R$$

and hence

$$\iint_{Q_\rho^0} |Dw - (Dw)_\rho|^2 dxdt$$

$$\le 2 \iint_{Q_\rho^0} |Dw|^2 dxdt + 2 \iint_{Q_\rho^0} |(Dw)_\rho|^2 dxdt$$

$$\le C\rho^{n+2+2\alpha} M_R.$$

□

Theorem 7.2.11 *Let u be a solution of equation (7.2.1) in Q_R^0 satisfying (7.2.3). Then*

$$[D^2 u]_{\alpha,\alpha/2;Q_{R/2}^0} \le C \left(\frac{1}{R^{2+\alpha}} |u|_{0;Q_R^0} + \frac{1}{R^\alpha} |f|_{0;Q_R^0} + [f]_{\alpha,\alpha/2;Q_R^0} \right),$$

where C is a positive constant depending only on n.

Proof.   Similar to the case of elliptic equations (Theorem 6.2.11), we need only to use Theorem 7.2.10 and the corresponding result about the interior estimate (Theorem 7.2.4).   □

7.2.4   *Near lateral estimate*

Similar to the case of Poisson's equation, in establishing the near lateral estimate, we can estimate the tangential derivatives directly, while for the normal derivatives, we need to apply the equation and the results for tangential derivatives. Since in equation (7.2.1) there is a term of time derivative $D_t u$, we need to estimate it too.

First, we establish the following Caccioppoli's inequalities.

Theorem 7.2.12   *Let u be a solution of equation (7.2.1) in Q_R^+ satisfying (7.2.5), $w = D_i u\,(1 \leq i < n)$. Then for any $0 < \rho < R$ and any $\vec{\lambda} \in \mathbb{R}^n$, we have*

$$\sup_{I_\rho} \int_{B_\rho^+} u^2 dx + \iint_{Q_\rho^+} |Du|^2 dxdt$$

$$\leq C\Big[\frac{1}{(R-\rho)^2}\iint_{Q_R^+} u^2 dxdt + (R-\rho)^2 \iint_{Q_R^+} f^2 dxdt\Big], \qquad (7.2.19)$$

$$\sup_{I_\rho} \int_{B_\rho^+} w^2 dx + \iint_{Q_\rho^+} |Dw|^2 dxdt$$

$$\leq C\Big[\frac{1}{(R-\rho)^2}\iint_{Q_R^+} w^2 dxdt + \iint_{Q_R^+} (f - f_R)^2 dxdt\Big], \qquad (7.2.20)$$

$$\sup_{I_\rho} \int_{B_\rho^+} |Du - \vec{\lambda}|^2 dx + \iint_{Q_\rho^+} |D_t u|^2 dxdt$$

$$\leq C\Big[\frac{1}{(R-\rho)^2}\iint_{Q_R^+} |Du - \vec{\lambda}|^2 dxdt + \iint_{Q_R^+} f^2 dxdt\Big], \qquad (7.2.21)$$

where $f_R = \dfrac{1}{|Q_R^+|}\iint_{Q_R^+} f(x,t)dxdt$ and C is a positive constant depending only on n.

Proof.   Since the proofs of (7.2.19) and (7.2.20) are similar, we merely show (7.2.20), which is similar to that of Theorem 7.2.1, the only difference is that here we multiply the equation for w

$$\frac{\partial w}{\partial t} - \Delta w = D_i f(x,t) = D_i(f(x,t) - f_R),$$

by $\eta^2\xi^2 w$ and integrate over $Q_R^{+,s} \equiv B_R^+ \times (t_0 - R^2, s)$ $(s \in I_R)$. Then

$$\iint_{Q_R^{+,s}} \eta^2\xi^2 w \frac{\partial w}{\partial t} dxdt$$

$$= \iint_{Q_R^{+,s}} \eta^2\xi^2 w \Delta w dxdt + \iint_{Q_R^{+,s}} \eta^2\xi^2 w D_i(f - f_R) dxdt.$$

Since $\eta \in C_0^\infty(B_R)$ and $w\big|_{x_n=0} = 0$, the boundary term resulting from integrating by parts with respect to the spatial variables is equal to zero.

Now we show (7.2.21). Rewrite equation (7.2.1) as

$$\frac{\partial u}{\partial t} - \operatorname{div}(Du - \vec{\lambda}) = f(x, t).$$

Let η and ξ be the cut-off functions in the proof of Theorem 7.2.1. Multiply both side of the above equation by $\eta^2\xi^2 D_t u$ and integrate over $Q_R^{+,s} \equiv B_R^+ \times (t_0 - R^2, s)$ $(s \in I_R)$. Then integrating by parts and noticing that $D_t u\big|_{x_n=0} = 0$, we obtain

$$\iint_{Q_R^{+,s}} \eta^2\xi^2 |D_t u|^2 dxdt$$

$$= -\iint_{Q_R^{+,s}} (Du - \vec{\lambda}) \cdot D(\eta^2\xi^2 D_t u) dxdt + \iint_{Q_R^{+,s}} \eta^2\xi^2 D_t u f dxdt$$

$$= -\iint_{Q_R^{+,s}} \eta^2\xi^2 (Du - \vec{\lambda}) \cdot DD_t u dxdt$$

$$\quad - 2\iint_{Q_R^{+,s}} \eta\xi^2 D_t u (Du - \vec{\lambda}) \cdot D\eta dxdt + \iint_{Q_R^{+,s}} \eta^2\xi^2 D_t u f dxdt$$

$$= -\frac{1}{2}\iint_{Q_R^{+,s}} \frac{\partial}{\partial t}(\eta^2\xi^2 |Du - \vec{\lambda}|^2) dxdt + \iint_{Q_R^{+,s}} \eta^2\xi\xi' |Du - \vec{\lambda}|^2 dxdt$$

$$\quad - 2\iint_{Q_R^{+,s}} \eta\xi^2 D_t u (Du - \vec{\lambda}) \cdot D\eta dxdt + \iint_{Q_R^{+,s}} \eta^2\xi^2 D_t u f dxdt.$$

Utilizing Cauchy's inequality with ε, we further have

$$\iint_{Q_R^{+,s}} \eta^2\xi^2 |D_t u|^2 dxdt + \frac{1}{2}\int_{B_R^+} \eta^2\xi^2 |Du - \vec{\lambda}|^2 \bigg|_{t=s} dx$$

$$\leq \frac{1}{2}\int_{B_R^+} \eta^2\xi^2 |Du - \vec{\lambda}|^2 \bigg|_{t=t_0-R^2} dx$$

$$+ \iint_{Q_R^{+,s}} \eta^2 \xi \xi' |Du - \vec{\lambda}|^2 dxdt + \frac{1}{2} \iint_{Q_R^{+,s}} \eta^2 \xi^2 |D_t u|^2 dxdt$$

$$+ 4 \iint_{Q_R^{+,s}} \xi^2 |D\eta|^2 |Du - \vec{\lambda}|^2 dxdt + 2 \iint_{Q_R^{+,s}} \eta^2 \xi^2 f^2 dxdt$$

$$\leq \frac{1}{2} \iint_{Q_R^{+,s}} \eta^2 \xi^2 |D_t u|^2 dxdt$$

$$+ \frac{C}{(R-\rho)^2} \iint_{Q_R^{+,s}} |Du - \vec{\lambda}|^2 dxdt + 2 \iint_{Q_R^{+,s}} f^2 dxdt,$$

from which (7.2.21) follows at once. $\qquad\qquad\qquad\qquad\qquad\qquad$ $\square$

Combining the inequalities (7.2.19), (7.2.20), (7.2.21) with equation (7.2.1) we obtain

Corollary 7.2.7 *Let u be a solution of equation (7.2.1) in Q_R^+ satisfying (7.2.5), $w = D_i u \, (1 \leq i \leq n)$. Then*

$$\sup_{I_{R/2}} \int_{B_{R/2}^+} w^2 dx + \iint_{Q_{R/2}^+} (|D_t u|^2 + |Dw|^2) dxdt$$

$$\leq \frac{C}{R^4} \iint_{Q_R^+} u^2 dxdt + CR^{n+2} |f|_{0;Q_R^+}^2 + CR^{n+2+2\alpha} [f]_{\alpha,\alpha/2;Q_R^+}^2,$$

where C is a positive constant depending only on n.

Corollary 7.2.8 *If $f \equiv 0$ in Q_R^+, then for any nonnegative integer k and any $0 < \rho < R$, we have*

$$\sup_{I_\rho} \int_{B_\rho^+} |D^k u|^2 dx + \iint_{Q_\rho^+} |D^{k+1} u|^2 dxdt$$

$$\leq \frac{C}{(R-\rho)^2} \iint_{Q_R^+} |D^k u|^2 dxdt, \qquad\qquad (7.2.22)$$

where C is a positive constant depending only on n.

Proof. We present the proof by considering the following four cases.

i) The case $k = 0$. In this case, the conclusion can be obtained by using Caccioppoli's inequality (7.2.19) directly.

ii) The case $k = 1$. For $1 \leq j < n$, $D_j u$ still satisfies the homogeneous equation and the zero lateral boundary value condition. It follows from the conclusion for the case $k = 0$ that

$$\iint_{Q_\rho^+} |DD_j u|^2 dxdt \leq \frac{C}{(R-\rho)^2} \iint_{Q_R^+} (D_j u)^2 dxdt$$

$$\leq \frac{C}{(R-\rho)^2} \iint_{Q_R^+} |Du|^2 dx dt. \qquad (7.2.23)$$

Caccioppoli's inequality (7.2.21) then implies

$$\sup_{I_\rho} \int_{B_\rho^+} |Du|^2 dx + \iint_{Q_\rho^+} |D_t u|^2 dx dt$$

$$\leq \frac{C}{(R-\rho)^2} \iint_{Q_R^+} |Du|^2 dx dt. \qquad (7.2.24)$$

Using the equation $D_{nn} u = D_t u - \sum_{j=1}^{n-1} D_{jj} u$ and the inequalities (7.2.23) and (7.2.24), we obtain

$$\iint_{Q_\rho^+} |D_{nn} u|^2 dx dt \leq \frac{C}{(R-\rho)^2} \iint_{Q_R^+} |Du|^2 dx dt,$$

which together with (7.2.23) and (7.2.24) implies (7.2.22) for the case of $k = 1$.

iii) The case $k = 2$. For $1 \leq j < n$, $D_j u$ still satisfies the homogeneous equation and zero lateral boundary value condition. It follows from the conclusion for the case $k = 1$ that

$$\sup_{I_\rho} \int_{B_\rho^+} |DD_j u|^2 dx + \iint_{Q_\rho^+} |D^2 D_j u|^2 dx dt$$

$$\leq \frac{C}{(R-\rho)^2} \iint_{Q_R^+} |DD_j u|^2 dx dt$$

$$\leq \frac{C}{(R-\rho)^2} \iint_{Q_R^+} |D^2 u|^2 dx dt.$$

Since $D_t u$ still satisfies the homogeneous equation and zero lateral boundary value condition, it follows from the conclusion for the case $k = 0$ and the equation $D_t u = \Delta u$ that

$$\sup_{I_\rho} \int_{B_\rho^+} |D_t u|^2 dx + \iint_{Q_\rho^+} |DD_t u|^2 dx dt$$

$$\leq \frac{C}{(R-\rho)^2} \iint_{Q_R^+} |D_t u|^2 dx dt$$

$$= \frac{C}{(R-\rho)^2} \iint_{Q_R^+} |\Delta u|^2 dx dt$$

$$\leq \frac{C}{(R-\rho)^2} \iint_{Q_R^+} |D^2 u|^2 dx dt.$$

Combining the above two inequalities and using the equation $D_{nn}u = D_t u - \sum_{j=1}^{n-1} D_{jj}u$ and the equality $D_{nnn}u = D_n D_t u - \sum_{j=1}^{n-1} D_{jjn}u$, we get the conclusion (7.2.22) for the case $k=2$.

iv) The case $k > 2$ can be deduced accordingly.   □

From Corollary 7.2.8, we obtain

Corollary 7.2.9   *If $f \equiv 0$ in Q_1^+, then for any nonnegative integer k and i, we have*

$$\sup_{I_{1/2}} \int_{B_{1/2}^+} |D^{k+i}u|^2 dx + \iint_{Q_{1/2}^+} |D^{k+i+1}u|^2 dx dt \leq C \iint_{Q_1^+} |D^i u|^2 dx dt,$$

where C is a positive constant depending only on n and k.

Using Corollary 7.2.9 and the embedding theorem, similar to the proof of Corollary 7.2.3, we obtain

Corollary 7.2.10   *If $f \equiv 0$ in Q_R^+, then for any nonnegative integer i, we have*

$$\sup_{Q_{R/2}^+} |D^i u| \leq C \left(\frac{1}{R^{n+2}} \iint_{Q_R^+} |D^i u|^2 dx dt \right)^{1/2},$$

where C is a positive constant depending only on n.

By virtue of Corollary 7.2.10, similar to the proof of Theorem 7.2.2, we obtain

Theorem 7.2.13   *Let u be a solution of equation (7.2.1) in Q_R^+ satisfying (7.2.5) and $f \equiv 0$ in Q_R^+. Then for any nonnegative integer i and any $0 < \rho \leq R$, we have*

$$\iint_{Q_\rho^+} |D^i u|^2 dx dt \leq C \left(\frac{\rho}{R} \right)^{n+2} \int_{Q_R^+} |D^i u|^2 dx dt, \tag{7.2.25}$$

where C is a positive constant depending only on n.

Furthermore, we have

Theorem 7.2.14 *Let u be a solution of equation (7.2.1) in Q_R^+ satisfying (7.2.5) and $f \equiv 0$ in Q_R^+. Then for any $0 < \rho \leq R$, we have*

$$\iint_{Q_\rho^+} u^2 \, dxdt \leq C \left(\frac{\rho}{R}\right)^{n+4} \iint_{Q_R^+} u^2 \, dxdt, \qquad (7.2.26)$$

where C is a positive constant depending only on n.

Proof. Noticing that $u\big|_{x_n=0} = 0$ and using (7.2.25), we see that if $0 < \rho < R/2$, then

$$\iint_{Q_\rho^+} u^2 \, dxdt \leq C\rho^2 \iint_{Q_\rho^+} |D_n u|^2 \, dxdt$$

$$\leq C\rho^2 \iint_{Q_\rho^+} |Du|^2 \, dxdt$$

$$\leq C\rho^2 \left(\frac{\rho}{R}\right)^{n+2} \iint_{Q_{R/2}^+} |Du|^2 \, dxdt.$$

In addition, using Caccioppoli's inequality (7.2.19), we obtain

$$\iint_{Q_{R/2}^+} |Du|^2 \, dxdt \leq \frac{C}{R^2} \iint_{Q_R^+} u^2 \, dxdt.$$

Therefore

$$\iint_{Q_\rho^+} u^2 \, dxdt \leq C \left(\frac{\rho}{R}\right)^{n+4} \int_{Q_R^+} u^2 \, dxdt, \quad 0 < \rho < R/2,$$

that is, (7.2.25) is valid for $0 < \rho < R/2$. For $R/2 \leq \rho \leq R$, (7.2.25) holds obviously with $C \geq 2^{n+4}$. □

Applying the near lateral boundary estimate (7.2.26) for the homogeneous heat equation and the iteration lemma (Lemma 6.2.1), we may obtain the near lateral boundary estimate for the nonhomogeneous heat equation.

Theorem 7.2.15 *Let u be a solution of equation (7.2.1) in $Q_{R_0}^+$ satisfying (7.2.5), $w = D_i u \, (1 \leq i < n)$. Set*

$$F_\rho(x,t) = |D_t u|^2 + \sum_{j=1}^{n-1} |D_j w|^2 + |D_n w - (D_n w)_\rho|^2.$$

Then for any $0 < \rho \leq R \leq R_0$, we have

$$\frac{1}{\rho^{n+2+2\alpha}} \iint_{Q_\rho^+} F_\rho(x,t) \, dxdt$$

$$\leq \frac{C}{R^{n+2+2\alpha}} \iint_{Q_R^+} F_R(x,t)dxdt + C[f]_{\alpha,\alpha/2;Q_R^+}^2,$$

where C is a positive constant depending only on n.

Proof.   We divide the proof into four steps.

Step 1   Estimate $D_t u$.

Decompose u into $u = u_1 + u_2$ with u_1 and u_2 satisfying

$$\begin{cases} \dfrac{\partial u_1}{\partial t} - \Delta u_1 = f_R, & \text{in } Q_R^+, \\[2mm] u_1\Big|_{\partial_p Q_R^+} = u \end{cases}$$

and

$$\begin{cases} \dfrac{\partial u_2}{\partial t} - \Delta u_2 = f - f_R, & \text{in } Q_R^+, \\[2mm] u_2\Big|_{\partial_p Q_R^+} = 0. \end{cases}$$

It is obvious that $D_t u_1$ satisfies the homogeneous equation and zero lateral boundary value condition. So, (7.2.26) is valid for $D_t u_1$, that is

$$\iint_{Q_\rho^+} |D_t u_1|^2 dxdt \leq C\left(\frac{\rho}{R}\right)^{n+4} \iint_{Q_R^+} |D_t u_1|^2 dxdt.$$

Hence, for any $0 < \rho \leq R \leq R_0$, we have

$$\begin{aligned} \iint_{Q_\rho^+} |D_t u|^2 dxdt &\leq 2\iint_{Q_\rho^+} |D_t u_1|^2 dxdt + 2\iint_{Q_\rho^+} |D_t u_2|^2 dxdt \\ &\leq C\left(\frac{\rho}{R}\right)^{n+4} \iint_{Q_R^+} |D_t u_1|^2 dxdt + 2\iint_{Q_R^+} |D_t u_2|^2 dxdt \\ &\leq C\left(\frac{\rho}{R}\right)^{n+4} \iint_{Q_R^+} |D_t u|^2 dxdt + C\iint_{Q_R^+} |D_t u_2|^2 dxdt. \end{aligned}$$

Multiplying the equation for u_2 by $D_t u_2$ and integrating over Q_R^+, we obtain

$$\iint_{Q_R^+} |D_t u_2|^2 dxdt - \iint_{Q_R^+} \Delta u_2 D_t u_2 dxdt = \iint_{Q_R^+} (f - f_R)D_t u_2 dxdt.$$

Integrating by parts for the left hand side of the above equality with respect to spatial variables, and using Cauchy's inequality to the right hand side, we see that

$$\iint_{Q_R^+} |D_t u_2|^2 \, dx dt + \frac{1}{2} \iint_{Q_R^+} \frac{\partial}{\partial t} |Du_2|^2 \, dx dt$$

$$\leq \frac{1}{2} \iint_{Q_R^+} |D_t u_2|^2 \, dx dt + \frac{1}{2} \iint_{Q_R^+} (f - f_R)^2 \, dx dt.$$

Since $u_2\big|_{\partial_p Q_R^+} = 0$, it follows that

$$\iint_{Q_R^+} \frac{\partial}{\partial t} |Du_2|^2 \, dx dt = \int_{B_R^+} |Du_2|^2 \bigg|_{t=t_0+R^2} \, dx \geq 0.$$

Therefore

$$\iint_{Q_R^+} |D_t u_2|^2 \, dx dt \leq \iint_{Q_R^+} (f - f_R)^2 \, dx dt \leq C R^{n+2+2\alpha} [f]^2_{\alpha, \alpha/2; Q_R^+}.$$

So, for any $0 < \rho \leq R \leq R_0$, we have

$$\iint_{Q_\rho^+} |D_t u|^2 \, dx dt$$

$$\leq C \left(\frac{\rho}{R}\right)^{n+4} \iint_{Q_R^+} |D_t u|^2 \, dx dt + C R^{n+2+2\alpha} [f]^2_{\alpha, \alpha/2; Q_R^+}.$$

Step 2 Estimate $D_j w \, (1 \leq j < n)$.

Decompose w into $w = w_1 + w_2$ with w_1 and w_2 satisfying

$$\begin{cases} \dfrac{\partial w_1}{\partial t} - \Delta w_1 = 0, & \text{in } Q_R^+, \\ w_1\big|_{\partial_p Q_R^+} = w \end{cases}$$

and

$$\begin{cases} \dfrac{\partial w_2}{\partial t} - \Delta w_2 = D_i f = D_i(f - f_R), & \text{in } Q_R^+, \\ w_2\big|_{\partial_p Q_R^+} = 0. \end{cases}$$

When $j = 1, 2, \cdots, n-1$, from (7.2.26), we have

$$\iint_{Q_\rho^+} |D_j w_1|^2 \, dx dt \leq C \left(\frac{\rho}{R}\right)^{n+4} \int_{Q_R^+} |D_j w_1|^2 \, dx dt.$$

So, for any $0 < \rho \le R \le R_0$,

$$\iint_{Q_\rho^+} |D_j w|^2 dxdt \le 2 \iint_{Q_\rho^+} |D_j w_1|^2 dxdt + 2 \iint_{Q_\rho^+} |D_j w_2|^2 dxdt$$

$$\le C \left(\frac{\rho}{R}\right)^{n+4} \iint_{Q_R^+} |D_j w_1|^2 dxdt + 2 \iint_{Q_R^+} |D_j w_2|^2 dxdt$$

$$\le C \left(\frac{\rho}{R}\right)^{n+4} \iint_{Q_R^+} |D_j w|^2 dxdt + C \iint_{Q_R^+} |D_j w_2|^2 dxdt.$$

Similar to the proof of Theorem 7.2.3, we multiply the equation for w_2 by w_2 and integrate over Q_R^+ to obtain

$$\iint_{Q_R^+} |D w_2|^2 dxdt \le \iint_{Q_R^+} (f - f_R)^2 dxdt$$

$$\le C R^{n+2+2\alpha} [f]_{\alpha,\alpha/2;Q_R^+}^2. \qquad (7.2.27)$$

Then, for $j = 1, 2, \cdots, n-1$ and any $0 < \rho \le R \le R_0$, we have

$$\iint_{Q_\rho^+} |D_j w|^2 dxdt$$

$$\le C \left(\frac{\rho}{R}\right)^{n+4} \iint_{Q_R^+} |D_j w|^2 dxdt + C R^{n+2+2\alpha} [f]_{\alpha,\alpha/2;Q_R^+}^2.$$

Step 3   Estimate $D_n w$.

From the t-anisotropic Poincaré's inequality (Theorem 1.3.4), we have

$$\iint_{Q_\rho^+} |D_n w_1 - (D_n w_1)_\rho|^2 dxdt$$

$$\le C \left(\rho^2 \iint_{Q_\rho^+} |DD_n w_1|^2 dxdt + \rho^4 \iint_{Q_\rho^+} |D_t D_n w_1|^2 dxdt \right).$$

By virtue of the equality $D_t D_n w_1 = \Delta D_n w_1$, we see that

$$\iint_{Q_\rho^+} |D_n w_1 - (D_n w_1)_\rho|^2 dxdt$$

$$\le C \left(\rho^2 \iint_{Q_\rho^+} |DD_n w_1|^2 dxdt + \rho^4 \iint_{Q_\rho^+} |\Delta D_n w_1|^2 dxdt \right)$$

$$\le C \left(\rho^2 \iint_{Q_\rho^+} |D^2 w_1|^2 dxdt + \rho^4 \iint_{Q_\rho^+} |D^3 w_1|^2 dxdt \right).$$

In addition from (7.2.22), we have

$$\iint_{Q_\rho^+} |D^3 w_1|^2 dxdt \le \frac{C}{\rho^2} \iint_{Q_{2\rho}^+} |D^2 w_1|^2 dxdt.$$

So, for any $0 < \rho \le R_0/2$,

$$\iint_{Q_\rho^+} |D_n w_1 - (D_n w_1)_\rho|^2 dxdt$$

$$\le C \left(\rho^2 \iint_{Q_\rho^+} |D^2 w_1|^2 dxdt + \rho^2 \iint_{Q_{2\rho}^+} |D^2 w_1|^2 dxdt \right)$$

$$\le C\rho^2 \iint_{Q_{2\rho}^+} |D^2 w_1|^2 dxdt.$$

From this, it follows by using (7.2.25) for w_1, we derive for $0 < \rho < 2\rho \le R \le R_0$,

$$\iint_{Q_\rho^+} |D_n w_1 - (D_n w_1)_\rho|^2 dxdt$$

$$\le C\rho^2 \iint_{Q_{2\rho}^+} |D^2 w_1|^2 dxdt$$

$$\le C\rho^2 \left(\frac{\rho}{R} \right)^{n+2} \iint_{Q_R^+} |D^2 w_1|^2 dxdt.$$

So, for $0 < \rho < 2\rho \le R/2 \le R_0/2$, we have

$$\iint_{Q_\rho^+} |D_n w - (D_n w)_\rho|^2 dxdt$$

$$\le 2 \iint_{Q_\rho^+} |D_n w_1 - (D_n w_1)_\rho|^2 dxdt + 2 \iint_{Q_\rho^+} |D_n w_2 - (D_n w_2)_\rho|^2 dxdt$$

$$\le C\rho^2 \left(\frac{\rho}{R} \right)^{n+2} \iint_{Q_{R/2}^+} |D^2 w_1|^2 dxdt + C \iint_{Q_{R/2}^+} |D w_2|^2 dxdt. \qquad (7.2.28)$$

By virtue of the equation $D_t w_1 = \Delta w_1$, it follows that

$$\iint_{Q_{R/2}^+} |D^2 w_1|^2 dxdt$$

$$\le \iint_{Q_{R/2}^+} |D_{nn} w_1|^2 dxdt + 2 \sum_{j=1}^{n-1} \iint_{Q_{R/2}^+} |D D_j w_1|^2 dxdt$$

$$\leq C \iint_{Q_{R/2}^+} |D_t w_1|^2 dxdt + C \sum_{j=1}^{n-1} \iint_{Q_{R/2}^+} |DD_j w_1|^2 dxdt. \qquad (7.2.29)$$

Using Caccioppoli's inequality (7.2.21) for w_1 with

$$\vec{\lambda} = (0, \cdots, 0, (D_n w)_R),$$

we obtain

$$\iint_{Q_{R/2}^+} |D_t w_1|^2 dxdt$$

$$\leq \frac{1}{R^2} \iint_{Q_R^+} \left(\sum_{j=1}^{n-1} |D_j w_1|^2 + |D_n w_1 - (D_n w)_R|^2 \right) dxdt$$

$$\leq \frac{2}{R^2} \iint_{Q_R^+} \left(\sum_{j=1}^{n-1} |D_j w|^2 + |D_n w - (D_n w)_R|^2 \right) dxdt$$

$$+ \frac{2}{R^2} \iint_{Q_R^+} \left(\sum_{j=1}^{n-1} |D_j w_2|^2 + |D_n w_2|^2 \right) dxdt$$

$$= \frac{2}{R^2} \iint_{Q_R^+} \left(\sum_{j=1}^{n-1} |D_j w|^2 + |D_n w - (D_n w)_R|^2 \right) dxdt$$

$$+ \frac{2}{R^2} \iint_{Q_R^+} |Dw_2|^2 dxdt, \qquad (7.2.30)$$

and using Caccioppoli's inequality (7.2.20) for $D_j w_1$ $(1 \leq j < n)$, we obtain

$$\sum_{j=1}^{n-1} \iint_{Q_{R/2}^+} |DD_j w_1|^2 dxdt$$

$$\leq \frac{1}{R^2} \sum_{j=1}^{n-1} \iint_{Q_R^+} |D_j w_1|^2 dxdt$$

$$\leq \frac{2}{R^2} \sum_{j=1}^{n-1} \iint_{Q_R^+} (|D_j w|^2 + |D_j w_2|^2) dxdt$$

$$\leq \frac{2}{R^2} \sum_{j=1}^{n-1} \iint_{Q_R^+} |D_j w|^2 dxdt + \frac{2}{R^2} \iint_{Q_R^+} |Dw_2|^2 dxdt. \qquad (7.2.31)$$

Combining (7.2.29), (7.2.30) and (7.2.31), we have

$$\iint_{Q_{R/2}^+} |D^2 w_1|^2 dxdt$$

$$\leq \frac{C}{R^2} \iint_{Q_R^+} \left(\sum_{j=1}^{n-1} |D_j w|^2 + |D_n w - (D_n w)_R|^2 \right) dxdt$$

$$+ \frac{C}{R^2} \iint_{Q_R^+} |Dw_2|^2 dxdt,$$

which together with (7.2.27), (7.2.28) implies that for $0 < \rho < 2\rho \leq R/2 \leq R_0/2$, there holds

$$\iint_{Q_\rho^+} |D_n w - (D_n w)_\rho|^2 dxdt$$

$$\leq C \left(\frac{\rho}{R} \right)^{n+4} \iint_{Q_R^+} \left(\sum_{j=1}^{n-1} |D_j w|^2 + |D_n w - (D_n w)_R|^2 \right) dxdt$$

$$+ C R^{n+2+2\alpha} [f]_{\alpha,\alpha/2;Q_R^+}^2.$$

The above inequality is obviously valid for $R/4 \leq \rho \leq R \leq R_0$ (with $C \geq 4^{n+4}$).

Step 4 Estimate $D_t u$ and Dw.

Combining the estimates obtained from the above three steps, we see that for any $0 < \rho \leq R \leq R_0$,

$$\iint_{Q_\rho^+} F_\rho(x,t) dxdt$$

$$\leq C \left(\frac{\rho}{R} \right)^{n+4} \iint_{Q_R^+} F_R(x,t) dxdt + C R^{n+2+2\alpha} [f]_{\alpha,\alpha/2;Q_R^+}^2.$$

Finally, we use the iteration lemma (Lemma 6.2.1 of Chapter 6) and immediately obtain the conclusion of the theorem. $\qquad\square$

Theorem 7.2.16 *Let u be a solution of equation (7.2.1) in Q_R^+ satisfying (7.2.5), $w = D_i u \, (1 \leq i \leq n)$. Then for any $0 < \rho \leq \dfrac{R}{2}$, we have*

$$\iint_{Q_\rho^+} |Dw - (Dw)_\rho|^2 dxdt \leq C\rho^{n+2+2\alpha} M_R, \qquad (7.2.32)$$

where C is a positive constant depending only on n, and

$$M_R = \frac{1}{R^{4+2\alpha}}|u|_{0;Q_R^+}^2 + \frac{1}{R^{2\alpha}}|f|_{0;Q_R^+}^2 + [f]_{\alpha,\alpha/2;Q_R^+}^2.$$

Proof. Let $i = 1, 2, \cdots, n-1$. According to Theorem 7.2.15 and Corollary 7.2.7, we have

$$\iint_{Q_\rho^+}\left(|D_t u|^2 + \sum_{j=1}^{n-1}|D_j w|^2 + |D_n w - (D_n w)_\rho|^2\right)dxdt$$

$$\leq C\rho^{n+2+2\alpha}\left(\frac{1}{R^{n+2+2\alpha}}\iint_{Q_{R/2}^+}\left(|D_t u|^2 + \sum_{j=1}^{n-1}|D_j w|^2\right.\right.$$

$$+ |D_n w - (D_n w)_{R/2}|^2\Big)dxdt + [f]_{\alpha,\alpha/2;Q_{R/2}^+}^2\Big)$$

$$\leq C\rho^{n+2+2\alpha}\left(\frac{1}{R^{n+2+2\alpha}}\iint_{Q_{R/2}^+}\left(|D_t u|^2 + |Dw|^2\right)dxdt + [f]_{\alpha,\alpha/2;Q_{R/2}^+}^2\right)$$

$$\leq C\rho^{n+2+2\alpha}\left(\frac{1}{R^{n+6+2\alpha}}\iint_{Q_R^+}u^2 dxdt + \frac{1}{R^{2\alpha}}|f|_{0;Q_R^+}^2 + [f]_{\alpha,\alpha/2;Q_R^+}^2\right)$$

$$\leq C\rho^{n+2+2\alpha}M_R. \tag{7.2.33}$$

In particular, the above inequality implies that for $j = 1, 2, \cdots, n-1$,

$$\iint_{Q_\rho^+}|D_j w|^2 dxdt \leq C\rho^{n+2+2\alpha}M_R.$$

Thus

$$(D_j w)_\rho^2 = \frac{1}{|Q_\rho^+|^2}\left(\iint_{Q_\rho^+}|D_j w|dxdt\right)^2$$

$$\leq \frac{1}{|Q_\rho^+|}\iint_{Q_\rho^+}|D_j w|^2 dxdt \leq C\rho^{2\alpha}M_R.$$

So, for $j = 1, 2, \cdots, n-1$, we have

$$\iint_{Q_\rho^+}|D_j w - (D_j w)_\rho|^2 dxdt$$

$$\leq 2\iint_{Q_\rho^+}|D_j w|^2 dxdt + 2\iint_{Q_\rho^+}|(D_j w)_\rho|^2 dxdt$$

$$\leq C\rho^{n+2+2\alpha}M_R.$$

In addition, (7.2.33) implies

$$\iint_{Q_\rho^+} |D_n w - (D_n w)_\rho|^2 dxdt \le C\rho^{n+2+2\alpha} M_R.$$

Therefore,

$$\iint_{Q_\rho^+} |Dw - (Dw)_\rho|^2 dxdt \le C\rho^{n+2+2\alpha} M_R.$$

To sum up, we have proved (7.2.32) for $w = D_i u \, (i = 1, 2, \cdots, n-1)$. Similarly, using (7.2.33), we have

$$\iint_{Q_\rho^+} |D_t u - (D_t u)_\rho|^2 dxdt$$
$$\le 2 \iint_{Q_\rho^+} |D_t u|^2 dxdt + 2 \iint_{Q_\rho^+} |(D_t u)_\rho|^2 dxdt$$
$$\le C\rho^{n+2+2\alpha} M_R.$$

Therefore, using the equation $D_{nn} u = D_t u - \sum_{i=1}^{n-1} D_{ii} u - f$ again, we see that (7.2.32) holds for $w = D_n u$. $\qquad \square$

Theorem 7.2.17 *Let u be a solution of equation (7.2.1) in Q_R^+ satisfying (7.2.5). Then*

$$[D^2 u]_{\alpha,\alpha/2;Q_{R/2}^+} \le C \left(\frac{1}{R^{2+\alpha}} |u|_{0;Q_R^+} + \frac{1}{R^\alpha} |f|_{0;Q_R^+} + [f]_{\alpha,\alpha/2;Q_R^+} \right),$$

where C is a positive constant depending only on n.

Proof. The proof is similar to that of Theorem 7.2.11, here we need to use Theorem 7.2.16 and the interior estimate (Theorem 7.2.4). $\qquad \square$

7.2.5 *Near lateral-bottom estimate*

Since on the bottom, all the spatial derivatives are tangential derivatives, we may establish the estimate near the lateral-bottom just as we did in establishing the near lateral estimate. Here, we only list the conclusions of such kind of estimate, whose proof are similar to the corresponding parts in the proof of the near lateral estimate.

Theorem 7.2.18   *Let u be a solution of equation (7.2.1) in Q_R^{0+} satisfying (7.2.3), (7.2.5), $w = D_i u\,(1 \le i < n)$. Then for any $0 < \rho < R$, we have*

$$\sup_{I_\rho^0} \int_{B_\rho^+} u^2 dx + \iint_{Q_\rho^{0+}} |Du|^2 dxdt$$

$$\le C\Big[\frac{1}{(R-\rho)^2} \iint_{Q_R^{0+}} u^2 dxdt + (R-\rho)^2 \iint_{Q_R^{0+}} f^2 dxdt\Big],$$

$$\sup_{I_\rho^0} \int_{B_\rho^+} w^2 dx + \iint_{Q_\rho^{0+}} |Dw|^2 dxdt$$

$$\le C\Big[\frac{1}{(R-\rho)^2} \iint_{Q_R^{0+}} w^2 dxdt + \iint_{Q_R^{0+}} (f - f_R)^2 dxdt\Big],$$

$$\sup_{I_\rho^0} \int_{B_\rho^+} |Du|^2 dx + \iint_{Q_\rho^{0+}} |D_t u|^2 dxdt$$

$$\le C\Big[\frac{1}{(R-\rho)^2} \iint_{Q_R^{0+}} |Du|^2 dxdt + \iint_{Q_R^{0+}} f^2 dxdt\Big],$$

where $f_R = \dfrac{1}{|Q_R^{0+}|} \iint_{Q_R^{0+}} f(x,t) dxdt$ and C is a positive constant depending only on n.

Corollary 7.2.11   *Let u be a solution of equation (7.2.1) in Q_R^{0+} satisfying (7.2.3), (7.2.5), $w = D_i u\,(1 \le i < n)$. Then*

$$\sup_{I_{R/2}^0} \int_{B_{R/2}^+} w^2 dx + \iint_{Q_{R/2}^{0+}} (|D_t u|^2 + |Dw|^2) dxdt$$

$$\le \frac{C}{R^4} \iint_{Q_R^{0+}} u^2 dxdt + CR^{n+2}|f|_{0;Q_R^{0+}}^2 + CR^{n+2+2\alpha}[f]_{\alpha,\alpha/2;Q_R^{0+}}^2,$$

where C is a positive constant depending only on n.

Corollary 7.2.12   *If $f \equiv 0$ in Q_R^{0+}, then for any nonnegative integer k and any $0 < \rho < R$, we have*

$$\sup_{I_\rho^0} \int_{B_\rho^+} |D^k u|^2 dx + \iint_{Q_\rho^{0+}} |D^{k+1} u|^2 dxdt \le \frac{C}{(R-\rho)^2} \iint_{Q_R^{0+}} |D^k u|^2 dxdt,$$

where C is a positive constant depending only on n.

Proof.   The proof is similar to that of Corollary 7.2.8. Here, we need to use the following fact: if u is appropriately smooth in $\overline{Q}_R^{0+}$ and satisfies the homogeneous heat equation and the conditions (7.2.3), (7.2.5), then $D_t u$ satisfies the same equation and boundary value condition. In fact, it is obvious that $D_t u$ satisfies the equation and the lateral boundary value condition. In addition, using the smoothness of u and the equation, we have

$$D_t u(x, 0) = \lim_{t \to 0^+} D_t u(x, t) = \lim_{t \to 0^+} \Delta u(x, t) = \Delta u(x, 0) = 0,$$

that is $D_t u$ satisfies the bottom boundary value condition. $\qquad\square$

Corollary 7.2.13 *If $f \equiv 0$ in Q_1^{0+}, then for any nonnegative integers k and i, we have*

$$\sup_{I_{1/2}^0} \int_{B_{1/2}^+} |D^{k+i} u|^2 dx + \iint_{Q_{1/2}^{0+}} |D^{k+i+1} u|^2 dx dt \le C \iint_{Q_1^{0+}} |D^i u|^2 dx dt,$$

where C is a positive constant depending only on n and k.

Corollary 7.2.14 *If $f \equiv 0$ in Q_R^{0+}, then for any nonnegative integer i, we have*

$$\sup_{Q_{R/2}^{0+}} |D^i u| \le C \left(\frac{1}{R^{n+2}} \iint_{Q_R^{0+}} |D^i u|^2 dx dt \right)^{1/2},$$

where C is a positive constant depending only on n.

Theorem 7.2.19 *Let u be a solution of equation (7.2.1) in Q_R^{0+} satisfying (7.2.3), (7.2.5), $f \equiv 0$ in Q_R^{0+}. Then for any nonnegative integer i and any $0 < \rho \le R$, we have*

$$\iint_{Q_\rho^{0+}} |D^i u|^2 dx dt \le C \left(\frac{\rho}{R} \right)^{n+2} \iint_{Q_R^{0+}} |D^i u|^2 dx dt,$$

where C is a positive constant depending only on n.

Theorem 7.2.20 *Let u be a solution of equation (7.2.1) in Q_R^{0+} satisfying (7.2.3), (7.2.5), $f \equiv 0$ in Q_R^{0+}. Then for any nonnegative integer i and any $0 < \rho \le R$, we have*

$$\iint_{Q_\rho^{0+}} |D^i u|^2 dx dt \le C \left(\frac{\rho}{R} \right)^{n+4} \iint_{Q_R^{0+}} |D^i u|^2 dx dt, \qquad (7.2.34)$$

where C is a positive constant depending only on n.

Proof. Noticing that $D^i u \big|_{t=0} = 0$ and using equation (7.2.1) and Theorem 7.2.19, we see that if $0 < \rho < R/2$, then

$$\iint_{Q_\rho^{0+}} |D^i u|^2 dx dt \le C \rho^4 \iint_{Q_\rho^{0+}} |D_t D^i u|^2 dx dt$$

$$=C\rho^4 \iint_{Q_\rho^{0+}} |\Delta D^i u|^2 dxdt$$

$$\leq C\rho^4 \iint_{Q_\rho^{0+}} |D^{i+2}u|^2 dxdt$$

$$\leq C\rho^4 \left(\frac{\rho}{R}\right)^{n+2} \iint_{Q_{R/2}^{0+}} |D^{i+2}u|^2 dxdt.$$

In addition, using Corollary 7.2.12, we obtain

$$\iint_{Q_{R/2}^{0+}} |D^{i+2}u|^2 dxdt \leq \frac{C}{R^2} \iint_{Q_{3R/4}^{0+}} |D^{i+1}u|^2 dxdt \leq \frac{C}{R^4} \iint_{Q_R^{0+}} |D^i u|^2 dxdt.$$

Therefore

$$\iint_{Q_\rho^+} |D^i u|^2 dxdt \leq C \left(\frac{\rho}{R}\right)^{n+6} \int_{Q_R^+} |D^i u|^2 dxdt$$

$$\leq C \left(\frac{\rho}{R}\right)^{n+4} \int_{Q_R^+} |D^i u|^2 dxdt, \quad 0 < \rho < R/2,$$

that is, (7.2.34) is valid for $0 < \rho < R/2$. For $R/2 \leq \rho \leq R$, (7.2.34) holds obviously with $C \geq 2^{n+4}$. □

Theorem 7.2.21 *Let u be a solution of equation (7.2.1) in Q_R^{0+} satisfying (7.2.3), (7.2.5), $w = D_i u\, (1 \leq i < n)$. Then for any $0 < \rho \leq R \leq R_0$, we have*

$$\frac{1}{\rho^{n+2+2\alpha}} \iint_{Q_\rho^{0+}} (|D_t u|^2 + |Dw|^2) dxdt$$

$$\leq \frac{C}{R^{n+2+2\alpha}} \iint_{Q_R^{0+}} (|D_t u|^2 + |Dw|^2) dxdt + C[f]_{\alpha,\alpha/2;Q_R^{0+}}^2,$$

where C is a positive constant depending only on n.

Proof. Similar to the proof of Theorem 7.2.15, we decompose u and w in the same manner. The estimates on $D_t u$ and $D_j w\, (1 \leq j < n)$ are quite similar to the corresponding estimates in Theorem 7.2.15. To estimate $D_t u_1$, we need the fact used in the proof of Corollary 7.2.12. The estimate on $D_n w$ is even easier than the corresponding estimate in the proof of Theorem 7.2.15. In fact, according to (7.2.34), we may estimate $D_n u$ in the same way as what we do on $D_j w\, (1 \leq j < n)$. □

Theorem 7.2.22 *Let u be a solution of equation (7.2.1) in Q_R^{0+} satisfying (7.2.3), (7.2.5), $w = D_i u \, (1 \leq i \leq n)$. Then for any $0 < \rho \leq \dfrac{R}{2}$, we have*

$$\iint_{Q_\rho^{0+}} |Dw - (Dw)_\rho|^2 dxdt \leq C\rho^{n+2+2\alpha} M_R,$$

where C is a positive constant depending only on n, and

$$M_R = \frac{1}{R^{4+2\alpha}} |u|^2_{0;Q_R^{0+}} + \frac{1}{R^{2\alpha}} |f|^2_{0;Q_R^{0+}} + [f]^2_{\alpha,\alpha/2;Q_R^{0+}}.$$

Theorem 7.2.23 *Let u be a solution of equation (7.2.1) in Q_R^{0+} satisfying (7.2.3), (7.2.5). Then*

$$[D^2 u]_{\alpha,\alpha/2;Q_{R/2}^{0+}} \leq C \left(\frac{1}{R^{2+\alpha}} |u|_{0;Q_R^{0+}} + \frac{1}{R^\alpha} |f|_{0;Q_R^{0+}} + [f]_{\alpha,\alpha/2;Q_R^{0+}} \right),$$

where C is a positive constant depending only on n.

7.2.6 Schauder's estimates for general linear parabolic equations

Now we turn to the general linear parabolic equation and consider the corresponding first boundary value problem

$$\frac{\partial u}{\partial t} - a_{ij}(x,t)D_{ij}u + b_i(x,t)D_i u + c(x,t)u = f(x,t), \quad (x,t) \in Q_T, \tag{7.2.35}$$

$$u\Big|_{\partial_p Q_T} = \varphi(x,t), \tag{7.2.36}$$

where $Q_T = \Omega \times (0,T)$, $\Omega \subset \mathbb{R}^n$ is a bounded domain, $T > 0$. Similar to the case of elliptic equations, we may use the interior estimates and the near boundary estimates (including near bottom estimates, near lateral boundary estimates and near lateral-bottom estimates) for the heat equations, to establish the corresponding estimates for (7.2.35), and then use the finite covering technique to derive the global Schauder's estimates. Exactly speaking, we have the following theorem.

Theorem 7.2.24 *Let $0 < \alpha < 1$, $\partial\Omega \in C^{2,\alpha}$, $a_{ij}, b_i, c \in C^{\alpha,\alpha/2}(\overline{Q}_T)$, $a_{ij} = a_{ji}$ and equation (7.2.35) satisfies the uniform parabolicity conditions, that is, for some constants $0 < \lambda \leq \Lambda$,*

$$\lambda|\xi|^2 \leq a_{ij}(x,t)\xi_i\xi_j \leq \Lambda|\xi|^2, \quad \forall \xi \in \mathbb{R}^n, \, (x,t) \in Q_T.$$

In addition, assume that $f \in C^{\alpha,\alpha/2}(\overline{Q}_T)$, $\varphi \in C^{2+\alpha,1+\alpha/2}(\overline{Q}_T)$. If $u \in C^{2+\alpha,1+\alpha/2}(\overline{Q}_T)$ is the solution of the first initial-boundary value problem (7.2.35), (7.2.36), then

$$|u|_{2+\alpha,1+\alpha/2;Q_T} \le C(|f|_{\alpha,\alpha/2;Q_T} + |\varphi|_{2+\alpha,1+\alpha/2;Q_T} + |u|_{0;Q_T}), \quad (7.2.37)$$

where C is a positive constant depending only on n, α, λ, Λ, Ω, T and the $C^{2+\alpha,1+\alpha/2}(\overline{Q}_T)$ norm of a_{ij}, b_i, c.

Remark 7.2.5   *Note that in order the estimate holds, it suffices to require $u \in C^{2+\alpha,1+\alpha/2}(Q_T) \cap C(\overline{Q}_T)$ instead of $u \in C^{2+\alpha,1+\alpha/2}(\overline{Q}_T)$.*

Exercises

1. Prove Theorem 7.1.1 and Remark 7.1.2.
2. Prove Remark 7.2.3.
3. Establish the near lateral-bottom estimate for the heat equation.
4. Prove Theorem 7.2.24 and Remark 7.2.6.
5. Let $u \in C^{2+\alpha,1+\alpha/2}(\overline{B}_T)$ be a solution of the following initial-boundary value problem

$$\begin{cases} \dfrac{\partial u}{\partial t} - \Delta u + u^p = f, & (x,t) \in B_T = B \times (0,T), \\[2mm] u\Big|_{\partial_p B_T} = 0, & \end{cases}$$

where $0 < \alpha < 1$, B is the unit ball of $\mathbb{R}^n$, $p \ge \alpha$. Prove that

$$|u|_{2,\alpha;\Omega} \le C,$$

where $C > 0$ depending only on n, p, $|f|_{\alpha,\alpha/2;B_T}$ and $|u|_{0;B_T}$.

6. Establish Schauder's estimates for solutions of the following initial-boundary value problem of fourth order parabolic equation

$$\begin{cases} \dfrac{\partial u}{\partial t} + \Delta^2 u = f, & (x,t) \in Q_T = \Omega \times (0,T), \\[2mm] u = \dfrac{\partial u}{\partial \nu} = 0, & (x,t) \in \partial\Omega \times (0,T), \\[2mm] u(x,0) = 0, & x \in \Omega. \end{cases}$$

Chapter 8

Existence of Classical Solutions for Linear Equations

In this chapter, we establish the existence theory of classical solutions for linear elliptic and parabolic equations of second order.

8.1 Maximum Principle and Comparison Principle

The existence of classical solutions is based on Schauder's estimates. In addition, the L^∞ norm estimate on solutions is also needed. In this section, we introduce the maximum principle, which will be used to establish the L^∞ norm estimate and comparison principle on classical solutions.

8.1.1 The case of elliptic equations

Consider the following linear elliptic equation

$$Lu \equiv -a_{ij}(x)D_{ij}u + b_i(x)D_iu + c(x)u = f(x), \quad x \in \Omega, \qquad (8.1.1)$$

where $\Omega \subset \mathbb{R}^n$ is a bounded domain, $a_{ij} = a_{ji}$ and for some constant $\lambda > 0$,

$$a_{ij}(x)\xi_i\xi_j \geq \lambda|\xi|^2, \quad \forall \xi \in \mathbb{R}^n, \ x \in \Omega.$$

Theorem 8.1.1 *(Maximum Principle) Let $c(x) \geq 0$, $b_i(x)$ and $c(x)$ be bounded in Ω, $u \in C^2(\Omega) \cap C(\overline{\Omega})$ satisfy $Lu = f \leq 0 \ (\geq 0)$ in Ω. Then*

$$\sup_\Omega u(x) \leq \sup_{\partial\Omega} u_+(x) \quad \left(\inf_\Omega u(x) \geq \inf_{\partial\Omega} u_-(x)\right),$$

where $u_+ = \max\{u, 0\}$, $u_- = \min\{u, 0\}$.

Proof. We first show that if $f < 0$, then the conclusion holds. If the conclusion were not true, then there would exist $x^0 \in \Omega$, such that

$$u(x^0) = \max_{\overline{\Omega}} u(x) > 0.$$

Thus

$$(D_{ij}u(x^0))_{n \times n} \leq 0, \quad D_i u(x^0) = 0.$$

On the other hand, since $(a_{ij}(x^0))_{n \times n} \geq 0$, $c(x^0) \geq 0$, we have

$$Lu(x^0) = -a_{ij}(x^0)D_{ij}u(x^0) + b_i(x^0)D_i u(x^0) + c(x^0)u(x^0) \geq 0,$$

which contradicts $f(x^0) < 0$ and hence the conclusion is valid for the case $f < 0$.

Now, we turn to the general case $f \leq 0$. If we may find an auxiliary function $h \in C^2(\Omega) \cap C(\overline{\Omega})$, which satisfies

$$h \geq 0, \quad Lh < 0, \quad \text{in } \Omega,$$

then for any $\varepsilon > 0$, there holds

$$L(u + \varepsilon h) = Lu + \varepsilon Lh < 0 \quad \text{in } \Omega.$$

So, according to the above proved conclusion, we infer

$$\sup_{\Omega}\{u(x) + \varepsilon h(x)\} \leq \sup_{\partial\Omega}\{u(x) + \varepsilon h(x)\}_+.$$

Thus

$$\sup_{\Omega} u(x) \leq \sup_{\Omega}\{u(x) + \varepsilon h(x)\}$$
$$\leq \sup_{\partial\Omega}\{u(x) + \varepsilon h(x)\}_+$$
$$\leq \sup_{\partial\Omega} u_+(x) + \varepsilon \sup_{\partial\Omega} h(x).$$

Letting $\varepsilon \to 0$, we get the desired conclusion. There are many functions with the above properties, for example, we may take $h(x) = e^{\alpha x_1}$, where $\alpha > 0$ is a constant to be determined. Noticing that

$$Lh(x) = e^{\alpha x_1}(-\alpha^2 a_{11}(x) + \alpha b_1(x) + c(x))$$
$$\leq e^{\alpha x_1}(-\alpha^2 \lambda + \alpha b_1(x) + c(x)), \quad x \in \Omega,$$

and $b_i(x)$ and $c(x)$ are bounded in Ω, we need only to take α to be sufficiently large, such that $Lh < 0$ in Ω.

As for the case of $f \geq 0$, we may consider $-u$ instead of u, and get the desired conclusion. □

Remark 8.1.1 *From the proof of the theorem, we see that the condition that $b_i(x)$ $(i = 1, \cdots, n)$ are bounded in Ω can be replaced by the boundedness of $b_i(x)$ for some i.*

Remark 8.1.2 *Theorem 8.1.1 can also be proved by the following approach. Let $v = u - \sup_{\partial\Omega} u_+$ and first show that $v \leq 0$ if $c > 0$. As for the general case $c \geq 0$, we may let $v = hw$, and consider the equation for w, where h is an auxiliary function to be determined.*

Using the maximum principle, we can now establish the comparison principle.

Theorem 8.1.2 *(Comparison Principle) Let $c(x) \geq 0$, $b_i(x)$ and $c(x)$ be bounded in Ω, $v, w \in C^2(\Omega) \cap C(\overline{\Omega})$ satisfy $Lv \leq Lw$ in Ω and $v\big|_{\partial\Omega} \leq w\big|_{\partial\Omega}$. Then*

$$v(x) \leq w(x), \quad \forall x \in \Omega.$$

Proof. It suffices to use Theorem 8.1.1 by taking $u = v - w$. □

By a suitable choice of the functions v and w in the comparison principle, we may obtain the a priori bound of solutions of the Dirichlet problem for equation (8.1.1).

Theorem 8.1.3 *Let $c(x) \geq 0$, $b_i(x)$ and $c(x)$ be bounded in Ω, $u \in C^2(\Omega) \cap C(\overline{\Omega})$ satisfy $Lu = f$ in Ω. Then*

$$\sup_{\Omega} |u| \leq \sup_{\partial\Omega} |u| + C \sup_{\Omega} |f|,$$

where C depends only on λ, $\text{diam}\,\Omega$ and the bound of $b_i(x)$ in Ω.

Proof. Without loss of generality, we assume that f is bounded in Ω; otherwise the conclusion is obvious. Set

$$d = \text{diam}\,\Omega, \quad \beta = \sup_{\Omega} |b_1|.$$

Take a fixed point $x^0 = (x_1^0, x_2^0, \cdots, x_n^0) \in \overline{\Omega}$, such that

$$x_1^0 < x_1, \quad \forall x = (x_1, x_2, \cdots, x_n) \in \Omega.$$

Then

$$0 \leq x_1 - x_1^0 \leq d, \quad \forall x = (x_1, x_2, \cdots, x_n) \in \Omega.$$

Let

$$g(x_1) = \left(e^{\alpha d} - e^{\alpha(x_1 - x_1^0)}\right) \sup_{\Omega} |f|,$$

$$w(x) = \sup_{\partial\Omega} |u| + g(x_1), \quad x \in \overline{\Omega},$$

where $\alpha > 0$ is a constant to be determined. Then $w \in C^2(\Omega) \cap C(\overline{\Omega})$, and for any $x \in \Omega$, w satisfies

$$w(x) \geq g(x_1) \geq 0,$$

$$\begin{aligned}
Lw(x) &= -a_{11}(x)g''(x_1) + b_1(x)g'(x_1) + c(x)w(x) \\
&\geq -a_{11}(x)g''(x_1) + b_1(x)g'(x_1) \\
&= e^{\alpha(x_1 - x_1^0)}(\alpha^2 a_{11}(x) - \alpha b_1(x)) \sup_{\Omega} |f| \\
&\geq \alpha(\alpha\lambda - \beta) \sup_{\Omega} |f|.
\end{aligned}$$

Choosing $\alpha = (\beta + 1)/\lambda + 1$ yields

$$Lw(x) \geq \sup_{\Omega} |f|, \quad \forall x \in \Omega.$$

Similarly, we have

$$L\{-w(x)\} \leq -\sup_{\Omega} |f|, \quad \forall x \in \Omega.$$

Thus

$$L\{-w(x)\} \leq Lu(x) \leq Lw(x), \quad \forall x \in \Omega.$$

In addition, it is obvious that

$$-w(x) \leq u(x) \leq w(x), \quad \forall x \in \partial\Omega,$$

and so, from the comparison principle, the desired conclusion is valid for $C = e^{((\beta+1)/\lambda+1)d}$. $\qquad\square$

8.1.2 *The case of parabolic equations*

Consider the following linear parabolic equation

$$Lu \equiv \frac{\partial u}{\partial t} - a_{ij}(x,t)D_{ij}u + b_i(x,t)D_i u + c(x,t)u$$

$$= f(x,t), \quad (x,t) \in Q_T, \tag{8.1.2}$$

where $Q_T = \Omega \times (0, T)$, $\Omega \subset \mathbb{R}^n$ is a bounded domain, $a_{ij} = a_{ji}$ and

$$a_{ij}(x, t)\xi_i\xi_j \geq 0, \quad \forall \xi \in \mathbb{R}^n, \ (x, t) \in Q_T.$$

Theorem 8.1.4 *(Maximum Principle) Let $c(x, t) \geq 0$ be bounded in Q_T, $u \in C^2(Q_T) \cap C(\overline{Q}_T)$ satisfy $Lu = f \leq 0 (\geq 0)$ in Q_T. Then*

$$\sup_{Q_T} u(x, t) \leq \sup_{\partial_p Q_T} u_+(x, t) \quad \left(\inf_{Q_T} u(x, t) \geq \inf_{\partial_p Q_T} u_-(x, t)\right).$$

Proof. We first show the conclusion when $f < 0$. If the conclusion were false, then there would exist a point $(x^0, t_0) \in \overline{Q}_T \backslash \partial_p Q_T$ such that

$$u(x^0, t_0) = \max_{\overline{Q}_T} u(x, t) > 0.$$

Thus

$$\frac{\partial u(x^0, t_0)}{\partial t} \geq 0, \quad (D_{ij}u(x^0, t_0))_{n \times n} \leq 0,$$

$$D_i u(x^0, t_0) = 0, \quad u(x^0, t_0) \geq 0.$$

In addition, since $(a_{ij}(x^0, t_0))_{n \times n} \geq 0$, $c(x^0, t_0) \geq 0$, we have

$$Lu(x^0, t_0) = \frac{\partial u(x^0, t_0)}{\partial t} - a_{ij}(x^0, t_0)D_{ij}u(x^0, t_0)$$
$$+ b_i(x^0, t_0)D_i u(x^0, t_0) + c(x^0, t_0)u(x^0, t_0) \geq 0,$$

which contradicts $f(x^0, t_0) < 0$, and hence the conclusion is valid if $f < 0$.

Next we consider the general case of $f \leq 0$. Let $h(x) = e^{-\alpha t}$, where $\alpha = \sup_{Q_T} c(x, t) + 1$. Then $h \in C^2(Q_T) \cap C(\overline{Q}_T)$ and

$$h \geq 0, \quad Lh = \frac{\partial h}{\partial t} + ch = (-\alpha + c)e^{-\alpha t} < 0, \quad \text{in } Q_T.$$

Hence, for any $\varepsilon > 0$, we have

$$L(u + \varepsilon h) = Lu + \varepsilon Lh < 0 \quad \text{in } Q_T.$$

According to the proved conclusion, it follows that

$$\sup_{Q_T}\{u(x, t) + \varepsilon h(x, t)\} \leq \sup_{\partial_p Q_T}\{u(x, t) + \varepsilon h(x, t)\}_+.$$

Therefore

$$\sup_{Q_T} u(x, t) \leq \sup_{Q_T}\{u(x, t) + \varepsilon h(x, t)\}$$

$$\leq \sup_{\partial_p Q_T} \{u(x,t) + \varepsilon h(x,t)\}+$$

$$\leq \sup_{\partial_p Q_T} u_+(x,t) + \varepsilon \sup_{\partial_p Q_T} h(x,t).$$

Letting $\varepsilon \to 0$, we get the desired conclusion.

For the case of $f \geq 0$, we need only to consider $-u$ instead of u. □

Remark 8.1.3 *In Theorem 8.1.2, we merely assume the parabolicity condition rather than the uniform parabolicity condition for equation (8.1.2), and no boundedness condition for b_i is assumed.*

If we do not assume the condition $c(x,t) \geq 0$, then the above maximum principle is invalid. However, we still have the following useful result.

Theorem 8.1.5 *Let $c(x,t)$ be bounded in Q_T, $u \in C^2(Q_T) \cap C(\overline{Q}_T)$ satisfy $Lu = f \leq 0$ in Q_T and $\sup\limits_{\partial_p Q_T} u(x,t) \leq 0$. Then*

$$\sup_{Q_T} u(x,t) \leq 0.$$

Proof. Let $c_0 = \inf\limits_{Q_T} c(x,t)$ and set

$$v(x,t) = e^{c_0 t} u(x,t).$$

Then v satisfies

$$\frac{\partial v}{\partial t} - a_{ij}(x,t)D_{ij}v + b_i(x,t)D_i v + (c(x,t) - c_0)v = e^{c_0 t} f(x,t) \leq 0,$$

$$(x,t) \in Q_T.$$

Noticing that $c(x,t) - c_0 \geq 0$ in Q_T, from Theorem 8.1.4, we obtain

$$\sup_{Q_T} v(x,t) \leq \sup_{\partial_p Q_T} v_+(x,t) = \sup_{\partial_p Q_T} e^{c_0 t} u_+(x,t) = 0,$$

which leads to the conclusion of the theorem. □

Applying Theorem 8.1.5, we may establish the following comparison principle.

Theorem 8.1.6 *(Comparison Principle) Assume that $c(x,t)$ is bounded in Q_T, $v, w \in C^2(Q_T) \cap C(\overline{Q}_T)$ satisfy $Lv \leq Lw$ in Q_T and $v\big|_{\partial_p Q_T} \leq w\big|_{\partial_p Q_T}$. Then*

$$v(x,t) \leq w(x,t), \quad \forall (x,t) \in Q_T.$$

Proof. We may take $u = v - w$ in Theorem 8.1.5 to obtain the desired conclusion. $\square$

Similar to the case of elliptic equations, by suitably choosing the functions v and w in the comparison principle, we may obtain the a priori bound for solutions of the first initial-boundary value problem for equation (8.1.2).

Theorem 8.1.7 *Let $c(x, t) \geq 0$ and be bounded in Q_T, $u \in C^2(Q_T) \cap C(\overline{Q}_T)$ satisfy $Lu = f$ in Q_T. Then*

$$\sup_{Q_T} |u| \leq \sup_{\partial_p Q_T} |u| + T \sup_{Q_T} |f|.$$

Proof. Without loss of generality, we assume that f is bounded in Q_T; otherwise the conclusion is obvious. Let

$$w(x, t) = \sup_{\partial_p Q_T} |u| + t \sup_{Q_T} |f|, \quad (x, t) \in \overline{Q}_T.$$

Then $w \in C^2(Q_T) \cap C(\overline{Q}_T)$ and for any $(x, t) \in Q_T$, we have

$$Lw(x, t) = \sup_{Q_T} |f| + c(x, t)w(x, t) \geq \sup_{Q_T} |f|.$$

Similarly, for any $(x, t) \in Q_T$, we also have

$$L\{-w(x, t)\} \leq - \sup_{\Omega} |f|.$$

Thus

$$L\{-w(x, t)\} \leq Lu(x, t) \leq Lw(x, t), \quad \forall (x, t) \in Q_T.$$

In addition, it is obvious that

$$-w(x, t) \leq u(x, t) \leq w(x, t), \quad \forall (x, t) \in \partial_p Q_T,$$

from which and the comparison principle we get the desired conclusion. $\square$

Theorem 8.1.8 *Let $c(x, t)$ be bounded in Q_T, $c_0 = \min\{0, \inf_{Q_T} c(x, t)\}$, $u \in C^2(Q_T) \cap C(\overline{Q}_T)$ satisfy $Lu = f$ in Q_T. Then*

$$\sup_{Q_T} |u| \leq e^{-c_0 T} \left(\sup_{\partial_p Q_T} |u| + T \sup_{Q_T} |f| \right).$$

Proof. Let

$$v(x,t) = e^{c_0 t} u(x,t), \quad (x,t) \in \overline{Q}_T.$$

Then v satisfies

$$\frac{\partial v}{\partial t} - a_{ij}(x,t)D_{ij}v + b_i(x,t)D_i v + (c(x,t) - c_0)v = e^{c_0 t} f(x,t),$$

$$(x,t) \in Q_T.$$

Noticing that $c(x,t) - c_0 \geq 0$ in Q_T, from Theorem 8.1.7, we have

$$\sup_{Q_T} |v| \leq \sup_{\partial_p Q_T} |v| + T \sup_{Q_T} |e^{c_0 t} f|,$$

that is

$$\sup_{Q_T} |e^{c_0 t} u| \leq \sup_{\partial_p Q_T} |e^{c_0 t} u| + T \sup_{Q_T} |e^{c_0 t} f|.$$

It follows from $c_0 \leq 0$ that

$$\sup_{Q_T} |u| \leq e^{-c_0 T} \left(\sup_{\partial_p Q_T} |u| + T \sup_{Q_T} |f| \right).$$

$\square$

8.2 Existence and Uniqueness of Classical Solutions for Linear Elliptic Equations

In this section, we first investigate the existence of $C^{2,\alpha}(\overline{\Omega})$ solutions and $C^{2,\alpha}(\Omega) \cap C(\overline{\Omega})$ solutions for Poisson's equation and then investigate the existence of the same kinds of solutions for general linear elliptic equations.

8.2.1 Existence and uniqueness of the classical solution for Poisson's equation

Consider the Dirichlet problem for Poisson's equation

$$-\Delta u = f, \quad x \in \Omega, \tag{8.2.1}$$

$$u\big|_{\partial\Omega} = \varphi, \tag{8.2.2}$$

where $\Omega \subset \mathbb{R}^n$ is a bounded domain. We first prove the existence and uniqueness of its $C^{2,\alpha}(\overline{\Omega})$ solution.

Theorem 8.2.1 *Let $\partial\Omega \in C^\infty$, $0 < \alpha < 1$, $f \in C^\alpha(\overline{\Omega})$, $\varphi \in C^{2,\alpha}(\overline{\Omega})$. Then problem (8.2.1), (8.2.2) admits a unique solution $u \in C^{2,\alpha}(\overline{\Omega})$.*

Proof. Without loss of generality, we assume that $\varphi \equiv 0$. Otherwise, we may consider the equation for $w = u - \varphi$. Using the standard approximation technique, we may choose a function $f_\varepsilon \in C^\infty(\overline{\Omega})$, such that

$$|f_\varepsilon|_{\alpha;\Omega} \le 2|f|_{\alpha;\Omega}$$

and f_ε converges to f uniformly on $\overline{\Omega}$ as $\varepsilon \to 0$. Consider the approximate problem

$$\begin{cases} -\Delta u = f_\varepsilon(x), & x \in \Omega, \\ u\Big|_{\partial\Omega} = 0. \end{cases}$$

By the L^2 theory (Theorem 2.2.5 of Chapter 2), we see that the above problem admits a unique solution $u_\varepsilon \in C^\infty(\overline{\Omega})$. From the global Schauder's estimate (Theorem 6.3.1 of Chapter 6), we have

$$|u_\varepsilon|_{2,\alpha;\Omega} \le C(|f_\varepsilon|_{\alpha;\Omega} + |u_\varepsilon|_{0;\Omega}).$$

Using the maximum principle (Theorem 8.1.3) yields

$$|u_\varepsilon|_{0;\Omega} \le C|f_\varepsilon|_{0;\Omega} \le C|f_\varepsilon|_{\alpha;\Omega} \le C|f|_{\alpha;\Omega}.$$

Thus

$$|u_\varepsilon|_{2,\alpha;\Omega} \le C|f|_{\alpha;\Omega}.$$

The constant C in the above formula is independent of ε. By Arzela-Ascoli's theorem, there exists a subsequence of $\{u_\varepsilon\}$, denoted by itself, and a function $u \in C^{2,\alpha}(\overline{\Omega})$, such that

$$u_\varepsilon \to u, \quad Du_\varepsilon \to Du, \quad D^2 u_\varepsilon \to D^2 u$$

uniformly on $\overline{\Omega}$ as $\varepsilon \to 0$. Letting $\varepsilon \to 0$ in the approximate problem, we see that u satisfies equation (8.2.1) and the boundary value condition $u\Big|_{\partial\Omega} = 0$. So, we have proved the existence of solutions. The uniqueness follows from the maximum principle. $\qquad\square$

In Theorem 8.2.1, it is assumed that the domain Ω has C^∞ smooth boundary, which means that the theorem could not be applied even to Poisson's equation in square domain. In what follows, we will relax the restriction on the domain, but the solution space is enlarged to be $C^{2,\alpha}(\Omega) \cap C(\overline{\Omega})$ in the same time.

Definition 8.2.1   We call a domain Ω to have exterior ball property, if for any $x^0 \in \partial\Omega$, there exists $R > 0$ and $y \in \mathbb{R}^n \backslash \overline{\Omega}$ such that $\overline{B}_R(y) \cap \overline{\Omega} = \{x^0\}$. If such R can be chosen to be independent of x^0, then the domain Ω is said to have uniform exterior ball property.

Theorem 8.2.2   *Assume that Ω has the exterior ball property, and there exists a sequence of subdomains $\{\Omega_k\}$ with C^∞ boundary, such that $\overline{\Omega}_k \subset \Omega_{k+1}$ and $\partial\Omega_k$ converges to $\partial\Omega$ uniformly. Let $0 < \alpha < 1$, $f \in C^\alpha(\overline{\Omega})$, $\varphi \in C^{2,\alpha}(\overline{\Omega})$. Then problem (8.2.1), (8.2.2) admits a unique solution $u \in C^{2,\alpha}(\Omega) \cap C(\overline{\Omega})$.*

Proof.   Without loss of generality, we may assume that $\varphi \equiv 0$; otherwise, we set $w = u - \varphi$ and consider the equation for w. Consider the approximate problem

$$\begin{cases} -\Delta u = f(x), & x \in \Omega_k, \\ u\Big|_{\partial\Omega_k} = 0, \end{cases}$$

which admits a unique solution $u_k \in C^{2,\alpha}(\overline{\Omega}_k)$ by Theorem 8.2.1. For fixed positive integer m, according to Schauder's interior estimate (Theorem 6.2.8) and the maximum norm estimate, we have

$$\begin{aligned} |u_k|_{2,\alpha;\Omega_m} &\leq C_1(|f|_{\alpha;\Omega_k} + |u_k|_{0;\Omega_k}) \\ &\leq C_2|f|_{\alpha;\Omega_k} \leq C_2|f|_{\alpha;\Omega}, \quad \forall k > m, \end{aligned}$$

where C_1 and C_2 are independent of k. By a diagonal process, we may obtain a subsequence $\{u_{k_i}\}_{i=1}^\infty$ of $\{u_k\}_{k=1}^\infty$ and a function $u \in C^{2,\alpha}(\Omega)$, such that for any fixed $m \geq 1$,

$$u_{k_i} \to u, \quad Du_{k_i} \to Du, \quad D^2 u_{k_i} \to D^2 u$$

uniformly on Ω_m as $i \to \infty$. Therefore, u satisfies equation (8.2.1) in Ω.

Now, we use the barrier function technique to show that $u\Big|_{\partial\Omega} = 0$, that is for any fixed $x^0 \in \partial\Omega$, $u(x^0) = \lim\limits_{\substack{x \in \Omega \\ x \to x^0}} u(x) = 0$. For this purpose, it suffices to construct a continuous function $w(x) \geq 0$, such that $w(x^0) = 0$ and

$$|u(x)| \leq Cw(x), \quad x \in \Omega \cap B_\delta(x^0).$$

Such a function is called an **exterior barrier function**.

Set

$$w(x) = M\big(e^{-\beta R^2} - e^{-\beta|x-y|^2}\big), \quad x \in \overline{\Omega},$$

where R and y are the radius and the center of the exterior ball at the point x^0 respectively, $\beta > 0$ and $M > 0$ are constants to be determined. It is easily seen that the function $w(x)$ has the following properties:

i) $w(x^0) = 0$, $w(x) > 0$ for all $x \in \overline{\Omega} \backslash \{x^0\}$;

ii) $w \in C^2(\overline{\Omega})$ and for appropriate large $\beta > 0$ and sufficiently large $M > 0$, $-\Delta w \geq 1$ in Ω. In fact,

$$
\begin{aligned}
-\Delta w(x) =&\, M\big(4\beta^2|x-y|^2 e^{-\beta|x-y|^2} - 2n\beta e^{-\beta|x-y|^2}\big) \\
\geq&\, M e^{-\beta|x-y|^2}(4\beta^2 R^2 - 2n\beta), \quad x \in \Omega.
\end{aligned}
$$

Next, we set

$$v_k(x) = u_k(x) - |f|_{0;\Omega} w(x), \quad x \in \overline{\Omega}_k$$

and proceed to show $v_k(x) \leq 0$ in Ω_k. In fact,

$$-\Delta v_k(x) = -\Delta u_k(x) + |f|_{0;\Omega}\Delta w(x) \leq f(x) - |f|_{0;\Omega} \leq 0, \quad x \in \Omega_k.$$

In addition,

$$v_k\Big|_{\partial\Omega_k} = u_k\Big|_{\partial\Omega_k} - |f|_{0;\Omega} w\Big|_{\partial\Omega_k} \leq 0,$$

and so, from the comparison principle, we have

$$v_k(x) \leq 0, \quad \forall x \in \Omega_k,$$

that is

$$u_k(x) \leq |f|_{0;\Omega} w(x), \quad \forall x \in \Omega_k.$$

For any fixed $x \in \Omega$, choosing m sufficiently large such that $x \in \Omega_m$, we have

$$u_k(x) \leq |f|_{0;\Omega} w(x), \quad \forall k \geq m.$$

Taking $k = k_i$ and setting $i \to \infty$ lead to

$$u(x) \leq |f|_{0;\Omega} w(x), \quad \forall x \in \Omega.$$

Similarly,

$$u(x) \geq -|f|_{0;\Omega} w(x), \quad \forall x \in \Omega.$$

Summing up, we have

$$|u(x)| \leq |f|_{0;\Omega} w(x), \quad \forall x \in \Omega.$$

Therefore, $u(x^0) = 0$ and the existence of solutions is proved. The uniqueness follows from the maximum principle. $\qquad\square$

Remark 8.2.1 *If Ω is a rectangle, then Ω satisfies the assumptions of Theorem 8.2.2.*

In Theorem 8.2.2, the boundary function $\varphi \in C^{2,\alpha}(\overline{\Omega})$, but the solution obtained is only continuous up to the boundary. Is it possible to weaken the restriction on the boundary function φ? The answer is positive. We have

Theorem 8.2.3 *If $\varphi \in C(\overline{\Omega})$, then the conclusion of Theorem 8.2.2 is still valid.*

Proof. Choose $\varphi_k \in C^\infty(\overline{\Omega})$, such that

$$|\varphi_k(x) - \varphi(x)| \leq \frac{1}{k}, \quad \forall x \in \overline{\Omega}, \quad k = 1, 2, \cdots.$$

Consider the approximate problem

$$\begin{cases} -\Delta u(x) = f(x), & x \in \Omega_k, \\ u\big|_{\partial\Omega_k} = \varphi_k. \end{cases}$$

By Theorem 8.2.2, the above problem admits a unique solution $u_k \in C^{2,\alpha}(\overline{\Omega}_k)$. Using the interior estimate (Theorem 6.2.8 of Chapter 6) and a diagonal process, it is easy to prove that there exists a subsequence of $\{u_k\}$, denoted still by $\{u_k\}$, and a function $u \in C^{2,\alpha}(\Omega)$, such that for any fixed $m \geq 1$,

$$u_k \to u, \quad Du_k \to Du, \quad D^2 u_k \to D^2 u$$

uniformly on Ω_m as $k \to \infty$, from which it follows that u satisfies equation (8.2.1).

Now, we verify $u\big|_{\partial\Omega} = \varphi$. Let $x^0 \in \partial\Omega$. For any $\varepsilon > 0$, by the continuity of $\varphi(x)$ we see that, there exists $\delta > 0$, such that

$$|\varphi(x) - \varphi(x^0)| < \varepsilon, \quad \forall x \in B_\delta(x^0) \cap \Omega.$$

Choose $C_\varepsilon > |f|_{0;\Omega} + 1$ such that

$$|\varphi(x) - \varphi(x^0)| < \varepsilon + C_\varepsilon w(x), \quad \forall x \in \Omega,$$

where w is the barrier function defined in the proof of Theorem 8.2.2. Thus

$$|\varphi_k(x) - \varphi(x^0)| < \varepsilon + C_\varepsilon w(x) + \frac{1}{k}, \quad \forall x \in \Omega, \quad k = 1, 2, \cdots.$$

Set

$$v_k(x) = u_k(x) - C_\varepsilon w(x) - \varepsilon - \frac{1}{k} - \varphi(x^0), \quad x \in \overline{\Omega}_k.$$

Then

$$-\Delta v_k = -\Delta u_k + C_\varepsilon \Delta w = f + C_\varepsilon \Delta w \le f - C_\varepsilon \le 0, \quad x \in \Omega_k,$$

$$\begin{aligned}
v_k\big|_{\partial\Omega_k} &= u_k\big|_{\partial\Omega_k} - C_\varepsilon w\big|_{\partial\Omega_k} - \varepsilon - \frac{1}{k} - \varphi(x^0) \\
&= \varphi_k\big|_{\partial\Omega_k} - C_\varepsilon w\big|_{\partial\Omega_k} - \varepsilon - \frac{1}{k} - \varphi(x^0) \le 0.
\end{aligned}$$

The comparison principle gives

$$v_k(x) \le 0, \quad \forall x \in \Omega_k,$$

that is

$$u_k(x) \le C_\varepsilon w(x) + \varepsilon + \frac{1}{k} + \varphi(x^0), \quad \forall x \in \Omega_k.$$

For fixed $x \in \Omega$, we choose m sufficiently large, such that $x \in \Omega_m$. Then

$$u_k(x) \le C_\varepsilon w(x) + \varepsilon + \frac{1}{k} + \varphi(x^0), \quad \forall k \ge m.$$

Letting $k \to \infty$ yields

$$u(x) \le C_\varepsilon w(x) + \varepsilon + \varphi(x^0), \quad \forall x \in \Omega.$$

So

$$\overline{\lim_{x \to x^0}} \, u(x) \le \varepsilon + \varphi(x^0).$$

By the arbitrariness of $\varepsilon > 0$, we see that

$$\overline{\lim_{x \to x^0}} \, u(x) \le \varphi(x^0).$$

Similarly, we obtain

$$\lim_{x \to x^0} u(x) \geq \varphi(x^0).$$

Summing up, we have

$$\lim_{x \to x^0} u(x) = \varphi(x^0).$$

Thus the existence is proved. The uniqueness of solutions follows from the maximum principle.   □

8.2.2   *The method of continuity*

Contraction Mapping Principle   *Let T be a contraction mapping on the Banach space B, that is, there exists $0 < \theta < 1$, such that*

$$\|Tu - Tv\| \leq \theta \|u - v\|, \quad \forall u, v \in B. \tag{8.2.3}$$

Then T admits a unique fixed point, that is, the operator equation

$$Tu = u$$

admits a unique solution $u \in B$.

Proof.   For fixed $u_0 \in B$, set

$$u_i = Tu_{i-1}, \quad i = 1, 2, \cdots.$$

For any positive integer $1 \leq i \leq j$, using (8.2.3), we obtain

$$\|u_j - u_i\| \leq \sum_{k=i+1}^{j} \|u_k - u_{k-1}\|$$

$$= \sum_{k=i+1}^{j} \|Tu_{k-1} - Tu_{k-2}\|$$

$$\leq \sum_{k=i+1}^{j} \theta^{k-1} \|u_1 - u_0\|$$

$$\leq \frac{\theta^i}{1 - \theta} \|u_1 - u_0\| \to 0 \quad (i \to \infty).$$

So, $\{u_i\}$ is a Cauchy sequence, and from the completeness of B, it converges to some $u \in B$. From (8.2.3) we see that T is continuous, and so

$$Tu = \lim_{i \to \infty} Tu_i = \lim_{i \to \infty} u_{i+1} = u.$$

The uniqueness follows directly from (8.2.3). □

Remark 8.2.2 *From the proof of the theorem, we see that the conclusion is still valid if we replace B by any closed subset of B.*

The Method of Continuity *Let B be a Banach space, V be a normed linear space, T_0 and T_1 be bounded linear operators from B to V. Set*

$$T_\tau = (1 - \tau)T_0 + \tau T_1, \quad \tau \in [0, 1].$$

If there exists some constant $C > 0$, such that

$$\|u\|_B \le C\|T_\tau u\|_V, \quad u \in B, \tau \in [0, 1], \tag{8.2.4}$$

then T_1 maps B onto V if and only if T_0 maps B onto V.

Proof. Let $s \in [0, 1]$ and T_s maps B onto V. By (8.2.4), we see that T_s is injective and so the inverse map $T_s^{-1} : V \to B$ exists. For $\tau \in [0, 1]$, $v \in V$, the operator equation $T_\tau u = v$ is equivalent to the equation

$$T_s u = v + (T_s - T_\tau)u = v + (\tau - s)(T_0 - T_1)u.$$

Furthermore, since T_s^{-1} exists, it is also equivalent to

$$u = T_s^{-1}v + (\tau - s)T_s^{-1}(T_0 - T_1)u \equiv Tu.$$

If

$$|\tau - s| < \delta \equiv \frac{1}{C(\|T_0\| + \|T_1\| + 1)},$$

then from (8.2.4), we see that $T : B \to B$ is a contraction map. According to the contraction mapping principle, for any $s \in [0, 1]$ satisfying $|\tau - s| < \delta$, the map T_τ is bijective. We decompose $[0, 1]$ into several intervals with their length less than δ. It is easy to see that if for some fixed $\tau_0 \in [0, 1]$ (in particular for $\tau_0 = 0$ or $\tau_0 = 1$), T_{τ_0} is bijective, then for all $\tau \in [0, 1]$, T_τ is bijective too. □

Remark 8.2.3 *The method of continuity shows that the invertibility of a bounded linear operator can be deduced from the invertibility of another similar kind of operators.*

8.2.3   Existence and uniqueness of classical solutions for general linear elliptic equations

By the method of continuity, we may extend the above results about the Dirichlet problem (8.2.1), (8.2.2) for Poisson's equation to the Dirichlet problem for general linear elliptic equation

$$-a_{ij}(x)D_{ij}u + b_i(x)D_i u + c(x)u = f(x), \quad x \in \Omega, \tag{8.2.5}$$

$$u\Big|_{\partial\Omega} = \varphi, \tag{8.2.6}$$

where $\Omega \subset \mathbb{R}^n$ is a bounded domain, $a_{ij}, b_i, c \in C^\alpha(\overline{\Omega})$, $c \geq 0$, $a_{ij} = a_{ji}$, and there exists some constants $0 < \lambda \leq \Lambda$, such that

$$\lambda|\xi|^2 \leq a_{ij}(x)\xi_i\xi_j \leq \Lambda|\xi|^2, \quad \forall \xi \in \mathbb{R}^n, \ x \in \Omega. \tag{8.2.7}$$

Theorem 8.2.4   Let $\partial\Omega \in C^\infty$, $0 < \alpha < 1$, $a_{ij}, b_i, c, f \in C^\alpha(\overline{\Omega})$, $c \geq 0$, $\varphi \in C^{2,\alpha}(\overline{\Omega})$, $a_{ij} = a_{ji}$ satisfy (8.2.7). Then problem (8.2.5), (8.2.6) admits a unique solution $u \in C^{2,\alpha}(\overline{\Omega})$.

Proof.   Without loss of generality, we may assume that $\varphi \equiv 0$. Otherwise, we consider the equation for $w = u - \varphi$.

Let

$$L_0 u = -\Delta u,$$
$$L_1 u = -a_{ij}(x)D_{ij}u + b_i(x)D_i u + c(x)u.$$

Consider the family of elliptic equations with a parameter τ,

$$L_\tau u = (1 - \tau)L_0 u + \tau L_1 u = f, \quad 0 \leq \tau \leq 1, \tag{8.2.8}$$

where the coefficients of second order term satisfy (8.2.7) with λ, Λ taken as

$$\lambda_\tau = \min\{1, \lambda\}, \quad \Lambda_\tau = \max\{1, \Lambda\}.$$

L_τ can be regarded as a linear operator from the Banach space $B = \{u \in C^{2,\alpha}(\overline{\Omega}) : u\big|_{\partial\Omega} = 0\}$ to the normed linear space $V = C^\alpha(\overline{\Omega})$. So the solvability of problem (8.2.8), (8.2.6) is equivalent to the invertibility of the operator L_τ. Let $u \in B$ be a solution of problem (8.2.8), (8.2.6). According to Schauder's estimates (Theorem 6.3.1 of Chapter 6) and the maximum norm estimates (Theorem 8.1.3), and noticing the assumption $\varphi \equiv 0$, we have

$$|u|_{2,\alpha;\Omega} \leq C(|f|_{\alpha;\Omega} + |u|_{0;\Omega}) \leq C|f|_{\alpha;\Omega}, \quad \tau \in [0, 1],$$

that is

$$\|u\|_B \le C\|L_\tau u\|_V, \quad u \in B, \ \tau \in [0,1],$$

where C is a constant independent of τ. When $\tau = 0$, problem (8.2.8), (8.2.6) is just problem (8.2.1), (8.2.2), which admits a unique solution $u \in B$ according to Theorem 8.2.1. This means that L_0 maps B onto V. Using the method of continuity, we see that L_1 maps B onto V too, and so problem (8.2.5), (8.2.6) admits a solution $u \in C^{2,\alpha}(\overline{\Omega})$. The uniqueness can be proved by the maximum principle. □

Using Theorem 8.2.4 and the barrier function technique, similar to the proof of Theorem 8.2.2, we obtain the following

Theorem 8.2.5 *Assume that Ω has the exterior ball property, and there exists a sequence of subdomains $\{\Omega_k\}$ with C^∞ boundary, such that $\overline{\Omega}_k \subset \Omega_{k+1}$ and $\partial\Omega_k$ approximates $\partial\Omega$ uniformly. Let $0 < \alpha < 1$, $a_{ij}, b_i, c, f \in C^\alpha(\overline{\Omega})$, $c \ge 0$, $\varphi \in C^{2,\alpha}(\overline{\Omega})$, $a_{ij} = a_{ji}$ satisfy (8.2.7). Then problem (8.2.1), (8.2.2) admits a unique solution $u \in C^{2,\alpha}(\Omega) \cap C(\overline{\Omega})$.*

Using Theorem 8.2.4, Theorem 8.2.5 and the barrier function technique, similar to the proof of Theorem 8.2.3, we obtain

Theorem 8.2.6 *If $\varphi \in C(\overline{\Omega})$, then the conclusion of Theorem 8.2.5 is still valid.*

Furthermore, we may prove the following theorem.

Theorem 8.2.7 *Let $0 < \alpha < 1$, $\partial\Omega \in C^{2,\alpha}$, $a_{ij}, b_i, c, f \in C^\alpha(\overline{\Omega})$, $c \ge 0$, $\varphi \in C^{2,\alpha}(\overline{\Omega})$, $a_{ij} = a_{ji}$ satisfy (8.2.7). Then problem (8.2.5), (8.2.6) admits a unique solution $u \in C^{2,\alpha}(\overline{\Omega})$.*

Proof. Since $\partial\Omega \in C^{2,\alpha}$, Ω has the exterior ball property, and all the conditions for Ω in Theorem 8.2.5 are satisfied. So, according to Theorem 8.2.5, we see that problem (8.2.5), (8.2.6) admits a unique solution $u \in C^{2,\alpha}(\Omega) \cap C(\overline{\Omega})$. An application of Schauder's estimates (Theorem 6.3.1 and Remark 6.3.1 of Chapter 6) shows that $u \in C^{2,\alpha}(\overline{\Omega})$. □

8.3 Existence and Uniqueness of Classical Solutions for Linear Parabolic Equations

In this section, we introduce the theory parallel to the second section for linear parabolic equations.

8.3.1 Existence and uniqueness of the classical solution for the heat equation

Consider the first initial-boundary value problem

$$\frac{\partial u}{\partial t} - \Delta u = f(x,t), \quad (x,t) \in Q_T, \tag{8.3.1}$$

$$u(x,t) = \varphi(x,t), \quad (x,t) \in \partial_p Q_T, \tag{8.3.2}$$

where $Q_T = \Omega \times (0,T)$, $\Omega \subset \mathbb{R}^n$ is a bounded domain, $T > 0$.

Theorem 8.3.1 *Let* $\partial\Omega \in C^\infty$, $0 < \alpha < 1$, $f \in C^{\alpha,\alpha/2}(\overline{Q}_T)$, $\varphi \in C^{2+\alpha,1+\alpha/2}(\overline{Q}_T)$. *Then the first initial-boundary value problem (8.3.1), (8.3.2) admits a unique solution* $u \in C^{2+\alpha,1+\alpha/2}(\overline{Q}_T)$.

The proof is similar to that of Theorem 8.2.1, and we leave the details to the reader.

Theorem 8.3.2 *Assume that* Ω *has the exterior ball property, and there exists a sequence* $\{\Omega_k\}$ *with* C^∞ *smooth boundary, such that* $\overline{\Omega}_k \subset \Omega_{k+1}$ *and* $\partial\Omega_k$ *approximates* $\partial\Omega$ *uniformly. Let* $0 < \alpha < 1$, $f \in C^{\alpha,\alpha/2}(\overline{Q}_T)$, $\varphi \in C^{2+\alpha,1+\alpha/2}(\overline{Q}_T)$. *Then problem (8.3.1), (8.3.2) admits a unique solution* $u \in C^{2+\alpha,1+\alpha/2}(Q_T) \cap C(\overline{Q}_T)$.

Proof. Without loss of generality, we assume that $\varphi \equiv 0$. Otherwise, we consider the equation for $w = u - \varphi$. Similar to the proof of Theorem 8.2.2, consider the approximation problems of (8.3.1), (8.3.2). We first prove that the limit of solutions of the approximation problems satisfies equation (8.3.1), and then apply the barrier function technique to check that $u\big|_{\partial_p Q_T} = 0$. Here, we only point out the construction of the barrier function $w(x,t)$. Let $(x^0, t_0) \in \partial_p Q_T$. The barrier function $w(x,t)$ should have the following properties:

   i) $w(x^0, t_0) = 0$, $w(x,t) > 0$ for all $x \in \overline{Q}_T \backslash \{x^0, t_0\}$;

   ii) $w \in C^{2,1}(\overline{Q}_T)$, $\dfrac{\partial w}{\partial t} - \Delta w \geq 1$ in Q_T.

Now, for the point (x^0, t_0) at the lateral boundary, we choose $w(x,t) = w(x)$, the barrier function constructed in the proof of Theorem 8.2.2, and for the point $(x^0, 0)$ at the bottom, we choose $w(x,t) = t$. Clearly the function thus defined possesses the above properties. $\qquad\square$

Theorem 8.3.3 *If* $\varphi \in C(\overline{Q}_T)$, *then the conclusion of Theorem 8.3.2 is still valid.*

The proof is similar to that of Theorem 8.2.3 and the details are left to the reader.

8.3.2 Existence and uniqueness of classical solutions for general linear parabolic equations

Using the method of continuity, we may extend Theorem 8.3.1 for the heat equation to the first initial-boundary value problem

$$\frac{\partial u}{\partial t} - a_{ij}(x,t)D_{ij}u + b_i(x,t)D_i u + c(x,t)u$$
$$= f(x,t), \quad (x,t) \in Q_T, \tag{8.3.3}$$

$$u(x,t) = \varphi(x,t), \quad (x,t) \in \partial_p Q_T, \tag{8.3.4}$$

where $\Omega \subset \mathbb{R}^n$ is a bounded domain, $a_{ij}, b_i, c \in C^{\alpha,\alpha/2}(\overline{Q}_T)$, $a_{ij} = a_{ji}$ and for some constants $0 < \lambda \leq \Lambda$, such that

$$\lambda|\xi|^2 \leq a_{ij}(x,t)\xi_i\xi_j \leq \Lambda|\xi|^2, \quad \forall \xi \in \mathbb{R}^n, \ (x,t) \in Q_T. \tag{8.3.5}$$

Theorem 8.3.4 *Let $\partial\Omega \in C^\infty$, $0 < \alpha < 1$, $a_{ij}, b_i, c, f \in C^{\alpha,\alpha/2}(\overline{Q}_T)$, $\varphi \in C^{2+\alpha,1+\alpha/2}(\overline{Q}_T)$, $a_{ij} = a_{ji}$ satisfy (8.3.5). Then problem (8.3.3), (8.3.4) admits a unique solution $u \in C^{2+\alpha,1+\alpha/2}(\overline{Q}_T)$.*

Using the barrier function technique, we may further obtain

Theorem 8.3.5 *Assume that Ω has the exterior ball property, and there exists a sequence of subdomains $\{\Omega_k\}$ with C^∞ boundary, such that $\overline{\Omega}_k \subset \Omega_{k+1}$ and $\partial\Omega_k$ approximates $\partial\Omega$ uniformly. Let $0 < \alpha < 1$, $a_{ij}, b_i, c, f \in C^{\alpha,\alpha/2}(\overline{Q}_T)$, $\varphi \in C^{2+\alpha,1+\alpha/2}(\overline{Q}_T)$, $a_{ij} = a_{ji}$ satisfy (8.3.5). Then problem (8.3.1), (8.3.2) admits a unique solution $u \in C^{2+\alpha,1+\alpha/2}(Q_T) \cap C(\overline{Q}_T)$.*

Theorem 8.3.6 *If $\varphi \in C(\overline{Q}_T)$, then the conclusion of Theorem 8.3.5 is still valid.*

We may further establish the following theorem.

Theorem 8.3.7 *Let $0 < \alpha < 1$, $\partial\Omega \in C^{2,\alpha}$, $a_{ij}, b_i, c, f \in C^{\alpha,\alpha/2}(\overline{Q}_T)$, $\varphi \in C^{2+\alpha,1+\alpha/2}(\overline{Q}_T)$, $a_{ij} = a_{ji}$ satisfy (8.3.5). Then problem (8.3.3), (8.3.4) admits a unique solution $u \in C^{2+\alpha,1+\alpha/2}(\overline{Q}_T)$.*

Proof. Since $\partial\Omega \in C^{2,\alpha}$, Ω has the exterior ball property, and all the conditions for Ω in Theorem 8.3.5 are satisfied. So, according to Theorem 8.3.5 we see that problem (8.3.3), (8.3.4) admits a unique solution $u \in C^{2+\alpha,1+\alpha/2}(Q_T) \cap C(\overline{Q}_T)$. Then an application of Schauder's

estimates (Theorem 7.2.24 and Remark 7.2.5 of Chapter 7) shows that
$u \in C^{2+\alpha, 1+\alpha/2}(\overline{Q}_T)$.   □

Remark 8.3.1   *Different from the case of elliptic equations, to ensure the solvability of the first initial-boundary value problem for equation (8.3.3), we need not require $c \geq 0$.*

Exercises

1. Prove Theorem 8.1.1 by the method mentioned in the proof of Remark 8.1.2.

2. Let B be the unit ball in $\mathbb{R}^n$. Assume that $u \in C^2(B) \cap C(\overline{B})$ satisfies

$$-\Delta u(x) \leq 0, \quad x \in B.$$

Let $x^0 \in \partial B$ with

$$u(x) \leq u(x^0), \quad x \in B.$$

Prove that

$$\frac{\partial u}{\partial \nu}(x^0) > 0,$$

where ν is the unit normal vector outward to ∂B.

3. Prove Theorems 8.2.5 and 8.2.6.

4. Let $\Omega \subset \mathbb{R}^n$ be a bounded domain with appropriately smooth boundary, $\lambda \in \mathbb{R}$, $0 < \alpha < 1$. Prove that there is exact one of the following alternatives:

i) The homogeneous boundary value problem

$$-\Delta u + \lambda u = 0 \text{in } \Omega, \quad u\Big|_{\partial \Omega} = 0$$

admits a nontrivial classical solution $u \in C^{2,\alpha}(\overline{\Omega})$;

ii) For any $f \in C^\alpha(\overline{\Omega})$, the nonhomogeneous boundary value problem

$$-\Delta u + \lambda u = f \text{in } \Omega, \quad u\Big|_{\partial \Omega} = 0$$

admits a unique classical solution $u \in C^{2,\alpha}(\overline{\Omega})$.

5. Consider the second initial-boundary value problem

$$
\begin{cases}
\dfrac{\partial u}{\partial t} - \Delta u + \lambda u = f, & (x,t) \in Q_T = \Omega \times (0,T), \\[2mm]
\dfrac{\partial u}{\partial \nu} = 0, & (x,t) \in \partial\Omega \times (0,T), \\[2mm]
u(x,0) = u_0(x), & x \in \Omega,
\end{cases}
$$

where $\lambda \in \mathbb{R}$, $\Omega \subset \mathbb{R}^n$ is a bounded domain with appropriately smooth boundary, ν is the unit normal vector outward to $\partial\Omega$. Prove that the problem admits at most a smooth solution $u \in C^{2,1}(\overline{Q}_T)$.

6. Prove Theorems 8.3.1 and 8.3.3.

7. Prove Theorems 8.3.4–8.3.6.

Chapter 9

L^p Estimates for Linear Equations and Existence of Strong Solutions

In the previous chapters we have investigated two classes of solutions, that is weak solutions and classical solutions of linear elliptic and parabolic equations. In this chapter, we consider another kind of solutions with intermediate regularity, called strong solutions. For this purpose, we need to establish the L^p estimates. Just as the existence of classical solutions is based on Schauder's estimates, the existence of strong solutions is based on the L^p estimates.

We will first apply Stampacchia's interpolation theorem and the results on Schauder's estimates, to establish the L^p estimates for Poisson's equation and the heat equation. On the basis of these estimates, we establish the L^p estimates for general linear elliptic and parabolic equations, and establish the existence theory of strong solutions. It is worthy noting that the L^p estimates can be established for equations in nondivergence form, but a crucial condition, i.e. the continuity assumption on the coefficients of second order terms is required.

9.1 L^p Estimates for Linear Elliptic Equations and Existence and Uniqueness of Strong Solutions

In this section, we first introduce the L^p estimates on solutions of Poisson's equation in cubes, and then apply these estimates to establish the L^p estimates for general linear elliptic equations, and further establish the existence theory of strong solutions.

9.1.1 L^p estimates for Poisson's equation in cubes

Consider the homogeneous Dirichlet problem for Poisson's equation

$$-\Delta u = f, \quad x \in Q_0, \tag{9.1.1}$$

$$u\Big|_{\partial Q_0} = 0, \tag{9.1.2}$$

where Q_0 is a cube in $\mathbb{R}^n$ with its edges parallel to the axes.

To obtain the L^p estimate on a solution u of problem (9.1.1), (9.1.2) in the cube Q_0, we need to establish the estimate on D^2u in the Campanato space $\mathcal{L}^{2,n}(Q_0)$. We first establish the interior estimate.

Proposition 9.1.1   *Let $f \in L^\infty(Q_0)$, $u \in H^2(Q_0) \cap H_0^1(Q_0)$ be a weak solution of equation (9.1.1), $x^0 \in Q_0$, $B_{2R_0}(x^0) \subset\subset Q_0$. Then*

$$[D^2u]_{2,n;B_{R_0}(x^0)} \le C \left(\|D^2u\|_{L^2(B_{2R_0}(x^0))} + \|f\|_{L^\infty(B_{2R_0}(x^0))} \right), \tag{9.1.3}$$

where C is a constant depending only on n and R_0.

Proof.   Let f_ε be the standard smooth approximation of f and u_ε be the solution of the problem

$$\begin{cases} -\Delta u_\varepsilon = f_\varepsilon, & x \in Q_0, \\ u_\varepsilon\Big|_{\partial Q_0} = 0. \end{cases}$$

By the L^2 theory, u_ε is sufficiently smooth in $\overline{B}_{2R_0}(x^0)$ and

$$u_\varepsilon \to u, \quad \text{in } H^2(B_{2R_0}(x^0)) \text{ as } \varepsilon \to 0.$$

Therefore, to show (9.1.3), we need only to prove

$$[D^2u_\varepsilon]_{2,n;B_{R_0}(x^0)} \le C \left(\|D^2u_\varepsilon\|_{L^2(B_{2R_0}(x^0))} + \|f_\varepsilon\|_{L^\infty(B_{2R_0}(x^0))} \right),$$

where the constant C is independent of ε. Owing to this reason, in what follows, we may assume that u is sufficiently smooth in $\overline{B}_{2R_0}(x^0)$.

For any $x \in B_{R_0}(x^0)$, we have $B_{R_0}(x) \subset B_{2R_0}(x^0)$. From the proof of Schauder's interior estimate (Theorem 6.2.6 of Chapter 6), we see that for any $0 < \rho \le R \le R_0$,

$$\int_{B_\rho(x)} |Dw(y) - (Dw)_{x,\rho}|^2 dy$$

$$\le C \left(\frac{\rho}{R}\right)^{n+2} \int_{B_R(x)} |Dw(y) - (Dw)_{x,R}|^2 dy + C \int_{B_R(x)} (f(y) - f_{x,R})^2 dy$$

$$\le C \left(\frac{\rho}{R}\right)^{n+2} \int_{B_R(x)} |Dw(y) - (Dw)_{x,R}|^2 dy + CR^n \|f\|_{L^\infty(B_R(x))}^2,$$

where $w = D_i u \, (1 \leq i \leq n)$. So, the iteration lemma (Lemma 6.2.1 of Chapter 6) yields that for any $0 < \rho \leq R \leq R_0$,

$$
\int_{B_\rho(x)} |Dw(y) - (Dw)_{x,\rho}|^2 dy
$$

$$
\leq C \left(\frac{\rho}{R}\right)^n \left(\int_{B_R(x)} |Dw(y) - (Dw)_{x,R}|^2 dy + R^n \|f\|^2_{L^\infty(B_R(x))}\right)
$$

$$
\leq C\rho^n \left(\frac{1}{R^n}\|Dw\|^2_{L^2(B_R(x))} + \|f\|^2_{L^\infty(B_R(x))}\right).
$$

Thus, for any $0 < \rho \leq R_0$,

$$
\int_{B_\rho(x) \cap B_{R_0}(x^0)} |D^2 u(y) - (D^2 u)_{B_\rho(x) \cap B_{R_0}(x^0)}|^2 dy
$$

$$
\leq \int_{B_\rho(x)} |D^2 u(y) - (D^2 u)_{x,\rho}|^2 dy
$$

$$
\leq C\rho^n \left(\frac{1}{R_0^n}\|D^2 u\|^2_{L^2(B_{R_0}(x))} + \|f\|^2_{L^\infty(B_{R_0}(x))}\right)
$$

$$
\leq C\rho^n \left(\frac{1}{R_0^n}\|D^2 u\|^2_{L^2(B_{2R_0}(x^0))} + \|f\|^2_{L^\infty(B_{2R_0}(x^0))}\right),
$$

where

$$
(D^2 u)_{B_\rho(x) \cap B_{R_0}(x^0)} = \frac{1}{|B_\rho(x) \cap B_{R_0}(x^0)|} \int_{B_\rho(x) \cap B_{R_0}(x^0)} D^2 u(y) dy.
$$

Therefore

$$
[D^2 u]^{(1/2)}_{2,n;B_{R_0}(x^0)} \leq C \left(\|D^2 u\|_{L^2(B_{2R_0}(x^0))} + \|f\|_{L^\infty(B_{2R_0}(x^0))}\right),
$$

where C is a constant depending only on n and R_0. The notation $[\,\cdot\,]^{(1/2)}_{2,n}$ is defined in Remark 6.1.2 of Chapter 6. In addition, by virtue of $[\,\cdot\,]_{2,n} \leq [\,\cdot\,]^{(1/2)}_{2,n} + C\|\cdot\|_{L^2}$, we obtain (9.1.3). $\qquad\square$

Having the interior estimate in hand, we may further obtain the global estimate by the extension of solution.

Proposition 9.1.2   *Let $f \in L^\infty(Q_0)$ and $u \in H^2(Q_0) \cap H_0^1(Q_0)$ be the weak solution of the Dirichlet problem (9.1.1), (9.1.2). Then*

$$
\|D^2 u\|_{\mathcal{L}^{2,n}(Q_0)} \leq C\|f\|_{L^\infty(Q_0)}, \tag{9.1.4}
$$

where C is a constant depending only on n and the length of the edge of Q_0.

Proof. Let

$$Q_0 = Q(x^0, R) = \{x \in \mathbb{R}^n; |x_i - x_i^0| < R, i = 1, 2, \cdots, n\}.$$

To prove (9.1.4), we extend the definition of u in the following manner, and denote the extended function by $\tilde{u}$. First, we make the antisymmetric extension of u with respect to the super planes $x_1 = x_1^0 + R$ and $x_1 = x_1^0 - R$, namely, define

$$\tilde{u}(x) = \begin{cases} u(x), & \text{if } x \in \overline{Q}_0, \\ -u(2x_1^0 + 2R - x_1, x_2, \cdots, x_n), \\ \quad \text{if } x_1 \in (x_1^0 + R, x_1^0 + 3R], |x_i| < R \, (2 \le i \le n), \\ -u(2x_1^0 - 2R - x_1, x_2, \cdots, x_n), \\ \quad \text{if } x_1 \in [x_1^0 - 3R, x_1^0 - R), |x_i| < R \, (2 \le i \le n). \end{cases}$$

Next, we make the antisymmetric extension with respect to the super planes $x_2 = x_2^0 + R$ and $x_2 = x_2^0 - R$, $\cdots$, $x_n = x_n^0 + R$ and $x_n = x_n^0 - R$. Then we obtain a function $\tilde{u}$ defined on $\overline{Q}(x^0, 3R)$. Repeating the above procedure for n times then yields a function $\tilde{u}$ defined on $\overline{Q}(x^0, 3^n R)$. Similarly, we get the antisymmetric extension $\tilde{f}$ of f on $\overline{Q}(x^0, 3^n R)$.

Obviously, $\tilde{f} \in L^\infty(Q(x^0, 3^n R))$. It is not difficult to check that $\tilde{u} \in H^2(Q(x^0, 3^n R)) \cap H_0^1(Q(x^0, 3^n R))$, and $\tilde{u}$ is a weak solution of the equation

$$-\Delta \tilde{u} = \tilde{f}, \quad x \in Q(x^0, 3^n R).$$

Owing to

$$Q_0 = Q(x^0, R) \subset B_{\sqrt{n}R}(x^0) \subset B_{2\sqrt{n}R}(x^0) \subset\subset Q(x^0, 3^n R),$$

using Proposition 9.1.1 in $Q(x^0, 3^n R)$ leads to

$$\begin{aligned} [D^2 u]_{2,n;Q_0} &\le [D^2 \tilde{u}]_{2,n;B_{\sqrt{n}R}(x^0)} \\ &\le C \left(\|D^2 \tilde{u}\|_{L^2(B_{2\sqrt{n}R}(x^0))} + \|\tilde{f}\|_{L^\infty(B_{2\sqrt{n}R}(x^0))} \right) \\ &\le C \left(\|D^2 \tilde{u}\|_{L^2(Q(x^0, 3^n R))} + \|\tilde{f}\|_{L^\infty(Q(x^0, 3^n R))} \right). \end{aligned}$$

So, by changing a constant C depending only on n and R, we derive

$$[D^2 u]_{2,n;Q_0} \le C \left(\|D^2 u\|_{L^2(Q_0)} + \|f\|_{L^\infty(Q_0)} \right). \tag{9.1.5}$$

According to the L^2 theory (Remark 2.2.2 of Chapter 2, which is still valid for cubes and can be proved by using the method similar to the proof of

Theorem 8.2.2 of Chapter 8), we have

$$\|u\|_{H^2(Q_0)} \leq C\|f\|_{L^2(Q_0)}. \tag{9.1.6}$$

Combining (9.1.5) with (9.1.6) yields (9.1.4). □

To obtain the L^p estimate for Poisson's equation in the cube Q_0, we need Stampacchia's interpolation theorem.

Definition 9.1.1 Let Q_0 be a cube in $\mathbb{R}^n$ with its edges parallel to the axes. For $u \in L^1(Q_0)$, if

$$|u|_{*,Q_0} \equiv \sup \left\{ \frac{1}{|Q \cap Q_0|} \int_{Q \cap Q_0} |u - u_{Q \cap Q_0}| dx; Q \text{ is a cube parallel to } Q_0 \right\}$$
$$< +\infty,$$

then we say that $u \in \mathrm{BMO}(Q_0)$.

We define the norm in $\mathrm{BMO}(Q_0)$ by

$$\|\cdot\|_{\mathrm{BMO}(Q_0)} = \|\cdot\|_{L^1(Q_0)} + |\cdot|_{*,Q_0}.$$

It is easily shown that $\mathrm{BMO}(Q_0)$ is a Banach space.

From the definition of Campanato spaces and the space $\mathrm{BMO}(Q_0)$, we see that

$$\mathcal{L}^{2,n}(Q_0) \subset \mathcal{L}^{1,n}(Q_0) \simeq \mathrm{BMO}(Q_0).$$

Define the operator

$$T_i : L^\infty(Q_0) \to \mathrm{BMO}(Q_0), \quad f \mapsto DD_i u,$$

where $1 \leq i \leq n$, and u is a weak solution of problem (9.1.1), (9.1.2). The estimate (9.1.4) shows that $T_i (1 \leq i \leq n)$ is a bounded linear operator from $L^\infty(Q_0)$ to $\mathcal{L}^{2,n}(Q_0) \subset BMO(Q_0)$.

Stampacchia's Interpolation Theorem *Let $1 < q < +\infty$. If T is a bounded linear operator both from $L^q(Q_0)$ to $L^q(Q_0)$ and from $L^\infty(Q_0)$ to $BMO(Q_0)$, namely*

$$\|Tu\|_{L^q(Q_0)} \leq C_1 \|u\|_{L^q(Q_0)}, \quad \forall u \in L^q(Q_0),$$
$$\|Tu\|_{BMO(Q_0)} \leq C_2 \|u\|_{L^\infty(Q_0)}, \quad \forall u \in L^\infty(Q_0),$$

then T is a bounded linear operator from $L^p(Q_0)$ to $L^p(Q_0)$, namely

$$\|Tu\|_{L^p(Q_0)} \leq C\|u\|_{L^p(Q_0)}, \quad \forall u \in L^p(Q_0),$$

where $q \leq p < +\infty$, and C depends only on n, q, p, C_1 and C_2.

For the proof of the theorem, we refer to [Chen and Wu (1997)] Appendix 4.

Using Stampacchia's interpolation theorem, we may obtain

Theorem 9.1.1　*Let $p \geq 2$, $f \in L^p(Q_0)$, $u \in W^{2,p}(Q_0) \cap W_0^{1,p}(Q_0)$ satisfy Poisson's equation (9.1.1) in Q_0 almost everywhere. Then*

$$\|D^2 u\|_{L^p(Q_0)} \leq C\|f\|_{L^p(Q_0)},$$

where C depends only on n, p and the length of the edge of Q_0.

Proof.　By the L^2 theory and the estimate (9.1.4), T_i $(1 \leq i \leq n)$ is a bounded linear operator both from $L^2(Q_0)$ to $L^2(Q_0)$ and from $L^\infty(Q_0)$ to $\mathrm{BMO}(Q_0)$. Then, Stampacchia's interpolation theorem shows that T_i is a bounded linear operator from $L^p(Q_0)$ to $L^p(Q_0)$. □

Remark 9.1.1　*Under the conditions of Theorem 9.1.1, we may further use Ehrling–Nirenberg–Gagliardo's interpolation inequality to obtain*

$$\|u\|_{W^{2,p}(Q_0)} \leq C(\|f\|_{L^p(Q_0)} + \|u\|_{L^p(Q_0)}).$$

Remark 9.1.2　*The conclusion of Theorem 9.1.1 is still valid for the case $1 < p < 2$. For the proof we refer to [Chen and Wu (1997)] Chapter 3.*

Remark 9.1.3　*The conclusion of Theorem 9.1.1 can be extended to the general elliptic equations with the coefficients matrix of second order being constant and positive definite and throwing lower order terms. This is because Schauder's interior estimates which have been applied to prove (9.1.3) is still valid for such kind of equations, while the extension technique using in the proof of (9.1.4) is available to such equations too.*

9.1.2　L^p estimates for general linear elliptic equations

Now, we turn to the general linear elliptic equations

$$-a_{ij}(x)D_{ij}u + b_i(x)D_i u + c(x)u = f(x), \quad x \in \Omega, \tag{9.1.7}$$

where $\Omega \subset \mathbb{R}^n$ is a bounded domain, $a_{ij} = a_{ji}$, and for some constants $\lambda > 0$, $M > 0$

$$a_{ij}(x)\xi_i\xi_j \geq \lambda|\xi|^2, \quad \forall \xi \in \mathbb{R}^n, \ x \in \Omega, \tag{9.1.8}$$

$$\sum_{i,j=1}^{n} \|a_{ij}\|_{L^{\infty}(\Omega)} + \sum_{i=1}^{n} \|b_i\|_{L^{\infty}(\Omega)} + \|c\|_{L^{\infty}(\Omega)} \leq M. \tag{9.1.9}$$

Definition 9.1.2 Let $\omega(R)$ be a nondecreasing and continuous function defined on $[0, +\infty)$ and $\omega(R) = \lim_{R \to 0} \omega(R) = 0$. For a function $a(x)$ on $\overline{\Omega}$, we say that $a(x)$ has continuity module $\omega(R)$ on $\overline{\Omega}$, if

$$|a(x) - a(y)| \leq \omega(|x - y|), \quad \forall x, y \in \overline{\Omega}.$$

It is easy to verify that if $a \in C(\overline{\Omega})$, then $a(x)$ has continuity module on $\overline{\Omega}$.

For the general linear elliptic equation (9.1.7), we have

Theorem 9.1.2 *Let $\partial\Omega \in C^2$, $p > 1$. Assume that the coefficients of equation (9.1.7) satisfy (9.1.8) and (9.1.9) and $a_{ij} \in C(\overline{\Omega})$. If $u \in W^{2,p}(\Omega) \cap W_0^{1,p}(\Omega)$ satisfies equation (9.1.7) almost everywhere in Ω, then*

$$\|u\|_{W^{2,p}(\Omega)} \leq C(\|f\|_{L^p(\Omega)} + \|u\|_{L^p(\Omega)}),$$

where C is a positive constant depending only on n, p, λ, M, Ω and the continuity module of a_{ij}.

Proof. Similar to Schauder's estimates, we proceed to prove the theorem by three steps.

Step 1 Establish the interior estimate.

For any $x^0 \in \Omega$, choose $\hat{R}_0 > 0$ such that $Q_{\hat{R}_0} = Q(x^0, \hat{R}_0) \subset \Omega$. Let $0 < R \leq \hat{R}_0$ and η be a cut-off function on Q_R relative to $B_{R/2}$, that is, $\eta \in C_0^{\infty}(Q_R)$ satisfying $\eta(x) = 1$ in $B_{R/2}$ and

$$0 \leq \eta(x) \leq 1, \quad |\nabla \eta(x)| \leq \frac{C}{R}, \quad |D_{ij}\eta(x)| \leq \frac{C}{R^2}, \quad x \in Q_R.$$

Let $v = \eta u$. Then $v \in W_0^{2,p}(Q_R)$ and from equation (9.1.7) we obtain

$$-a_{ij}(x)D_{ij}v = g(x), \quad x \in Q_R,$$

where

$$g = -(b_i\eta + 2a_{ij}D_j\eta)D_iu(x) - (c\eta + a_{ij}D_{ij}\eta)u + \eta f \quad \text{in } Q_R.$$

Rewrite the above equation as

$$-a_{ij}(x^0)D_{ij}v = F(x) = g(x) + h(x), \quad x \in Q_R, \tag{9.1.10}$$

where

$$h(x) = (a_{ij}(x) - a_{ij}(x^0))D_{ij}v(x), \quad x \in Q_R.$$

Using the conclusion of Theorem 9.1.1 and Remark 9.1.2 to equation (9.1.10) (as pointed in Remark 9.1.3, for such kind of equations, the conclusion of Theorem 9.1.1 is still valid), we obtain

$$\|D^2v\|_{L^p(Q_R)} \le C_0 \|F\|_{L^p(Q_R)},$$

where C_0 depends only on n, p and R. In addition,

$$\|g\|_{L^p(Q_R)} \le M\left(\frac{C}{R^2} + 1\right)\|u\|_{W^{1,p}(Q_R)} + \|f\|_{L^p(Q_R)},$$

$$\|h\|_{L^p(Q_R)} \le \omega(\sqrt{n}R)\|D^2v\|_{L^p(Q_R)},$$

where $\omega(\cdot)$ is the common continuity module of each a_{ij} $(i, j = 1, 2, \cdots, n)$. Choose $0 < R_0 \le \hat{R}_0$ such that $C_0\omega(\sqrt{n}R_0) \le 1/2$. Then

$$\|D^2v\|_{L^p(Q_{R_0})} \le C\left(\left(\frac{1}{R_0^2} + 1\right)\|u\|_{W^{1,p}(Q_{R_0})} + \|f\|_{L^p(Q_{R_0})}\right).$$

Therefore, from the definition of v, we get that for any $0 < R \le R_0$,

$$\|D^2u\|_{L^p(B_{R/2})} \le C(\|f\|_{L^p(\Omega)} + \|u\|_{W^{1,p}(\Omega)}). \tag{9.1.11}$$

Step 2   Establish the near boundary estimate.

Let $x^0 \in \partial\Omega$. Since $\partial\Omega \in C^2$, there exists a neighborhood V of the point x^0 and a C^2 mapping $\Psi : V \to B_1 = B_1(0)$ such that

$$\Psi(V) = B_1, \quad \Psi(\Omega \cap V) = B_1^+, \quad \Psi(\partial\Omega \cap V) = \partial B_1^+ \cap B_1.$$

Denote by Q_0 the maximal cube contained in B_1^+ with its edges parallel to the axes. After a coordinate transformation $y = \Psi(x)$, equation (9.1.7) in $\Psi^{-1}(Q_0)$ is transformed into the equation in Q_0

$$-\hat{a}_{ij}(y)\hat{D}_{ij}\hat{u} + \hat{b}_i(y)\hat{D}_i\hat{u} + \hat{c}(y)\hat{u} = \hat{f}(y), \tag{9.1.12}$$

where $\hat{D}_i = \dfrac{\partial}{\partial y_i}$, $\hat{D}_{ij} = \dfrac{\partial^2}{\partial y_i \partial y_j}$, $\hat{u} = u(\Psi^{-1}(y))$ and the meaning of $\hat{a}_{ij}$, $\hat{b}_i$, $\hat{c}$, $\hat{f}$ are understood similarly. It is easy to check that the coefficients and the right hand side function of equation (9.1.12) have the same properties as the corresponding ones of equation (9.1.7). To establish the near boundary estimate for equation (9.1.12), as we did for the interior estimate, we first cut off the function $\hat{u}$, consider the equation for the function after cutting off and throwing off the lower order terms, adopt the method of solidifying coefficients, and use the conclusion of Theorem 9.1.1 and Remark 9.1.2 (as pointed out in Remark 9.1.3, for such equation, the conclusion of Theorem 9.1.1 is still valid). After returning to the original equation with respect to the variable x, we finally obtain

$$\|D^2 u\|_{L^p(O_R)} \leq C \left(\|f\|_{L^p(\Omega)} + \|u\|_{W^{1,p}(\Omega)} \right), \tag{9.1.13}$$

where $0 < R \leq R_0$, and R_0 is a given constant, $O_R \subset \Omega$ is such a domain which depends on R and for some constant $\sigma > 0$ independent of R such that $\Omega \cap B_{\sigma R}(x^0) \subset O_R$.

Step 3 Establish the global estimate.

Combining the interior estimate (9.1.11) and the near boundary estimate (9.1.13) and using the finite covering argument, we get the following global estimate

$$\|D^2 u\|_{L^p(\Omega)} \leq C(\|f\|_{L^p(\Omega)} + \|u\|_{W^{1,p}(\Omega)}).$$

Then, Ehrling–Nirenberg–Gagliardo's interpolation inequality implies

$$\|u\|_{W^{2,p}(\Omega)} \leq C(\|f\|_{L^p(\Omega)} + \|u\|_{L^p(\Omega)}).$$

$\square$

Remark 9.1.4 *From the proof of Theorem 9.1.2, we see that the establishment of $W^{2,p}$ estimates on solutions depends essentially on the conditions $a_{ij} \in C(\overline{\Omega})$. For this reason, it should be careful in applications.*

In Theorem 9.1.1 and Theorem 9.1.2, we have mentioned a new kind of solutions which have weak derivatives up to the second order and satisfy the equation almost everywhere. Such kind of solutions will be called strong solutions.

More general, consider the following nonhomogeneous boundary value condition

$$u\Big|_{\partial\Omega} = \varphi, \tag{9.1.14}$$

where $\varphi \in W^{2,p}(\Omega)$.

Definition 9.1.3 A function $u \in W^{2,p}(\Omega)$ is called a strong solution of equation (9.1.7), if u satisfies equation (9.1.7) almost everywhere in Ω. If, in addition, $u - \varphi \in W_0^{1,p}(\Omega)$, then u is called a strong solution of the Dirichlet problem (9.1.7), (9.1.14).

From Theorem 9.1.2 and Definition 9.1.3, it is easy to prove

Theorem 9.1.3   *Let $\partial\Omega \in C^2$, $p > 1$. Assume that the coefficients of equation (9.1.7) satisfy (9.1.8), (9.1.9) and $a_{ij} \in C(\overline{\Omega})$. If $u \in W^{2,p}(\Omega)$ is a strong solution of the Dirichlet problem (9.1.7), (9.1.14), then*

$$\|u\|_{W^{2,p}(\Omega)} \le C \left(\|f\|_{L^p(\Omega)} + \|\varphi\|_{W^{2,p}(\Omega)} + \|u\|_{L^p(\Omega)} \right),$$

where C is a constant depending only on n, p, λ, M, Ω and the continuity module of a_{ij}.

9.1.3   Existence and uniqueness of strong solutions for linear elliptic equations

Now, we discuss the existence and uniqueness of strong solutions. As shown in Chapter 8, the existence and uniqueness of classical solutions are based on not only Schauder's estimates but also the L^∞ norm estimates on solutions themselves. Similarly, the existence and uniqueness of strong solutions are based not only the L^p estimates but also the L^p norm estimates on solutions themselves. For general equations, the L^p norm estimates can be established by Aleksandrov's maximum principle (see [Chen and Wu (1997)]). While for special equations, for example, for the equation

$$-\Delta u + c(x)u = f(x), \quad x \in \Omega, \tag{9.1.15}$$

the L^p norm estimate can be obtained by the methods similar to those in establishing the L^2 norm estimate (see the following Theorem 9.1.4). However, such kind of methods can only be generalized to equations in divergence form and can only be used to treat the case $p \ge 2$.

Theorem 9.1.4   *Let $\partial\Omega \in C^{2,\alpha}$, $p \ge 2$, $c \in L^\infty(\Omega)$ and $c \ge 0$. Then for any $f \in L^p(\Omega)$, equation (9.1.15) admits a unique strong solution $u \in W^{2,p}(\Omega) \cap W_0^{1,p}(\Omega)$.*

Proof.   We first establish the L^p norm estimates on strong solutions of equation (9.1.15). Let $u \in W^{2,p}(\Omega) \cap W_0^{1,p}(\Omega)$ be a strong solution of equation (9.1.15). Multiplying both sides of equation (9.1.15) by $|u|^{p-2}u$ and integrating the resulting relation over Ω, we have

$$-\int_\Omega |u|^{p-2}u\Delta u dx + \int_\Omega c|u|^p dx = \int_\Omega f|u|^{p-2}u dx.$$

Integrating by parts yields

$$\frac{4(p-1)}{p^2} \int_\Omega |\nabla(|u|^{p/2-1}u)|^2 dx + \int_\Omega c|u|^p dx = \int_\Omega f|u|^{p-2}u dx.$$

Then, using Poincaré's inequality, Hölder's inequality and Young's inequality with ε leads to

$$\frac{4(p-1)}{\mu p^2} \int_\Omega |u|^p dx + \int_\Omega c|u|^p dx \leq \int_\Omega f|u|^{p-1} dx$$

$$\leq \|f\|_{L^p(\Omega)} \|u\|_{L^p(\Omega)}^{p-1}$$

$$\leq \varepsilon \int_\Omega |u|^p dx + \varepsilon^{-1/(p-1)} \int_\Omega |f|^p dx,$$

where $\mu > 0$ is the constant in Poincaré's inequality, $\varepsilon > 0$ is an arbitrary constant. Owing to $c \geq 0$, the above estimate yields

$$\int_\Omega |u|^p dx \leq C \int_\Omega |f|^p dx$$

with C depending only on μ and p. Thus by virtue of Theorem 9.1.2, we obtain

$$\|u\|_{W^{2,p}(\Omega)} \leq C\|f\|_{L^p(\Omega)}, \tag{9.1.16}$$

where C depends only on n, p, μ, $\|c\|_{L^\infty(\Omega)}$ and Ω.

Using the a priori estimate (9.1.16), we immediately obtain the uniqueness of the strong solution. In fact, let $u_1, u_2 \in W^{2,p}(\Omega) \cap W_0^{1,p}(\Omega)$ be two strong solutions of equation (9.1.15) and set $v = u_1 - u_2$. Then $v \in W^{2,p}(\Omega) \cap W_0^{1,p}(\Omega)$ is a strong solution of the homogeneous equation

$$-\Delta v + cv = 0, \quad x \in \Omega.$$

The estimate (9.1.16) gives $\|v\|_{W^{2,p}(\Omega)} \leq 0$, which implies that $v = 0$ a.e. in Ω, that is $u_1 = u_2$ a.e. in Ω.

Finally, we show the existence of strong solutions. Let $c_k, f_k \in C^\alpha(\overline{\Omega})$, $c_k \geq 0$, and c_k converges to c weakly star in $L^\infty(\Omega)$, f_k converges to f in $L^p(\Omega)$. Consider the approximate problem

$$\begin{cases} -\Delta u_k + c_k u = f_k(x), & x \in \Omega, \\ u_k\big|_{\partial\Omega} = 0. \end{cases}$$

According to Theorem 8.2.7 of Chapter 8, the above problem admits a solution $u_k \in C^{2,\alpha}(\overline{\Omega}) \subset W^{2,p}(\Omega) \cap W_0^{1,p}(\Omega)$. The estimate (9.1.16) implies

$$\|u_k\|_{W^{2,p}(\Omega)} \leq C\|f_k\|_{L^p(\Omega)},$$

which shows that $\{u_k\}$ is uniformly bounded in $W^{2,p}(\Omega)$. From the weak compactness of the bounded set in $W^{2,p}(\Omega)$ and the compactly embedding theorem, there exists a subsequence of $\{u_k\}$, which converges weakly in $W^{2,p}(\Omega)$ and converges strongly in $W^{1,p}(\Omega)$. Let u be the limit function. Then $u \in W^{2,p}(\Omega) \cap W_0^{1,p}(\Omega)$ and it is easy to verify that u satisfies equation (9.1.15) almost everywhere. $\qquad\square$

Remark 9.1.5 *The conclusion of Theorem 9.1.4 is still valid for the case $1 < p < 2$. In addition, the smoothness condition on $\partial\Omega$ can be relaxed to $\partial\Omega \in C^2$.*

For general linear elliptic equation (9.1.7), we have the following theorem, whose proof can be found in [Chen and Wu (1997)] Chapter 3.

Theorem 9.1.5 *Let $\partial\Omega \in C^2$, $p > 1$. Assume that the coefficients of equation (9.1.7) satisfy (9.1.8), (9.1.9), $c \geq 0$ and $a_{ij} \in C(\overline{\Omega})$. Then for any $f \in L^p(\Omega)$, the Dirichlet problem (9.1.7), (9.1.14) admits a unique strong solution $u \in W^{2,p}(\Omega)$.*

9.2 L^p Estimates for Linear Parabolic Equations and Existence and Uniqueness of Strong Solutions

In this section, we introduce a parallel theory to the one in the first section for linear parabolic equations.

9.2.1 L^p estimates for the heat equation in cubes

Consider the first initial-boundary value problem for the heat equation

$$\frac{\partial u}{\partial t} - \Delta u = f, \quad (x,t) \in Q_T, \tag{9.2.1}$$

$$u\Big|_{\partial_p Q_T} = 0, \tag{9.2.2}$$

where $Q_T = Q_0 \times (0,T)$, Q_0 is a cube in $\mathbb{R}^n$ with edges parallel to its axes.

Denote

$$Q_R = Q_R(x^0, t_0) = B_R(x^0) \times (t_0 - R^2, t_0 + R^2),$$

$$Q_R^0 = Q_R^0(x^0, 0) = B_R(x^0) \times (0, R^2).$$

Proposition 9.2.1   *Let $f \in L^\infty(Q_T)$, $u \in W_2^{2,1}(Q_T) \cap \overset{\bullet}{W}{}_2^{1,1}(Q_T)$ be a weak solution of equation (9.2.1), $x^0 \in Q_0$, $B_{2R_0}(x^0) \subset\subset Q_0$. Then*

$$[D^2 u]_{2, n+2; B_{R_0}(x^0) \times (0,T)}$$

$$\leq C \left(\|D^2 u\|_{L^2(B_{2R_0}(x^0) \times (0,T))} + \|f\|_{L^\infty(B_{2R_0}(x^0) \times (0,T))} \right), \qquad (9.2.3)$$

where C depends only on n, T and R_0.

Proof.   Let f_ε be a smooth approximation of f, u_ε be the solution of the problem

$$\begin{cases} \dfrac{\partial u_\varepsilon}{\partial t} - \Delta u_\varepsilon = f_\varepsilon, & (x, t) \in Q_T, \\[2mm] u_\varepsilon \big|_{\partial_p Q_T} = 0. \end{cases}$$

From the L^2 theory, u_ε is sufficiently smooth in $\overline{B}_{2R_0}(x^0) \times [0, T]$ and

$$u_\varepsilon \to u, \quad \text{in } W_2^{2,1}(B_{2R_0}(x^0) \times (0, T)) \text{ as } \varepsilon \to 0.$$

So, in order to prove (9.2.3), it suffices to show

$$[D^2 u_\varepsilon]_{2, n+2; B_{R_0}(x^0) \times (0,T)}$$

$$\leq C \left(\|D^2 u_\varepsilon\|_{L^2(B_{2R_0}(x^0) \times (0,T))} + \|f_\varepsilon\|_{L^\infty(B_{2R_0}(x^0) \times (0,T))} \right),$$

where the constant C is independent of ε. Owing to this reason, in what follows, we may suppose that u is sufficiently smooth in $\overline{B}_{2R_0}(x^0) \times [0, T]$.

First, we establish the interior estimate. For any $x \in \overline{B}_{R_0}(x^0)$, $0 < t < T$, we choose $0 < \hat{R}_0 \leq R_0$ such that $Q_{\hat{R}_0}(x, t) \subset B_{2R_0}(x^0) \times (0, T) \subset Q_T$. From the proof of Schauder's interior estimate (Theorem 7.2.3 of Chapter 7), we see that, for any $0 < \rho \leq R \leq \hat{R}_0$,

$$\iint_{Q_\rho(x,t)} |Dw(y, s) - (Dw)_{x,t,\rho}|^2 \, dyds$$

$$\leq C \left(\frac{\rho}{R} \right)^{n+4} \iint_{Q_R(x,t)} |Dw(y, s) - (Dw)_{x,t,R}|^2 \, dyds$$

$$+ C \iint_{Q_R(x,t)} (f(y, s) - f_{x,t,R})^2 \, dyds$$

$$\leq C \left(\frac{\rho}{R} \right)^{n+4} \iint_{Q_R(x,t)} |Dw(y, s) - (Dw)_{x,t,R}|^2 \, dyds$$

$$+ CR^{n+2}\|f\|^2_{L^\infty(Q_R(x,t))},$$

where $w = D_i u\,(1 \le i \le n)$. By virtue of this and the iteration lemma (Lemma 6.2.1 of Chapter 6), we deduce that for any $0 < \rho \le R \le \hat{R}_0$,

$$\iint_{Q_\rho(x,t)} |Dw(y,s) - (Dw)_{x,t,\rho}|^2 dyds$$

$$\le C\left(\frac{\rho}{R}\right)^{n+2} \left(\iint_{Q_R(x,t)} |Dw(y,s) - (Dw)_{x,t,R}|^2 dyds \right.$$

$$\left. + R^{n+2}\|f\|^2_{L^\infty(Q_R(x,t))}\right)$$

$$\le C\rho^{n+2} \left(\frac{1}{R^{n+2}}\|Dw\|^2_{L^2(Q_R(x,t))} + \|f\|^2_{L^\infty(Q_R(x,t))}\right).$$

Similar to the proofs of Theorem 7.2.4 and Theorem 7.2.5 of Chapter 7, we obtain

$$[D^2 u]^{(1/2)}_{2,n+2;Q_{\hat{R}_0/2}(x,t)}$$

$$\le C\left(\|D^2 u\|_{L^2(Q_{\hat{R}_0}(x,t))} + \|f\|_{L^\infty(Q_{\hat{R}_0}(x,t))}\right)$$

$$\le C\left(\|D^2 u\|_{L^2(B_{2R_0}(x^0)\times(0,T))} + \|f\|_{L^\infty(B_{2R_0}(x^0)\times(0,T))}\right), \qquad (9.2.4)$$

where C depends only on n and $\hat{R}_0$.

Similarly, we may obtain the near bottom estimate. For any $x \in \overline{B}_{R_0}(x^0)$, we choose $0 < \hat{R}_0 \le R_0$ such that $Q^0_{\hat{R}_0}(x,0) \subset B_{2R_0}(x^0)\times(0,T) \subset Q_T$. From the proof of the near bottom Schauder's estimate (Theorem 7.2.9 of Chapter 7), we see that for any $0 < \rho \le R \le \hat{R}_0$,

$$\iint_{Q^0_\rho(x,0)} |Dw(y,s)|^2 dyds$$

$$\le C\left(\frac{\rho}{R}\right)^{n+4} \iint_{Q^0_R(x,0)} |Dw(y,s)|^2 dyds$$

$$+ C\iint_{Q^0_R(x,0)} (f(y,s) - f_{x,0,R})^2 dyds$$

$$\le C\left(\frac{\rho}{R}\right)^{n+4} \iint_{Q^0_R(x,0)} |Dw(y,s)|^2 dyds + CR^{n+2}\|f\|^2_{L^\infty(Q^0_R(x,0))},$$

where $w = D_i u\,(1 \le i \le n)$ and

$$v_{x,0,R} = \frac{1}{|Q^0_R(x,0)|} \iint_{Q^0_R(x,0)} v(y,s)dyds.$$

Therefore, the iteration lemma (Lemma 6.2.1 of Chapter 6) leads to that for any $0 < \rho \leq R \leq \hat{R}_0$,

$$\iint_{Q^0_\rho(x,0)} |Dw(y,s)|^2 dyds$$

$$\leq C \left(\frac{\rho}{R}\right)^{n+2} \left(\iint_{Q^0_R(x,0)} |Dw(y,s)|^2 dyds + R^{n+2}\|f\|^2_{L^\infty(Q^0_R(x,0))}\right)$$

$$\leq C\rho^{n+2} \left(\frac{1}{R^{n+2}}\|Dw\|^2_{L^2(Q^0_R(x,0))} + \|f\|^2_{L^\infty(Q^0_R(x,0))}\right).$$

Similar to the proofs of Theorem 7.2.10 and Theorem 7.2.11, we obtain

$$[D^2u]^{(1/2)}_{2,n+2;Q^0_{\hat{R}_0/2}(x,0)}$$

$$\leq C \left(\|D^2u\|_{L^2(Q^0_{\hat{R}_0}(x,0))} + \|f\|_{L^\infty(Q^0_{\hat{R}_0}(x,0))}\right)$$

$$\leq C \left(\|D^2u\|_{L^2(B_{2R_0}(x^0)\times(0,T))} + \|f\|_{L^\infty(B_{2R_0}(x^0)\times(0,T))}\right), \qquad (9.2.5)$$

where C depends only on n and $\hat{R}_0$.

Combining the interior estimate (9.2.4) with the near bottom estimate (9.2.5), using the finite covering argument and the relation $[\cdot]_{2,n} \leq [\cdot]^{(1/2)}_{2,n} + C\|\cdot\|_{L^2}$, we obtain (9.2.3) immediately. $\qquad\square$

Proposition 9.2.2 *Let $f \in L^\infty(Q_T)$ and $u \in W^{2,1}_2(Q_T) \cap \overset{\bullet}{W}^{1,1}_2(Q_T)$ be a weak solution of the first initial-boundary value problem (9.2.1), (9.2.2). Then*

$$\|D^2u\|_{\mathcal{L}^{2,n+2}(Q_T)} \leq C\|f\|_{L^\infty(Q_T)}, \qquad (9.2.6)$$

where C depends only on n, T and the length of the edge of Q_0.

Proof. Let

$$Q_0 = Q(x^0, R) = \{x \in \mathbb{R}^n; |x_i - x^0_i| < R, i = 1, 2, \cdots, n\}.$$

Similar to the proof of the corresponding conclusion for elliptic equations (Proposition 9.1.2), for fixed $t \in [0,T]$, we extend the definition of $u(\cdot, t)$ and $f(\cdot, t)$ to $\overline{Q}(x^0, 3^n R)$ antisymmetrically, and obtain functions $\tilde{u}$ and $\tilde{f}$ in $\overline{Q}_T(x^0, 3^n R)$, where

$$Q_T(x^0, 3^n R) = Q(x^0, 3^n R) \times (0, T).$$

Obviously $\tilde{f} \in L^\infty(Q_T(x^0, 3^n R))$. It is not difficult to check that $\tilde{u} \in W^{2,1}_2(Q_T(x^0, 3^n R)) \cap \overset{\bullet}{W}^{1,1}_2(Q_T(x^0, 3^n R))$ and $\tilde{u}$ is a weak solution of the equation

$$\frac{\partial \tilde{u}}{\partial t} - \Delta \tilde{u} = \tilde{f}, \quad (x,t) \in Q_T(x^0, 3^n R).$$

Owing to

$$Q_0 = Q(x^0, R) \subset B_{\sqrt{n}R}(x^0) \subset B_{2\sqrt{n}R}(x^0) \subset\subset Q(x^0, 3^n R),$$

use Proposition 9.2.1 in $Q_T(x^0, 3^n R)$ to get

$$\begin{aligned}
[D^2 u]_{2,n+2;Q_T} &\leq [D^2 \tilde{u}]_{2,n+2;B_{\sqrt{n}R}(x^0)\times(0,T)} \\
&\leq C\left(\|D^2\tilde{u}\|_{L^2(B_{2\sqrt{n}R}(x^0)\times(0,T))} + \|\tilde{f}\|_{L^\infty(B_{2\sqrt{n}R}(x^0)\times(0,T))}\right) \\
&\leq C\left(\|D^2\tilde{u}\|_{L^2(Q_T(x^0,3^n R))} + \|\tilde{f}\|_{L^\infty(Q_T(x^0,3^n R))}\right),
\end{aligned}$$

which implies

$$[D^2 u]_{2,n+2;Q_T} \leq C\left(\|D^2 u\|_{L^2(Q_T)} + \|f\|_{L^\infty(Q_T)}\right) \tag{9.2.7}$$

with another constant C depending only on n and R. According to the L^2 theory (Remark 3.4.1 of Chapter 3, although there the spatial domain is assumed to have C^2 smoothness, the conclusion is still valid for cubes, which can be proved by the methods similar to those in Theorem 8.3.2 of Chapter 8), we have

$$\|u\|_{W_2^{2,1}(Q_T)} \leq C\|f\|_{L^2(Q_T)}. \tag{9.2.8}$$

Combining (9.2.7) with (9.2.8) leads to (9.2.6).   □

Define an operator

$$T_i : L^\infty(Q_T) \to \mathrm{BMO}(Q_T), \quad f \mapsto DD_i u,$$

where $1 \leq i \leq n$ and u is a weak solution of problem (9.2.1), (9.2.2). The estimate (9.2.6) shows that T_i $(1 \leq i \leq n)$ is an bounded linear operator from $L^\infty(Q_T)$ to $BMO(Q_T)$.

Theorem 9.2.1 *Let $p \geq 2$ and $u \in W_p^{2,1}(Q_T) \cap \overset{\bullet}{W}{}_p^{1,1}(Q_T)$ satisfy the heat equation (9.2.1) almost everywhere in Q_T. Then*

$$\|D^2 u\|_{L^p(Q_T)} + \left\|\frac{\partial u}{\partial t}\right\|_{L^p(Q_T)} \leq C\|f\|_{L^p(Q_T)},$$

where C depends only on n, p, T and the length of the edge of Q_0.

Proof. From the L^2 theory and the estimate (9.2.6), T_i $(1 \leq i \leq n)$ is a bounded linear operator both from $L^2(Q_T)$ to $L^2(Q_T)$ and from $L^\infty(Q_T)$ to BMO(Q_T). Then, by Stampacchia's interpolation theorem, T_i is a bounded linear operator from $L^p(Q_T)$ to $L^p(Q_T)$ and

$$\|D^2 u\|_{L^p(Q_T)} \leq C\|f\|_{L^p(Q_T)}.$$

In addition, by virtue of this estimate, equation (9.2.1) gives

$$\left\|\frac{\partial u}{\partial t}\right\|_{L^p(Q_T)} \leq C\|f\|_{L^p(Q_T)}.$$

$\square$

Remark 9.2.1 *Under the conditions of Theorem 9.2.1, using Ehrling–Nirenberg–Gagliardo's interpolation inequality, we further obtain*

$$\|u\|_{W_p^{2,1}(Q_T)} \leq C(\|f\|_{L^p(Q_T)} + \|u\|_{L^p(Q_T)}).$$

Remark 9.2.2 *The conclusion of Theorem 9.2.1 is also valid for the case $1 < p < 2$, whose proof can be found in [Gu (1995)] Chapter 7.*

Remark 9.2.3 *The conclusion of Theorem 9.2.1 can be extended to parabolic equations with the coefficients matrix of second order derivatives being constant and positive definite and throwing all lower order terms. This is because Schauder's interior estimates, near bottom estimates and the extension used to prove (9.2.6) is also available to such kind of equations.*

9.2.2 L^p estimates for general linear parabolic equations

Now we turn to the general linear parabolic equations

$$\frac{\partial u}{\partial t} - a_{ij}(x,t)D_{ij}u + b_i(x,t)D_iu + c(x,t)u = f(x,t), \quad (x,t) \in Q_T, \quad (9.2.9)$$

where $Q_T = \Omega \times (0,T)$, $\Omega \subset \mathbb{R}^n$ is a bounded domain, $a_{ij} = a_{ji}$ and for some constants $\lambda > 0$, $M > 0$

$$a_{ij}(x,t)\xi_i\xi_j \geq \lambda|\xi|^2, \quad \forall \xi \in \mathbb{R}^n, \ (x,t) \in Q_T, \quad (9.2.10)$$

$$\sum_{i,j=1}^n \|a_{ij}\|_{L^\infty(Q_T)} + \sum_{i=1}^n \|b_i\|_{L^\infty(Q_T)} + \|c\|_{L^\infty(Q_T)} \leq M. \quad (9.2.11)$$

Similar to the case of elliptic equations, we may obtain the L^p estimates for general linear parabolic equations.

Theorem 9.2.2 *Let $\Omega \in C^2$, $p > 1$. Assume that the coefficients of equation (9.2.9) satisfy (9.2.10) and (9.2.11) and $a_{ij} \in C(\overline{Q}_T)$. If $u \in W_p^{2,1}(Q_T) \cap \overset{\bullet}{W}_p^{1,1}(Q_T)$ satisfies equation (9.2.9) almost everywhere in Q_T, then*

$$\|u\|_{W_p^{2,1}(Q_T)} \leq C \left(\|f\|_{L^p(Q_T)} + \|u\|_{L^p(Q_T)} \right),$$

where C depends only on n, p, λ, M, T, Ω and the continuity module of a_{ij}.

Remark 9.2.4 *The establishment of $W_p^{2,1}$ estimates on solutions is essentially depending on the condition $a_{ij} \in C(\overline{Q}_T)$, and so, in applying such estimates one should take special care.*

Now, we consider the first initial-boundary value problem of equation (9.2.9) with the following boundary and initial value condition

$$u\Big|_{\partial_p Q_T} = \varphi, \tag{9.2.12}$$

where $\varphi \in W_p^{2,1}(Q_T)$.

Definition 9.2.1 A function $u \in W_p^{2,1}(Q_T)$ is called a strong solution of equation (9.2.9), if u satisfies equation (9.2.9) almost everywhere in Q_T. If, in addition, $u - \varphi \in \overset{\bullet}{W}_p^{1,1}(Q_T)$, then u is called a strong solution of the first initial-boundary value problem (9.2.9), (9.2.12).

From Theorem 9.2.2 and Definition 9.2.1, we easily obtain the following

Theorem 9.2.3 *Let $\Omega \in C^2$, $p > 1$. Assume that the coefficients of equation (9.2.9) satisfy (9.2.10) and (9.2.11) and $a_{ij} \in C(\overline{Q}_T)$. If $u \in W_p^{2,1}(Q_T)$ is a strong solution of the first initial-boundary value problem (9.2.9), (9.2.12), then*

$$\|u\|_{W_p^{2,1}(Q_T)} \leq C \left(\|f\|_{L^p(Q_T)} + \|\varphi\|_{W_p^{2,1}(Q_T)} + \|u\|_{L^p(Q_T)} \right),$$

where C depends only on n, p, λ, M, T, Ω and the continuity module of a_{ij}.

9.2.3 *Existence and uniqueness of strong solutions for linear parabolic equations*

To show the existence and uniqueness of strong solutions, besides the L^p estimates established in Theorem 9.2.2, we also need the L^p norm estimates on solutions themselves. Similar to the case of elliptic equations, for general equations, the L^p norm estimates can be established by using the Aleksandrov's maximum principle (see [Gu (1995)]). While for some special equations, for example, for the equation

$$\frac{\partial u}{\partial t} - \Delta u + c(x,t)u = f(x,t), \quad (x,t) \in Q_T, \quad (9.2.13)$$

we may utilize the methods similar to those utilized in establishing the L^2 norm estimates (see the following Theorem 9.2.4). However, such kind of methods can only be generalized to the equations with divergence form and can only be used to treat the case $p \geq 2$.

Theorem 9.2.4 *Let $\partial\Omega \in C^{2,\alpha}$, $p \geq 2$ and $c \in L^\infty(Q_T)$. Then for any $f \in L^p(Q_T)$, equation (9.2.13) admits a unique strong solution $u \in W_p^{2,1}(Q_T) \cap \overset{\bullet}{W}_p^{1,1}(Q_T)$.*

Proof. Let $u \in W_p^{2,1}(Q_T) \cap \overset{\bullet}{W}_p^{1,1}(Q_T)$ be a strong solution of equation (9.2.13). Setting

$$w(x,t) = e^{-Mt}u(x,t), \quad (x,t) \in Q_T,$$

where $M = \|c\|_{L^\infty(Q_T)}$. Then $w \in W_p^{2,1}(Q_T) \cap \overset{\bullet}{W}_p^{1,1}(Q_T)$ and w satisfies

$$\frac{\partial w}{\partial t} - \Delta w + (M+c)w$$

$$= e^{-Mt}\left(\frac{\partial u}{\partial t} - \Delta u + cu\right) = e^{-Mt}f, \quad \text{a.e. in } Q_T$$

with $M + c \geq 0$. Hence we may assume that $c \geq 0$ in equation (9.2.13).

We first establish the L^p norm estimates on strong solutions for equation (9.2.13). Assume that $u \in W_p^{2,1}(Q_T) \cap \overset{\bullet}{W}_p^{1,1}(Q_T)$ is a strong solution of equation (9.2.13). Multiplying both sides of equation (9.2.13) by $|u|^{p-2}u$ and integrating over Q_T, we obtain

$$\iint_{Q_T} |u|^{p-2}u\frac{\partial u}{\partial t}dxdt - \iint_{Q_T} |u|^{p-2}u\Delta u\,dxdt + \iint_{Q_T} c|u|^p dxdt$$

$$= \iint_{Q_T} f|u|^{p-2}u\,dxdt.$$

Integrating by parts with respect to the spatial variable yields

$$\frac{1}{p}\iint_{Q_T} \frac{\partial |u|^p}{\partial t}dxdt + \frac{4(p-1)}{p^2}\iint_{Q_T} |\nabla(|u|^{p/2-1}u)|^2 dxdt + \iint_{Q_T} c|u|^p dxdt$$

$$= \iint_{Q_T} f|u|^{p-2}u\,dxdt.$$

Using Poincaré's inequality, Hölder's inequality and Young's inequality with ε, we get

$$\frac{1}{p}\int_{\Omega} |u|^p \Big|_0^T dx + \frac{4(p-1)}{\mu p^2} \iint_{Q_T} |u|^p dxdt + \iint_{Q_T} c|u|^p dxdt$$

$$\leq \iint_{Q_T} f|u|^{p-1} dxdt$$

$$\leq \|f\|_{L^p(Q_T)} \|u\|_{L^p(Q_T)}^{p-1}$$

$$\leq \varepsilon \iint_{Q_T} |u|^p dxdt + \varepsilon^{-1/(p-1)} \iint_{Q_T} |f|^p dxdt,$$

where $\mu > 0$ is the constant in Poincaré's inequality, $\varepsilon > 0$ is an arbitrary constant. Owing to $u \in \overset{\bullet}{W}_p^{1,1}(Q_T)$ and $c \geq 0$, the above estimate yields

$$\iint_{Q_T} |u|^p dxdt \leq C \iint_{Q_T} |f|^p dxdt$$

with C depends only on μ and p. Combine this with Theorem 9.2.2 to get

$$\|u\|_{W_p^{2,1}(Q_T)} \leq C\|f\|_{L^p(Q_T)} \tag{9.2.14}$$

with C depending only on n, p, μ, $\|c\|_{L^\infty(Q_T)}$, T and Ω.

From (9.2.14), we immediately obtain the uniqueness of the strong solution. In fact, assume $u_1, u_2 \in W_p^{2,1}(Q_T) \cap \overset{\bullet}{W}_p^{1,1}(Q_T)$ are two strong solutions of equation (9.2.13). Set $v = u_1 - u_2$. Then $v \in W_p^{2,1}(Q_T) \cap \overset{\bullet}{W}_p^{1,1}(Q_T)$ is a strong solution of the homogeneous equation

$$\frac{\partial v}{\partial t} - \Delta v + c(x,t)v = 0, \quad (x,t) \in Q_T.$$

According to the estimate (9.2.14), we have $\|v\|_{W_p^{2,1}(Q_T)} \leq 0$. Therefore $v = 0$ a.e. in Q_T, that is $u_1 = u_2$ a.e. in Q_T.

Finally, we prove the existence of strong solutions. Let $c_k, f_k \in C^{\alpha,\alpha/2}(\overline{Q}_T)$, $c_k \geq 0$, and c_k converges to c weakly star in $L^\infty(Q_T)$, f_k converges to f in $L^p(Q_T)$. Consider the approximate problem

$$\begin{cases} \dfrac{\partial u_k}{\partial t} - \Delta u_k + c_k(x,t)u = f_k(x,t), & (x,t) \in Q_T, \\[2mm] u_k\Big|_{\partial Q_T} = 0. \end{cases}$$

According to Theorem 8.3.7 of Chapter 8, the above problem admits a solution $u_k \in C^{2,\alpha}(\overline{Q}_T)$. Thus the estimate (9.2.14) gives

$$\|u_k\|_{W_p^{2,1}(Q_T)} \leq C\|f_k\|_{L^p(Q_T)},$$

which implies that $\{u_k\}$ is uniformly bounded in $W_p^{2,1}(Q_T)$. From the weak compactness of the bounded set in $W_p^{2,1}(Q_T)$ and $\overset{\bullet}{W}\,_p^{1,1}(Q_T)$ and the compactly embedding theorem, there exists a subsequence of $\{u_k\}$, which converges weakly in $W_p^{2,1}(Q_T)$ and $\overset{\bullet}{W}_p^{1,1}(Q_T)$, and converges strongly in $L^p(Q_T)$. Let u be the limit function. It is easy to verify that $u \in W_p^{2,1}(Q_T) \cap \overset{\bullet}{W}\,_p^{1,1}(Q_T)$ and u satisfies equation (9.2.13) almost everywhere in Q_T. $\qquad\square$

Remark 9.2.5 *Different from the case of elliptic equations (Theorem 9.2.4), the restriction condition $c \geq 0$ is not required.*

Remark 9.2.6 *The conclusion of Theorem 9.2.4 is still valid for the case $1 < p < 2$. In addition, it suffices to require $\partial\Omega \in C^2$.*

For general linear parabolic equation (9.2.9), we have the following theorem, whose proof can be found in [Gu (1995)] Chapter 7.

Theorem 9.2.5 *Let $\partial\Omega \in C^2$, $p > 1$. Assume that the coefficients of equation (9.2.9) satisfy (9.2.10) and (9.2.11), and $a_{ij} \in C(\overline{Q}_T)$. Then for any $f \in L^p(Q_T)$, the first initial-boundary value problem (9.2.9), (9.2.12) admits a unique strong solution $u \in W_p^{2,1}(Q_T)$.*

Exercises

1. Check $\tilde{u} \in H^2(Q(x^0, 3^n R))$ in Proposition 9.1.2, and judge whether the following arguments are valid:
   i) If $u \in C^{1,\alpha}(\overline{Q}(x^0, R))$ and $u\big|_{\partial Q(x^0,R)} = 0$, then $\tilde{u} \in C^{1,\alpha}(\overline{Q}(x^0, 3^n R))$;
   ii) If $u \in C^2(\overline{Q}(x^0, R))$ and $u\big|_{\partial Q(x^0,R)} = 0$, then $\tilde{u} \in H^2(\overline{Q}(x^0, 3^n R))$.

2. Prove that $a(x)$ has a continuity module in $\overline{\Omega}$ if and only if $a(x) \in C(\overline{\Omega})$, where Ω is an bounded open set in $\mathbb{R}^n$.

3. Let $\Omega \subset \mathbb{R}^n$ be a bounded domain with appropriately smooth boundary, $\lambda \in \mathbb{R}$, $p > 1$. Then there is one and only one of the following alternatives:

i) The boundary value problem for the homogeneous equation

$$-\Delta u + \lambda u = 0 \text{ in } \Omega, \quad u\big|_{\partial \Omega} = 0$$

admits a nontrivial strong solution $u \in W^{2,p}(\Omega) \cap W_0^{1,p}(\Omega)$;

ii) For any $f \in L^p(\Omega)$, the boundary value problem for the inhomogeneous equation

$$-\Delta u + \lambda u = f \text{ in } \Omega, \quad u\big|_{\partial \Omega} = 0$$

admits a unique strong solution $u \in W^{2,p}(\Omega) \cap W_0^{1,p}(\Omega)$.

Chapter 10

Fixed Point Method

The approaches based on fixed point theorems have very important applications in the investigation of partial differential equations, especially the nonlinear differential equations. In this chapter, as an example, we apply such a method to the solvability of quasilinear elliptic equations.

10.1 Framework of Solving Quasilinear Equations via Fixed Point Method

In this section, we describe the basic framework of fixed point method in solving quasilinear equations.

10.1.1 *Leray-Schauder's fixed point theorem*

Leray-Schauder's Fixed Point Theorem *Let U be a Banach space, $T(u, \sigma)$ be a mapping from $U \times [0, 1]$ to U satisfying the following conditions:*

i) T is a compact mapping;

ii) $T(u, 0) = 0$, $\forall u \in U$;

iii) There exists a constant $M > 0$, such that for any $u \in U$, if $u = T(u, \sigma)$ for some $\sigma \in [0, 1]$, then $\|u\|_U \leq M$.

Then the mapping $T(\cdot, 1)$ has a fixed point, that is, there exists $u \in U$, such that $u = T(u, 1)$.

10.1.2 *Solvability of quasilinear elliptic equations*

We first consider the Dirichlet problem for quasilinear elliptic equations

$$- \operatorname{div} \vec{a}(x, u, \nabla u) + b(x, u, \nabla u) = 0, \quad x \in \Omega, \tag{10.1.1}$$

$$u\Big|_{\partial\Omega} = \varphi, \tag{10.1.2}$$

where $\vec{a} = (a_1, a_2, \cdots, a_n)$ and $\Omega \subset \mathbb{R}^n$ is a bounded domain. Assume that $\vec{a}(x, z, \eta)$, $b(x, z, \eta)$ satisfy the following structure conditions:

$$\frac{\partial a_i}{\partial \eta_j} \xi_i \xi_j \geq \lambda |\xi|^2, \quad \forall \xi \in \mathbb{R}^n, \tag{10.1.3}$$

$$|a_i(x, z, 0)| \leq g(x), \tag{10.1.4}$$

$$\left| \frac{\partial a_i}{\partial \eta_j} \right| \leq \mu(|z|), \tag{10.1.5}$$

$$\left| \frac{\partial a_i}{\partial z} \right| + |a_i| \leq \mu(|z|)(1 + |\eta|), \tag{10.1.6}$$

$$\left| \frac{\partial a_i}{\partial x_j} \right| + |b| \leq \mu(|z|)(1 + |\eta|^2), \tag{10.1.7}$$

$$-b(x, z, \eta)\mathrm{sgn}z \leq \Lambda(|\eta| + h(x)), \tag{10.1.8}$$

where $i, j = 1, 2, \cdots, n$, $\lambda, \Lambda > 0$, $g \in L^q(\Omega)$ $(q > n)$, $h \in L^{q_*}(\Omega)$, $q_* = nq/(n+q)$ and $\mu(s)$ is a nondecreasing function on $[0, +\infty)$.

A typical example of such equations satisfying (10.1.3)–(10.1.8) is

$$-\mathrm{div}(a(u)\nabla u) + b(u) = f(x), \quad x \in \Omega, \tag{10.1.9}$$

where

$$a(u) = (u^2 + 1)^{m/2} \ (m > 0), \quad b(u) = |u|^{\gamma-1}u \ (\gamma \geq 1)$$

and $f \in C^\alpha(\overline{\Omega})$. For simplicity, we consider only the Dirichlet problem with the homogeneous boundary value condition

$$u\Big|_{\partial\Omega} = 0. \tag{10.1.10}$$

Theorem 10.1.1 *Let $0 < \alpha < 1$, $\partial\Omega \in C^{2,\alpha}$, $\vec{a} \in C^{1,\alpha}$, $b \in C^\alpha$, $\varphi \in C^{2,\alpha}(\overline{\Omega})$. Assume that the structure conditions (10.1.3)–(10.1.8) hold. Then*

i) There exist a constant $0 < \beta < 1$ and a constant $M > 0$ depending only on n, λ, Λ, $\|g\|_{L^q(\Omega)}$, $\|h\|_{L^{q_*}(\Omega)}$, $\mu(s)$, $|\varphi|_{2,\alpha;\Omega}$, $|\vec{a}|_{1,\alpha}$, $|b|_\alpha$ and Ω, such that for any solution $u \in C^{2,\alpha}(\overline{\Omega})$ of problem (10.1.1), (10.1.2),

$$|u|_{1,\beta;\Omega} \leq M;$$

ii) Problem (10.1.1), (10.1.2) admits a solution $u \in C^{2,\alpha}(\overline{\Omega})$.

Proof. We do not intend to give a detailed proof for such a general theorem. For simplicity, we merely consider problem (10.1.9), (10.1.10). The proof of the conclusion i) is quite lengthy, which will be completed in the subsequent sections. Here we merely give the proof of the conclusion ii) by assuming that the conclusion i) is valid.

Choose $U = C^{1,\alpha}(\overline{\Omega})$. For any $v \in U$, $0 \le \sigma \le 1$, consider the problem

$$- \sigma \operatorname{div}(a(v)\nabla u) - (1 - \sigma)\Delta u + \sigma b(v) = \sigma f(x), \quad x \in \Omega, \qquad (10.1.11)$$

$$u\Big|_{\partial\Omega} = 0. \qquad (10.1.12)$$

By the Schauder theory for linear equations, this problem admits a unique solution $u \in C^{2,\alpha}(\overline{\Omega})$. Define the mapping

$$T : U \times [0, 1] \to U,$$
$$(v, \sigma) \mapsto u.$$

In what follows, we check the properties of the mapping T.

Firstly, since $C^{2,\alpha}(\overline{\Omega})$ can be compactly embedded into $C^{1,\alpha}(\overline{\Omega})$, T is compact.

Secondly, when $\sigma = 0$, (10.1.11), (10.1.12) reduces to the homogeneous Dirichlet problem for Laplace's equation

$$\begin{cases} -\Delta u = 0, & x \in \Omega, \\ u\Big|_{\partial\Omega} = 0, \end{cases}$$

which has only a trivial solution. Therefore

$$T(v, 0) = 0, \quad \forall v \in U.$$

Finally, assume that u is a fixed point of the mapping T for some $\sigma \in [0, 1]$, namely, u is a solution of the problem

$$\begin{cases} -\sigma \operatorname{div}(a(u)\nabla u) - (1 - \sigma)\Delta u + \sigma b(u) = \sigma f(x), & x \in \Omega, \\ u\Big|_{\partial\Omega} = 0. \end{cases}$$

Then according to the conclusion i), there exist $0 < \beta < 1$ and a constant $M > 0$ independent of u and σ, such that

$$|u|_{1,\beta;\Omega} \le M. \qquad (10.1.13)$$

In applying the conclusion i), it should be noted that, all elements appeared in the structure conditions which the constant M depends on can be chosen to be independent of $\sigma \in [0, 1]$. Now, we rewrite the equation of u into the nondivergence form

$$-(\sigma a(u) + 1 - \sigma)\Delta u = \sigma(a'(u)|\nabla u|^2 - b(u) + f(x)). \qquad (10.1.14)$$

Owing to the estimate (10.1.13), the coefficients of equation (10.1.14) belong to $C^{\alpha\beta}(\overline{\Omega})$. So, according to the Schauder theory, we conclude that $u \in C^{2,\alpha\beta}(\overline{\Omega})$, and there exists a constant C independent of u and σ, such that

$$|u|_{2,\alpha\beta;\Omega} \leq C,$$

which implies

$$\|u\|_U = |u|_{1,\alpha;\Omega} \leq C.$$

Summing up, we have proved that the mapping $T(u, \sigma)$ satisfies all conditions of Leray-Schauder's fixed point theorem, and so, there exists $u \in U$ such that $T(u, 1) = u$. Then, from the definition of the operator T and the classical theory for linear elliptic equations, we get further that $u \in C^{2,\alpha}(\overline{\Omega})$. $\qquad \square$

10.1.3 *Solvability of quasilinear parabolic equations*

Now we turn to the first initial-boundary value problem for quasilinear parabolic equations

$$\frac{\partial u}{\partial t} - \operatorname{div}\vec{a}(x, t, u, \nabla u) + b(x, t, u, \nabla u) = 0, \quad (x, t) \in Q_T, \qquad (10.1.15)$$

$$u\Big|_{\partial_p Q_T} = \varphi, \qquad (10.1.16)$$

where $\vec{a} = (a_1, a_2, \cdots, a_n)$, $\Omega \subset \mathbb{R}^n$ is a bounded domain, $T > 0$, $Q_T = \Omega \times (0, T)$. Assume that $\vec{a}(x, t, z, \eta)$, $b(x, t, z, \eta)$ satisfy the following structure conditions:

$$\lambda(|z|)|\xi|^2 \leq \frac{\partial a_i}{\partial \eta_j}\xi_i\xi_j \leq \Lambda(|z|)|\xi|^2, \quad \forall \xi \in \mathbb{R}^n, \qquad (10.1.17)$$

$$-zb(x, t, z, 0) \leq b_1|z|^2 + b_2, \qquad (10.1.18)$$

$$\left|\frac{\partial a_i}{\partial z}\right| + |a_i| \leq \mu(|z|)(1 + |\eta|), \qquad (10.1.19)$$

$$\left|\frac{\partial a_i}{\partial x_j}\right| + |b| \le \mu(|z|)(1 + |\eta|^2), \qquad (10.1.20)$$

where $i, j = 1, 2, \cdots, n$, $\lambda(s), \Lambda(s)$ are positive continuous functions on $[0, +\infty)$, b_1, b_2 are positive constants and $\mu(s)$ is a nondecreasing function on $[0, +\infty)$.

The equation

$$\frac{\partial u}{\partial t} - \operatorname{div}(a(u)\nabla u) + b(u) + c(x, t)u = f(x, t), \quad (x, t) \in Q_T$$

satisfies all the structure conditions (10.1.17)–(10.1.20), where

$$a(u) = (u^2 + 1)^{m/2} \; (m > 0), \quad b(u) = |u|^{\gamma - 1}u \; (\gamma \ge 1)$$

and $c, f \in C^{\alpha, \alpha/2}(\overline{Q}_T)$.

Theorem 10.1.2 *Let* $0 < \alpha < 1$, $\partial\Omega \in C^{2,\alpha}$, $\vec{a} \in C^{1,\alpha}$, $b \in C^1$, $\varphi \in C^{2+\alpha, 1+\alpha/2}(\overline{Q}_T)$ *and satisfy the first order compatibility condition*

$$\frac{\partial\varphi}{\partial t} - \operatorname{div}\vec{a}(x, t, \varphi, \nabla\varphi) + b(x, t, \varphi, \nabla\varphi) = 0, \quad \text{when } x \in \partial\Omega, \; t = 0.$$

Assume that equation (10.1.15) satisfies the structure conditions (10.1.17)–(10.1.20). Then

i) There exist a constant $0 < \beta < 1$ *and a constant* $M > 0$ *depending only on* n, b_1, b_2, $\lambda(s)$, $\Lambda(s)$, $\mu(s)$, $|\varphi|_{2+\alpha, 1+\alpha/2; Q_T}$, $|\vec{a}|_{1,\alpha}$, $|b|_\alpha$, T *and* Ω, *such that for any solution* $u \in C^{2+\alpha, 1+\alpha/2}(\overline{Q}_T)$ *of problem (10.1.15), (10.1.16),*

$$|u|_{\beta, \beta/2; Q_T} \le M, \quad |\nabla u|_{\beta, \beta/2; Q_T} \le M;$$

ii) Problem (10.1.15), (10.1.16) admits a solution $u \in C^{2+\alpha, 1+\alpha/2}(\overline{Q}_T)$.

Proof. The proof of the conclusion i) is much complicated, here we do not intend to present. Assuming the conclusion i), the proof of the conclusion ii) is similar to that for elliptic equations, in which, instead of the space $U = C^{1,\alpha}(\overline{\Omega})$ we choose

$$U = \{u; u, \nabla u \in C^{\alpha, \alpha/2}(\overline{Q}_T)\}.$$

We omit the details. $\qquad\qquad\qquad\qquad\qquad\qquad\qquad\qquad\qquad\square$

10.1.4   *The procedures of the a priori estimates*

As shown in the previous section, to prove the existence of solutions for quasilinear equations, by means of the fixed point method, it suffices to establish the a priori estimates stated in Theorem 10.1.1 i) for solutions in $C^{2,\alpha}(\overline{\Omega})$ and Theorem 10.1.2 ii) for solutions in $C^{2+\alpha,1+\alpha/2}(\overline{Q}_T)$. We will do this in the subsequent sections. However, merely elliptic equations will be considered and for clarity of the expression, we merely discuss equation (10.1.9) with the homogeneous boundary value condition (10.1.10). Moreover, all arguments are presented for the case $n \geq 2$; the discussion of the case $n = 1$ needs some modification, although it is much simpler on the whole.

Let $u \in C^{2,\alpha}(\overline{\Omega})$ be a solution of problem (10.1.9), (10.1.10). To obtain the required a priori estimate, we proceed to establish the following estimates successively.

i) Maximum estimate $\|u\|_{L^\infty(\Omega)} \leq M$. It can be obtained by the maximum principle, the Moser iteration or the De Giorgi iteration;

ii) Hölder's estimate $[u]_{\alpha;\Omega} \leq M$. The main approaches are based on Harnack's inequality and Morrey's theorem;

iii) Boundary gradient estimate $\sup\limits_{\partial\Omega} |\nabla u| \leq M$. The main approaches are based on barrier function technique;

iv) Global gradient estimate $\sup\limits_{\Omega} |\nabla u| \leq M$. It can be derived by Bernstein approach;

v) Hölder's estimate for gradients $[\nabla u]_{\alpha;\Omega} \leq M$. The main methods are based on Harnack's inequality and Morrey's theorem.

10.2   Maximum Estimate

Several methods can be applied to establish the maximum norm estimate. We present one of them, which is based on the De Giorgi iteration technique.

Theorem 10.2.1   *Let* $u \in C^{2,\alpha}(\overline{\Omega})$ *be a solution of problem (10.1.9), (10.1.10). Then*

$$\sup_{\Omega} |u| \leq C\|f\|_{L^\infty(\Omega)},$$

where the constant C depends only on n, m, γ and Ω.

Proof. We need only to show

$$\sup_{\Omega} u \leq C\|f\|_{L^\infty(\Omega)},$$

since another part of the estimate can be obtained by considering $-u$.

We adopt the De Giorgi technique. Set $\varphi = (u - k)_+$, $A(k) = \{x \in \Omega; u(x) > k\}$. Multiplying both sides of equation (10.1.9) by φ, and integrating over Ω, we obtain

$$-\int_\Omega \varphi \operatorname{div}(a(u)\nabla u)dx + \int_\Omega \varphi b(u)dx = \int_\Omega \varphi f dx.$$

Owing to $(u - k)_+\big|_{\partial\Omega} = 0$, integrating by parts yields

$$\int_\Omega a(u)|\nabla \varphi|^2 dx + \int_\Omega \varphi b(u)dx = \int_\Omega \varphi f dx.$$

Therefore

$$\int_\Omega |\nabla \varphi|^2 dx \leq \int_\Omega \varphi f dx \leq \|\varphi\|_{L^p(A(k))}\|f\|_{L^q(A(k))},$$

where

$$2 < p < \begin{cases} +\infty, & n = 1, 2, \\ \dfrac{2n}{n-2}, & n \geq 3, \end{cases} \qquad \frac{1}{p} + \frac{1}{q} = 1.$$

The embedding theorem then implies that

$$\|\varphi\|_{L^p(A(k))}^2 \leq C\int_\Omega |\nabla \varphi|^2 dx \leq C\|f\|_{L^q(A(k))}\|\varphi\|_{L^p(A(k))},$$

that is

$$\|\varphi\|_{L^p(A(k))} \leq C\|f\|_{L^q(A(k))} \leq C\|f\|_{L^\infty(\Omega)}|A(k)|^{1/q}.$$

So, for any $h > k$,

$$(h - k)|A(h)|^{1/p} \leq \|\varphi\|_{L^p(A(k))} \leq C\|f\|_{L^\infty(\Omega)}|A(k)|^{1/q}$$

or

$$|A(h)| \leq \left(\frac{C\|f\|_{L^\infty(\Omega)}}{h - k}\right)^p |A(k)|^{p/q}.$$

Similar to the case of linear equations, by using the iteration lemma (Lemma 4.1.1 of Chapter 4), we achieve

$$\sup_{\Omega} u \leq C\|f\|_{L^\infty(\Omega)}.$$

$\square$

10.3 Interior Hölder's Estimate

In this section, we apply Harnack's inequality to estimate the interior Hölder norm of solutions. Since the proof is much complicated, we will divide it into several steps.

i) Estimate $\sup\limits_{B_{\theta R}} u$;

ii) Estimate $\inf\limits_{B_{\theta R}} u$ (weak Harnack's inequality);

iii) Prove Harnack's inequality;

iv) Estimate $[u]_\alpha$.

The following theorems present the details of the above steps.

Theorem 10.3.1 *Let $u \in C^{2,\alpha}(\overline{B}_R)$ be a nonnegative solution of equation (10.1.9) in B_R, $\tilde{u} = u + F_0$, $F_0 = R^2\|f\|_{L^\infty(B_R)}$. Then for any $p > 0$, $0 < \theta < 1$, we have*

$$\sup_{B_{\theta R}} \tilde{u} \leq C \left(\frac{1}{|B_R|} \int_{B_R} \tilde{u}^p dx \right)^{1/p},$$

where the constant C depends only on n, m, γ, $(1-\theta)^{-1}$ and $\|u\|_{L^\infty(B_R)}$.

Proof. Without loss of generality, we assume that $R = 1$. First consider the case $p \geq 2$. Let $\zeta(x)$ be the cut-off function on B_1. Multiplying both sides of equation (10.1.9) by $\zeta^2\tilde{u}^{p-1}$, integrating the resulting relation over Ω and then integrating by parts, we have

$$\int_{B_1} a(u)\nabla u \cdot \nabla(\zeta^2\tilde{u}^{p-1})dx + \int_{B_1} b(u)\zeta^2\tilde{u}^{p-1}dx = \int_{B_1} f\zeta^2\tilde{u}^{p-1}dx.$$

Noticing the structure of $a(u), b(u)$ and using the boundedness of u and Cauchy's inequality with ε, we obtain

$$(p-1)\int_{B_1} \zeta^2\tilde{u}^{p-2}|\nabla u|^2 dx$$

$$\leq -2\int_{B_1} \zeta a(u)\tilde{u}^{p-1}\nabla u \cdot \nabla\zeta dx - \int_{B_1} b(u)\zeta^2\tilde{u}^{p-1}dx + \int_{B_1} f\zeta^2\tilde{u}^{p-1}dx$$

$$\leq \varepsilon \int_{B_1} \zeta^2 \tilde{u}^{p-2} |\nabla u|^2 dx + \frac{C}{\varepsilon} \int_{B_1} |\nabla \zeta|^2 \tilde{u}^p dx + \frac{1}{2} \int_{B_1} \zeta^2 \tilde{u}^p dx$$
$$+ \frac{1}{2} \int_{B_1} |f|^2 \zeta^2 \tilde{u}^{p-2} dx,$$

where $\varepsilon > 0$ is an arbitrary constant. Thus

$$\int_{B_1} \zeta^2 \tilde{u}^{p-2} |\nabla u|^2 dx \leq C \int_{B_1} |\nabla \zeta|^2 \tilde{u}^p dx + C \int_{B_1} \zeta^2 \tilde{u}^p dx$$
$$+ C \int_{B_1} |f|^2 \zeta^2 \tilde{u}^{p-2} dx. \qquad (10.3.1)$$

Noticing that $\tilde{u} \geq F_0 = \|f\|_{L^\infty(B_1)}$ implies

$$\int_{B_1} |f|^2 \zeta^2 \tilde{u}^{p-2} dx \leq \int_{B_1} \zeta^2 \tilde{u}^p dx,$$

from (10.3.1), we see that

$$\int_{B_1} \zeta^2 \tilde{u}^{p-2} |\nabla u|^2 dx \leq C(1 + \sup_{B_1} |\nabla \zeta|^2) \int_{B_1} \tilde{u}^p dx.$$

Therefore

$$\int_{B_1} |\nabla(\zeta \tilde{u}^{p/2})|^2 dx \leq C(1 + \sup_{B_1} |\nabla \zeta|^2) \int_{B_1} \tilde{u}^p dx.$$

The embedding theorem then implies

$$\left(\int_{B_1} \zeta^{2q} \tilde{u}^{pq} dx \right)^{1/q} \leq C(1 + \sup_{B_1} |\nabla \zeta|^2) \int_{B_1} \tilde{u}^p dx,$$

where

$$1 < q < \begin{cases} +\infty, & n = 1, 2, \\ \dfrac{n}{n-2}, & n \geq 3. \end{cases}$$

Thus, applying the standard Moser iteration technique, we get the conclusion for the case $p \geq 2$. As for the case $0 < p < 2$, similar to the proof of the corresponding conclusion about Harnack's inequality for solutions of Laplace's equation (Theorem 5.1.3 of Chapter 5), we may use the result for $p = 2$ to get the desired conclusion. $\qquad \square$

Theorem 10.3.2 *Let $u \in C^{2,\alpha}(\overline{B}_R)$ be a nonnegative solution of equation (10.1.9) in B_R, $\tilde{u} = u + F_0$, $F_0 = R^2\|f\|_{L^\infty(B_R)}$. Then there exists $p_0 > 0$, such that for any $0 < \theta < 1$, we have*

$$\inf_{B_{\theta R}} \tilde{u} \geq C \left(\frac{1}{|B_R|} \int_{B_R} \tilde{u}^{p_0}\, dx \right)^{1/p_0},$$

where C depends only on n, m, γ, $(1-\theta)^{-1}$ and $\|u\|_{L^\infty(B_R)}$.

Proof. Without loss of generality, we assume that $F_0 > 0$, otherwise, we replace F_0 by $F_0 + \varepsilon$. Let $R = 1$, and ζ be a cut-off function on $B_{(\theta+1)/2}$. Multiplying both sides of equation (10.1.9) by $\zeta^2 \tilde{u}^{-(p+1)}$, similar to the proof of Theorem 10.3.1, we may obtain

$$\sup_{B_\theta} \tilde{u}^{-p} \leq C \int_{B_{(\theta+1)/2}} \tilde{u}^{-p} dx.$$

Thus

$$\inf_{B_\theta} \tilde{u} \geq \frac{1}{C^{1/p}} \left(\int_{B_{(\theta+1)/2}} \tilde{u}^{-p} dx \right)^{-1/p}$$

$$= \frac{1}{C^{1/p}} \left(\int_{B_{(\theta+1)/2}} \tilde{u}^{-p} dx \int_{B_{(\theta+1)/2}} \tilde{u}^{p} dx \right)^{-1/p} \left(\int_{B_{(\theta+1)/2}} \tilde{u}^{p} dx \right)^{1/p}.$$

So, to show the conclusion of the theorem, we need only to prove that for some $p_0 > 0$,

$$\int_{B_{(\theta+1)/2}} e^{p_0|w|} dx \leq C, \quad \text{where} \quad w = \ln \tilde{u} - \frac{1}{|B_{(\theta+3)/4}|} \int_{B_{(\theta+3)/4}} \ln \tilde{u}\, dx.$$

The remainder of the proof is almost similar to the linear case with some modifications. $\square$

Combine Theorem 10.3.1 and Theorem 10.3.2 to get Harnack's inequality.

Theorem 10.3.3 *Let $u \in C^{2,\alpha}(\overline{B}_R)$ be a nonnegative solution of equation (10.1.9) in B_R, $\tilde{u} = u + F_0$, $F_0 = R^2\|f\|_{L^\infty(B_R)}$. Then for any $0 < \theta < 1$,*

$$\sup_{B_{\theta R}} \tilde{u} \leq C \inf_{B_{\theta R}} \tilde{u},$$

where the constant C depends only on n, m, γ, $(1-\theta)^{-1}$ and $\|u\|_{L^\infty(B_R)}$.

Similar to the linear case, by virtue of Harnack's inequality we immediately deduce Hölder's estimate.

Theorem 10.3.4 *Let $u \in C^{2,\alpha}(\overline{\Omega})$ be a solution of equation (10.1.9). Then for any $\Omega' \subset\subset \Omega$, there exists $0 < \beta < 1$ such that*

$$[u]_{\beta;\Omega'} \leq C,$$

where the constant C depends only on n, m, γ, $\|u\|_{L^\infty(\Omega)}$, Ω' and Ω.

10.4 Boundary Hölder's Estimate and Boundary Gradient Estimate for Solutions of Poisson's Equation

The derivation of the boundary estimates for quasilinear equations is much delicate and complicated. In this section, we center our efforts on Poisson's equation to present the key ideas in getting such estimates. A discussion for general equations will be given in the next section.

Theorem 10.4.1 *Let $\Omega \subset \mathbb{R}^n$ be a bounded domain with uniform exterior ball property. Assume that $u \in C^{2,\alpha}(\overline{\Omega})$ satisfies*

$$-\Delta u = f \ in \ \Omega, \quad u\Big|_{\partial\Omega} = 0.$$

Then

$$\sup_{\partial\Omega} |\nabla u| \leq C|f|_{0;\Omega},$$

where the constant C depends only on n, $\operatorname{diam}\Omega$ and the uniform radius of the exterior ball of Ω.

Proof. For fixed $x^0 \in \partial\Omega$, we try to construct a continuously differentiable function $w^\pm(x)$, such that $w^\pm(x^0) = 0$ and

$$w^-(x) \leq u(x) \leq w^+(x), \quad x \in D, \tag{10.4.1}$$

where D is the set of some neighborhood of x^0 intersecting with Ω. Assume that, for the moment, such a function $w^\pm(x)$ exists. Then

$$\frac{w^-(x) - w^-(x^0)}{|x - x^0|} \leq \frac{u(x) - u(x^0)}{|x - x^0|} \leq \frac{w^+(x) - w^+(x^0)}{|x - x^0|}, \quad \forall x \in D.$$

Letting x tend to x^0 along the normal direction of x^0 gives

$$\frac{\partial w^-}{\partial \vec{\nu}}\bigg|_{x^0} \leq \frac{\partial u}{\partial \vec{\nu}}\bigg|_{x^0} \leq \frac{\partial w^+}{\partial \vec{\nu}}\bigg|_{x^0}, \tag{10.4.2}$$

where $\vec{\nu}$ is the unit normal vector inward to $\partial\Omega$. If $w^{\pm}(x)$ satisfies

$$-C|f|_{0;\Omega} \leq \left.\frac{\partial w^-}{\partial \vec{\nu}}\right|_{x^0}, \quad \left.\frac{\partial w^+}{\partial \vec{\nu}}\right|_{x^0} \leq C|f|_{0;\Omega} \qquad (10.4.3)$$

with the constant C independent of x^0, then from (10.4.2) we have

$$\left|\frac{\partial u}{\partial \vec{\nu}}\right|_{x^0} \leq C|f|_{0;\Omega}.$$

Since $\left.u\right|_{\partial\Omega} = 0$ implies that all the tangential derivatives of u are zero, we obtain

$$\left|\nabla u\right|_{x^0} \leq C|f|_{0;\Omega}.$$

By virtue of the maximum principle, to get (10.4.1), it suffices to require

$$-\Delta w^- \leq -\Delta u \leq -\Delta w^+, \quad \text{in } D, \qquad (10.4.4)$$

$$\left.w^-\right|_{\partial D} \leq \left.u\right|_{\partial D} \leq \left.w^+\right|_{\partial D}. \qquad (10.4.5)$$

We will seek such a function from the family of functions of the from

$$w^+(x) = A\left(\frac{1}{r^{n-1}} - \frac{1}{|x-y|^{n-1}}\right),$$

where y is the center of the exterior ball at the point x^0 with uniform radius r, and $A > 0$ is to be specified below.

For any $A > 0$, we have

$$\left.u\right|_{\partial\Omega} \leq \left.w^+\right|_{\partial\Omega}.$$

Since

$$-\Delta w^+ = \frac{A(n-1)}{|x-y|^{n+1}} \geq \frac{A(n-1)}{(r+\text{diam}\Omega)^{n+1}},$$

if we choose A such that

$$\frac{A(n-1)}{(r+\text{diam}\Omega)^{n+1}} \geq |f|_{0;\Omega},$$

i.e.

$$A \geq \frac{(r+\text{diam}\Omega)^{n+1}}{n-1}|f|_{0;\Omega},$$

then we obtain

$$-\Delta w^+ \geq |f|_{0;\Omega} \geq f = -\Delta u, \quad \text{in } \Omega.$$

So, we choose

$$A = \frac{(r + \text{diam}\Omega)^{n+1}}{n-1}|f|_{0;\Omega}. \tag{10.4.6}$$

With such a choice of A, we have, moreover,

$$\frac{\partial w^+}{\partial \vec{\nu}}\bigg|_{x^0} = \frac{A(n-1)}{r^n} = \frac{(r + \text{diam}\Omega)^{n+1}}{r^n}|f|_{0;\Omega},$$

which shows that $w^+(x)$ satisfies (10.4.3) too. Similarly, if we choose

$$w^-(x) = -A\left(\frac{1}{r^{n-1}} - \frac{1}{|x-y|^{n-1}}\right),$$

where A is the constant in (10.4.6), then $w^-(x)$ satisfies (10.4.4), (10.4.5) and also (10.4.3) with Ω in place of D. □

Remark 10.4.1 *The above method can be used to derive the boundary Hölder's estimate. In fact, it follows from (10.4.1) that*

$$|u(x) - u(x^0)| \leq |w^\pm(x) - w^\pm(x^0)| \leq [w^\pm]_\alpha |x - x^0|^\alpha, \quad x \in \Omega.$$

10.5 Boundary Hölder's Estimate and Boundary Gradient Estimate

In the previous section, we have used the barrier function technique to establish the boundary estimate for Poisson's equation. Such an important technique can also be applied to the general quasilinear equations with divergence structure.

Now, we consider problem (10.1.9), (10.1.2). First establish the boundary Hölder's estimate.

Theorem 10.5.1 *Let $\Omega \subset \mathbb{R}^n$ be a bounded domain with uniform exterior ball property. Assume that $u \in C^{2,\alpha}(\overline{\Omega})$ is a solution of problem (10.1.9), (10.1.2). Then for any $x^0 \in \partial\Omega$ and $x^1 \in \Omega$, we have*

$$|u(x^1) - u(x^0)| \leq C|x^1 - x^0|^{\alpha/(\alpha+1)}([u]_{\alpha;\partial\Omega} + 1),$$

where the constant C depends only on n, m, γ, $|f|_{0;\Omega}$, $|u|_{0;\Omega}$, α, diamΩ and the uniform radius ρ of the exterior ball of Ω.

Proof. We merely prove

$$u(x^1) - u(x^0) \le C|x^1 - x^0|^{\alpha/(\alpha+1)}([u]_{\alpha;\partial\Omega} + 1); \qquad (10.5.1)$$

another part of the theorem can be proved similarly. Without loss of generality, we make the following two assumptions:

i) Assume

$$|x^1 - x^0| = \text{dist}(x^1, \partial\Omega).$$

Otherwise, we may choose a point $y \in \partial\Omega$ such that $|x^1 - y| = \text{dist}(x^1, \partial\Omega)$. In this case,

$$|x^1 - y| \le |x^1 - x^0|,$$
$$|y - x^0| \le |y - x^1| + |x^1 - x^0| \le 2|x^1 - x^0|.$$

If we have proved

$$u(x^1) - u(y) \le C|x^1 - y|^{\alpha/(\alpha+1)}([u]_{\alpha;\partial\Omega} + 1),$$

then

$$
\begin{aligned}
u(x^1) - u(x^0) =& u(x^1) - u(y) + u(y) - u(x^0) \\
\le& C|x^1 - y|^{\alpha/(\alpha+1)}([u]_{\alpha;\partial\Omega} + 1) + C|y - x^0|^\alpha [u]_{\alpha;\partial\Omega} \\
\le& C|x^1 - x^0|^{\alpha/(\alpha+1)}([u]_{\alpha;\partial\Omega} + 1).
\end{aligned}
$$

ii) Assume $|x^1 - x^0| \le \rho$. This is because the constant C in (10.5.1) is allowed to depend on ρ, if $|x^1 - x^0| > \rho$, then the desired conclusion can be derived directly from $|u|_{0;\Omega} \le C$.

The basic idea of the proof of (10.5.1) is similar to the one for Poisson's equation in §10.4, namely, to construct a barrier function $w(x)$ such that $w(x^0) = u(x^0)$ and

$$u(x) \le w(x), \quad x \in D, \qquad (10.5.2)$$

where D is the set of some neighborhood of x^0 intersecting with Ω. According to the maximum principle, to get (10.5.2), it suffices to construct a second order elliptic operator L such that

$$Lu \le Lw, \quad \text{in } D, \qquad (10.5.3)$$

$$u\Big|_{\partial D} \le w\Big|_{\partial D}. \qquad (10.5.4)$$

We expect to seek a suitable barrier function from the family of functions of the form

$$w(x) = \psi(|x - y| - r),\qquad (10.5.5)$$

where $r \in (0, \rho]$ is a constant to be specified, y is the center of the exterior ball at the point x^0 with radius r, ψ is a function to be determined. We choose

$$D = \{x \in \mathbb{R}^n; |x - y| < r + \delta\} \cap \Omega$$

with $\delta > 0$ to be determined.

The proof of (10.5.1) will be completed in the following steps.

Step 1 Construction of the operator L.

Rewrite equation (10.1.9) as

$$-a(u)\Delta u - a'(u)|\nabla u|^2 + b(x, u) = 0,\qquad (10.5.6)$$

where $b(x, u) = b(u) - f(x)$. Owing to $|u|_{0;\Omega} \leq C$, there exist constants $\mu_0, \mu_1 \geq 1$ depending only on m, γ, $|f|_{0;\Omega}$ and $|u|_{0;\Omega}$, such that

$$1 \leq \bar{a}(x) \leq \mu_0,\qquad (10.5.7)$$

$$-\bar{a}(x)\Delta u \leq \mu_1(|\nabla u|^2 + 1),\qquad (10.5.8)$$

where

$$\bar{a}(x) = a(u(x)), \quad x \in \Omega.$$

Define

$$Lv = -\bar{a}(x)\Delta v - \mu_1(|\nabla v|^2 + 1).$$

Then from (10.5.6) and (10.5.8) we have

$$Lu \leq 0, \quad \text{in } D.\qquad (10.5.9)$$

Step 2 Construction of the function ψ.

A direct calculation shows that

$$\frac{\partial w}{\partial x_i} = \frac{x_i - y_i}{|x - y|}\psi',$$

$$\frac{\partial^2 w}{\partial x_i^2} = \frac{(x_i - y_i)^2}{|x - y|^2}\psi'' + \left(\frac{1}{|x - y|} - \frac{(x_i - y_i)^2}{|x - y|^3}\right)\psi',$$

$$\Delta w = \psi'' + \frac{n - 1}{|x - y|}\psi',$$

$$|\nabla w|^2 = (\psi')^2.$$

Thus

$$Lw = -\bar{a}(x)\psi'' - \bar{a}(x)\frac{n-1}{|x-y|}\psi' - \mu_1[(\psi')^2 + 1].$$

If we require

$$\psi' > 0, \quad \psi'' < 0, \tag{10.5.10}$$

then by (10.5.7)

$$Lw \geq -\psi'' - \frac{(n-1)\mu_0}{r}\psi' - \mu_1[(\psi')^2 + 1]$$
$$= -(\psi')^2\left(\frac{\psi''}{(\psi')^2} + \frac{(n-1)\mu_0}{r\psi'} + \mu_1 + \frac{\mu_1}{(\psi')^2}\right)$$

with r to be determined. If, in addition, we require

$$\psi' \geq \frac{(n-1)\mu_0}{\mu_1 r} + 1, \tag{10.5.11}$$

then

$$Lw > -(\psi')^2\left(\frac{\psi''}{(\psi')^2} + 2\mu_1\right).$$

Therefore, in order that w satisfies (10.5.3), since (10.5.9) holds, we only need $Lw \geq 0$, which is valid if we require ψ to satisfy

$$\psi'' + 2\mu_1(\psi')^2 \leq 0. \tag{10.5.12}$$

Summing up, in order to get (10.5.3), it suffices to construct a function $\psi(d)$ on $[0, \delta]$ satisfying (10.5.10), (10.5.11) and (10.5.12).

It is not difficult to check that, for any $k > 0$, the function

$$\psi(d) = \frac{1}{2\mu_1}\ln(1 + kd), \quad 0 \leq d \leq \delta \tag{10.5.13}$$

satisfies (10.5.10) and (10.5.12).

In what follows, we will show that we may choose suitable r, δ and k, such that (10.5.11) holds. Since for $d < \delta$,

$$\psi'(d) = \frac{k}{2\mu_1(1 + kd)} \geq \frac{1}{2\mu_1\delta}\cdot\frac{k\delta}{(1 + k\delta)},$$

we have $\psi' \geq \dfrac{1}{4\mu_1\delta}$, if

$$k \geq \frac{1}{\delta}. \qquad (10.5.14)$$

If we choose

$$0 < \delta \leq \frac{r}{4[(n-1)\mu_0 + \mu_1 r]}, \qquad (10.5.15)$$

then ψ satisfies (10.5.11). Therefore, if we choose $r \in (0, \rho]$, δ satisfying (10.5.15) and k satisfying (10.5.14) in turn, then the function ψ given by (10.5.13) satisfies (10.5.10), (10.5.11) and (10.5.12), and hence the function w given by (10.5.5) satisfies (10.5.3).

Step 3 Construction of the required barrier function.

The function

$$w(x) = \psi(d(x)) = \frac{1}{2\mu_1} \ln(1 + kd(x)), \quad d(x) = |x - y| - r$$

constructed above may not satisfy (10.5.4). In order to get a function satisfying both (10.5.3) and (10.5.4), we consider

$$\tilde{w}(x) = w(x) + u(x^0) + (3r)^\alpha [u]_{\alpha;\partial\Omega}.$$

Since $L\tilde{w} = Lw \geq 0$, we have

$$Lu \leq L\tilde{w}, \quad \text{in } D. \qquad (10.5.16)$$

Set

$$v(x) = \tilde{w}(x) - u(x) = w(x) - (u(x) - u(x^0)) + (3r)^\alpha [u]_{\alpha;\partial\Omega}.$$

If we require $\delta \leq r$, then

$$|x - x^0| \leq |x - y| + |y - x^0| \leq 3r, \quad \forall x \in D.$$

Hence

$$v(x) \geq -[u(x) - u(x^0)] + (3r)^\alpha [u]_{\alpha;\partial\Omega} \geq 0, \quad \forall x \in \partial\Omega \cap \overline{D}.$$

In order that $v(x) \geq 0$ in $\partial D \cap \Omega$, it suffices to require $w\big|_{\partial D \cap \Omega} \geq 2|u|_{0;\Omega}$, i.e.

$$\psi(\delta) = \frac{1}{2\mu_1} \ln(1 + k\delta) \geq 2|u|_{0;\Omega}$$

or

$$k\delta \geq e^{4\mu_1|u|_{0;\Omega}} - 1.$$

Therefore, if δ satisfies (10.5.15) (and hence $\delta \leq r$) and

$$k \geq \frac{1}{\delta}\left(e^{4\mu_1|u|_{0;\Omega}} - 1\right), \qquad (10.5.17)$$

then $v\big|_{\partial D} \geq 0$ namely

$$u\big|_{\partial D} \leq \tilde{w}\big|_{\partial D}. \qquad (10.5.18)$$

By virtue of (10.5.16) and (10.5.18), we have

$$\begin{cases} \tilde{L}v = -\bar{a}(x)(\Delta\tilde{w} - \Delta u) - \mu_1(|\nabla\tilde{w}|^2 - |\nabla u|^2) \\ \quad = -\bar{a}(x)\Delta v - \mu_1\nabla(\tilde{w} + u)\cdot\nabla v \geq 0, \quad \text{in } D, \\ v\big|_{\partial D} \geq 0. \end{cases}$$

The maximum principle for linear equations then yields

$$v(x) \geq 0, \quad x \in D,$$

that is

$$u(x) - u(x^0) \leq \frac{1}{2\mu_1}\ln(1+k(|x-y|-r)) + (3r)^\alpha[u]_{\alpha;\partial\Omega}, \quad \forall x \in D. \quad (10.5.19)$$

Step 4 Establishing Hölder's estimate.

For fixed $x^1 \in \Omega$, the choice of D depends on r and δ in the above discussion. We will further see in the following argument, that δ may depend on r, while r depends on x^1. Hence D depends on x^1. However, at the end of the proof we will prove that under the condition

$$|x^1 - x^0| < C(\rho) \qquad (10.5.20)$$

with some constant $C(\rho)$ depending only on ρ, there holds $x^1 \in D$. So, it suffices to prove (10.5.1) in the case $x^1 \in D$. This is because, if $x_1 \notin D$, then $|x^1 - x^0| \geq C(\rho)$, and since the constant in the right hand side of (10.5.1) is allowed to depend on ρ and $|u|_{0;\Omega}$, (10.5.1) holds clearly.

Let $x_1 \in D$ and take $x = x^1$ in (10.5.19). Since we have assumed that $|x^1 - x^0| = \text{dist}(x^1, \partial\Omega)$ at the beginning of the proof, $|x^1 - y| - r = |x^1 - x^0|$, and hence

$$u(x^1) - u(x^0) \leq \frac{1}{2\mu_1} \ln(1 + k|x^1 - x^0|) + (3r)^\alpha [u]_{\alpha;\partial\Omega}$$
$$\leq \frac{k}{2\mu_1} |x^1 - x^0| + (3r)^\alpha [u]_{\alpha;\partial\Omega}.$$

If we choose

$$k = \frac{1}{\delta} e^{4\mu_1 |u|_{0;\Omega}}, \tag{10.5.21}$$

then

$$u(x^1) - u(x^0) \leq \frac{C}{\delta} |x^1 - x^0| + (3r)^\alpha [u]_{\alpha;\partial\Omega}.$$

If, in addition,

$$\delta = k_1 r, \tag{10.5.22}$$

then the above inequality becomes

$$u(x^1) - u(x^0) \leq \frac{C}{k_1 r} |x^1 - x^0| + (3r)^\alpha [u]_{\alpha;\partial\Omega}.$$

Furthermore, if

$$r = k_2 |x^1 - x^0|^{1/(\alpha+1)}, \tag{10.5.23}$$

then we may further obtain

$$u(x^1) - u(x^0) \leq \frac{C}{k_1 k_2} |x^1 - x^0|^{\alpha/(\alpha+1)} ([u]_{\alpha;\partial\Omega} + 1),$$

which is just the desired conclusion (10.5.1).

Now, we first choose $k_2 = \rho^{\alpha/(\alpha+1)}$. Then $r \in (0, \rho]$ can be determined by (10.5.23), since we have assumed that $|x^1 - x^0| \leq \rho$ at the beginning of the proof. Next we take $k_1 = \dfrac{1}{4[(n-1)\mu_0 + \mu_1\rho]}$. Then δ determined by (10.5.22) satisfies (10.5.15). Finally, we determine the constant k by (10.5.21), which satisfies both (10.5.14) and (10.5.17) obviously. Summing up, if we choose k_2, r, k_1, δ and k according to the above procedures, then we may derive (10.5.1).

Finally, we verify that the condition (10.5.20) implies $x^1 \in D$. Since we have assumed that $|x^1 - x^0| = \text{dist}(x^1, \partial\Omega)$, it suffices to verify $|x^1 - x^0| < \delta$. By the choice of r and δ,

$$\delta = k_1 r = k_1 k_2 |x^1 - x^0|^{1/(\alpha+1)}.$$

Thus $|x^1 - x^0| < (k_1 k_2)^{(\alpha+1)/\alpha}$ implies $|x^1 - x^0| < \delta$. So, if we choose $C(\rho) = (k_1 k_2)^{(\alpha+1)/\alpha}$ in (10.5.20), then the condition (10.5.20) implies $x^1 \in D$. The proof of (10.5.1) is complete.   □

Similarly, we may establish the boundary gradient estimate.

Theorem 10.5.2   *Let $\Omega \subset \mathbb{R}^n$ be a bounded domain with uniform exterior ball property. Assume that $u \in C^{2,\alpha}(\overline{\Omega})$ is a solution of problem (10.1.9), (10.1.10). Then*

$$\left| \frac{\partial u}{\partial \vec{\nu}} \right|_{\partial \Omega} \le C,$$

where $\vec{\nu}$ is the unit normal vector outward to $\partial \Omega$, C depends only on n, m, γ, $|f|_{0;\Omega}$, distΩ and the uniform radius of the exterior ball of Ω.

Proof.   By (10.5.19), we have

$$u(x) - u(x^0) \le \frac{1}{2\mu_1} \ln(1 + k(|x - y| - r)) + (3r)^\alpha [u]_{\alpha;\partial\Omega}, \quad \forall x \in D.$$

Since $u\big|_{\partial\Omega} = 0$ implies $[u]_{\alpha;\partial\Omega} = 0$, we see that

$$u(x) - u(x^0) \le \frac{1}{2\mu_1} \ln(1 + k(|x - y| - r)) \le C(|x - y| - r), \quad \forall x \in D.$$

Similarly we may estimate the lower bound of $u(x) - u(x_0)$ to obtain an opposite inequality, and hence

$$\left| \frac{u(x) - u(x^0)}{x - x^0} \right| \le C \frac{|x - y| - r}{|x - x^0|}, \quad \forall x \in D.$$

Letting x tend to x^0 along the normal direction at the point x^0, and noticing that for such x, $|x - y| - r = |x - x^0|$, we immediately get the conclusion of the theorem.   □

10.6   Global Gradient Estimate

In this section, we apply the Bernstein approach to get the global gradient estimate. To present the main idea of this approach clearly, we first discuss Poisson's equation.

Theorem 10.6.1 *Let $\Omega \subset \mathbb{R}^n$ be a bounded domain with uniform exterior ball property. Assume that $u \in C^{2,\alpha}(\overline{\Omega})$ satisfies*

$$-\Delta u = f \text{ in } \Omega, \quad u\Big|_{\partial\Omega} = 0.$$

Then

$$\sup_{\Omega} |\nabla u| \leq C(|f|_{0;\Omega} + |\nabla f|_{0;\Omega} + |u|_{0;\Omega}),$$

where C depends only on n and the uniform radius of the exterior ball of Ω.

Proof. The main idea is to apply the sign rule to the function

$$w(x) = |\nabla u|^2 + u^2, \quad x \in \overline{\Omega}$$

to estimate its maximum. This is the so called Bernstein approach.

Since $u \in C^{2,\alpha}(\overline{\Omega})$, there exists $x^0 \in \overline{\Omega}$, such that $w(x^0) = \max_{\overline{\Omega}} w$. If $x^0 \in \partial\Omega$, then due to the boundary gradient estimate, the desired conclusion is obviously valid. So, without loss of generality, we assume that $x^0 \in \Omega$. Then $\Delta w(x^0) \leq 0$. A direct calculation shows that

$$\begin{aligned}
- \Delta w \\
= -\Delta \left(\sum_{i=1}^{n} (D_i u)^2 + u^2 \right) \\
= -2 \sum_{i=1}^{n} D_i u D_i \Delta u - 2 \sum_{i,j=1}^{n} (D_{ij} u)^2 - 2u\Delta u - 2|\nabla u|^2 \\
= 2 \sum_{i=1}^{n} D_i u D_i f - 2 \sum_{i,j=1}^{n} (D_{ij} u)^2 + 2uf - 2|\nabla u|^2.
\end{aligned}$$

Thus

$$\begin{aligned}
|\nabla u(x^0)|^2 \\
\leq \sum_{i=1}^{n} D_i u(x^0) D_i f(x^0) + u(x^0) f(x^0) \\
\leq \frac{1}{2} |\nabla u(x^0)|^2 + \frac{1}{2} |\nabla f(x^0)|^2 + \frac{1}{2} u^2(x^0) + \frac{1}{2} f^2(x^0)
\end{aligned}$$

and hence

$$|\nabla u(x^0)|^2 \leq |\nabla f(x^0)|^2 + u^2(x^0) + f^2(x^0).$$

Therefore, for any $x \in \Omega$, we have

$$
\begin{aligned}
|\nabla u(x)|^2 \le w(x) \le & w(x^0) \\
= & |\nabla u(x^0)|^2 + u^2(x^0) \\
\le & |\nabla f(x^0)|^2 + 2u^2(x^0) + f^2(x^0) \\
\le & |\nabla f|_{0;\Omega}^2 + 2|u|_{0;\Omega}^2 + |f|_{0;\Omega}^2,
\end{aligned}
$$

from which we get the conclusion of the theorem. $\qquad\square$

Now, we turn to problem (10.1.9), (10.1.10).

Theorem 10.6.2 *Let $\Omega \subset \mathbb{R}^n$ be a bounded domain with uniform exterior ball property. Assume that $u \in C^{2,\alpha}(\overline{\Omega})$ is a solution of problem (10.1.9), (10.1.10). Then*

$$
\sup_{\Omega} |\nabla u| \le C,
$$

where C depends only on n, m, γ, $|f|_{0;\Omega}$, $|u|_{\alpha;\Omega}$, distΩ and the uniform radius of the exterior ball of Ω.

Proof. Let $\varphi \in C_0^\infty(\Omega)$. Multiplying both sides of equation (10.1.9) by $D_k\varphi$ $(k = 1, 2, \cdots, n)$, and integrating over Ω, we obtain

$$
-\int_\Omega D_i(a(u)D_i u)D_k\varphi dx + \int_\Omega (b(u) - f(x))D_k\varphi dx = 0.
$$

Integrating the first term by parts, we further have

$$
-\int_\Omega D_k(a(u)D_i u)D_i\varphi dx + \int_\Omega (b(u) - f(x))D_k\varphi dx = 0,
$$

i.e.

$$
\int_\Omega a(u)D_{ik}u D_i\varphi dx + \int_\Omega f_k^i D_i\varphi dx = 0, \quad \forall \varphi \in C_0^\infty(\Omega), \tag{10.6.1}
$$

where

$$
f_k^i = a'(u)D_k u D_i u - \delta_{ik}(b(u) - f), \quad \text{in } \Omega.
$$

Since $C_0^\infty(\Omega)$ is dense in $H_0^1(\Omega)$, (10.6.1) is valid for any $\varphi \in H_0^1(\Omega)$. By the structure conditions on $a(u)$ and $b(u)$, the maximum estimate on u and the boundedness of f, we get

$$
|f_k^i| \le C(1 + |Du|^2), \quad \text{in } \Omega.
$$

Let $v = |Du|^2$. By virtue of $u \in C^{2,\alpha}(\overline{\Omega})$, there exists $x^0 \in \overline{\Omega}$, such that

$$N = \sqrt{v(x^0)} = \max_{\overline{\Omega}} |Du|.$$

If $x^0 \in \partial\Omega$, then due to the boundary gradient estimate, the desired conclusion is obviously valid. So, without loss of generality, we may assume that $x^0 \in \Omega$ and $N > 1$. Let $R = 1/N$ and ζ be a cut-off function on $B_R(x^0)$ satisfying

$$\zeta \in C_0^\infty(B_R(x^0)), \quad \zeta(x^0) = 1, \quad 0 \le \zeta(x) \le 1, \quad |D\zeta(x)| \le \frac{2}{R} = 2N.$$

Let $0 \le \varphi \in C_0^\infty(\Omega)$. Choose $\zeta^2\varphi D_k u$ as the test function in (10.6.1), and sum with respect to k, we have

$$\int_\Omega \zeta^2 \varphi a(u) D_{ik}u D_{ik}u dx + \int_\Omega \zeta^2 \varphi f_k^i D_{ik} u dx$$
$$+ \int_\Omega \left(\frac{1}{2}a(u)D_i v + f_k^i D_k u\right)(2\zeta\varphi D_i\zeta + \zeta^2 D_i\varphi)dx = 0.$$

Applying the structure conditions of $a(u)$ and setting $w = \zeta^2 v$, we obtain

$$\int_\Omega \zeta^2 \varphi |D^2 u|^2 dx + \int_\Omega \left(\zeta^2 f_k^i D_{ik}u + 2(a(u)D_k u D_{ik}u + f_k^i D_k u)\zeta D_i\zeta\right)\varphi dx$$
$$+ \int_\Omega \left(\frac{1}{2}a(u)D_i w - \zeta a(u)v D_i\zeta + \zeta^2 f_k^i D_k u\right) D_i\varphi dx \le 0.$$

Thus, by Cauchy's inequality with ε, we have

$$\int_\Omega \left(a(u)D_i w - 2\zeta a(u)v D_i\zeta + 2\zeta^2 f_k^i D_k u\right) D_i\varphi dx$$
$$\le C \int_\Omega \left(\zeta^2 |f_k^i|^2 + |D\zeta|^2(1 + |Du|^2)\right)\varphi dx, \quad \forall 0 \le \varphi \in C_0^\infty(\Omega),$$

which shows that w is a weak subsolution of some linear equation in $\Omega_R = B_R(x^0) \cap \Omega$. According to the maximum principle for weak subsolutions (Theorem 4.1.2 of Chapter 4. Although in that theorem the conclusion is established for Poisson's equation, the similar proof is available to general elliptic equations with divergence structure), for $p > n$, we have

$$\sup_{\Omega_R} w \le \sup_{\partial\Omega_R} w + C(\|\tilde{f}\|_{L^{p*}(\Omega_R)} + \|\vec{g}\|_{L^p(\Omega_R)})|\Omega_R|^{1/n - 1/p},$$

where $p_* = \dfrac{np}{n+p}$, $\vec{g} = (g_1, g_2, \cdots, g_n)$,

$$g_i = 2\zeta a(u)vD_i\zeta - 2\zeta^2 f_k^i D_k u, \quad \text{in } \Omega, \quad (i = 1, 2, \cdots, n)$$

$$\tilde{f} = \zeta^2 |f_k^i|^2 + |D\zeta|^2(1 + |Du|^2), \quad \text{in } \Omega.$$

Since

$$
\begin{aligned}
\|\tilde{f}\|_{L^{p_*}(\Omega_R)} &\leq \||\zeta^2|f_k^i|^2\|_{L^{p_*}(\Omega_R)} + \||D\zeta|^2(1 + |Du|^2)\|_{L^{p_*}(\Omega_R)} \\
&\leq C\|(1 + |Du|^2)^2\|_{L^{p_*}(\Omega_R)} + \|4N^2(1 + |Du|^2)\|_{L^{p_*}(\Omega_R)} \\
&\leq CN^2\|1 + |Du|^2\|_{L^{p_*}(\Omega_R)} \\
&\leq CN^2|\Omega_R|^{1/p_*} + CN^2\||Du|^2\|_{L^{p_*}(\Omega_R)},
\end{aligned}
$$

$$
\begin{aligned}
\|\vec{g}\|_{L^p(\Omega_R)} &\leq 2\|\zeta a(u)vD_i\zeta\|_{L^p(\Omega_R)} + 2\|\zeta^2 f_k^i D_k u\|_{L^p(\Omega_R)} \\
&\leq CN\||Du|^2\|_{L^p(\Omega_R)} + CN\|1 + |Du|^2\|_{L^p(\Omega_R)} \\
&\leq CN\|1 + |Du|^2\|_{L^p(\Omega_R)} \\
&\leq CN|\Omega_R|^{1/p} + CN\||Du|^2\|_{L^p(\Omega_R)},
\end{aligned}
$$

it follows that

$$
\begin{aligned}
\sup_{\Omega_R} w \leq \sup_{\partial\Omega \cap B_R(x^0)} w &+ C\Big(1 + N^{1+n/p}\||Du|^2\|_{L^{p_*}(\Omega_R)} \\
&+ N^{n/p}\||Du|^2\|_{L^p(\Omega_R)}\Big).
\end{aligned}
\tag{10.6.2}
$$

Now, we estimate $\|Du\|_{L^2(\Omega_R)}$. Let ξ be a cut-off function on $B_{2R}(x^0)$ relative to $B_R(x^0)$, that is $\xi \in C_0^\infty(B_{2R}(x^0))$, $\xi = 1$ in $B_R(x^0)$, and

$$0 \leq \xi(x) \leq 1, \quad |D\xi(x)| \leq \frac{C}{R} = CN, \quad x \in B_{2R}(x^0).$$

Multiplying both sides of equation (10.1.9) by $\xi^2(x)(u(x) - u(x^0))$, integrating over Ω and then integrating by parts, we obtain

$$\int_{\Omega_{2R}} a(u)\xi^2|Du|^2 dx + 2\int_{\Omega_{2R}} a(u)\xi(u(x) - u(x^0))Du \cdot D\xi dx$$

$$-\int_{\partial\Omega_{2R}} a(u)\xi^2(u(x) - u(x^0))Du \cdot \vec{\nu} ds$$

$$+\int_{\Omega_{2R}} (b(u) - f)\xi^2(u(x) - u(x^0)) dx = 0,$$

where $\vec{\nu}$ is the unit normal vector outward to $\partial\Omega_{2R}$. Applying Cauchy's inequality with ε and using

$$|u(x) - u(x^0)| \leq [u]_{\alpha;\Omega}(2R)^\alpha, \quad \forall x \in \Omega_{2R},$$

we have

$$\|Du\|^2_{L^2(\Omega_R)} \leq \int_{\Omega_{2R}} \xi^2 |Du|^2 dx$$
$$\leq C(N^2 R^{n+2\alpha} + NR^{n-1+\alpha} + R^{n+\alpha}) \leq CN^{2-n-\alpha}.$$

Thus

$$\||Du|^2\|_{L^{p_*}(\Omega_R)} \leq N^{2(p_*-1)/p_*}\|Du\|^{2/p_*}_{L^2(\Omega_R)} \leq CN^{2-(n+\alpha)/p_*},$$

$$\||Du|^2\|_{L^p(\Omega_R)} \leq N^{2(p-1)/p}\|Du\|^{2/p}_{L^2(\Omega_R)} \leq CN^{2-(n+\alpha)/p}.$$

Substituting these into (10.6.2) gives

$$N^2 = w(x^0) = \sup_{\Omega_R} w \leq \sup_{\partial\Omega \cap B_R(x^0)} w + C(1 + N^{2-\alpha/p_*} + N^{2-\alpha/p}).$$

Using Young's inequality with ε, we obtain

$$N^2 \leq \sup_{\partial\Omega \cap B_R(x^0)} w + C,$$

which implies

$$\sup_{\Omega} |Du| = N \leq C$$

with another constant C. $\qquad\qquad\qquad\qquad\qquad\qquad\square$

10.7 Hölder's Estimate for a Linear Equation

To establish Hölder's estimate for gradients of solutions of equation (10.1.9), we first investigate a special linear equation with divergence structure.

10.7.1 *An iteration lemma*

We first introduce a useful iteration lemma.

Lemma 10.7.1 *Let $\Phi(\rho)$ be a nonnegative and nondecreasing function defined on $[0, R_0]$ satisfying*

$$\Phi(\rho) \leq A\left[\left(\frac{\rho}{R}\right)^\alpha + \varepsilon\right]\Phi(R) + BR^\beta, \quad \forall 0 < \rho \leq R \leq R_0,$$

where $A, B > 0$, $0 < \beta < \alpha$. Then there exist constants $\varepsilon_0 > 0$ and $C > 0$ depending only on A, α, β, such that for any $0 \leq \varepsilon \leq \varepsilon_0$, there holds

$$\Phi(\rho) \leq C\left[\left(\frac{\rho}{R}\right)^{\beta}\Phi(R) + B\rho^{\beta}\right], \quad \forall 0 < \rho \leq R \leq R_0.$$

Proof. By assumption, for any $\tau \in (0,1)$,

$$\Phi(\tau R) \leq A\tau^{\alpha}(1 + \varepsilon\tau^{-\alpha})\Phi(R) + BR^{\beta}, \quad \forall 0 < R \leq R_0.$$

Without loss of generality we may assume $A \geq 1$.

First, choose a real number $\nu \in (\beta, \alpha)$ and then choose τ such that $2A\tau^{\alpha} = \tau^{\nu}$ or

$$\tau = \exp\left\{-\frac{\ln(2A)}{\alpha - \nu}\right\} \in (0,1).$$

Finally, choose $\varepsilon_0 > 0$ such that $\varepsilon_0\tau^{-\alpha} \leq 1$ or

$$0 < \varepsilon_0 \leq \exp\left\{-\alpha\frac{\ln(2A)}{\alpha - \nu}\right\}.$$

For such selected ν, τ and ε_0, when $0 \leq \varepsilon \leq \varepsilon_0$, we have

$$\phi(\tau R) \leq 2A\tau^{\alpha}\Phi(R) + BR^{\beta}$$
$$\leq \tau^{\nu}\Phi(R) + BR^{\beta}, \quad \forall 0 < R \leq R_0.$$

The remainder of the proof is completely similar to that of Lemma 6.2.1 of Chapter 6. $\qquad\square$

Remark 10.7.1 *When $\varepsilon = 0$, Lemma 10.7.1 reduces to Lemma 6.2.1 of Chapter 6.*

10.7.2 *Morrey's theorem*

In Chapter 6, we have applied the Campanato spaces to describe the integral characteristic of Hölder continuous functions (Theorem 6.1.1 of Chapter 6). Now, we introduce Morrey's theorem, which can also be used to describe the integral characteristic of Hölder continuous functions.

Morrey's Theorem *Let $p > 1$, $0 < \alpha < 1$, and $\Omega \subset \mathbb{R}^n$ be a bounded domain with appropriately smooth boundary (such as $\partial\Omega \in C^{1,\alpha}$).*

i) If $u \in W^{1,p}_{loc}(\mathbb{R}^n)$ and for any $x \in \mathbb{R}^n$ and $\rho > 0$, there holds

$$\int_{B_{\rho}(x)} |\nabla u(y)|^p dy \leq C\rho^{n-p+p\alpha},$$

then $u \in C^\alpha(\mathbb{R}^n)$.

ii) If $u \in W^{1,p}(\Omega)$ *and for any* $x \in \Omega$ *and* $0 < \rho < \mathrm{diam}\Omega$, *there holds*

$$\int_{\Omega_\rho(x)} |\nabla u(y)|^p dy \leq C\rho^{n-p+p\alpha},$$

where $\Omega_\rho(x) = B_\rho(x) \cap \Omega$, *then* $u \in C^\alpha(\overline{\Omega})$.

Proof. The proof of the above two conclusions are quite similar, and we only prove the second one. Let $x \in \Omega$, $0 < \rho < \mathrm{diam}\Omega$. From Poincaré's inequality, we have

$$\int_{\Omega_\rho(x)} |u(y) - u_{x,\rho}|^p dy \leq C\rho^p \int_{\Omega_\rho(x)} |\nabla u(y)|^p dy \leq C\rho^{n+p\alpha},$$

which implies $u \in \mathcal{L}^{p,n+p\alpha}(\Omega)$, and so $u \in C^\alpha(\overline{\Omega})$ according to Theorem 6.1.1 of Chapter 6. $\qquad\qquad\square$

10.7.3 *Hölder's estimate*

Consider the linear equation

$$-\mathrm{div}(\bar{a}(x)\nabla u) = \mathrm{div}\vec{f}, \quad x \in \mathbb{R}^n_+, \qquad (10.7.1)$$

where $\bar{a} \in C^\alpha(\overline{\mathbb{R}}^n_+)$, $\vec{f} \in L^\infty(\mathbb{R}^n_+, \mathbb{R}^n)$ and $\bar{a}(x) \geq \lambda > 0$.

Theorem 10.7.1 *Let* $u \in C^{1,\alpha}(\overline{\mathbb{R}}^n_+)$ *be a weak solution of equation (10.7.1),* $\beta \in (0,1)$. *Then for any bounded domain* $\Omega \subset\subset \mathbb{R}^n_+$, *we have*

$$[u]_{\beta;\Omega} \leq C \left(|\vec{f}|_{0;\mathbb{R}^n_+} + |u|_{0;\mathbb{R}^n_+} \right),$$

where the constant C *depends only on* n, λ, β, $|\bar{a}|_{\alpha;\mathbb{R}^n_+}$ *and* Ω.

Proof. Let

$$0 < R_0 < \frac{1}{2}\mathrm{dist}(\Omega, \partial\mathbb{R}^n_+)$$

and fix $x^0 \in \Omega$.

First consider the equation

$$-\mathrm{div}(\bar{a}(x^0)\nabla u) = \mathrm{div}\vec{f}, \quad x \in \mathbb{R}^n_+. \qquad (10.7.2)$$

Let $u \in C^{1,\alpha}(\overline{\mathbb{R}}^n_+)$ be its weak solution. Without loss of generality, we may assume that u is smooth in $\mathbb{R}^n_+$. Decompose u into $u = u_1 + u_2$ with u_1

and u_2 satisfying

$$\begin{cases} -\mathrm{div}(\bar{a}(x^0)\nabla u_1) = 0, & x \in B_R, \\ u_1\big|_{\partial B_R} = u \end{cases}$$

and

$$\begin{cases} -\mathrm{div}(\bar{a}(x^0)\nabla u_2) = \mathrm{div}\vec{f}, & x \in B_R, \\ u_2\big|_{\partial B_R} = 0, \end{cases}$$

where $0 < R \le R_0$ and $B_R = B_R(x^0)$. From Schauder's estimate for linear equations, for u_1, we have

$$\int_{B_\rho} |\nabla u_1|^2 dx \le C \left(\frac{\rho}{R}\right)^n \int_{B_R} |\nabla u_1|^2 dx, \quad 0 < \rho \le R \le R_0.$$

To estimate u_2, we multiply the equation of u_2 by u_2, integrate the resulting relation over B_R, integrate by parts and use Cauchy's inequality with ε. Then we have

$$\bar{a}(x^0) \int_{B_R} |\nabla u_2|^2 dx = -\int_{B_R} \vec{f} \cdot \nabla u_2 dx$$

$$\le \frac{\varepsilon}{2} \int_{B_R} |\nabla u_2|^2 dx + \frac{1}{2\varepsilon} \int_{B_R} |\vec{f}|^2 dx,$$

where $\varepsilon > 0$ is an arbitrary constant. From this and $\bar{a}(x^0) \ge \lambda$, it follows that

$$\int_{B_R} |\nabla u_2|^2 dx \le C \int_{B_R} |\vec{f}|^2 dx \le C |\vec{f}|_{0;\mathbb{R}_+^n}^2 R^n.$$

Thus, for any $0 < \rho \le R \le R_0$, we have

$$\int_{B_\rho} |\nabla u|^2 dx \le 2 \int_{B_\rho} |\nabla u_1|^2 dx + 2 \int_{B_\rho} |\nabla u_2|^2 dx$$

$$\le C \left(\frac{\rho}{R}\right)^n \int_{B_R} |\nabla u_1|^2 dx + 2 \int_{B_R} |\nabla u_2|^2 dx$$

$$\le C \left(\frac{\rho}{R}\right)^n \int_{B_R} |\nabla u|^2 dx + C \int_{B_R} |\nabla u_2|^2 dx$$

$$\le C \left(\frac{\rho}{R}\right)^n \int_{B_R} |\nabla u|^2 dx + C |\vec{f}|_{0;\mathbb{R}_+^n}^2 R^n$$

$$\le C \left(\frac{\rho}{R}\right)^n \int_{B_R} |\nabla u|^2 dx + C |\vec{f}|_{0;\mathbb{R}_+^n}^2 R^{n-2+2\beta}.$$

Applying the iteration lemma (Lemma 6.2.1 of Chapter 6) we further derive that, for any $0 < \rho \le R \le R_0$,

$$\int_{B_\rho} |\nabla u|^2 dx \le C \left(\frac{\rho}{R}\right)^{n-2+2\beta} \left(\int_{B_R} |\nabla u|^2 dx + C|\vec{f}|^2_{0;\mathbb{R}^n_+} R^{n-2+2\beta} \right).$$

In particular, by setting $R = R_0$, we deduce that, for any $0 < \rho \le R_0$,

$$\int_{B_\rho} |\nabla u|^2 dx \le C\rho^{n-2+2\beta} \left(\frac{1}{R_0^{n-2+2\beta}} \int_{B_{R_0}} |\nabla u|^2 dx + |\vec{f}|^2_{0;\mathbb{R}^n_+} \right). \quad (10.7.3)$$

Now, we estimate $\int_{B_{R_0}} |\nabla u|^2 dx$. Let ξ be a cut-off function on B_{2R_0} relative to B_{R_0}, i.e., $\xi \in C_0^\infty(B_{2R_0})$, $\xi = 1$ in B_{R_0} and

$$0 \le \xi(x) \le 1, \quad |\nabla \xi(x)| \le \frac{C}{R_0}, \quad x \in B_{2R_0}.$$

Choosing $\xi^2 u$ as the test function in the definition of weak solutions of equation (10.7.2), we obtain

$$\int_{B_{2R_0}} \bar{a}(x^0)\xi^2 \nabla u \cdot \nabla u dx + 2 \int_{B_{2R_0}} \bar{a}(x^0)\xi u \nabla u \cdot \nabla \xi dx$$

$$= - \int_{B_{2R_0}} \xi^2 \vec{f} \cdot \nabla u dx - 2 \int_{B_{2R_0}} \xi u \vec{f} \cdot \nabla \xi dx.$$

Then, Cauchy's inequality with ε yields

$$\int_{B_{R_0}} |\nabla u|^2 dx \le C \left(\int_{B_{2R_0}} |\vec{f}|^2 dx + \int_{B_{2R_0}} u^2 dx \right)$$

$$\le C \left(|\vec{f}|^2_{0;\mathbb{R}^n_+} + |u|^2_{0;\mathbb{R}^n_+} \right). \quad (10.7.4)$$

Substituting this into (10.7.3) leads to

$$\int_{B_\rho} |\nabla u|^2 dx \le C\rho^{n-2+2\beta} \left(|\vec{f}|^2_{0;\mathbb{R}^n_+} + |u|^2_{0;\mathbb{R}^n_+} \right), \quad 0 < \rho \le R_0.$$

Therefore, Morrey's theorem gives

$$[u]_{\beta;B_{R_0}} \le C \left(|\vec{f}|_{0;\mathbb{R}^n_+} + |u|_{0;\mathbb{R}^n_+} \right).$$

Finally we adopt the method of solidifying coefficients to equation (10.7.1). For this purpose, we rewrite the equation as

$$-\mathrm{div}(\bar{a}(x^0)\nabla u) = \mathrm{div}\vec{g},$$

where

$$\vec{g}(x) = (\bar{a}(x) - \bar{a}(x^0))\nabla u(x) + \vec{f}(x), \quad x \in \Omega.$$

By the conclusion proved above, we see that for any $0 < \rho \le R \le R_0$,

$$\int_{B_\rho} |\nabla u|^2 dx \le C\left(\frac{\rho}{R}\right)^n \int_{B_R} |\nabla u|^2 dx + C \int_{B_R} |\vec{g}|^2 dx,$$

which, together with

$$|\vec{g}|^2 \le 2|(\bar{a}(x) - \bar{a}(x^0))\nabla u|^2 + 2|\vec{f}|^2$$

$$\le 2[\bar{a}]_{\alpha;\mathbb{R}^n_+}^2 R^{2\alpha}|\nabla u|^2 + 2|\vec{f}|^2_{0;\mathbb{R}^n_+}, \quad \forall x \in B_R,$$

leads to that for any $0 < \rho \le R \le R_0$,

$$\int_{B_\rho} |\nabla u|^2 dx$$

$$\le C\left(\frac{\rho}{R}\right)^n \int_{B_R} |\nabla u|^2 dx + CR^{2\alpha}\int_{B_R} |\nabla u|^2 dx + C|\vec{f}|^2_{0;\mathbb{R}^n_+} R^n$$

$$\le C\left[\left(\frac{\rho}{R}\right)^n + R_0^{2\alpha}\right]\int_{B_R} |\nabla u|^2 dx + C|\vec{f}|^2_{0;\mathbb{R}^n_+} R^{n-2+2\beta}.$$

Let $\Phi(\rho) = \int_{B_\rho} |\nabla u|^2 dx$. Then the above inequality can be rewritten as

$$\Phi(\rho) \le C\left[\left(\frac{\rho}{R}\right)^n + R_0^{2\alpha}\right]\Phi(R) + C|\vec{f}|^2_{0;\mathbb{R}^n_+} R^{n-2+2\beta}, \quad \forall\, 0 < \rho \le R \le R_0.$$

Thus, Lemma 10.7.1 yields

$$\Phi(\rho) \le C\left[\left(\frac{\rho}{R}\right)^{n-2+2\beta}\Phi(R) + |\vec{f}|^2_{0;\mathbb{R}^n_+}\rho^{n-2+2\beta}\right], \quad \forall\, 0 < \rho \le R \le R_0$$

provided that $R_0 > 0$ is small enough. In particular, setting $R = R_0$ we conclude that for any $0 < \rho \le R_0$,

$$\int_{B_\rho} |\nabla u|^2 dx \le C\rho^{n-2+2\beta}\left(\frac{1}{R_0^{n-2+2\beta}}\int_{B_{R_0}} |\nabla u|^2 dx + |\vec{f}|^2_{0;\mathbb{R}^n_+}\right). \quad (10.7.5)$$

Similar to the proof of (10.7.4), we may obtain

$$\int_{B_{R_0}} |\nabla u|^2 dx \le C\left(|\vec{f}|^2_{0;\mathbb{R}^n_+} + |u|^2_{0;\mathbb{R}^n_+}\right).$$

Substituting this into (10.7.5) yields

$$\int_{B_\rho(x^0)} |\nabla u|^2 dx \le C\rho^{n-2+2\beta} \left(|\vec{f}|^2_{0;\mathbb{R}^n_+} + |u|^2_{0;\mathbb{R}^n_+} \right), \quad 0 < \rho \le R_0, \quad x^0 \in \Omega.$$

From the arbitrariness of $x^0 \in \Omega$ and by using Morrey's theorem, we deduce

$$[u]_{\beta;B_{R_0/2}(x^0)} \le C \left(|\vec{f}|_{0;\mathbb{R}^n_+} + |u|_{0;\mathbb{R}^n_+} \right), \quad \forall x^0 \in \Omega.$$

Then, we may use the finite covering argument to complete the proof of the theorem. □

Similarly, we may establish Hölder's boundary estimate, and prove the following

Theorem 10.7.2 *Let $u \in C^{1,\alpha}(\overline{\mathbb{R}}^n_+)$ be a weak solution of equation (10.7.1) with $u\big|_{\partial\mathbb{R}^n_+} = 0$ and $\beta \in (0,1)$. Then*

$$[u]_{\beta;\mathbb{R}^n_+} \le C \left(|\vec{f}|_{0;\mathbb{R}^n_+} + |u|_{0;\mathbb{R}^n_+} \right),$$

where C depends only on $|a|_{\alpha;\mathbb{R}^n_+}$, λ, n and β.

10.8 Hölder's Estimate for Gradients

In the previous sections, we have obtained Hölder's estimate for solutions and the maximum estimate for gradients of solutions of the Dirichlet problem (10.1.9), (10.1.10). In this section, we further establish Hölder's estimate for gradients of solutions.

10.8.1 *Interior Hölder's estimate for gradients of solutions*

Theorem 10.8.1 *Let $u \in C^{2,\alpha}(\overline{\Omega})$ be a solution of equation (10.1.9). Then for any $\Omega' \subset\subset \Omega$, we have*

$$[D_k u]_{\alpha;\Omega'} \le C, \quad k = 1, 2, \cdots, n,$$

where C depends only on n, m, γ, $|f|_{0;\Omega}$, $|u|_{\alpha;\Omega}$, $|\nabla u|_{0;\Omega}$, Ω' and Ω.

Proof. Let $\varphi \in C_0^\infty(\Omega)$. Multiplying both sides of equation (10.1.9) by $D_k\varphi \, (k = 1, 2, \cdots, n)$ and integrating the resulting relation over Ω, we obtain

$$-\int_\Omega D_i(a(u)D_iu)D_k\varphi dx + \int_\Omega (b(u) - f(x))D_k\varphi dx = 0.$$

Integrate the first term on the left hand side by parts to get

$$-\int_\Omega D_k(a(u)D_iu)D_i\varphi dx + \int_\Omega (b(u) - f(x))D_k\varphi dx = 0,$$

i.e.

$$\int_\Omega a(u)D_{ik}uD_i\varphi dx + \int_\Omega f_k^i D_i\varphi dx = 0, \quad \forall \varphi \in C_0^\infty(\Omega),$$

where

$$f_k^i = a'(u)D_kuD_iu - \delta_{ik}(b(u) - f), \quad \text{in } \Omega.$$

This shows that $D_ku \in C^{1,\alpha}(\overline{\Omega})$ is a weak solution of the equation

$$-\text{div}(\bar{a}(x)\nabla v) = \text{div}\vec{f}(x), \quad x \in \Omega,$$

where $\bar{a}(x) = a(u(x)) \in C^\alpha(\overline{\Omega})$, $\vec{f} = (f_k^1, f_k^2 \cdots, f_k^n) \in L^\infty(\Omega)$. Therefore, by interior Hölder's estimate obtained in §10.7, we immediately obtain the conclusion of the theorem. $\qquad\square$

10.8.2 Boundary Hölder's estimate for gradients of solutions

Theorem 10.8.2 *Let $\partial\Omega \in C^{2,\alpha}$ and $u \in C^{2,\alpha}(\overline{\Omega})$ be a solution of problem (10.1.9), (10.1.10). Then for any $x^0 \in \partial\Omega$, there exists $R > 0$, such that*

$$[D_ku]_{\alpha;\Omega_R} \le C, \quad k = 1, 2, \cdots, n,$$

where $\Omega_R = \Omega \cap B_R(x^0)$, C depends only on n, m, γ, $|f|_{0;\Omega}$, $|u|_{\alpha;\Omega}$, $|\nabla u|_{0;\Omega}$ and Ω.

Proof. We divide the proof into four steps.

Step 1 Local flatting.

From $\partial\Omega \in C^{2,\alpha}$, for fixed $x^0 \in \partial\Omega$, there exists a neighborhood U of x^0 and a $C^{2,\alpha}$ invertible mapping $\Psi : U \to B_1(0)$, such that

$$\Psi(U \cap \Omega) = B_1^+(0) = \{y \in B_1(0); y_n > 0\},$$

$$\Psi(U \cap \partial\Omega) = \partial B_1^+ \cap \{y; y_n = 0\}.$$

Then problem (10.1.9), (10.1.10) reduces locally in $U \cap \Omega$ to

$$- \hat{D}_j(a_{ij}(y, \hat{u})\hat{D}_i\hat{u}) = \tilde{f}(y, \hat{u}, \hat{D}\hat{u}), \quad y \in B_1^+(0), \tag{10.8.1}$$

$$\hat{u}\Big|_{\partial B_1^+ \cap \{y; y_n = 0\}} = 0, \tag{10.8.2}$$

where $\hat{D}_i = \dfrac{\partial}{\partial y_i}$, $\hat{u}(y) = u(\Psi^{-1}(y))$. It is easily seen that the metrics in x-space and in y-space are equivalent.

Step 2 Hölder's estimate for tangential derivatives.

Let $1 \le k < n$. From (10.8.1) and (10.8.2), it can be proved that $\hat{D}_k\hat{u} \in C^{1,\alpha}(\overline{B}_1^+)$ is a weak solution of the linear equation

$$-\hat{D}_j(a_{ij}(y, \hat{u}(y))\hat{D}_i\hat{D}_k\hat{u}) = \text{div}\vec{g}(y), \quad y \in B_1^+$$

with the boundary value condition

$$\hat{D}_k\hat{u}\Big|_{\partial B_1^+ \cap \{y; y_n = 0\}} = 0.$$

It is easy to verify $|a_{ij}(y, \hat{u}(y))|_{\alpha; B_1^+(0)} \le C$ and $|\vec{g}|_{0; B_1^+(0)} \le C$. By the boundary Hölder's estimate obtained in §10.7, we obtain

$$[\hat{D}_k\hat{u}]_{\alpha; B_1^+(0)} \le C, \quad k = 1, 2, \cdots, n-1.$$

Step 3 Hölder's estimate for normal derivative.

From the conclusion of Step 2, we have

$$\sum_{i+j<2n} \int_{B_R^+} |\hat{D}_{ij}\hat{u}|^2 dy \le CR^{n-2+2\alpha}, \quad \forall 0 < R \le 1,$$

where $\hat{D}_{ij} = \dfrac{\partial^2}{\partial y_i \partial y_j}$. Using equation (10.8.1), we further have

$$\int_{B_R^+} |\hat{D}_{nn}\hat{u}|^2 dy \le CR^{n-2+2\alpha}, \quad \forall 0 < R \le 1.$$

Therefore

$$\int_{B_R^+} |\hat{D}\hat{D}_n\hat{u}|^2 dy \le CR^{n-2+2\alpha}, \quad \forall 0 < R \le 1.$$

Morrey's theorem then yields

$$[\hat{D}_n\hat{u}]_{\alpha; B_1^+} \le C.$$

Step 4 Returning to the original coordinate.

Returning to the x coordinate, we obtain

$$[D_k u]_{\alpha; U \cap \Omega} \leq C, \quad k = 1, 2, \cdots, n,$$

from which we get the conclusion of the theorem. $\qquad\qquad\square$

10.8.3   Global Hölder's estimate for gradients of solutions

Combining the interior and the boundary Hölder's estimates for gradients
(Theorem 10.8.1 and Theorem 10.8.2), and using the finite covering argument, we obtain the following

Theorem 10.8.3   *Let $\partial\Omega \in C^{2,\alpha}$ and $u \in C^{2,\alpha}(\overline{\Omega})$ be a solution of problem (10.1.9), (10.1.10). Then*

$$[D_k u]_{\alpha; \Omega} \leq C, \quad k = 1, 2, \cdots, n,$$

where C depends only on n, m, γ, $|f|_{0;\Omega}$, $|u|_{\alpha;\Omega}$, $|\nabla u|_{0;\Omega}$ and Ω.

10.9   Solvability of More General Quasilinear Equations

In the previous discussion, we have investigated the solvability of quasilinear elliptic equations (10.1.1) with structure conditions (10.1.3)–(10.1.8)
and quasilinear parabolic equations (10.1.15) with structure conditions
(10.1.17)–(10.1.20). However, there are many important quasilinear equations which do not satisfy such kind of structure conditions, for example,
the quasilinear elliptic equation

$$-\mathrm{div}((|\nabla u|^2 + 1)^{p/2-1}\nabla u) + c(x)u = f(x), \quad x \in \Omega \qquad (10.9.1)$$

and the quasilinear parabolic equation

$$\frac{\partial u}{\partial t} - \mathrm{div}((|\nabla u|^2 + 1)^{p/2-1}\nabla u) + c(x,t)u = f(x,t), \quad (x,t) \in Q_T, \quad (10.9.2)$$

where $p > 1$. In this section, we will illustrate the solvability of a class of
more general quasilinear equations without proof.

10.9.1   Solvability of more general quasilinear elliptic equa-
tions

Consider the following Dirichlet problem for quasilinear elliptic equations

$$-\mathrm{div}\vec{a}(x, u, \nabla u) + b(x, u, \nabla u) = 0, \quad x \in \Omega, \qquad (10.9.3)$$

$$u\Big|_{\partial\Omega} = \varphi, \qquad\qquad (10.9.4)$$

where $\vec{a} = (a_1, a_2, \cdots, a_n)$, $\Omega \subset \mathbb{R}^n$ is a bounded domain. Assume that $\vec{a}(x, z, \eta)$, $b(x, z, \eta)$ satisfy the following structure conditions:

$$\lambda(|z|)(1 + |\eta|^{p-2})|\xi|^2 \leq \frac{\partial a_i}{\partial \eta_j}\xi_i\xi_j \leq \Lambda(|z|)(1 + |\eta|^{p-2})|\xi|^2, \quad \forall \xi \in \mathbb{R}^n,$$

$$(10.9.5)$$

$$|a_i(x, z, 0)| \leq g(x), \qquad\qquad (10.9.6)$$

$$\left|\frac{\partial a_i}{\partial z}\right| + |a_i| \leq \mu(|z|)(1 + |\eta|^{p-1}), \qquad\qquad (10.9.7)$$

$$\left|\frac{\partial a_i}{\partial x_j}\right| + |b| \leq \mu(|z|)(1 + |\eta|^p), \qquad\qquad (10.9.8)$$

$$-b(x, z, \eta)\text{sgn}z \leq b_0(|\eta|^{p-1} + h(x)), \qquad\qquad (10.9.9)$$

where $i, j = 1, 2, \cdots, n$, $p > 1$, $\lambda(s), \Lambda(s)$ are positive continuous functions on $[0, +\infty)$, $\mu(s)$ is a nondecreasing function on $[0, +\infty)$, $g \in L^q(\Omega)$, $q > n/(p-1)$, $h \in L^{q_*}(\Omega)$, $q_* = nq/(n+q)$, b_0 is a positive constant.

Equation (10.9.1) satisfies all of the above structure conditions provided $c, f \in C^\alpha(\overline{\Omega})$ and $c \geq 0$.

Theorem 10.9.1 *Let* $0 < \alpha < 1$, $\partial\Omega \in C^{2,\alpha}$, $\vec{a} \in C^{1,\alpha}$, $b \in C^\alpha$, $\varphi \in C^{2,\alpha}(\overline{\Omega})$. *Assume that equation (10.9.3) satisfies the structure conditions (10.9.5)–(10.9.9). Then problem (10.9.3), (10.9.4) admits a solution* $u \in C^{2,\alpha}(\overline{\Omega})$.

10.9.2 Solvability of more general quasilinear parabolic equations

Consider the following first initial-boundary value problem for quasilinear parabolic equations

$$\frac{\partial u}{\partial t} - \text{div}\vec{a}(x, t, u, \nabla u) + b(x, t, u, \nabla u) = 0, \quad (x, t) \in Q_T, \qquad (10.9.10)$$

$$u\Big|_{\partial_p Q_T} = \varphi, \qquad\qquad (10.9.11)$$

where $\vec{a} = (a_1, a_2, \cdots, a_n)$, $Q_T = \Omega \times (0, T)$, $\Omega \subset \mathbb{R}^n$ is a bounded domain, $T > 0$. Assume that $\vec{a}(x, t, z, \eta)$, $b(x, t, z, \eta)$ satisfy the following structure conditions:

$$\lambda(|z|)(1 + |\eta|^{p-2})|\xi|^2 \le \frac{\partial a_i}{\partial \eta_j}\xi_i\xi_j \le \Lambda(|z|)(1 + |\eta|^{p-2})|\xi|^2, \quad \xi \in \mathbb{R}^n,$$

$$(10.9.12)$$

$$-zb(x, t, z, 0) \le b_1|z|^2 + b_2, \qquad (10.9.13)$$

$$\left(\left|\frac{\partial a_i}{\partial z}\right| + |a_i|\right)(1 + |\eta|) + \left|\frac{\partial a_i}{\partial x_j}\right| \le \Psi(|z|, |\eta|)(1 + |\eta|^p), \qquad (10.9.14)$$

$$\left|\frac{\partial b}{\partial \eta_j}\right|(1 + |\eta|) + |b| \le \mu(|z|)(1 + |\eta|^p), \qquad (10.9.15)$$

$$\left|\frac{\partial b}{\partial x_j} - \frac{\partial b}{\partial z}(1 + |\eta|)\right| \le \Psi(|z|, |\eta|)(1 + |\eta|^{p+1}), \qquad (10.9.16)$$

where $i, j = 1, 2, \cdots, n$, $p > 1$, $\lambda(s), \Lambda(s)$ are positive continuous functions on $[0, +\infty)$, b_1, b_2 are positive constants, $\mu(s)$ is a nondecreasing function on $[0, +\infty)$, $\Psi(\tau, \rho)$ is a continuous function on $[0, +\infty) \times [0, +\infty)$ such that for any $\rho \in [0, +\infty)$, $\Psi(\cdot, \rho)$ is nondecreasing on $[0, +\infty)$, and as $\rho \to +\infty$, $\Psi(\tau, \rho)$ locally uniformly converges to 0 with respect to τ.

Equation (10.9.2) satisfies all of the above structure conditions provided that $c, f \in C^{\alpha, \alpha/2}(\overline{Q}_T)$.

Theorem 10.9.2 *Let $0 < \alpha < 1$, $\partial\Omega \in C^{2,\alpha}$, $\vec{a} \in C^{1,\alpha}$, $b \in C^1$, $\varphi \in C^{2+\alpha, 1+\alpha/2}(\overline{Q}_T)$ and satisfy the first order compatibility condition*

$$\frac{\partial\varphi}{\partial t} - \mathrm{div}\,\vec{a}(x, t, \varphi, \nabla\varphi) + b(x, t, \varphi, \nabla\varphi) = 0, \quad when \ x \in \partial\Omega, \ t = 0.$$

Assume that equation (10.9.10) satisfies the structure conditions (10.9.12)–(10.9.16). Then problem (10.9.10), (10.9.11) admits a solution $u \in C^{2+\alpha, 1+\alpha/2}(\overline{Q}_T)$.

Exercises

1. Prove Theorem 10.1.2 ii).
2. Complete the proof of Theorem 10.3.2.
3. Prove Theorem 10.3.4.
4. Prove Theorem 10.7.2.
5. Prove Theorem 10.8.3.
6. Establish the solvability of quasilinear parabolic equations.
7. Prove Theorems 10.9.1 and 10.9.2.

Chapter 11

Topological Degree Method

The concept of topological degree was first introduced by L. E. J. Brouwer for continuous mapping in finite dimensional space. It was J. Leray and J. Schauder who generalized such a concept to the completely continuous fields in Banach spaces, and developed a complete theory of topological degree, which has been applied extensively to the investigation of partial differential equations and integral equations. In this chapter, we will illustrate the application of the topological degree method by a heat equation with strongly nonlinear source.

11.1 Topological Degree

In this section, we introduce the definition of topological degree and present its basic properties without proof. For the detailed theory we refer to [Zhong, Fan and Chen (1998)].

11.1.1 *Brouwer degree*

Let Ω be an open set of $\mathbb{R}^n$ and f be a mapping from Ω to $\mathbb{R}^n$. Roughly speaking, the Brouwer degree is an integer valued function related to f and Ω. We first define it for $f \in C^1(\Omega; \mathbb{R}^n)$ and then extend to $f \in C(\Omega; \mathbb{R}^n)$. Assume that $f = (f^1, f^2, \cdots, f^n) \in C^1(\Omega; \mathbb{R}^n)$. Then for any $x \in \Omega$, the Frechet derivative operator of f at x, $f'(x) : \mathbb{R}^n \to \mathbb{R}^n$ is a linear operator, and

$$f'(x) = \left(\frac{\partial f^i(x)}{\partial x_j} \right)_{n \times n}.$$

Definition 11.1.1 $x \in \Omega$ is called a regular point of f, if the Frechet derivative operator $f'(x)$ is of full rank; otherwise, we call x a critical point of f. $y \in \mathbb{R}^n$ is called a critical value of f, if there exists a critical point $x \in \Omega$ of f such that $f(x) = y$; otherwise, we call y a regular value of f.

Theorem 11.1.1 *Let Ω be an open set of $\mathbb{R}^n$, $f \in C^1(\Omega; \mathbb{R}^n)$. Then the Lebesgue measure of the set of critical values of f in $\mathbb{R}^n$ is equal to zero.*

Definition 11.1.2 Let Ω be a bounded open set of $\mathbb{R}^n$, $f \in C^2(\overline{\Omega}; \mathbb{R}^n)$, $p \in \mathbb{R}^n \backslash f(\partial\Omega)$. We define the Brouwer degree $\deg(f, \Omega, p)$ of the mapping f in Ω at the point p in the following way:

i) If p is a regular value of f, then set

$$\deg(f, \Omega, p) = \sum_{x \in f^{-1}(p)} \mathrm{sgn} J_f(x),$$

where $J_f(x)$ is the determinant of $f'(x)$;

ii) If p is a critical value of f, then choose a regular value p_1 of f with $\|p_1 - p\| < \mathrm{dist}(p, f(\partial\Omega))$ and set

$$\deg(f, \Omega, p) = \deg(f, \Omega, p_1).$$

It can be proved that $\deg(f, \Omega, p)$ is independent of the choice of p_1.

Definition 11.1.3 Let Ω be a bounded open set of $\mathbb{R}^n$, $f \in C(\overline{\Omega}; \mathbb{R}^n)$, $p \in \mathbb{R}^n \backslash f(\partial\Omega)$. Choose $f_1 \in C^2(\overline{\Omega}; \mathbb{R}^n)$ such that

$$\sup_{x \in \Omega} \|f(x) - f_1(x)\| < \mathrm{dist}(p, f(\partial\Omega)).$$

Define the Brouwer degree of f in Ω at the point p by

$$\deg(f, \Omega, p) = \deg(f_1, \Omega, p).$$

It can be proved that $\deg(f, \Omega, p)$ is independent of the choice of f_1.

Some basic properties of the Brouwer degree are involved in the following theorem.

Theorem 11.1.2 *Let Ω be a bounded open set of $\mathbb{R}^n$, $f \in C(\overline{\Omega}; \mathbb{R}^n)$, $p \in \mathbb{R}^n \backslash f(\partial\Omega)$. The Brouwer degree $\deg(f, \Omega, p)$ has the following properties:*

i) (Normality)

$$\deg(\mathrm{id}, \Omega, p) = \begin{cases} 1, & p \in \Omega, \\ 0, & p \notin \overline{\Omega}, \end{cases}$$

where id *denotes the identity map;*

ii) (Domain Additivity) If Ω_1, Ω_2 are two open subsets of Ω with $\Omega_1 \cap \Omega_2 = \emptyset$ and $p \notin f(\overline{\Omega}\backslash(\Omega_1 \cup \Omega_2))$, then

$$\deg(f, \Omega, p) = \deg(f, \Omega_1, p) + \deg(f, \Omega_2, p);$$

iii) (Invariance of Homotopy) Let $H : \overline{\Omega} \times [0,1] \to \mathbb{R}^n$ be a continuous mapping and denote $h_t(x) = H(x,t)$. Assume the mapping $p : [0,1] \to \mathbb{R}^n$ is continuous and $p(t) \notin h_t(\partial\Omega)$ for each $t \in [0,1]$. Then $\deg(h_t, \Omega, p(t))$ is independent of t.

Based on these basic properties, we may derive a series of important properties of the degree. For example, we have the following theorem.

Theorem 11.1.3 *(Kronecker's Existence Theorem) Let Ω be a bounded open set of $\mathbb{R}^n$, $f \in C(\overline{\Omega}; \mathbb{R}^n)$, $p \in \mathbb{R}^n \backslash f(\partial\Omega)$. If $p \notin f(\overline{\Omega})$, then $\deg(f, \Omega, p) = 0$, and so, if $\deg(f, \Omega, p) \neq 0$, then the equation $f(x) = p$ must have at least one solution in Ω.*

11.1.2 Leray-Schauder degree

Since many problems in analysis are referred to infinite dimensional space, it is natural to extend the Brouwer degree theory to the infinite dimensional case. However, owing to the lack of compactness of the unit ball in the infinite dimensional space, one cannot establish the degree theory for general continuous mappings. It was Leray and Schauder who found an important class of mappings in the investigation of partial differential equations and integral equations, that is the completely continuous perturbations of the identity mappings (also called the compact continuous fields), and applied the method of finite dimensional approximation to establish the degree theory for this class of mappings.

Definition 11.1.4 Let X, Y be two normed linear spaces, $D \subset X$, A mapping $F : D \to Y$ is said to be compact, if for any bounded set $S \subset D$, $\overline{F(S)}$ is a compact set in Y. If, in addition, the mapping F is continuous, then we call F a completely continuous mapping or a compact continuous mapping.

Theorem 11.1.4 *Let X, Y be two normed linear spaces, M be a bounded closed subset of X. Assume that the mapping $F : M \to Y$ is continuous. If the mapping F is completely continuous, then for any $\varepsilon > 0$, there exists a bounded continuous mapping $F_k : M \to Y_k$ with finite dimensional range* *such that*

$$\sup_{x \in M} \|F(x) - F_k(x)\| < \varepsilon,$$

where $Y_k \subset Y$ is a finite dimensional space. Furthermore, if Y is complete, then the above condition is also sufficient.

Definition 11.1.5 Let X be a real normed linear space, $D \subset X$. If the mapping $F : D \to X$ is completely continuous, then $f = \text{id} - F$ is called a completely continuous field in D, or a compact continuous field in D.

Let X be a real normed linear space, Ω be a bounded open set in X, $F : \overline{\Omega} \to X$ be a completely continuous mapping, $f = \text{id} - F$ be a completely continuous field, $p \in X \backslash f(\partial \Omega)$. By Theorem 11.1.4, there exist a finite dimensional subspace $X_k \subset X$, $p_k \in X_k$ and a bounded continuous mapping $F_k : \overline{\Omega} \to X_k$, such that

$$\|p - p_k\| + \sup_{x \in \Omega} \|F(x) - F_k(x)\| < \text{dist}(p, f(\partial \Omega)).$$

Denote $\Omega_k = X_k \cap \Omega$, $f_k = \text{id} - F_k$. Then $f_k \in C(\overline{\Omega}_k, X_k)$, $p_k \in X_k \backslash f_k(\partial \Omega_k)$, and hence the Brouwer degree $\deg(f_k, \Omega_k, p_k)$ is well-defined.

Definition 11.1.6 Define the Leray–Schauder degree of the completely continuous field f in Ω at the point p by

$$\deg(f, \Omega, p) = \deg(f_k, \Omega_k, p_k).$$

It can be shown that $\deg(f, \Omega, p)$ is independent of the choice of X_k, p_k and F_k.

Since the Leray–Schauder degree is obtained by the approximation of the Brouwer degree, it can be proved that most properties of the Brouwer degree are retained.

Theorem 11.1.5 *Let Ω be a bounded open subset of X which is a real normed linear space, $f = \text{id} - F$ be a completely continuous field on $\overline{\Omega}$, $p \in X \backslash f(\partial \Omega)$. Then the Leray–Schauder degree $\deg(f, \Omega, p)$ has the following properties:*

i) (Normality)

$$\deg(\text{id}, \Omega, p) = \begin{cases} 1, & p \in \Omega, \\ 0, & p \notin \overline{\Omega}; \end{cases}$$

ii) (Domain Additivity) If Ω_1, Ω_2 are two open subsets of Ω with $\Omega_1 \cap \Omega_2 = \emptyset$ and $p \notin f(\overline{\Omega} \backslash (\Omega_1 \cup \Omega_2))$, then

$$\deg(f, \Omega, p) = \deg(f, \Omega_1, p) + \deg(f, \Omega_2, p);$$

iii) (Invariance of Compact Homotopy) Let $H : \overline{\Omega} \times [0, 1] \to X$ be a completely continuous mapping and denote $h_t(x) = x - H(x, t)$. Assume the mapping $p : [0, 1] \to X$ is continuous and $p(t) \notin h_t(\partial\Omega)$ for every $t \in [0, 1]$. Then $\deg(h_t, \Omega, p(t))$ is independent of t.

Theorem 11.1.6 *(Kronecker's Existence Theorem) Let X be a real normed linear space, Ω be a bounded open subset of X, and $f = \mathrm{id} - F$ be a completely continuous field defined on $\overline{\Omega}$, $p \in X \backslash f(\partial\Omega)$. If $p \notin f(\overline{\Omega})$, then $\deg(f, \Omega, p) = 0$. Thus, if $\deg(f, \Omega, p) \neq 0$, then the equation $f(x) = p$ admits at least one solution in Ω.*

11.2   Existence of a Heat Equation with Strong Nonlinear Source

As an example in applications of the topological degree method, let us consider the heat equation with the strong nonlinear source

$$\frac{\partial u}{\partial t} - \Delta u = |u|^p, \quad (x, t) \in Q_T \tag{11.2.1}$$

with the initial-boundary value condition

$$u(x, t) = \varphi, \quad (x, t) \in \partial_p Q_T, \tag{11.2.2}$$

where $p > 1$, $Q_T = \Omega \times (0, T)$, Ω is a bounded domain in $\mathbb{R}^n$ with $\partial\Omega \in C^{2,\alpha}$, $\alpha \in (0, 1)$, $T > 0$.

If the right hand side of (11.2.1) is a function $f(x, t)$ independent of u, namely, the equation

$$\frac{\partial u}{\partial t} - \Delta u = f, \quad (x, t) \in Q_T \tag{11.2.3}$$

is considered, then by the theory of classical solutions for the nonhomogeneous heat equation, the first initial-boundary value problem (11.2.3), (11.2.2) admits a unique solution $u \in C^{2+\alpha, 1+\alpha/2}(\overline{Q}_T)$ and

$$|u|_{2+\alpha, 1+\alpha/2; Q_T} \leq C_0 \left(|f|_{\alpha, \alpha/2; Q_T} + |\varphi|_{2+\alpha, 1+\alpha/2; Q_T} \right), \tag{11.2.4}$$

provided $f \in C^{\alpha,\alpha/2}(\overline{Q}_T)$, $\varphi \in C^{2+\alpha,1+\alpha/2}(\overline{Q}_T)$, where C_0 depends only on n, Ω and T. Our aim is to obtain the existence of $C^{2+\alpha,1+\alpha/2}(\overline{Q}_T)$ solutions of the first initial-boundary value problem (11.2.1), (11.2.2). To this purpose, we use the topological degree method.

Assume that $\varphi \in C^{2+\alpha,1+\alpha/2}(\overline{Q}_T)$. Define a mapping

$$F : C^{\alpha,\alpha/2}(\overline{Q}_T) \times [0,1] \to C^{\alpha,\alpha/2}(\overline{Q}_T),$$
$$(f,\sigma) \mapsto u,$$

where $u \in C^{2+\alpha,1+\alpha/2}(\overline{Q}_T)$ is the solution of the problem

$$\begin{cases} \dfrac{\partial u}{\partial t} - \Delta u = \sigma f, & (x,t) \in Q_T, \\ u(x,t) = \varphi, & (x,t) \in \partial_p Q_T. \end{cases}$$

We proceed to show that F is a completely continuous mapping.

Lemma 11.2.1   *The mapping F is compact.*

Proof.   Assume that $\{f_k\}_{k=1}^\infty \subset C^{\alpha,\alpha/2}(\overline{Q}_T)$, $\{\sigma_k\}_{k=1}^\infty \subset [0,1]$ and there exists a constant $M > 0$, such that

$$|f_k|_{\alpha,\alpha/2;Q_T} \leq M, \quad \forall k \geq 1.$$

Denote $u_k = F(f_k,\sigma_k)$, that is $u_k \in C^{2+\alpha,1+\alpha/2}(\overline{Q}_T)$ is the solution of the following problem

$$\begin{cases} \dfrac{\partial u_k}{\partial t} - \Delta u_k = \sigma_k f_k, & (x,t) \in Q_T, \\ u_k(x,t) = \varphi, & (x,t) \in \partial_p Q_T. \end{cases}$$

By virtue of the classical theory, we have

$$|u_k|_{2+\alpha,1+\alpha/2;Q_T} \leq C_0 \left(|\sigma_k f_k|_{\alpha,\alpha/2;Q_T} + |\varphi|_{2+\alpha,1+\alpha/2;Q_T} \right)$$

with C_0 given in (11.2.4), which implies that $\{u_k\}_{k=1}^\infty$ is uniformly bounded in $C^{2+\alpha,1+\alpha/2}(\overline{Q}_T)$. Therefore, there exists a convergent subsequence of $\{u_k\}_{k=1}^\infty$ in $C^{\alpha,\alpha/2}(\overline{Q}_T)$, this means that the mapping F is compact. $\square$

Lemma 11.2.2   *The mapping F is continuous.*

Proof.   Assume that $\{f_k\}_{k=1}^\infty \subset C^{\alpha,\alpha/2}(\overline{Q}_T)$, $\{\sigma_k\}_{k=1}^\infty \subset [0,1]$, $f \in C^{\alpha,\alpha/2}(\overline{Q}_T)$, $\sigma \in [0,1]$, and

$$\lim_{k\to\infty} |f_k - f|_{\alpha,\alpha/2;Q_T} = 0, \quad \lim_{k\to\infty} \sigma_k = \sigma.$$

Denote $u_k = F(f_k, \sigma_k)$, $u = F(f, \sigma)$. By the definition of F, $u_k - u \in C^{2+\alpha, 1+\alpha/2}(\overline{Q}_T)$ is the solution of the following problem

$$\begin{cases} \dfrac{\partial w}{\partial t} - \Delta w = (\sigma_k f_k - \sigma f), & (x, t) \in Q_T, \\ w(x, t) = 0, & (x, t) \in \partial_p Q_T. \end{cases}$$

By the classical theory, we have

$$|u_k - u|_{2+\alpha, 1+\alpha/2; Q_T}$$
$$\leq C_0 |\sigma_k f_k - \sigma f|_{\alpha, \alpha/2; Q_T}$$
$$\leq C_0 \left(\sigma_k |f_k - f|_{\alpha, \alpha/2; Q_T} + |\sigma_k - \sigma| |f|_{\alpha, \alpha/2; Q_T} \right)$$
$$\leq C_0 \left(|f_k - f|_{\alpha, \alpha/2; Q_T} + |\sigma_k - \sigma| |f|_{\alpha, \alpha/2; Q_T} \right),$$

where C_0 is the constant given in (11.2.4). Therefore

$$\lim_{k \to \infty} |u_k - u|_{2+\alpha, 1+\alpha/2; Q_T} = 0,$$

which implies that

$$\lim_{k \to \infty} |u_k - u|_{\alpha, \alpha/2; Q_T} = 0.$$

Thus, the mapping F is continuous. $\qquad\square$

Combining Lemma 11.2.1 and Lemma 11.2.2, we see that the mapping F is completely continuous.

Theorem 11.2.1 *Assume that $\varphi \in C^{2+\alpha, 1+\alpha/2}(\overline{Q}_T)$ and*

$$|\varphi|_{2+\alpha, 1+\alpha/2; Q_T} < \frac{1}{2C_0} (2(p+1)C_0)^{1/(1-p)},$$

where C_0 is the constant given in (11.2.4). Then problem (11.2.1), (11.2.2) admits at least one solution in $C^{2+\alpha, 1+\alpha/2}(\overline{Q}_T)$.

Proof. Denote $\Phi(v) = |v|^p$. Since F is completely continuous and $p > 1$, it is easy to see that

$$F(\Phi(\cdot), \cdot) : C^{\alpha, \alpha/2}(\overline{Q}_T) \times [0, 1] \to C^{\alpha, \alpha/2}(\overline{Q}_T)$$

is also completely continuous.

According to the classical theory, solving problem (11.2.1), (11.2.2) in $C^{2+\alpha, 1+\alpha/2}(\overline{Q}_T)$ is equivalent to solving the equation

$$u - F(\Phi(u), 1) = 0 \qquad\qquad (11.2.5)$$

in $C^{\alpha,\alpha/2}(\overline{Q}_T)$. The latter will be solved by using the Leray–Schauder topological degree theory. To this purpose, we first choose $R > 0$ such that

$$0 \neq (\mathrm{id} - F(\Phi(\cdot),\sigma))\,(\partial\hat{B}_R(0)), \quad \forall\sigma \in [0,1], \tag{11.2.6}$$

where $\hat{B}_R(0)$ is the ball of radius R centered at the origin in $C^{\alpha,\alpha/2}(\overline{Q}_T)$.

If (11.2.6) holds, then by Theorem 11.1.6, in order to show that (11.2.5) has at least one solution in $C^{\alpha,\alpha/2}(\overline{Q}_T)$, we need only to show that

$$\deg(\mathrm{id} - F(\Phi(\cdot),1), \hat{B}_R(0), 0) \neq 0. \tag{11.2.7}$$

Furthermore, if (11.2.6) holds, then Theorem 11.1.5 iii) yields

$$\deg(\mathrm{id} - F(\Phi(\cdot),1), \hat{B}_R(0), 0) = \deg(\mathrm{id} - F(\Phi(\cdot),0), \hat{B}_R(0), 0).$$

From the definition of F, it is seen that

$$F(\Phi(\cdot),0) : C^{\alpha,\alpha/2}(\overline{Q}_T) \to C^{\alpha,\alpha/2}(\overline{Q}_T)$$

is a constant mapping, that is

$$F(\Phi(v),0) \equiv \hat{u}, \quad \forall v \in C^{\alpha,\alpha/2}(\overline{Q}_T),$$

where $\hat{u} \in C^{2+\alpha,1+\alpha/2}(\overline{Q}_T)$ is the solution of the following problem

$$\frac{\partial\hat{u}}{\partial t} - \Delta\hat{u} = 0, \quad (x,t) \in Q_T, \tag{11.2.8}$$

$$\hat{u}(x,t) = \varphi, \quad (x,t) \in \partial_p Q_T. \tag{11.2.9}$$

Consider the following mapping

$$G(v,\sigma) = v - \sigma\hat{u}, \quad v \in C^{\alpha,\alpha/2}(\overline{Q}_T),\ \sigma \in [0,1].$$

If

$$0 \neq G(\partial\hat{B}_R(0),\sigma), \quad \forall\sigma \in [0,1], \tag{11.2.10}$$

then applying Theorem 11.1.5 iii) and i) to $G(v,\sigma)$ yields

$$\deg(\mathrm{id} - F(\Phi(\cdot),0), \hat{B}_R(0), 0) = \deg(\mathrm{id}, \hat{B}_R(0), 0) = 1.$$

Thus (11.2.7)holds.

Consequently, if we can find $R > 0$ satisfying (11.2.6) and (11.2.10), then the proof is complete.

We now show that if we choose

$$R = (2(p+1)C_0)^{1/(1-p)}$$

with C_0 given in (11.2.4), then both (11.2.6) and (11.2.10) are satisfied.

Assume that $v \in \partial \hat{B}_R(0)$, i.e. $v \in C^{\alpha, \alpha/2}(\overline{Q}_T)$ and $|v|_{\alpha, \alpha/2; Q_T} = R$. For any $\sigma \in [0, 1]$, the classical theory gives

$$|F(\Phi(v), \sigma)|_{2+\alpha, 1+\alpha/2; Q_T}$$
$$\leq C_0 \left(|\sigma \Phi(v)|_{\alpha, \alpha/2; Q_T} + |\varphi|_{2+\alpha, 1+\alpha/2; Q_T} \right)$$
$$\leq C_0 \left(|\, |v|^p \, |_{\alpha, \alpha/2; Q_T} + |\varphi|_{2+\alpha, 1+\alpha/2; Q_T} \right).$$

In addition,

$$|\, |v|^p \, |_{\alpha, \alpha/2; Q_T} = |\, |v|^p \, |_{0; Q_T} + [\, |v|^p \,]_{\alpha, \alpha/2; Q_T}$$
$$\leq |v|_{0; Q_T}^p + p |v|_{0; Q_T}^{p-1} [v]_{\alpha, \alpha/2; Q_T}$$
$$\leq R^p + p R^{p-1} R = (p+1) R^p = \frac{1}{2 C_0} R,$$

$$|\varphi|_{2+\alpha, 1+\alpha/2; \Omega} < \frac{1}{2 C_0} (2(p+1) C_0)^{1/(1-p)} = \frac{1}{2 C_0} R,$$

thus

$$|F(\Phi(v), \sigma)|_{2+\alpha, 1+\alpha/2; Q_T} < R.$$

Therefore

$$|F(\Phi(v), \sigma)|_{\alpha, \alpha/2; Q_T} < R, \quad \forall v \in \partial \hat{B}_R(0), \ \forall \sigma \in [0, 1].$$

This shows

$$F(\Phi(v), \sigma) \neq v, \quad \forall v \in \partial \hat{B}_R(0), \ \forall \sigma \in [0, 1]$$

and so (11.2.6) follows. Moreover, as the solution of problem (11.2.8), (11.2.9), the classical theory gives that $\hat{u} \in C^{2+\alpha, 1+\alpha/2}(\overline{Q}_T)$ and

$$|\hat{u}|_{\alpha, \alpha/2; Q_T} \leq |\hat{u}|_{2+\alpha, 1+\alpha/2; Q_T} \leq C_0 |\varphi|_{2+\alpha, 1+\alpha/2; Q_T} < \frac{R}{2}.$$

Therefore

$$|G(v, \sigma)|_{\alpha, \alpha/2; Q_T} = |v - \sigma \hat{u}|_{\alpha, \alpha/2; Q_T}$$
$$\geq |v|_{\alpha, \alpha/2; Q_T} - \sigma |\hat{u}|_{\alpha, \alpha/2; Q_T}$$
$$> R/2, \quad \forall v \in \partial \hat{B}_R(0), \ \forall \sigma \in [0, 1],$$

which implies (11.2.10). The proof of Theorem 11.2.1 is complete. □

Exercises

1. Consider the first initial-boundary value problem

$$\begin{cases} \dfrac{\partial u}{\partial t} - \Delta u = e^u, & (x,t) \in Q_T = \Omega \times (0,T), \\ u(x,t) = \varphi, & (x,t) \in \partial_p Q_T, \end{cases}$$

where Ω is a bounded domain in $\mathbb{R}^n$ with $\partial\Omega \in C^{2,\alpha}$, $\varphi \in C^{2+\alpha,1+\alpha/2}(\overline{Q}_T)$, $\alpha \in (0,1)$. Prove that the above problem admits at least one classical solution $u \in C^{2+\alpha,1+\alpha/2}(\overline{Q}_T)$ provided $|\varphi|_{2+\alpha,1+\alpha/2;Q_T}$ is sufficiently small.

2. Consider the first initial-boundary value problem

$$\begin{cases} \dfrac{\partial u}{\partial t} - \Delta u = |u|^p, & (x,t) \in Q_T, \\ u(x,t) = \varphi, & (x,t) \in \partial_p Q_T, \end{cases}$$

where $p > 1$, $Q_T = \Omega \times (0,T)$, Ω is a bounded domain in $\mathbb{R}^n$ with $\partial\Omega \in C^{2,\alpha}$, $\alpha \in (0,1)$. Prove that for any $\varphi \in C^{2+\alpha,1+\alpha/2}(\overline{Q}_T)$, there exists $T' \in (0,T]$, such that the above problem admits at least one classical solution $u \in C^{2+\alpha,1+\alpha/2}(\overline{Q}_{T'})$.

Chapter 12

Monotone Method

The method of supersolutions and subsolutions is a powerful tool in establishing existence results for differential equations. What is more, this method can also be applied to systems. The basic idea of this method is to use a supersolution or a subsolution as the initial iteration in a suitable iterative process, so that the resulting sequence of iterations is monotone and converges to a solution of the problem. The underlying monotone iterative scheme can also be used for the computation of numerical solutions when these equations are replaced by suitable finite difference equations.

In this chapter, the method of supersolutions and subsolutions and its associated monotone iteration are introduced for a scalar heat equation and a system of coupled heat equations as two typical examples. Similar argument can be applied to the general parabolic equations and systems and also to elliptic equations and systems.

12.1 Monotone Method for Parabolic Problems

We consider the following nonlinear parabolic problem

$$\frac{\partial u}{\partial t} - \Delta u = f(u), \qquad (x,t) \in Q_T = \Omega \times (0,T), \qquad (12.1.1)$$

$$u(x,t) = g(x,t), \qquad (x,t) \in \partial_p Q_T, \qquad (12.1.2)$$

where $\Omega \subset \mathbb{R}^n$ is a bounded domain with $\partial\Omega \in C^{2,\alpha}$, $f \in C^{\alpha}(\mathbb{R})$ and $g \in C^{2+\alpha,1+\alpha/2}(\overline{Q}_T)$ with some $0 < \alpha < 1$. In this chapter, we merely consider classical solutions in $C^{2,1}(Q_T) \cap C(\overline{Q}_T)$.

12.1.1   *Definition of supersolutions and subsolutions*

Definition 12.1.1   A function $\tilde{u} \in C^{2,1}(Q_T) \cap C(\overline{Q}_T)$ is called a super-solution of problem (12.1.1), (12.1.2), if

$$\frac{\partial \tilde{u}}{\partial t} - \Delta \tilde{u} \geq f(\tilde{u}), \qquad\qquad (x,t) \in Q_T,$$

$$\tilde{u}(x,t) \geq g(x,t), \qquad\qquad (x,t) \in \partial_p Q_T.$$

Similarly, a function $\underline{u} \in C^{2,1}(Q_T) \cap C(\overline{Q}_T)$ is called a subsolution of problem (12.1.1), (12.1.2) if

$$\frac{\partial \underline{u}}{\partial t} - \Delta \underline{u} \leq f(\underline{u}), \qquad\qquad (x,t) \in Q_T,$$

$$\underline{u}(x,t) \leq g(x,t), \qquad\qquad (x,t) \in \partial_p Q_T.$$

For a supersolution $\tilde{u}$ and a subsolution $\underline{u}$ of problem (12.1.1), (12.1.2), we say that the pair $\tilde{u}$, $\underline{u}$ are ordered if

$$\tilde{u}(x,t) \geq \underline{u}(x,t), \quad (x,t) \in \overline{Q}_T.$$

Definition 12.1.2   For any ordered supersolution and subsolution $\tilde{u}$, $\underline{u}$, we define the sector $\langle \underline{u}, \tilde{u} \rangle$ as the functional interval

$$\langle \underline{u}, \tilde{u} \rangle = \left\{ u \in C(\overline{Q}_T); \underline{u}(x,t) \leq u(x,t) \leq \tilde{u}(x,t), \quad (x,t) \in \overline{Q}_T \right\}.$$

12.1.2   *Iteration and monotone property*

It is clear that every solution of problem (12.1.1), (12.1.2) in $C^{2,1}(Q_T) \cap C(\overline{Q}_T)$ is a supersolution as well as a subsolution. Therefore, supersolutions and subsolutions exist unless the problem has no solution in $C^{2,1}(Q_T) \cap C(\overline{Q}_T)$. To ensure the existence of a solution it is necessary to impose more condition on the reaction function f. A basic assumption is the following one-sided Lipschitz condition

$$f(u_1) - f(u_2) \geq -\underline{c}(u_1 - u_2), \quad \underline{u} \leq u_2 \leq u_1 \leq \tilde{u}, \qquad (12.1.3)$$

where $\underline{c}$ is a constant and $\tilde{u}$, $\underline{u}$ are given ordered supersolution and subsolution. Clearly this condition is satisfied with $\underline{c} = 0$ when f is monotone nondecreasing in $\mathbb{R}$. In view of (12.1.3), the function

$$F(u) = \underline{c}u + f(u)$$

is monotone nondecreasing in u for $u \in \langle \underline{u}, \tilde{u} \rangle$.

Adding $\underline{c}u$ on both sides of equation (12.1.1) and choosing a suitable initial iteration $u^{(0)} \in C^{2,1}(Q_T) \cap C(\overline{Q}_T)$, we construct a sequence $\{u^{(k)}\}_{k=0}^{\infty}$ successively from the iteration process

$$\frac{\partial u^{(k)}}{\partial t} - \Delta u^{(k)} + \underline{c}u^{(k)} = F(u^{(k-1)}), \qquad (x,t) \in Q_T, \qquad (12.1.4)$$

$$u^{(k)}(x,t) = g(x,t), \qquad (x,t) \in \partial_p Q_T. \qquad (12.1.5)$$

Since for each $k \geq 1$ the right side of (12.1.4) is known, the L^2 theory and the maximum principle guarantee that the sequence $\{u^{(k)}\}_{k=0}^{\infty}$ is well defined. From the regularity of solutions of heat equations,

$$u^{(1)} \in C^{\alpha,\alpha/2}(\overline{Q}_T), \quad u^{(k)} \in C^{2+\alpha,1+\alpha/2}(\overline{Q}_T) \text{ for } k = 2, 3, \cdots.$$

A natural choice of $u^{(0)}$ is $u^{(0)} = \widetilde{u}$ and $u^{(0)} = \underline{u}$. Denote the sequences defined by (12.1.4), (12.1.5) with $u^{(0)} = \widetilde{u}$ and $u^{(0)} = \underline{u}$ by $\{\overline{u}^{(k)}\}_{k=0}^{\infty}$ and $\{\underline{u}^{(k)}\}_{k=0}^{\infty}$, and refer to them as the upper sequence and the lower sequence of (12.1.4), (12.1.5), respectively. The following lemma presents the monotone property of these two sequences.

Lemma 12.1.1 *Let $\widetilde{u}$, $\underline{u}$ be ordered supersolution and subsolution of problem (12.1.1), (12.1.2) and f satisfy (12.1.3). Then the sequences $\{\overline{u}^{(k)}\}_{k=0}^{\infty}$ and $\{\underline{u}^{(k)}\}_{k=0}^{\infty}$ possess the monotone property*

$$\underline{u}(x,t) = \underline{u}^{(0)}(x,t) \leq \underline{u}^{(k)}(x,t) \leq \underline{u}^{(k+1)}(x,t)$$

$$\leq \overline{u}^{(k+1)}(x,t) \leq \overline{u}^{(k)}(x,t) \leq \overline{u}^{(0)}(x,t) = \widetilde{u}(x,t), \quad (x,t) \in \overline{Q}_T \quad (12.1.6)$$

for every $k = 1, 2, \cdots$.

Proof. Let

$$w(x,t) = \overline{u}^{(0)}(x,t) - \overline{u}^{(1)}(x,t) = \widetilde{u}(x,t) - \overline{u}^{(1)}(x,t), \quad (x,t) \in \overline{Q}_T.$$

Then $w \in C^{2,1}(Q_T) \cap C(\overline{Q}_T)$ is a solution of the problem

$$\frac{\partial w}{\partial t} - \Delta w + \underline{c}w \geq F(\widetilde{u}) - F(\widetilde{u}) = 0, \qquad (x,t) \in Q_T,$$

$$w(x,t) \geq g(x,t) - g(x,t) = 0, \qquad (x,t) \in \partial_p Q_T.$$

The maximum principle leads to $w \geq 0$ on $\overline{Q}_T$, i.e.

$$\overline{u}^{(1)}(x,t) \leq \overline{u}^{(0)}(x,t) = \widetilde{u}(x,t), \quad (x,t) \in \overline{Q}_T.$$

Similarly

$$\underline{u}^{(1)}(x,t) \geq \underline{u}^{(0)}(x,t) = \underline{u}(x,t), \quad (x,t) \in \overline{Q}_T.$$

Let

$$w^{(1)}(x,t) = \overline{u}^{(1)}(x,t) - \underline{u}^{(1)}(x,t), \quad (x,t) \in \overline{Q}_T.$$

Then $w^{(1)} \in C^{2,1}(Q_T) \cap C(\overline{Q}_T)$ satisfies

$$\frac{\partial w^{(1)}}{\partial t} - \Delta w^{(1)} + \underline{c} w^{(1)} = F(\widetilde{u}) - F(\underline{u}) \geq 0, \qquad (x,t) \in Q_T,$$

$$w^{(1)}(x,t) = g(x,t) - g(x,t) = 0, \qquad (x,t) \in \partial_p Q_T.$$

Again, by the maximum principle, $w^{(1)} \geq 0$ on $\overline{Q}_T$, i.e.

$$\underline{u}^{(1)}(x,t) \leq \overline{u}^{(1)}(x,t), \qquad (x,t) \in \overline{Q}_T.$$

Then, we have

$$\underline{u}^{(0)}(x,t) = \underline{u}(x,t) \leq \underline{u}^{(1)}(x,t) \leq \overline{u}^{(1)}(x,t) \leq \widetilde{u}(x,t) = \overline{u}^{(0)}(x,t),$$
$$(x,t) \in \overline{Q}_T.$$

Suppose

$$\underline{u}^{(k-1)}(x,t) \leq \underline{u}^{(k)}(x,t) \leq \overline{u}^{(k)}(x,t) \leq \overline{u}^{(k-1)}(x,t), \quad (x,t) \in \overline{Q}_T$$

for some $k \geq 1$. Then the function

$$w^{(k)}(x,t) = \overline{u}^{(k)}(x,t) - \overline{u}^{(k+1)}(x,t), \quad (x,t) \in \overline{Q}_T$$

satisfies

$$\frac{\partial w^{(k)}}{\partial t} - \Delta w^{(k)} + \underline{c} w^{(k)} = F(\overline{u}^{(k-1)}) - F(\overline{u}^k) \geq 0, \qquad (x,t) \in Q_T,$$

$$w^{(k)}(x,t) = g(x,t) - g(x,t) = 0, \qquad (x,t) \in \partial_p Q_T.$$

The maximum principle implies that $w^{(k)} \geq 0$ on $\overline{Q}_T$, i.e.

$$\overline{u}^{(k+1)}(x,t) \leq \overline{u}^{(k)}(x,t), \quad (x,t) \in \overline{Q}_T.$$

Similar reasoning gives

$$\underline{u}^{(k+1)}(x,t) \leq \underline{u}^{(k)}(x,t), \quad \underline{u}^{(k+1)}(x,t) \leq \overline{u}^{(k+1)}(x,t), \quad (x,t) \in \overline{Q}_T.$$

Thus, by induction, the monotone property (12.1.6) follows. $\qquad \square$

12.1.3 Existence results

The relation (12.1.6) implies that the upper sequence $\{\overline{u}^{(k)}\}_{k=0}^{\infty}$ is monotone nonincreasing and is bounded from below and that the lower sequence $\{\underline{u}^{(k)}\}_{k=0}^{\infty}$ is monotone nondecreasing and is bounded from above. Hence the limits

$$\lim_{k \to \infty} \overline{u}^{(k)}(x,t) = \overline{u}(x,t), \quad \lim_{k \to \infty} \underline{u}^{(k)}(x,t) = \underline{u}(x,t), \quad (x,t) \in \overline{Q}_T \quad (12.1.7)$$

exist and satisfy

$$\underline{\widetilde{u}}(x,t) \leq \underline{u}(x,t) \leq \overline{u}(x,t) \leq \widetilde{u}(x,t), \quad (x,t) \in \overline{Q}_T.$$

We will show that both $\overline{u}$ and $\underline{u}$ are solutions of problem (12.1.1), (12.1.2). Furthermore, if there exists a constant $\overline{c} \leq \underline{c}$ such that

$$f(u_1) - f(u_2) \leq -\overline{c}(u_1 - u_2), \quad \underline{\widetilde{u}} \leq u_2 \leq u_1 \leq \widetilde{u}, \quad (12.1.8)$$

then the solution is also unique in $\langle \underline{\widetilde{u}}, \widetilde{u} \rangle$.

Theorem 12.1.1 *Let $\widetilde{u}$, $\underline{\widetilde{u}}$ be ordered supersolution and subsolution of problem (12.1.1), (12.1.2) and f satisfy (12.1.3). Then*

i) The upper sequence $\{\overline{u}^{(k)}\}_{k=0}^{\infty}$ converges monotonically from above to a solution $\overline{u}$ and the lower sequence $\{\underline{u}^{(k)}\}_{k=0}^{\infty}$ converges monotonically from below to a solution $\underline{u}$, and

$$\overline{u}(x,t) \geq \underline{u}(x,t), \quad (x,t) \in \overline{Q}_T; \quad (12.1.9)$$

ii) Any solution $u^* \in \langle \underline{\widetilde{u}}, \widetilde{u} \rangle$ of problem (12.1.1), (12.1.2) satisfies

$$\underline{u}(x,t) \leq u^*(x,t) \leq \overline{u}(x,t), \quad (x,t) \in \overline{Q}_T;$$

iii) If, in addition, the condition (12.1.8) holds, then $\overline{u} = \underline{u}$ and is the unique solution in $\langle \underline{\widetilde{u}}, \widetilde{u} \rangle$.

Proof. Let $\{u^{(k)}\}_{k=0}^{\infty}$ be either $\{\overline{u}^{(k)}\}_{k=0}^{\infty}$ or $\{\underline{u}^{(k)}\}_{k=0}^{\infty}$ and u be $\overline{u}$ or $\underline{u}$ respectively. Since F is Hölder continuous and monotone nondecreasing, the monotone convergence of $\{u^{(k)}\}_{k=0}^{\infty}$ to u implies that $\{F(u^{(k)})\}_{k=0}^{\infty}$ converges to $F(u)$ as $k \to \infty$. As indicated above, we have

$$u^{(1)} \in C^{\alpha, \alpha/2}(\overline{Q}_T), \quad u^{(k)} \in C^{2+\alpha, 1+\alpha/2}(\overline{Q}_T) \text{ for } k = 2, 3, \cdots.$$

Moreover, it follows from the maximum principle and Schauder's estimate that

$$|u^{(k)}|_{2+\alpha, 1+\alpha/2; Q_T} \leq C(|g|_{2+\alpha, 1+\alpha/2; Q_T} + |u^{(k-1)}|_{0; Q_T}), \quad k = 2, 3, \cdots,$$

where $C > 0$ is a constant depending only on α, Ω, T and f but independent of k. From the monotone property (12.1.6), $\{u^{(k)}\}_{k=1}^{\infty}$ is uniformly bounded in $C^{2+\alpha, 1+\alpha/2}(\overline{Q}_T)$. Therefore, $u \in C^{2+\alpha, 1+\alpha/2}(\overline{Q}_T)$ is a solution of problem (12.1.1), (12.1.2). And (12.1.9) follows from the monotone property (12.1.6).

If $u^* \in \langle \underline{u}, \widetilde{u} \rangle$ is a solution of problem (12.1.1), (12.1.2), then the functions u^*, $\underline{u}$ are ordered supersolution and subsolution. Since the sequence $\{u^{(k)}\}_{k=0}^{\infty}$ with $u^{(0)} = u^*$ consists of the same function u^* for every k, the above conclusion implies that $u^* \geq \underline{u}$. Similarly, by considering $\widetilde{u}$, u^* as ordered supersolution and subsolution, the same reasoning leads to $\overline{u} \geq u^*$. This proves ii). To prove iii), it suffices to show that

$$\overline{u}(x, t) \leq \underline{u}(x, t), \quad (x, t) \in \overline{Q}_T. \tag{12.1.10}$$

Indeed, the function

$$w(x, t) = \underline{u}(x, t) - \overline{u}(x, t), \quad (x, t) \in \overline{Q}_T$$

satisfies

$$\frac{\partial w}{\partial t} - \Delta w = f(\underline{u}) - f(\overline{u}) \geq -\overline{c}w, \qquad (x, t) \in Q_T,$$

$$w(x, t) = g(x, t) - g(x, t) = 0, \qquad (x, t) \in \partial_p Q_T$$

and hence, by the maximum principle, $w \geq 0$ on $\overline{Q}_T$ and (12.1.10) follows immediately. $\qquad \square$

In the conditions (12.1.3) and (12.1.8) the constants $\underline{c}$ and $\overline{c}$ are not necessarily nonnegative. This is different from the case of elliptic problem. When f is a C^1-function in $\langle \underline{u}, \widetilde{u} \rangle$, we may take these constants as

$$\underline{c} = -\min\{f'(u(x, t)); u \in \langle \underline{u}, \widetilde{u} \rangle, (x, t) \in \overline{Q}_T\}$$

and

$$\overline{c} = -\max\{f'(u(x, t)); u \in \langle \underline{u}, \widetilde{u} \rangle, (x, t) \in \overline{Q}_T\}.$$

If f is Lipschitz continuous in $\langle \underline{u}, \widetilde{u} \rangle$, namely there exists a constant $K \geq 0$ such that

$$|f(u_1) - f(u_2)| \leq K|u_1 - u_2|, \quad u_1, u_2 \in \langle \underline{u}, \widetilde{u} \rangle.$$

Then we may take $\underline{c} = K$ and $\overline{c} = -K$. This observation leads to

Corollary 12.1.1 *Let $\widetilde{u}$, $\underline{u}$ be ordered supersolution and subsolution of problem (12.1.1), (12.1.2) and f be a C^1-function in $\langle \underline{u}, \widetilde{u} \rangle$. Then problem*

(12.1.1), (12.1.2) *has a unique solution in* $\langle \underset{\sim}{u}, \widetilde{u} \rangle$. *Moreover this solution is the limit of the sequence defined by* (12.1.4), (12.1.5) *with either* $u^{(0)} = \widetilde{u}$ *or* $u^{(0)} = \underset{\sim}{u}$. *If* f *is Lipschitz continuous in* $\langle \underset{\sim}{u}, \widetilde{u} \rangle$, *the same conclusion holds.*

If both f and g are nonnegative functions, then the trivial function $\underset{\sim}{u} = 0$ is a subsolution of problem (12.1.1), (12.1.2). Hence the existence of solutions is valid provided that there is a nonnegative supersolution. A sufficient condition is that for some constant $\rho \geq 0$,

$$f(\rho) \leq 0, \quad \rho \geq g(x,t), \quad (x,t) \in \partial_p Q_T. \tag{12.1.11}$$

This follows immediately from Definition 12.1.1 with $\widetilde{u} = \rho$. By an application of Theorem 12.1.1 we have the following conclusion, which is quite useful in applications.

Theorem 12.1.2 *Let* $\widetilde{u}$ *be a nonnegative supersolution of problem* (12.1.1), (12.1.2) *and* f *be a* C^1-*function in* $\langle 0, \widetilde{u} \rangle$. *If*

$$f(0) \geq 0, \quad g(x,t) \geq 0, \quad (x,t) \in \partial_p Q_T,$$

then there exists a unique solution of problem (12.1.1), (12.1.2) *in* $\langle 0, \widetilde{u} \rangle$. *If* (12.1.11) *holds for some constant* $\rho \geq 0$, *then* $\widetilde{u} = \rho$ *is a nonnegative supersolution.*

To achieve the conclusion of Theorem 12.1.1, the existence of ordered supersolution and subsolution is necessary. In the following, we will show that under the conditions (12.1.3) and (12.1.8) any supersolution and subsolution of problem (12.1.1), (12.1.2) are ordered and $\overline{u}^{(k)}$, $\underline{u}^{(k)}$ are ordered supersolution and subsolution for each $k = 1, 2, \cdots$.

Theorem 12.1.3 *Let* $\widetilde{u}$ *and* $\underset{\sim}{u}$ *be a supersolution and a subsolution of problem* (12.1.1), (12.1.2) *respectively. Assume that* f *satisfies* (12.1.3) *and* (12.1.8) *for any* u_1 *and* u_2 *between* $\underset{\sim}{u}$ *and* $\widetilde{u}$ *with* $u_2 \leq u_1$. *Then*

$$\underset{\sim}{u}(x,t) \leq \widetilde{u}(x,t), \quad (x,t) \in \overline{Q}_T.$$

Thus, $\widetilde{u}$, $\underset{\sim}{u}$ *are ordered supersolution and subsolution of problem* (12.1.1), (12.1.2). *Moreover,* $\overline{u}^{(k)}$, $\underline{u}^{(k)}$ *are also ordered supersolution and subsolution for each* $k = 1, 2, \cdots$.

Proof. Let

$$c^* = \max\{|\underline{c}|, |\overline{c}|\}, \quad w(x,t) = \widetilde{u}(x,t) - \underset{\sim}{u}(x,t), \quad (x,t) \in \overline{Q}_T,$$

where $\underline{c}$ and $\overline{c}$ are the constants in (12.1.3) and (12.1.8). Define

$$c(x,t) = c^* \text{sgn} w(x,t), \quad (x,t) \in \overline{Q}_T,$$

where $\text{sgn}(\cdot)$ is the sign function. Then

$$\frac{\partial w}{\partial t} - \Delta w \geq f(\tilde{u}) - f(\underline{u}) \geq -cw, \qquad (x,t) \in Q_T,$$

$$w(x,t) \geq g(x,t) - g(x,t) = 0, \qquad (x,t) \in \partial_p Q_T.$$

Since c is bounded on $\overline{Q}_T$, the maximum principle implies $w \geq 0$ on $\overline{Q}_T$, i.e.

$$\underline{u}(x,t) \leq \tilde{u}(x,t), \quad (x,t) \in \overline{Q}_T.$$

For each $k = 1, 2, \cdots$, by (12.1.4) and (12.1.3),

$$\frac{\partial \overline{u}^{(k)}}{\partial t} - \Delta \overline{u}^{(k)} = -\underline{c}\,\overline{u}^{(k)} + F(\overline{u}^{(k-1)})$$

$$= \left[\underline{c}(\overline{u}^{(k-1)} - \overline{u}^{(k)}) + f(\overline{u}^{(k-1)}) - f(\overline{u}^{(k)}) \right] + f(\overline{u}^{(k)})$$

$$\geq f(\overline{u}^{(k)}), \qquad (x,t) \in Q_T,$$

$$\frac{\partial \underline{u}^{(k)}}{\partial t} - \Delta \underline{u}^{(k)} = -\underline{c}\,\underline{u}^{(k)} + F(\underline{u}^{(k-1)})$$

$$= -\left[\underline{c}(\underline{u}^{(k)} - \underline{u}^{(k-1)}) + f(\underline{u}^{(k)}) - f(\underline{u}^{(k-1)}) \right] + f(\underline{u}^{(k)})$$

$$\leq f(\underline{u}^{(k)}), \qquad (x,t) \in Q_T.$$

On the other hand, (12.1.5) and (12.1.6) imply

$$\overline{u}^{(k)}(x,t) \geq g(x,t), \quad \underline{u}^{(k)}(x,t) \leq g(x,t), \quad (x,t) \in \partial_p Q_T$$

and

$$\overline{u}^{(k)}(x,t) \geq \underline{u}^{(k)}(x,t), \quad (x,t) \in \overline{Q}_T.$$

Hence $\overline{u}^{(k)}, \underline{u}^{(k)}$ are ordered supersolution and subsolution. $\square$

12.1.4 *Application to more general parabolic equations*

The monotone method used above may be applied to the heat equations with more general reaction terms and even to uniformly parabolic equations of general form.

Remark 12.1.1 *For the nonlinear parabolic problem*

$$\frac{\partial u}{\partial t} - \Delta u = f(x,t,u), \qquad\qquad (x,t) \in Q_T, \qquad\qquad (12.1.12)$$

$$u(x,t) = g(x,t), \qquad\qquad (x,t) \in \partial_p Q_T, \qquad\qquad (12.1.13)$$

we may apply the monotone method to get the same results as those for problem (12.1.1), (12.1.2). Here $f \in C^{\alpha,\alpha/2,\alpha}(\overline{Q}_T \times \mathbb{R})$, *and the conditions (12.1.3) and (12.1.8) are replaced by*

$$f(x,t,u_1) - f(x,t,u_2) \geq -\underline{c}(u_1 - u_2),$$
$$(x,t) \in \overline{Q}_T, \ \underline{u} \leq u_2 \leq u_1 \leq \widetilde{u} \qquad (12.1.14)$$

and

$$f(x,t,u_1) - f(x,t,u_2) \leq -\overline{c}(u_1 - u_2),$$
$$(x,t) \in \overline{Q}_T, \ \underline{u} \leq u_2 \leq u_1 \leq \widetilde{u} \qquad (12.1.15)$$

respectively.

Remark 12.1.2 *The monotone method may also be applied to the uniformly parabolic equation of general form*

$$\frac{\partial u}{\partial t} - \sum_{i,j=1}^{n} a_{ij}(x,t)D_{ij}u + \sum_{i=1}^{n} b_i(x,t)D_i u + c(x,t)u = f(x,t,u), \quad (x,t) \in Q_T,$$

where $a_{ij}, b_i, c \in C^{\alpha,\alpha/2}(\overline{Q}_T)$, $a_{ij} = a_{ji}$ *and there exist two positive constants* λ, Λ, *such that*

$$\lambda|\xi|^2 \leq \sum_{i,j=1}^{n} a_{ij}(x,t)\xi_i\xi_j \leq \Lambda|\xi|^2, \quad \forall \xi \in \mathbb{R}^n, \ (x,t) \in Q_T.$$

We may also use the method of supersolutions and subsolutions to establish existence results for elliptic problems.

Remark 12.1.3 *The monotone method may be applied to the following nonlinear elliptic problem*

$$-\Delta u = f(x,u), \qquad\qquad x \in \Omega,$$
$$u(x) = g(x), \qquad\qquad x \in \partial\Omega,$$

where $\Omega \subset \mathbb{R}^n$ *is a bounded domain and* $\partial\Omega \in C^{2,\alpha}$, $f \in C^{\alpha}(\overline{\Omega} \times \mathbb{R})$ *and* $g \in C^{2,\alpha}(\overline{\Omega})$ *with some* $0 < \alpha < 1$. *The conditions corresponding to*

(12.1.12) and (12.1.13) are

$$f(x, u_1) - f(x, u_2) \geq -\underline{c}(u_1 - u_2), \quad x \in \overline{\Omega}, \ \underline{u} \leq u_2 \leq u_1 \leq \widetilde{u}$$

and

$$f(x, u_1) - f(x, u_2) \leq -\overline{c}(u_1 - u_2), \quad x \in \overline{\Omega}, \ \underline{u} \leq u_2 \leq u_1 \leq \widetilde{u}$$

respectively. However, the constants $\underline{c}$ and $\overline{c}$ should be nonnegative due to the same reason as that for linear elliptic equations. Furthermore, this method may also be applied to the uniformly elliptic equation of general form

$$-a_{ij}(x)D_{ij}u + b_i(x)D_i u + c(x)u = f(x, u), \quad x \in \Omega,$$

where $a_{ij}, b_i, c \in C^\alpha(\overline{\Omega})$, $a_{ij} = a_{ji}$ and there exist two positive constants λ, Λ, such that

$$\lambda |\xi|^2 \leq a_{ij}(x)\xi_i \xi_j \leq \Lambda |\xi|^2, \quad \forall \xi \in \mathbb{R}^n, \ x \in \Omega.$$

12.1.5   *Nonuniqueness of solutions*

The existence theorem shows that if f satisfies the left-hand side Lipschitz condition (12.1.14), then problem (12.1.12), (12.1.13) has at least one solution in the sector $\langle \underline{u}, \widetilde{u} \rangle$. This solution is unique if f also satisfies the right-hand side Lipschitz condition (12.1.15). In particular, the existence of a unique solution of problem (12.1.12), (12.1.13) is guaranteed if f is a C^1-function or a Lipschitz continuous function in $u \in \langle \underline{u}, \widetilde{u} \rangle$. However, this uniqueness result is ensured only with respect to the given supersolution and subsolution, and it does not rule out the possibility of other solutions outside the sector $\langle \underline{u}, \widetilde{u} \rangle$. Furthermore, the uniqueness result may not hold when f is not Lipschitz continuous in $u \in \langle \underline{u}, \widetilde{u} \rangle$. In the following discussion, we give some examples to show that if f satisfies the condition (12.1.14) but fails to satisfy (12.1.15) then problem (12.1.12), (12.1.13) may possess more than one solution.

Let us consider the one-dimensional problem

$$u_t - u_{xx} = f(x, u), \quad 0 < x < \pi, \ t > 0, \tag{12.1.16}$$

$$u(x, t) = 0, \quad (x, t) \in (\{0, \pi\} \times (0, +\infty)) \cup ((0, \pi) \times \{0\}). \tag{12.1.17}$$

Any nontrivial solution of problem (12.1.16), (12.1.17) must be spatially dependent. Consider the function

$$f(x, u) = u + 3\sin^{2/3}(x/2)u^{1/3}, \quad 0 \le x \le \pi, \, u \in \mathbb{R}.$$

Clearly this function is Hölder continuous and is nondecreasing in $u \in \mathbb{R}$. This implies that f satisfies (12.1.14) for all $-\infty < u_2 \le u_1 < +\infty$ with $\underline{c} = 0$. We seek some ordered supersolution and subsolution of the form

$$\widetilde{u}(x, t) = \rho t^{3/2} \sin x, \quad \underset{\sim}{u}(x, t) = -\rho t^{3/2} \sin x, \quad 0 \le x \le \pi, \, t \ge 0$$

with $\rho > 1$. Since $\widetilde{u}$ and $\underset{\sim}{u}$ satisfy the boundary condition (12.1.17), it suffices to verify the differential inequality. In view of the relation

$$\widetilde{u}_t - \widetilde{u}_{xx} = \rho\left(\frac{3}{2}t^{1/2} + t^{3/2}\right)\sin x = \frac{3}{2}\rho t^{1/2}\sin x + \widetilde{u}, \quad 0 < x < \pi, \, t > 0,$$

$\widetilde{u}$ is a supersolution if

$$\frac{3}{2}\rho t^{1/2}\sin x + \widetilde{u} \ge \widetilde{u} + 3\sin^{2/3}(x/2)(\rho t^{3/2}\sin x)^{1/3}, \quad 0 < x < \pi, \, t > 0.$$

This inequality is equivalent to $\rho \ge \rho^{1/3}$, which is clearly satisfied by any $\rho > 1$. The same argument shows that $\underset{\sim}{u}$ is a subsolution. Therefore, there exists at least one solution u of problem (12.1.16), (12.1.17) such that

$$-\rho t^{3/2}\sin x \le u(x, t) \le \rho t^{3/2}\sin x, \quad 0 \le x \le \pi, \, t \ge 0.$$

However, all the three functions

$$u_1(x, t) = -t^{3/2}\sin x, \, u_2(x, t) = 0, \, u_3(x, t) = t^{3/2}\sin x, \, 0 \le x \le \pi, \, t \ge 0$$

are true solutions of problem (12.1.16), (12.1.17) in the sector $\langle 0, \widetilde{u}\rangle$. In fact, for each $t_0 \ge 0$ the function

$$u(x, t) = \begin{cases} 0, & \text{when } 0 \le x \le \pi, \, 0 \le t \le t_0, \\ (t - t_0)^{3/2}\sin x, & \text{when } 0 \le x \le \pi, \, t > t_0 \end{cases}$$

is also a solution, so that the problem has infinitely many solutions. This nonuniqueness result is due to the fact that f does not satisfy a right-hand side Lipschitz condition (12.1.15) in $\langle 0, \widetilde{u}\rangle$. It should be noted that f is a C^1-function in each of the intervals $(-\infty, 0)$ and $(0, +\infty)$, so that the negative solution u_1 and the positive solution u_3 are unique in their respective sectors.

The nonuniqueness results for the one-dimensional model can be extended to problem (12.1.12), (12.1.13) in an arbitrary bounded domain $\Omega \subset \mathbb{R}^n$. Consider, for simplicity, the case where

$$f(x,0) = 0, \quad x \in \overline{\Omega} \tag{12.1.18}$$

and

$$g(x,t) = 0, \quad (x,t) \in \partial_p Q_T. \tag{12.1.19}$$

Then $u = 0$ is always a solution of problem (12.1.12), (12.1.13). To show the existence of another solution we define

$$\overline{f}(u) = \sup\{f(x,u); x \in \Omega\}$$

and consider the Cauchy problem

$$p'(t) = \overline{f}(p(t)), \quad p(0) = p_0 \tag{12.1.20}$$

with $p_0 > 0$. By the continuity of f there exists $T^* \leq +\infty$ such that this problem has at least one solution $p(t)$ in $[0, T^*)$. In the following theorem we give a sufficient condition on f for problem (12.1.12), (12.1.13) to have at least one positive solution in Q_T for any $T < T^*$.

Theorem 12.1.4 *Let f be Hölder continuous and satisfy (12.1.14) for $0 \leq u_2 \leq u_1$, and let (12.1.18) and (12.1.19) hold. If there exist a constant σ_0 and positive constants σ, γ with $\gamma < 1$ such that*

$$f(x,u) \geq -\sigma_0 u + \sigma u^\gamma, \quad x \in \overline{\Omega}, \ u \geq 0, \tag{12.1.21}$$

then for any $T < T^*$, problem (12.1.12), (12.1.13) has the trivial solution $u_1 = 0$ and a positive solution $u_2(x,t)$ in Q_T. In fact, there are infinitely many solutions to problem (12.1.12), (12.1.13).

Proof. Let

$$\underline{u}(x,t) = e^{-\beta t} q(t)\phi(x), \quad (x,t) \in \overline{Q}_T$$

with $\beta = \sigma_0 + \lambda_0$, where $\lambda_0 > 0$ is the smallest eigenvalue of the problem

$$\begin{cases} \Delta\phi(x) + \lambda\phi(x) = 0, & x \in \Omega, \\ \phi(x) = 0, & x \in \partial\Omega, \end{cases}$$

ϕ is its corresponding normalized eigenfunction, and q, determined below, is a positive function with $q(0) = 0$. Since $\underline{u}\big|_{\partial_p Q_T} = 0$, $\underline{u}$ is a subsolution if

$$e^{-\beta t}(q'(t) - \beta q(t))\phi(x) - e^{-\beta t}q(t)\Delta\phi(x) \leq f(x, e^{-\beta t}q(t)\phi(x)), \quad (x,t) \in Q_T,$$

which is equivalent to

$$(q'(t) - \sigma_0 q(t))\phi(x) \leq e^{\beta t}f(x, e^{-\beta t}q(t)\phi(x)), \quad (x,t) \in Q_T.$$

In view of the hypothesis (12.1.21), it suffices to find $q \geq 0$ such that

$$(q'(t) - \sigma_0 q(t))\phi(x) \leq e^{\beta t}\left[-\sigma_0 e^{-\beta t}q(t)\phi(x) + \sigma(e^{-\beta t}q(t)\phi(x))^\gamma\right], \quad (x,t) \in Q_T$$

or, equivalently,

$$q'(t)\phi^{1-\gamma}(x) \leq \sigma e^{\beta(1-\gamma)t}q^\gamma(t), \quad (x,t) \in Q_T.$$

Since $0 < \phi \leq 1$ and $\gamma < 1$, the above inequality is satisfied by any function $q \geq 0$ which is a solution of the Cauchy problem

$$q'(t) = \sigma q^\gamma(t), \quad q(0) = 0. \tag{12.1.22}$$

A positive solution of this problem is given by

$$q(t) = (\sigma(1-\gamma)t)^{1/(1-\gamma)}, \quad t \geq 0.$$

With this choice of q, $\underline{u}$ is a positive subsolution.

We next seek a positive supersolution by letting

$$\widetilde{u}(x,t) = p(t), \quad (x,t) \in \overline{Q}_T,$$

where p is the solution of problem (12.1.20). Clearly, $\widetilde{u}\big|_{\partial_p Q_T} \geq 0$ and

$$\widetilde{u}_t - \Delta\widetilde{u} = p'(t) = \overline{f}(p(t)) \geq f(x, \widetilde{u}), \quad (x,t) \in Q_T.$$

This implies that p is a supersolution. By (12.1.21), the function $z(t) = e^{\beta t}p(t)$ satisfies the relation

$$\begin{aligned} z'(t) &= e^{\beta t}(p'(t) + \beta p(t)) \geq e^{\beta t}(f(x, p(t)) + \beta p(t)) \\ &\geq \sigma e^{\beta t}p^\gamma(t) \geq \sigma z^\gamma(t), \quad t > 0. \end{aligned} \tag{12.1.23}$$

A comparison between (12.1.22) and (12.1.23) shows that $z(t) \geq q(t)$ in $[0, +\infty)$ and thus problem (12.1.20) has a positive solution p such that

$$e^{\beta t}p(t) \geq q(t), \quad t \geq 0.$$

Therefore, the pair $\widetilde{u}(x,t) = p(t)$, $\underset{\sim}{u}(x,t) = e^{-\beta t}q(t)\phi(x)$ are ordered supersolution and subsolution. Hence problem (12.1.12), (12.1.13) has at least one positive solution u_2 in the section $\langle e^{-\beta t}q(t)\phi(x), p(t)\rangle$. This proves the existence of two solutions of problem (12.1.12), (12.1.13), $u_1 = 0$ and u_2. It is easily seen that for each $t_0 \geq 0$ the function

$$u(x,t) = \begin{cases} 0, & \text{when } x \in \Omega,\ 0 \leq t \leq t_0, \\ u_2(x, t - t_0), & \text{when } x \in \Omega,\ t_0 < t < T^* \end{cases}$$

is also a solution. This shows that problem (12.1.12), (12.1.13) has infinitely many solutions. $\qquad\square$

It is seen form construction of the supersolution and subsolution in the proof of Theorem 12.1.4 that if the condition (12.1.21) is satisfied only for $u \in [0, \rho]$ with some $\rho > 0$, then there is a $T_\rho < T^*$ such that the solution q of problem (12.1.22) exists and is bounded by ρ on $[0, T_\rho]$. This implies that $\widetilde{u}(x,t) = p(t)$ and $\underset{\sim}{u}(x,t) = e^{-\beta t}q(t)\phi(x)$ are ordered supersolution and subsolution in Q_{T_ρ}. As a consequence, we have

Corollary 12.1.2 *Let the hypotheses of Theorem 12.1.4 be satisfied except that the condition (12.1.21) holds only for $u \in [0, \rho]$, where ρ is a positive constant. Then there exists $T_\rho < T^*$ such that all the conclusions in Theorem 12.1.4 hold in Q_{T_ρ}.*

12.2 Monotone Method for Coupled Parabolic Systems

The monotone method and its associated supersolution and subsolution for scalar equations, discussed in the previous section, can be extended to coupled systems of parabolic and elliptic equations. However, for coupled systems of equations, the definition of supersolutions and subsolutions and the construction of monotone sequences depend on the quasimonotone property of the reaction functions in the system. To illustrate the basic idea of the method, we consider a coupled system of two parabolic equations of the form

$$\frac{\partial u_1}{\partial t} - \Delta u_1 = f_1(u_1, u_2), \qquad (x,t) \in Q_T = \Omega \times (0,T), \qquad (12.2.1)$$

$$\frac{\partial u_2}{\partial t} - \Delta u_2 = f_2(u_1, u_2), \qquad (x,t) \in Q_T = \Omega \times (0,T), \qquad (12.2.2)$$

$$u_1(x,t) = g_1(x,t), \qquad (x,t) \in \partial_p Q_T, \qquad (12.2.3)$$

$$u_2(x,t) = g_2(x,t), \qquad (x,t) \in \partial_p Q_T, \qquad (12.2.4)$$

where $\Omega \subset \mathbb{R}^n$ is a bounded domain with $\partial\Omega \in C^{2,\alpha}$, $f_i \in C^\alpha(\mathbb{R}^2)$ and $g_i \in C^{2+\alpha,1+\alpha/2}(\overline{Q}_T)$ with some $0 < \alpha < 1$ for each $i = 1, 2$.

12.2.1 *Quasimonotone reaction functions*

Let J_i $(i = 1, 2)$ be open sets of $\mathbb{R}$.

Definition 12.2.1 A function $f_i = f_i(u_1, u_2)$ $(i = 1, 2)$ is said to be quasimonotone nondecreasing (quasimonotone nonincreasing) in $J_1 \times J_2$ if for any fixed $u_i \in J_i$, f_i is nondecreasing (nonincreasing) in $u_j \in J_j$ for $j \neq i$.

Definition 12.2.2 A vector function $\mathbf{f} = (f_1, f_2)$ is said to be quasimonotone nondecreasing (quasimonotone nonincreasing) in $J_1 \times J_2$, if both f_1 and f_2 are quasimonotone nondecreasing (quasimonotone nonincreasing) in $J_1 \times J_2$. If f_1 is quasimonotone nonincreasing and f_2 is quasimonotone nondecreasing in $J_1 \times J_2$ (or vice versa), then $\mathbf{f}$ is said to be mixed quasimonotone. The function $\mathbf{f}$ is said to be quasimonotone in $J_1 \times J_2$ if it has any one of the above quasimonotone properties.

As usual, we call $\mathbf{f}$ a C^γ-function $(0 \leq \gamma \leq 1)$ in $J_1 \times J_2$ if $f_1 \in C^\gamma$, $f_2 \in C^\gamma$. If $f_1(u_1, \cdot)$ is continuously differentiable in J_2 for any $u_1 \in J_1$ and $f_2(\cdot, u_2)$ is continuously differentiable in J_1 for any $u_2 \in J_2$, then we call $f = (f_1, f_2)$ a quasi C^1-function in $J_1 \times J_2$. If $\mathbf{f}$ is a C^1-function or a quasi C^1-function, then the three types of quasimonotone functions in Definition 12.2.2 are corresponding to

$$\frac{\partial f_1}{\partial u_2} \geq 0, \quad \frac{\partial f_2}{\partial u_1} \geq 0, \quad (u_1, u_2) \in J_1 \times J_2,$$

$$\frac{\partial f_1}{\partial u_2} \leq 0, \quad \frac{\partial f_2}{\partial u_1} \leq 0, \quad (u_1, u_2) \in J_1 \times J_2$$

and

$$\frac{\partial f_1}{\partial u_2} \leq 0, \quad \frac{\partial f_2}{\partial u_1} \geq 0, \quad (u_1, u_2) \in J_1 \times J_2 \quad \text{(or vice versa)}$$

respectively. These three types of reaction functions appear most often in many physical problems.

12.2.2 *Definition of supersolutions and subsolutions*

Suppose the reaction function $\mathbf{f} = (f_1, f_2)$ defined in $\mathbb{R}^2$ possesses the quasimonotone properties described in Definition 12.2.2. Then we can extend the monotone method for scalar equations to the coupled system (12.2.1)–(12.2.4) using a supersolution and subsolution as the initial iterations. The supersolution and subsolution, denoted by $\widetilde{\mathbf{u}} = (\widetilde{u}_1, \widetilde{u}_2)$ and $\underline{\mathbf{u}} = (\underline{u}_1, \underline{u}_2)$, respectively, are required to satisfy the boundary inequality

$$\widetilde{\mathbf{u}}(x,t) \geq \mathbf{g}(x,t) \geq \underline{\mathbf{u}}(x,t), \quad (x,t) \in \partial_p Q_T, \tag{12.2.5}$$

where $\mathbf{g} = (g_1, g_2)$. The inequality $\mathbf{u} = (u_1, u_2) \geq \mathbf{v} = (v_1, v_2)$ means that $u_1 \geq u_2$, $v_1 \geq v_2$.

Similar to scalar problems, the supersolution $\widetilde{\mathbf{u}}$ and subsolution $\underline{\mathbf{u}}$ are defined by differential inequalities. However, the form of differential inequalities for $\widetilde{\mathbf{u}}$ and $\underline{\mathbf{u}}$ depends on the different quasimonotone property of $\mathbf{f}$. For definiteness, we always consider the case that f_1 is quasimonotone nonincreasing and f_2 is quasimonotone nondecreasing when $\mathbf{f}$ is mixed quasimonotone.

Definition 12.2.3 A pair of functions $\widetilde{\mathbf{u}} = (\widetilde{u}_1, \widetilde{u}_2)$ and $\underline{\mathbf{u}} = (\underline{u}_1, \underline{u}_2)$ in $C^{2,1}(Q_T) \cap C(\overline{Q}_T)$ are called ordered supersolution and subsolution of problem (12.2.1)–(12.2.4), if they satisfy

$$\widetilde{\mathbf{u}}(x,t) \geq \underline{\mathbf{u}}(x,t), \quad (x,t) \in \overline{Q}_T$$

and (12.2.5) and if

$$\begin{cases} \dfrac{\partial \widetilde{u}_1}{\partial t} - \Delta \widetilde{u}_1 - f_1(\widetilde{u}_1, \widetilde{u}_2) \geq 0 \geq \dfrac{\partial \underline{u}_1}{\partial t} - \Delta \underline{u}_1 - f_1(\underline{u}_1, \underline{u}_2), \\ \dfrac{\partial \widetilde{u}_2}{\partial t} - \Delta \widetilde{u}_2 - f_2(\widetilde{u}_1, \widetilde{u}_2) \geq 0 \geq \dfrac{\partial \underline{u}_2}{\partial t} - \Delta \underline{u}_2 - f_2(\underline{u}_1, \underline{u}_2), \end{cases} \quad (x,t) \in Q_T \tag{12.2.6}$$

when (f_1, f_2) is quasimonotone nondecreasing;

$$\begin{cases} \dfrac{\partial \widetilde{u}_1}{\partial t} - \Delta \widetilde{u}_1 - f_1(\widetilde{u}_1, \underline{u}_2) \geq 0 \geq \dfrac{\partial \underline{u}_1}{\partial t} - \Delta \underline{u}_1 - f_1(\underline{u}_1, \widetilde{u}_2), \\ \dfrac{\partial \widetilde{u}_2}{\partial t} - \Delta \widetilde{u}_2 - f_2(\underline{u}_1, \widetilde{u}_2) \geq 0 \geq \dfrac{\partial \underline{u}_2}{\partial t} - \Delta \underline{u}_2 - f_2(\widetilde{u}_1, \underline{u}_2), \end{cases} \quad (x,t) \in Q_T \tag{12.2.7}$$

when (f_1, f_2) is quasimonotone nonincreasing; and

$$\begin{cases} \dfrac{\partial \widetilde{u}_1}{\partial t} - \Delta \widetilde{u}_1 - f_1(\widetilde{u}_1, \underline{u}_2) \geq 0 \geq \dfrac{\partial \underline{u}_1}{\partial t} - \Delta \underline{u}_1 - f_1(\underline{u}_1, \widetilde{u}_2), \\ \dfrac{\partial \widetilde{u}_2}{\partial t} - \Delta \widetilde{u}_2 - f_2(\widetilde{u}_1, \widetilde{u}_2) \geq 0 \geq \dfrac{\partial \underline{u}_2}{\partial t} - \Delta \underline{u}_2 - f_2(\underline{u}_1, \underline{u}_2), \end{cases} \quad (x,t) \in Q_T \tag{12.2.8}$$

when (f_1, f_2) is mixed quasimonotone.

Remark 12.2.1 *It is seen from this definition that when (f_1, f_2) is quasi-monotone nondecreasing, we can use the first and third inequalities in (12.2.6) to determine $\widetilde{\mathbf{u}}$ and use the second and the fourth inequalities in (12.2.6) to determine $\underline{\mathbf{u}}$ independently; when (f_1, f_2) is quasimonotone nonincreasing, we can use the first and fourth inequalities in (12.2.7) to determine $(\widetilde{u}_1, \underline{u}_2)$ and use the second and third inequalities in (12.2.7) to determine $(\underline{u}_1, \widetilde{u}_2)$ independently. Moreover, if (f_1, f_2) is mixed quasi-monotone, then $(\widetilde{u}_1, \widetilde{u}_2, \underline{u}_1, \underline{u}_2)$ must be determined simultaneously by all of the four inequalities in (12.2.8).*

Definition 12.2.4 For any ordered supersolution $\widetilde{\mathbf{u}} = (\widetilde{u}_1, \widetilde{u}_2)$ and sub-solution $\underline{\mathbf{u}} = (\underline{u}_1, \underline{u}_2)$, we define the sector

$$\langle \underline{\mathbf{u}}, \widetilde{\mathbf{u}} \rangle = \left\{ \mathbf{u} = (u_1, u_2) \in C(\overline{Q}_T); \underline{\mathbf{u}}(x,t) \leq \mathbf{u}(x,t) \leq \widetilde{\mathbf{u}}(x,t), \quad (x,t) \in \overline{Q}_T \right\}.$$

12.2.3 *Monotone sequences*

Suppose for a given type of quasimonotone reaction function there exist a pair of ordered supersolution $\widetilde{\mathbf{u}} = (\widetilde{u}_1, \widetilde{u}_2)$ and subsolution $\underline{\mathbf{u}} = (\underline{u}_1, \underline{u}_2)$. In the following discussion we consider each of the three types of reaction functions in the sector $\langle \underline{\mathbf{u}}, \widetilde{\mathbf{u}} \rangle$. In addition, we assume that there exist constants $\underline{c}_i$ $(i = 1, 2)$ such that for every $(u_1, u_2), (v_1, v_2) \in \langle \underline{\mathbf{u}}, \widetilde{\mathbf{u}} \rangle$, (f_1, f_2) satisfies the one-sided Lipschitz condition

$$\begin{cases} f_1(u_1, u_2) - f_1(v_1, u_2) \geq -\underline{c}_1(u_1 - v_1), & \text{when } u_1 \geq v_1, \\ f_2(u_1, u_2) - f_2(u_1, v_2) \geq -\underline{c}_2(u_2 - v_2), & \text{when } u_2 \geq v_2. \end{cases} \quad (12.2.9)$$

To ensure the uniqueness of the solution we also assume that there exist constants $\overline{c}_i \leq \underline{c}_i$ $(i = 1, 2)$ such that for every $(u_1, u_2), (v_1, v_2) \in \langle \underline{\mathbf{u}}, \widetilde{\mathbf{u}} \rangle$ with $(u_1, u_2) \geq (v_1, v_2)$,

$$\begin{cases} f_1(u_1, u_2) - f_1(v_1, v_2) \leq -\overline{c}_1((u_1 - v_1) + (u_2 - v_2)), \\ f_2(u_1, u_2) - f_2(v_1, v_2) \leq -\overline{c}_2((u_1 - v_1) + (u_2 - v_2)). \end{cases} \quad (12.2.10)$$

It is clear that if there exist constants $K_i \geq 0$ $(i = 1, 2)$ such that (f_1, f_2) satisfies the Lipschitz condition

$$|f_i(u_1, u_2) - f_i(v_1, v_2)| \leq K_i(|u_1 - v_1| + |u_2 - v_2|),$$
$$(u_1, u_2), (v_1, v_2) \in \langle \underline{\mathbf{u}}, \widetilde{\mathbf{u}} \rangle, \quad (i = 1, 2)$$

then both the conditions (12.2.9) and (12.2.10) hold with $\underline{c}_i = K_i$ and $\overline{c}_i = -K_i$. In particular, if (f_1, f_2) is a C^1-function in $\langle \underline{\mathbf{u}}, \widetilde{\mathbf{u}} \rangle$, then the conditions (12.2.9) and (12.2.10) are satisfied. Let

$$F_i(u_1, u_2) = \underline{c}_i u_i + f_i(u_1, u_2), \quad (u_1, u_2) \in \langle \underline{\mathbf{u}}, \widetilde{\mathbf{u}} \rangle \quad (i = 1, 2).$$

Then the condition (12.2.9) is equivalent to that F_i is monotone nondecreasing in u_i for $i = 1, 2$.

Starting from a suitable initial iteration $\mathbf{u}^{(0)} = (u_1^{(0)}, u_2^{(0)}) \in C^{2,1}(Q_T) \cap C(\overline{Q}_T)$, we construct a sequence $\{\mathbf{u}^{(k)}\}_{k=0}^{\infty} = \{(u_1^{(k)}, u_2^{(k)})\}_{k=0}^{\infty}$ from the iteration process

$$\frac{\partial u_1^{(k)}}{\partial t} - \Delta u_1^{(k)} + \underline{c}_1 u_1^{(k)} = F_1(u_1^{(k-1)}, u_2^{(k-1)}), \quad (x, t) \in Q_T, \quad (12.2.11)$$

$$\frac{\partial u_2^{(k)}}{\partial t} - \Delta u_2^{(k)} + \underline{c}_2 u_2^{(k)} = F_2(u_1^{(k-1)}, u_2^{(k-1)}), \quad (x, t) \in Q_T, \quad (12.2.12)$$

$$u_1^{(k)}(x, t) = g_1(x, t), \quad (x, t) \in \partial_p Q_T, \quad (12.2.13)$$

$$u_2^{(k)}(x, t) = g_2(x, t), \quad (x, t) \in \partial_p Q_T. \quad (12.2.14)$$

It is clear that for each $k = 1, 2, \cdots$, the above system consists of two linear uncoupled initial-boundary problems, and therefore the existence of $\mathbf{u}^{(k)} = (u_1^{(k)}, u_2^{(k)})$ is guaranteed by the L^2 theory and the maximum principle. Furthermore, from the regularity of solutions of heat equations,

$$\mathbf{u}^{(1)} \in C^{\alpha, \alpha/2}(\overline{Q}_T), \quad \mathbf{u}^{(k)} \in C^{2+\alpha, 1+\alpha/2}(\overline{Q}_T) \text{ for } k = 2, 3, \cdots.$$

Similar to the scalar case, to ensure that this sequence is monotone and converges to a solution of problem (12.2.1)–(12.2.4), it is necessary to choose a suitable initial iteration. The choice of this function depends on the type of quasimonotone property of (f_1, f_2).

(I) **Quasimonotone nondecreasing function.** For this type of quasimonotone function it suffices to take either $(\widetilde{u}_1, \widetilde{u}_2)$ or $(\underline{u}_1, \underline{u}_2)$ as the initial iteration $(u_1^{(0)}, u_2^{(0)})$. Denote these two sequences by $\{(\overline{u}_1^{(k)}, \overline{u}_2^{(k)})\}_{k=0}^{\infty}$ and $\{(\underline{u}_1^{(k)}, \underline{u}_2^{(k)})\}_{k=0}^{\infty}$, respectively. The following lemma presents the monotone property of these two sequences.

Lemma 12.2.1 *For quasimonotone nondecreasing (f_1, f_2), the two sequences $\{(\overline{u}_1^{(k)}, \overline{u}_2^{(k)})\}_{k=0}^{\infty}$ and $\{(\underline{u}_1^{(k)}, \underline{u}_2^{(k)})\}_{k=0}^{\infty}$ possess the monotone property*

$$\underline{u}^{(k)}(x, t) \leq \underline{u}^{(k+1)}(x, t) \leq \overline{u}^{(k+1)}(x, t) \leq \overline{u}^{(k)}(x, t), \quad (x, t) \in \overline{Q}_T \quad (12.2.15)$$

for every $k = 0, 1, \cdots$.

Proof. Let

$$w_i^{(0)}(x,t) = \overline{u}_i^{(0)}(x,t) - \overline{u}_i^{(1)}(x,t) = \widetilde{u}_i(x,t) - \overline{u}_i^{(1)}(x,t), \quad (x,t) \in \overline{Q}_T.$$

By (12.2.5), (12.2.6) and (12.2.11)–(12.2.14),

$$\frac{\partial w_i^{(0)}}{\partial t} - \Delta w_i^{(0)} + \underline{c}_i w_i^{(0)} \geq F_i(\widetilde{u}_1, \widetilde{u}_2) - F_i(\overline{u}_1^{(0)}, \overline{u}_2^{(0)}) = 0, \quad (x,t) \in Q_T,$$

$$w_i^{(0)}(x,t) \geq g_i(x,t) - g_i(x,t) = 0, \quad (x,t) \in \partial_p Q_T.$$

The maximum principle leads to $w_i^{(0)} \geq 0$ on $\overline{Q}_T$, i.e.

$$\overline{u}_i^{(1)}(x,t) \leq \overline{u}_i^{(0)}(x,t) = \widetilde{u}_i(x,t), \quad (x,t) \in \overline{Q}_T \quad (i = 1, 2).$$

Similarly

$$\underline{u}_i^{(1)}(x,t) \geq \underline{u}_i^{(0)}(x,t) = \underline{u}_i(x,t), \quad (x,t) \in \overline{Q}_T \quad (i = 1, 2).$$

Let

$$w_i^{(1)}(x,t) = \overline{u}_i^{(1)}(x,t) - \underline{u}_i^{(1)}(x,t), \quad (x,t) \in \overline{Q}_T \quad (i = 1, 2).$$

Then, by (12.2.11)–(12.2.14) and the monotone property of F_i,

$$\frac{\partial w_i^{(1)}}{\partial t} - \Delta w_i^{(1)} + \underline{c}_i w_i^{(1)} = F_i(\overline{u}_1^{(0)}, \overline{u}_2^{(0)}) - F_i(\underline{u}_1^{(0)}, \underline{u}_2^{(0)}) \geq 0, \quad (x,t) \in Q_T,$$

$$w_i^{(1)}(x,t) = g_i(x,t) - g_i(x,t) = 0, \quad (x,t) \in \partial_p Q_T.$$

Using the maximum principle gives $w_i^{(1)} \geq 0$ on $\overline{Q}_T$, i.e.

$$\underline{u}_i^{(1)}(x,t) \leq \overline{u}_i^{(1)}(x,t), \quad (x,t) \in \overline{Q}_T \quad (i = 1, 2).$$

Thus, we have

$$\underline{u}_i^{(0)}(x,t) \leq \underline{u}_i^{(1)}(x,t) \leq \overline{u}_i^{(1)}(x,t) \leq \overline{u}_i^{(0)}(x,t), \quad (x,t) \in \overline{Q}_T \quad (i = 1, 2).$$

Suppose

$$\underline{u}_i^{(k-1)}(x,t) \leq \underline{u}_i^{(k)}(x,t) \leq \overline{u}_i^{(k)}(x,t) \leq \overline{u}_i^{(k-1)}(x,t), \quad (x,t) \in \overline{Q}_T \quad (i = 1, 2)$$

for some $k \geq 1$. Then, by (12.2.11)–(12.2.14) and the monotone property of F_i, the function

$$w_i^{(k)}(x,t) = \overline{u}_i^{(k)}(x,t) - \overline{u}_i^{(k+1)}(x,t), \quad (x,t) \in \overline{Q}_T \quad (i = 1, 2)$$

satisfies the relation

$$\frac{\partial w_i^{(k)}}{\partial t} - \Delta w_i^{(k)} + \underline{c}_i w_i^{(k)} = F_i(\overline{u}_1^{(k-1)}, \overline{u}_2^{(k-1)}) - F_i(\overline{u}_1^{(k)}, \overline{u}_2^{(k)})$$

$$\geq 0, \qquad (x, t) \in Q_T,$$

$$w_i^{(k)}(x, t) = g_i(x, t) - g_i(x, t) = 0, \qquad (x, t) \in \partial_p Q_T.$$

This leads to the inequality

$$\overline{u}_i^{(k+1)}(x, t) \leq \overline{u}_i^{(k)}(x, t), \quad (x, t) \in \overline{Q}_T \quad (i = 1, 2).$$

A similar argument gives

$$\underline{u}_i^{(k+1)}(x, t) \geq \underline{u}_i^{(k)}(x, t), \, \underline{u}_i^{(k+1)}(x, t) \leq \overline{u}_i^{(k+1)}(x, t), \, (x, t) \in \overline{Q}_T \quad (i = 1, 2).$$

Thus (12.2.15) follows by induction. $\qquad \square$

Remark 12.2.2 *From the proof of this lemma, we see that in the absence of a supersolution, the monotone nondecreasing property of the sequence* $\{(\underline{u}_1^{(k)}, \underline{u}_2^{(k)})\}_{k=0}^{\infty}$ *remains true provided that the condition (12.2.9) holds for every bounded function* $(\widetilde{u}_1, \widetilde{u}_2)$. *In this situation the sequence* $\{(\underline{u}_1^{(k)}, \underline{u}_2^{(k)})\}_{k=0}^{\infty}$ *either converges to some limit as* $k \to \infty$ *or becomes unbounded at some point in* $\overline{Q}_T$. *A similar conclusion holds for the sequence* $\{(\overline{u}_1^{(k)}, \overline{u}_2^{(k)})\}_{k=0}^{\infty}$.

(II) **Quasimonotone nonincreasing function.** When the reaction function (f_1, f_2) is quasimonotone nonincreasing, we choose $(\widetilde{u}_1, \underline{u}_2)$ or $(\underline{u}_1, \widetilde{u}_2)$ as the initial iteration $(u_1^{(0)}, u_2^{(0)})$ in the iteration process (12.2.11)–(12.2.14) and denote the corresponding sequences by $\{(\overline{u}_1^{(k)}, \underline{u}_2^{(k)})\}_{k=0}^{\infty}$ and $\{(\underline{u}_1^{(k)}, \overline{u}_2^{(k)})\}_{k=0}^{\infty}$, respectively. The monotone property of these two sequences is presented in the following lemma.

Lemma 12.2.2 *For quasimonotone nonincreasing* (f_1, f_2), *the two sequences* $\{(\overline{u}_1^{(k)}, \underline{u}_2^{(k)})\}_{k=0}^{\infty}$ *and* $\{(\underline{u}_1^{(k)}, \overline{u}_2^{(k)})\}_{k=0}^{\infty}$ *possess the mixed monotone property in the sense that their components* $\overline{u}_i^{(k)}$ *and* $\underline{u}_i^{(k)}$ *satisfy the relation (12.2.15) for every* $k = 0, 1, \cdots$.

Proof. Let

$$w_1^{(0)}(x, t) = \overline{u}_1^{(0)}(x, t) - \overline{u}_1^{(1)}(x, t) = \widetilde{u}_1(x, t) - \overline{u}_1^{(1)}(x, t), \quad (x, t) \in \overline{Q}_T$$

and

$$w_2^{(0)}(x, t) = \underline{u}_2^{(1)}(x, t) - \underline{u}_2^{(0)}(x, t) = \underline{u}_2^{(1)}(x, t) - \underline{u}_2(x, t), \quad (x, t) \in \overline{Q}_T.$$

By (12.2.5), (12.2.7) and (12.2.11)–(12.2.14),

$$\frac{\partial w_1^{(0)}}{\partial t} - \Delta w_1^{(0)} + \underline{c}_1 w_1^{(0)} \geq F_1(\tilde{u}_1, \underline{u}_2) - F_1(\overline{u}_1^{(0)}, \underline{u}_2^{(0)}) = 0, \quad (x,t) \in Q_T,$$

$$\frac{\partial w_2^{(0)}}{\partial t} - \Delta w_2^{(0)} + \underline{c}_2 w_2^{(0)} \geq F_2(\overline{u}_1^{(0)}, \underline{u}_2^{(0)}) - F_2(\tilde{u}_1, \underline{u}_2) = 0, \quad (x,t) \in Q_T,$$

$$w_1^{(0)}(x,t) \geq g_1(x,t) - g_1(x,t) = 0, \qquad\qquad (x,t) \in \partial_p Q_T,$$

$$w_2^{(0)}(x,t) \geq g_2(x,t) - g_2(x,t) = 0, \qquad\qquad (x,t) \in \partial_p Q_T.$$

The maximum principle implies that $w_i^{(0)} \geq 0$ on $\overline{Q}_T$, i.e.

$$\overline{u}_1^{(1)}(x,t) \leq \overline{u}_1^{(0)}(x,t), \quad \underline{u}_2^{(1)}(x,t) \geq \underline{u}_2^{(0)}(x,t), \quad (x,t) \in \overline{Q}_T.$$

A similar argument gives

$$\underline{u}_1^{(1)}(x,t) \geq \underline{u}_1^{(0)}(x,t), \quad \overline{u}_2^{(1)}(x,t) \leq \overline{u}_2^{(0)}(x,t), \quad (x,t) \in \overline{Q}_T.$$

Let

$$w_i^{(1)}(x,t) = \overline{u}_i^{(1)}(x,t) - \underline{u}_i^{(1)}(x,t), \quad (x,t) \in \overline{Q}_T \quad (i=1,2).$$

Then, by (12.2.11)–(12.2.14), (12.2.9) and the quasimonotone property of f_i,

$$\frac{\partial w_1^{(1)}}{\partial t} - \Delta w_1^{(1)} + \underline{c}_1 w_1^{(1)} = F_1(\overline{u}_1^{(0)}, \underline{u}_2^{(0)}) - F_1(\underline{u}_1^{(0)}, \overline{u}_2^{(0)})$$

$$= \left[\underline{c}_1(\tilde{u}_1 - \underline{u}_1) + f_1(\tilde{u}_1, \underline{u}_2) - f_1(\underline{u}_1, \underline{u}_2) \right]$$

$$\qquad + \left[f_1(\underline{u}_1, \underline{u}_2) - f_1(\underline{u}_1, \tilde{u}_2) \right] \geq 0, \qquad (x,t) \in Q_T,$$

$$\frac{\partial w_2^{(1)}}{\partial t} - \Delta w_2^{(1)} + \underline{c}_2 w_2^{(1)} = F_2(\underline{u}_1^{(0)}, \overline{u}_2^{(0)}) - F_2(\overline{u}_1^{(0)}, \underline{u}_2^{(0)})$$

$$= \left[\underline{c}_2(\tilde{u}_2 - \underline{u}_2) + f_2(\underline{u}_1, \tilde{u}_2) - f_2(\underline{u}_1, \underline{u}_2) \right]$$

$$\qquad + \left[f_2(\underline{u}_1, \underline{u}_2) - f_2(\tilde{u}_1, \underline{u}_2) \right] \geq 0, \qquad (x,t) \in Q_T,$$

$$w_1^{(1)}(x,t) = g_1(x,t) - g_1(x,t) = 0, \qquad\qquad (x,t) \in \partial_p Q_T,$$

$$w_2^{(1)}(x,t) = g_2(x,t) - g_2(x,t) = 0, \qquad\qquad (x,t) \in \partial_p Q_T.$$

Using the maximum principle again gives $w_i^{(1)} \geq 0$ on $\overline{Q}_T$. Thus we obtain

$$\underline{u}_i^{(0)}(x,t) \leq \underline{u}_i^{(1)}(x,t) \leq \overline{u}_i^{(1)}(x,t) \leq \overline{u}_i^{(0)}(x,t), \quad (x,t) \in \overline{Q}_T \quad (i=1,2).$$

The proof of the monotone property (12.2.15) can be completed by a induction argument similar to that of Lemma 12.2.1. $\qquad\square$

(III) **Mixed quasimonotone function.** The construction of monotone sequences for mixed quasimonotone functions requires the use of both supersolution and subsolution simultaneously. When f_1 is quasimonotone nonincreasing and f_2 is quasimonotone nondecreasing, the monotone iteration process is given by

$$\frac{\partial \overline{u}_1^{(k)}}{\partial t} - \Delta \overline{u}_1^{(k)} + \underline{c}_1 \overline{u}_1^{(k)} = F_1(\overline{u}_1^{(k-1)}, \underline{u}_2^{(k-1)}), \qquad (x,t) \in Q_T, \qquad (12.2.16)$$

$$\frac{\partial \underline{u}_1^{(k)}}{\partial t} - \Delta \underline{u}_1^{(k)} + \underline{c}_1 \underline{u}_1^{(k)} = F_1(\underline{u}_1^{(k-1)}, \overline{u}_2^{(k-1)}), \qquad (x,t) \in Q_T, \qquad (12.2.17)$$

$$\frac{\partial \overline{u}_2^{(k)}}{\partial t} - \Delta \overline{u}_2^{(k)} + \underline{c}_2 \overline{u}_2^{(k)} = F_2(\overline{u}_1^{(k-1)}, \overline{u}_2^{(k-1)}), \qquad (x,t) \in Q_T, \qquad (12.2.18)$$

$$\frac{\partial \underline{u}_2^{(k)}}{\partial t} - \Delta \underline{u}_2^{(k)} + \underline{c}_2 \underline{u}_2^{(k)} = F_2(\underline{u}_1^{(k-1)}, \underline{u}_2^{(k-1)}), \qquad (x,t) \in Q_T, \qquad (12.2.19)$$

$$\overline{u}_1^{(k)}(x,t) = \underline{u}_1^{(k)}(x,t) = g_1(x,t), \qquad\qquad (x,t) \in \partial_p Q_T, \quad (12.2.20)$$

$$\overline{u}_2^{(k)}(x,t) = \underline{u}_2^{(k)}(x,t) = g_2(x,t), \qquad\qquad (x,t) \in \partial_p Q_T, \quad (12.2.21)$$

and

$$(\overline{u}_1^{(0)}, \overline{u}_2^{(0)}) = (\widetilde{u}_1, \widetilde{u}_2), \quad (\underline{u}_1^{(0)}, \underline{u}_2^{(0)}) = (\underline{u}_1, \underline{u}_2). \qquad (12.2.22)$$

It is seen from this iteration process that the equations in (12.2.16)–(12.2.19) are uncoupled but are interrelated in the sense that the k-th iteration $(\overline{u}_1^{(k)}, \overline{u}_2^{(k)})$ or $(\underline{u}_1^{(k)}, \underline{u}_2^{(k)})$ depends on all of the four components in the previous iteration. This kind of iteration is fundamental in its extension to coupled system with any finite number of equations. The idea of this construction is to obtain the monotone property of the sequences shown in the following lemma.

Lemma 12.2.3 *For mixed quasimonotone* (f_1, f_2)*, the two sequences* $\{(\overline{u}_1^{(k)}, \overline{u}_2^{(k)})\}_{k=0}^{\infty}$ *and* $\{(\underline{u}_1^{(k)}, \underline{u}_2^{(k)})\}_{k=0}^{\infty}$ *given by* (12.2.16)–(12.2.21) *with* (12.2.22) *possess the monotone property* (12.2.15).

Proof. Let

$$w_i^{(0)}(x,t) = \overline{u}_i^{(0)}(x,t) - \overline{u}_i^{(1)}(x,t) = \widetilde{u}_i(x,t) - \overline{u}_i^{(1)}(x,t), \quad (x,t) \in \overline{Q}_T.$$

By (12.2.8) and (12.2.16)–(12.2.22),

$$\frac{\partial w_1^{(0)}}{\partial t} - \Delta w_1^{(0)} + \underline{c}_1 w_1^{(0)} \geq F_1(\widetilde{u}_1, \underline{u}_2) - F_1(\overline{u}_1^{(0)}, \underline{u}_2^{(0)}) = 0, \quad (x,t) \in Q_T,$$

$$\frac{\partial w_2^{(0)}}{\partial t} - \Delta w_2^{(0)} + \underline{c}_2 w_2^{(0)} \geq F_2(\tilde{u}_1, \tilde{u}_2) - F_2(\overline{u}_1^{(0)}, \overline{u}_2^{(0)}) = 0, \quad (x,t) \in Q_T,$$

$$w_1^{(0)}(x,t) \geq g_1(x,t) - g_1(x,t) = 0, \qquad (x,t) \in \partial_p Q_T,$$

$$w_2^{(0)}(x,t) \geq g_2(x,t) - g_2(x,t) = 0, \qquad (x,t) \in \partial_p Q_T.$$

The maximum principle implies that $w_i^{(0)} \geq 0$ on $\overline{Q}_T$, i.e.

$$\overline{u}_i^{(1)}(x,t) \leq \overline{u}_i^{(0)}(x,t), \quad (x,t) \in \overline{Q}_T \quad (i=1,2).$$

A similar argument gives

$$\underline{u}_i^{(1)}(x,t) \geq \underline{u}_i^{(0)}(x,t), \quad (x,t) \in \overline{Q}_T \quad (i=1,2).$$

Let

$$w_i^{(1)}(x,t) = \overline{u}_i^{(1)}(x,t) - \underline{u}_i^{(1)}(x,t), \quad (x,t) \in \overline{Q}_T \quad (i=1,2).$$

Then, by (12.2.16)–(12.2.22), (12.2.9) and the mixed quasimonotone property of (f_1, f_2),

$$\frac{\partial w_1^{(1)}}{\partial t} - \Delta w_1^{(1)} + \underline{c}_1 w_1^{(1)} = F_1(\overline{u}_1^{(0)}, \underline{u}_2^{(0)}) - F_1(\underline{u}_1^{(0)}, \overline{u}_2^{(0)})$$

$$= [\underline{c}_1(\tilde{u}_1 - \underline{u}_1) + f_1(\tilde{u}_1, \underline{u}_2) - f_1(\underline{u}_1, \underline{u}_2)]$$

$$\quad + [f_1(\underline{u}_1, \underline{u}_2) - f_1(\underline{u}_1, \tilde{u}_2)] \geq 0, \qquad (x,t) \in Q_T,$$

$$\frac{\partial w_2^{(1)}}{\partial t} - \Delta w_2^{(1)} + \underline{c}_2 w_2^{(1)} = F_2(\overline{u}_1^{(0)}, \overline{u}_2^{(0)}) - F_2(\underline{u}_1^{(0)}, \underline{u}_2^{(0)})$$

$$= [\underline{c}_2(\tilde{u}_2 - \underline{u}_2) + f_2(\tilde{u}_1, \tilde{u}_2) - f_2(\tilde{u}_1, \underline{u}_2)]$$

$$\quad + [f_2(\tilde{u}_1, \underline{u}_2) - f_2(\underline{u}_1, \underline{u}_2)] \geq 0, \qquad (x,t) \in Q_T,$$

$$w_1^{(1)}(x,t) = g_1(x,t) - g_1(x,t) = 0, \qquad (x,t) \in \partial_p Q_T,$$

$$w_2^{(1)}(x,t) = g_2(x,t) - g_2(x,t) = 0, \qquad (x,t) \in \partial_p Q_T.$$

Using the maximum principle gives $w_i^{(1)} \geq 0$ on $\overline{Q}_T$. Thus we have

$$\underline{u}_i^{(0)}(x,t) \leq \underline{u}_i^{(1)}(x,t) \leq \overline{u}_i^{(1)}(x,t) \leq \overline{u}_i^{(0)}(x,t), \quad (x,t) \in \overline{Q}_T \quad (i=1,2).$$

Assume

$$\underline{u}_i^{(k-1)}(x,t) \leq \underline{u}_i^{(k)}(x,t) \leq \overline{u}_i^{(k)}(x,t) \leq \overline{u}_i^{(k-1)}(x,t), \quad (x,t) \in \overline{Q}_T \quad (i=1,2)$$

for some $k \geq 1$. Then, by (12.2.16)–(12.2.21), (12.2.9) and the mixed quasimonotone property of (f_1, f_2), the function

$$w_i^{(k)}(x, t) = \overline{u}_i^{(k)}(x, t) - \overline{u}_i^{(k+1)}(x, t), \quad (x, t) \in \overline{Q}_T \quad (i = 1, 2)$$

satisfies the relation

$$\frac{\partial w_1^{(k)}}{\partial t} - \Delta w_1^{(k)} + c_1 w_1^{(k)} = F_1(\overline{u}_1^{(k-1)}, \underline{u}_2^{(k-1)}) - F_1(\overline{u}_1^{(k)}, \underline{u}_2^{(k)})$$

$$= \left[c_1(\overline{u}_1^{(k-1)} - \overline{u}_1^{(k)}) + f_1(\overline{u}_1^{(k-1)}, \underline{u}_2^{(k-1)}) - f_1(\overline{u}_1^{(k)}, \underline{u}_2^{(k-1)}) \right]$$
$$+ \left[f_1(\overline{u}_1^{(k)}, \underline{u}_2^{(k-1)}) - f_1(\overline{u}_1^{(k)}, \underline{u}_2^{(k)}) \right] \geq 0, \qquad (x, t) \in Q_T,$$

$$\frac{\partial w_2^{(k)}}{\partial t} - \Delta w_2^{(k)} + c_2 w_2^{(k)} = F_2(\overline{u}_1^{(k-1)}, \overline{u}_2^{(k-1)}) - F_2(\overline{u}_1^{(k)}, \overline{u}_2^{(k)})$$

$$= \left[c_2(\overline{u}_2^{(k-1)} - \overline{u}_2^{(k)}) + f_2(\overline{u}_1^{(k-1)}, \overline{u}_2^{(k-1)}) - f_2(\overline{u}_1^{(k-1)}, \overline{u}_2^{(k)}) \right]$$
$$+ \left[f_2(\overline{u}_1^{(k-1)}, \overline{u}_2^{(k)}) - f_2(\overline{u}_1^{(k)}, \overline{u}_2^{(k)}) \right] \geq 0, \qquad (x, t) \in Q_T,$$

$$w_1^{(k)}(x, t) = g_1(x, t) - g_1(x, t) = 0, \qquad (x, t) \in \partial_p Q_T,$$

$$w_2^{(k)}(x, t) = g_2(x, t) - g_2(x, t) = 0, \qquad (x, t) \in \partial_p Q_T.$$

Using the maximum principle again leads to that $w_i^{(k)} \geq 0$ on $\overline{Q}_T$, i.e.

$$\overline{u}_i^{(k+1)}(x, t) \leq \overline{u}_i^{(k)}(x, t), \quad (x, t) \in \overline{Q}_T \quad (i = 1, 2).$$

A similar argument gives

$$\underline{u}_i^{(k+1)}(x, t) \geq \underline{u}_i^{(k)}(x, t), \ \underline{u}_i^{(k+1)}(x, t) \leq \overline{u}_i^{(k+1)}(x, t), \ (x, t) \in \overline{Q}_T \quad (i = 1, 2).$$

The conclusion of the lemma follows by induction. $\qquad \square$

The following lemma shows that the above construction of monotone sequences yields a sequence of ordered supersolutions and subsolutions for problem (12.2.1)–(12.2.4).

Lemma 12.2.4 *Let $(\tilde{u}_1, \tilde{u}_2)$, $(\underline{u}_1, \underline{u}_2)$ be ordered supersolution and subsolution of problem (12.2.1)–(12.2.4) and (f_1, f_2) be quasimonotone and satisfy (12.2.9). Then, for each type of quasimonotone (f_1, f_2), the corresponding iterations $(\overline{u}_1^{(k)}, \overline{u}_2^{(k)})$ and $(\underline{u}_1^{(k)}, \underline{u}_2^{(k)})$ $(k = 1, 2, \cdots)$ given by Lemmas 12.2.1–12.2.3 are ordered supersolution and subsolution.*

Proof. First, consider the case where (f_1, f_2) is quasimonotone nondecreasing. Then, by (12.2.9) and (12.2.11)–(12.2.14), we have, for $k =$

$1, 2 \cdots,$

$$\frac{\partial \overline{u}_1^{(k)}}{\partial t} - \Delta \overline{u}_1^{(k)} = -\underline{c}_1 \overline{u}_1^{(k)} + F_1(\overline{u}_1^{(k-1)}, \overline{u}_2^{(k-1)})$$

$$= [\underline{c}_1(\overline{u}_1^{(k-1)} - \overline{u}_1^{(k)}) + f_1(\overline{u}_1^{(k-1)}, \overline{u}_2^{(k-1)}) - f_1(\overline{u}_1^{(k)}, \overline{u}_2^{(k-1)})]$$

$$+ [f_1(\overline{u}_1^{(k)}, \overline{u}_2^{(k-1)}) - f_1(\overline{u}_1^{(k)}, \overline{u}_2^{(k)})] + f_1(\overline{u}_1^{(k)}, \overline{u}_2^{(k)})$$

$$\geq f_1(\overline{u}_1^{(k)}, \overline{u}_2^{(k)}), \qquad (x, t) \in Q_T,$$

$$\frac{\partial \overline{u}_2^{(k)}}{\partial t} - \Delta \overline{u}_2^{(k)} = -\underline{c}_2 \overline{u}_2^{(k)} + F_2(\overline{u}_1^{(k-1)}, \overline{u}_2^{(k-1)})$$

$$= [\underline{c}_2(\overline{u}_2^{(k-1)} - \overline{u}_2^{(k)}) + f_2(\overline{u}_1^{(k-1)}, \overline{u}_2^{(k-1)}) - f_2(\overline{u}_1^{(k-1)}, \overline{u}_2^{(k)})]$$

$$+ [f_2(\overline{u}_1^{(k-1)}, \overline{u}_2^{(k)}) - f_2(\overline{u}_1^{(k)}, \overline{u}_2^{(k)})] + f_2(\overline{u}_1^{(k)}, \overline{u}_2^{(k)})$$

$$\geq f_2(\overline{u}_1^{(k)}, \overline{u}_2^{(k)}), \qquad (x, t) \in Q_T,$$

$$\overline{u}_1^{(k)}(x, t) = g_1(x, t), \qquad (x, t) \in \partial_p Q_T,$$

$$\overline{u}_2^{(k)}(x, t) = g_2(x, t), \qquad (x, t) \in \partial_p Q_T,$$

which shows that $(\overline{u}_1^{(k)}, \overline{u}_2^{(k)})$ is a supersolution. The proof for the subsolution is similar.

If (f_1, f_2) is quasimonotone nonincreasing, then from the construction of the sequence and using (12.2.9) and the quasimonotone nonincreasingness of (f_1, f_2), for $k = 1, 2 \cdots,$

$$\frac{\partial \overline{u}_1^{(k)}}{\partial t} - \Delta \overline{u}_1^{(k)} = -\underline{c}_1 \overline{u}_1^{(k)} + F_1(\overline{u}_1^{(k-1)}, \underline{u}_2^{(k-1)})$$

$$= [\underline{c}_1(\overline{u}_1^{(k-1)} - \overline{u}_1^{(k)}) + f_1(\overline{u}_1^{(k-1)}, \underline{u}_2^{(k-1)}) - f_1(\overline{u}_1^{(k)}, \underline{u}_2^{(k-1)})]$$

$$+ [f_1(\overline{u}_1^{(k)}, \underline{u}_2^{(k-1)}) - f_1(\overline{u}_1^{(k)}, \underline{u}_2^{(k)})] + f_1(\overline{u}_1^{(k)}, \underline{u}_2^{(k)})$$

$$\geq f_1(\overline{u}_1^{(k)}, \underline{u}_2^{(k)}), \qquad (x, t) \in Q_T,$$

$$\frac{\partial \underline{u}_2^{(k)}}{\partial t} - \Delta \underline{u}_2^{(k)} = -\underline{c}_2 \underline{u}_2^{(k)} + F_2(\overline{u}_1^{(k-1)}, \underline{u}_2^{(k-1)})$$

$$= -[\underline{c}_2(\underline{u}_2^{(k)} - \underline{u}_2^{(k-1)}) + f_2(\overline{u}_1^{(k-1)}, \underline{u}_2^{(k)}) - f_2(\overline{u}_1^{(k-1)}, \underline{u}_2^{(k-1)})]$$

$$+ [f_2(\overline{u}_1^{(k-1)}, \underline{u}_2^{(k)}) - f_2(\overline{u}_1^{(k)}, \underline{u}_2^{(k)})] + f_2(\overline{u}_1^{(k)}, \underline{u}_2^{(k)})$$

$$\leq f_2(\overline{u}_1^{(k)}, \underline{u}_2^{(k)}), \qquad (x, t) \in Q_T.$$

A similar argument gives

$$\frac{\partial \underline{u}_1^{(k)}}{\partial t} - \Delta \underline{u}_1^{(k)} \leq f_1(\underline{u}_1^{(k)}, \overline{u}_2^{(k)}), \qquad\qquad (x,t) \in Q_T,$$

$$\frac{\partial \overline{u}_2^{(k)}}{\partial t} - \Delta \overline{u}_2^{(k)} \geq f_2(\underline{u}_1^{(k)}, \overline{u}_2^{(k)}), \qquad\qquad (x,t) \in Q_T$$

for $k = 1, 2, \cdots$. Therefore, $(\overline{u}_1^{(k)}, \overline{u}_2^{(k)})$ and $(\underline{u}_1^{(k)}, \underline{u}_2^{(k)})$ are ordered supersolutions and subsolutions.

Finally for mixed quasimonotone (f_1, f_2), $(\overline{u}_1^{(k)}, \overline{u}_2^{(k)})$ and $(\underline{u}_1^{(k)}, \underline{u}_2^{(k)})$ are determined by (12.2.16)–(12.2.22). In view of (12.2.9) and the mixed quasimonotone property of (f_1, f_2), for $k = 1, 2, \cdots$,

$$\frac{\partial \overline{u}_1^{(k)}}{\partial t} - \Delta \overline{u}_1^{(k)} = -\underline{c}_1 \overline{u}_1^{(k)} + F_1(\overline{u}_1^{(k-1)}, \underline{u}_2^{(k-1)})$$

$$= \left[\underline{c}_1(\overline{u}_1^{(k-1)} - \overline{u}_1^{(k)}) + f_1(\overline{u}_1^{(k-1)}, \underline{u}_2^{(k-1)}) - f_1(\overline{u}_1^{(k)}, \underline{u}_2^{(k-1)}) \right]$$

$$+ \left[f_1(\overline{u}_1^{(k)}, \underline{u}_2^{(k-1)}) - f_1(\overline{u}_1^{(k)}, \underline{u}_2^{(k)}) \right] + f_1(\overline{u}_1^{(k)}, \underline{u}_2^{(k)})$$

$$\geq f_1(\overline{u}_1^{(k)}, \underline{u}_2^{(k)}), \qquad (x,t) \in Q_T,$$

$$\frac{\partial \overline{u}_2^{(k)}}{\partial t} - \Delta \overline{u}_2^{(k)} = -\underline{c}_2 \overline{u}_2^{(k)} + F_2(\overline{u}_1^{(k-1)}, \overline{u}_2^{(k-1)})$$

$$= \left[\underline{c}_2(\overline{u}_2^{(k-1)} - \overline{u}_2^{(k)}) + f_2(\overline{u}_1^{(k-1)}, \overline{u}_2^{(k-1)}) - f_2(\overline{u}_1^{(k-1)}, \overline{u}_2^{(k)}) \right]$$

$$+ \left[f_2(\overline{u}_1^{(k-1)}, \overline{u}_2^{(k)}) - f_2(\overline{u}_1^{(k)}, \overline{u}_2^{(k)}) \right] + f_2(\overline{u}_1^{(k)}, \overline{u}_2^{(k)})$$

$$\geq f_2(\overline{u}_1^{(k)}, \overline{u}_2^{(k)}), \qquad (x,t) \in Q_T.$$

A similar argument gives

$$\frac{\partial \underline{u}_1^{(k)}}{\partial t} - \Delta \underline{u}_1^{(k)} \leq f_1(\underline{u}_1^{(k)}, \overline{u}_2^{(k)}), \qquad\qquad (x,t) \in Q_T,$$

$$\frac{\partial \underline{u}_2^{(k)}}{\partial t} - \Delta \underline{u}_2^{(k)} \leq f_2(\underline{u}_1^{(k)}, \underline{u}_2^{(k)}), \qquad\qquad (x,t) \in Q_T$$

for $k = 1, 2, \cdots$. Hence $(\overline{u}_1^{(k)}, \overline{u}_2^{(k)})$ and $(\underline{u}_1^{(k)}, \underline{u}_2^{(k)})$ are ordered supersolution and subsolution for mixed quasimonotone functions. This completes the proof of the lemma. $\qquad\square$

It is worthy noting that the iteration process stated above is not the only way to construct the monotone sequences. For example, for the case that (f_1, f_2) is quasimonotone nondecreasing in $\langle \underline{u}, \widetilde{u} \rangle$, a different iteration process is given by

$$\frac{\partial u_1^{(k)}}{\partial t} - \Delta u_1^{(k)} + \underline{c}_1 u_1^{(k)} = F_1(u_1^{(k-1)}, u_2^{(k-1)}), \qquad (x,t) \in Q_T, \qquad (12.2.23)$$

$$\frac{\partial u_2^{(k)}}{\partial t} - \Delta u_2^{(k)} + \underline{c}_2 u_2^{(k)} = F_2(u_1^{(k)}, u_2^{(k-1)}), \qquad (x,t) \in Q_T, \qquad (12.2.24)$$

$$u_1^{(k)}(x,t) = g_1(x,t), \qquad (x,t) \in \partial_p Q_T, \quad (12.2.25)$$

$$u_2^{(k)}(x,t) = g_2(x,t), \qquad (x,t) \in \partial_p Q_T. \quad (12.2.26)$$

Compared with (12.2.11), (12.2.12), the difference of the present iteration process is that in determining $u_2^{(k)}$ by (12.2.24), (12.2.26), we have to use $u_1^{(k)}$ in addition to $u_2^{(k-1)}$. This kind of iteration is similar to the Gauss-Seidal iterative method for algebraic systems, which has the advantage of obtaining faster convergent sequences. It may be shown that the sequences thus defined possess monotone property when the initial iteration is either a supersolution or a subsolution.

Lemma 12.2.5 *Let (f_1, f_2) be quasimonotone nondecreasing in $\langle \underline{u}, \widetilde{u} \rangle$. Then the sequences $\{(\overline{u}_1^{(k)}, \overline{u}_2^{(k)})\}_{k=0}^{\infty}$ and $\{(\underline{u}_1^{(k)}, \underline{u}_2^{(k)})\}_{k=0}^{\infty}$, obtained from (12.2.23)–(12.2.26) with*

$$(\overline{u}_1^{(0)}, \overline{u}_2^{(0)}) = (\widetilde{u}_1, \widetilde{u}_2) \quad and \quad (\underline{u}_1^{(0)}, \underline{u}_2^{(0)}) = (\underline{u}_1, \underline{u}_2),$$

possess the monotone property (12.2.15) for every $k = 0, 1, \cdots$.

Similarly, for the quasimonotone nondecreasing reaction function, we have

Lemma 12.2.6 *Let (f_1, f_2) be quasimonotone nonincreasing in $\langle \underline{u}, \widetilde{u} \rangle$. Then the sequences $\{(\overline{u}_1^{(k)}, \underline{u}_2^{(k)})\}_{k=0}^{\infty}$ and $\{(\underline{u}_1^{(k)}, \overline{u}_2^{(k)})\}_{k=0}^{\infty}$, obtained from (12.2.23)–(12.2.26) with*

$$(\overline{u}_1^{(0)}, \underline{u}_2^{(0)}) = (\widetilde{u}_1, \underline{u}_2) \quad and \quad (\underline{u}_1^{(0)}, \overline{u}_2^{(0)}) = (\underline{u}_1, \widetilde{u}_2),$$

possess the monotone property (12.2.15) for every $k = 0, 1, \cdots$.

In the case of mixed quasimonotone (f_1, f_2), a modified iteration process for $\{(\overline{u}_1^{(k)}, \overline{u}_2^{(k)})\}_{k=0}^{\infty}$ and $\{(\underline{u}_1^{(k)}, \underline{u}_2^{(k)})\}_{k=0}^{\infty}$ is given by

$$\frac{\partial \overline{u}_1^{(k)}}{\partial t} - \Delta \overline{u}_1^{(k)} + \underline{c}_1 \overline{u}_1^{(k)} = F_1(\overline{u}_1^{(k-1)}, \underline{u}_2^{(k-1)}), \qquad (x,t) \in Q_T, \qquad (12.2.27)$$

$$\frac{\partial \underline{u}_1^{(k)}}{\partial t} - \Delta \underline{u}_1^{(k)} + \underline{c}_1 \underline{u}_1^{(k)} = F_1(\underline{u}_1^{(k-1)}, \overline{u}_2^{(k-1)}), \qquad (x,t) \in Q_T, \qquad (12.2.28)$$

$$\frac{\partial \overline{u}_2^{(k)}}{\partial t} - \Delta \overline{u}_2^{(k)} + \underline{c}_2 \overline{u}_2^{(k)} = F_2(\overline{u}_1^{(k)}, \overline{u}_2^{(k-1)}), \qquad (x,t) \in Q_T, \qquad (12.2.29)$$

$$\frac{\partial \underline{u}_2^{(k)}}{\partial t} - \Delta \underline{u}_2^{(k)} + \underline{c}_2 \underline{u}_2^{(k)} = F_2(\underline{u}_1^{(k)}, \underline{u}_2^{(k-1)}), \qquad (x,t) \in Q_T, \qquad (12.2.30)$$

$$\overline{u}_1^{(k)}(x,t) = \underline{u}_1^{(k)}(x,t) = g_1(x,t), \qquad (x,t) \in \partial_p Q_T, \quad (12.2.31)$$

$$\overline{u}_2^{(k)}(x,t) = \underline{u}_2^{(k)}(x,t) = g_2(x,t), \qquad (x,t) \in \partial_p Q_T. \quad (12.2.32)$$

Lemma 12.2.7 *For mixed quasimonotone (f_1, f_2), the two sequences $\{(\overline{u}_1^{(k)}, \overline{u}_2^{(k)})\}_{k=0}^{\infty}$ and $\{(\underline{u}_1^{(k)}, \underline{u}_2^{(k)})\}_{k=0}^{\infty}$ given by (12.2.27)–(12.2.32) with*

$$(\overline{u}_1^{(0)}, \overline{u}_2^{(0)}) = (\tilde{u}_1, \tilde{u}_2) \quad \text{and} \quad (\underline{u}_1^{(0)}, \underline{u}_2^{(0)}) = (\underline{u}_1, \underline{u}_2)$$

possess the monotone property (12.2.15) for every $k = 0, 1, 2, \cdots$.

The proofs of these three lemmas are similar to those of Lemmas 12.2.1–12.2.3 and we leave them to the interested readers.

12.2.4   Existence results

Lemmas 12.2.1 to 12.2.3 imply that for each of the three types of quasimonotone functions, the corresponding sequence obtained from (12.2.11)–(12.2.14) and (12.2.16)–(12.2.21) converges monotonically to some limit function. The same is true for the sequences given by (12.2.23)–(12.2.26) and (12.2.27)–(12.2.32). Define

$$\lim_{k \to \infty} \overline{u}_i^{(k)}(x,t) = \overline{u}_i(x,t), \quad \lim_{k \to \infty} \underline{u}_i^{(k)}(x,t) = \underline{u}_i(x,t),$$
$$(x,t) \in \overline{Q}_T \quad (i = 1, 2). \qquad (12.2.33)$$

Following the same argument as in the proof of Theorem 12.1.1, we will show that under the conditions (12.2.9) and (12.2.10),

$$\overline{u}_i(x,t) = \underline{u}_i(x,t) = u_i(x,t), \quad (x,t) \in \overline{Q}_T \quad (i = 1, 2)$$

and $\mathbf{u} = (u_1, u_2)$ is the unique solution of problem (12.2.1)–(12.2.4) for each of the three types of quasimonotone reaction functions.

Theorem 12.2.1   *Let $(\tilde{u}_1, \tilde{u}_2)$, $(\underline{u}_1, \underline{u}_2)$ be ordered supersolution and subsolution of problem (12.2.1)–(12.2.4), and (f_1, f_2) be quasimonotone nondecreasing in $\langle \underline{u}, \tilde{u} \rangle$ and satisfy the conditions (12.2.9) and (12.2.10). Then*

problem (12.2.1)–(12.2.4) *has a unique solution* $\mathbf{u} = (u_1, u_2)$ *in* $\langle \underline{\mathbf{u}}, \tilde{\mathbf{u}} \rangle$.
Moreover, the sequences $\{(\overline{u}_1^{(k)}, \overline{u}_2^{(k)})\}_{k=0}^{\infty}$ *and* $\{(\underline{u}_1^{(k)}, \underline{u}_2^{(k)})\}_{k=0}^{\infty}$, *obtained from* (12.2.11)–(12.2.14) *with*

$$(\overline{u}_1^{(0)}, \overline{u}_2^{(0)}) = (\tilde{u}_1, \tilde{u}_2) \quad and \quad (\underline{u}_1^{(0)}, \underline{u}_2^{(0)}) = (\underline{u}_1, \underline{u}_2),$$

converge monotonically to (u_1, u_2) *and satisfy the relation*

$$(\underline{u}_1, \underline{u}_2) \le (\underline{u}_1^{(k)}, \underline{u}_2^{(k)}) \le (u_1, u_2) \le (\overline{u}_1^{(k)}, \overline{u}_2^{(k)}) \le (\tilde{u}_1, \tilde{u}_2) \quad on \quad \overline{Q}_T$$

$$(12.2.34)$$

for every $k = 1, 2, \cdots$.

Proof. Consider problem (12.2.11)–(12.2.14) where the sequence $\{u^{(k)}\}_{k=0}^{\infty}$ represents either $\{\overline{u}^{(k)}\}_{k=0}^{\infty}$ or $\{\underline{u}^{(k)}\}_{k=0}^{\infty}$. Since by Lemma 12.2.1 this sequence converges monotonically to some limit (u_1, u_2) as $k \to \infty$, the continuity and monotonicity property of F_i imply that $F(u_1^{(k)}, u_2^{(k)})$ converges monotonically to $F_i(u_1, u_2)$ for $i = 1, 2$. From the regularity of solutions of heat equations,

$$u_i^{(1)} \in C^{\alpha, \alpha/2}(\overline{Q}_T), \quad u_i^{(k)} \in C^{2+\alpha, 1+\alpha/2}(\overline{Q}_T), \quad k = 2, 3, \cdots$$

and

$$|u_i^{(k)}|_{2+\alpha, 1+\alpha/2; Q_T} \le C_i (|g_i|_{2+\alpha, 1+\alpha/2; Q_T}$$
$$+ |u_1^{(k-1)}|_{0; Q_T} + |u_2^{(k-1)}|_{0; Q_T}), \quad k = 2, 3, \cdots,$$

where $i = 1, 2$, and $C_i > 0$ is a constant depending only on α, Ω, T and f_i but independent of k. From the monotone property (12.2.15), $\{u_i^{(k)}\}_{k=2}^{\infty}$ $(i = 1, 2)$ is uniformly bounded in $C^{2+\alpha, 1+\alpha/2}(\overline{Q}_T)$. Therefore, $\mathbf{u} \in C^{2+\alpha, 1+\alpha/2}(\overline{Q}_T)$ is a solution of problem (12.2.1)–(12.2.4). And (12.2.34) follows from the monotone property (12.2.15).

Now we show that

$$\overline{u}_i(x, t) = \underline{u}_i(x, t), \quad (x, t) \in \overline{Q}_T \quad (i = 1, 2). \tag{12.2.35}$$

Let

$$w_i(x, t) = \underline{u}_i(x, t) - \overline{u}_i(x, t), \quad (x, t) \in \overline{Q}_T \quad (i = 1, 2).$$

Then, from the monotone property (12.2.15),

$$w_i(x, t) \le 0, \quad (x, t) \in \overline{Q}_T \quad (i = 1, 2). \tag{12.2.36}$$

By (12.2.1)–(12.2.4) and the condition (12.2.10), $w_i \in C^{2,1}(Q_T) \cap C(\overline{Q}_T)$ $(i = 1, 2)$ satisfies

$$\frac{\partial w_i}{\partial t} - \Delta w_i = f_i(\underline{u}_1, \underline{u}_2) - f_i(\overline{u}_1, \overline{u}_2)$$

$$\geq \overline{c}_i((\overline{u}_1 - \underline{u}_1) + (\overline{u}_2 - \underline{u}_2)) = -\overline{c}_i(w_1 + w_2), \qquad (x, t) \in Q_T,$$

$$w_i(x, t) = g_i(x, t) - g_i(x, t) = 0, \qquad (x, t) \in \partial_p Q_T.$$

Therefore, $w_1 + w_2$ satisfies the relation

$$\frac{\partial(w_1 + w_2)}{\partial t} - \Delta(w_1 + w_2) \geq -(\overline{c}_1 + \overline{c}_2)(w_1 + w_2), \qquad (x, t) \in Q_T,$$

$$(w_1 + w_2)(x, t) = 0, \qquad (x, t) \in \partial_p Q_T.$$

The maximum principle guarantees that

$$w_1(x, t) + w_2(x, t) \geq 0, \quad (x, t) \in \overline{Q}_T.$$

This and (12.2.36) lead to (12.2.35).

From (12.2.35), to show the uniqueness of the solution of problem (12.2.1)–(12.2.4) in $\langle \underline{u}, \widetilde{u} \rangle$, it suffices to verify that any solution $\mathbf{u}^* \in \langle \underline{u}, \widetilde{u} \rangle$ to problem (12.2.1)–(12.2.4) satisfies the relation

$$\underline{u}(x, t) \leq \mathbf{u}^*(x, t) \leq \overline{u}(x, t), \quad (x, t) \in \overline{Q}_T.$$

This may be proved by the same argument as in the proof of Theorem 12.1.1 ii) and we leave the details to the reader. $\qquad \square$

For the other two types of quasimonotone reaction functions, we may prove similarly the following theorems.

Theorem 12.2.2 *Let* $(\widetilde{u}_1, \widetilde{u}_2)$, $(\underline{u}_1, \underline{u}_2)$ *be ordered supersolution and subsolution of problem* (12.2.1)–(12.2.4), *and* (f_1, f_2) *be quasimonotone nonincreasing in* $\langle \underline{u}, \widetilde{u} \rangle$ *and satisfy the conditions* (12.2.9) *and* (12.2.10). *Then problem* (12.2.1)–(12.2.4) *has a unique solution* $\mathbf{u} = (u_1, u_2)$ *in* $\langle \underline{u}, \widetilde{u} \rangle$. *Moreover, the sequences* $\{(\overline{u}_1^{(k)}, \underline{u}_2^{(k)})\}_{k=0}^{\infty}$ *and* $\{(\underline{u}_1^{(k)}, \overline{u}_2^{(k)})\}_{k=0}^{\infty}$, *obtained from* (12.2.11)–(12.2.14) *with*

$$(\overline{u}_1^{(0)}, \underline{u}_2^{(0)}) = (\widetilde{u}_1, \underline{u}_2) \quad and \quad (\underline{u}_1^{(0)}, \overline{u}_2^{(0)}) = (\underline{u}_1, \widetilde{u}_2),$$

converge monotonically to (u_1, u_2) *and satisfy the relation* (12.2.34).

Theorem 12.2.3 *Let* $(\widetilde{u}_1, \widetilde{u}_2)$, $(\underline{u}_1, \underline{u}_2)$ *be ordered supersolution and subsolution of problem* (12.2.1)–(12.2.4), *and* (f_1, f_2) *be quasimonotone nonincreasing in* $\langle \underline{u}, \widetilde{u} \rangle$ *and satisfy the conditions* (12.2.9) *and* (12.2.10). *Then* *problem* (12.2.1)–(12.2.4) *has a unique solution* $\mathbf{u} = (u_1, u_2)$ *in* $\langle \underline{\mathbf{u}}, \tilde{\mathbf{u}} \rangle$. *Moreover, the sequences* $\{(\overline{u}_1^{(k)}, \overline{u}_2^{(k)})\}_{k=0}^{\infty}$ *and* $\{(\underline{u}_1^{(k)}, \underline{u}_2^{(k)})\}_{k=0}^{\infty}$, *obtained from* (12.2.16)–(12.2.21) *with*

$$(\overline{u}_1^{(0)}, \overline{u}_2^{(0)}) = (\tilde{u}_1, \tilde{u}_2) \quad \text{and} \quad (\underline{u}_1^{(0)}, \underline{u}_2^{(0)}) = (\underline{u}_1, \underline{u}_2),$$

converge monotonically to (u_1, u_2) *and satisfy the relation* (12.2.34).

When the iteration processes (12.2.11)–(12.2.14) and (12.2.16)–(12.2.21) are replaced by (12.2.23)–(12.2.26) and (12.2.27)–(12.2.32), respectively, the results of Lemmas 12.2.5 to 12.2.7 imply that the corresponding sequences converge to some limit functions in the same fashion as in (12.2.33). It is easy to prove by an argument similar to the proof of Theorems 12.2.1 to 12.2.3 that these limits are also solutions of problem (12.2.1)–(12.2.4) in accordance with the quasimonotone property of (f_1, f_2). This observation leads to the following conclusion.

Theorem 12.2.4 *Under the hypothesis of Theorems 12.2.1–12.2.3, except that the iteration processes* (12.2.11)–(12.2.14) *and* (12.2.16)–(12.2.21) *are replaced by* (12.2.23)–(12.2.26) *and* (12.2.27)–(12.2.32) *respectively, all conclusions in the corresponding theorem remain true.*

12.2.5 Extension

As the scalar equations, the monotone method used above may be applied to coupled uniformly parabolic systems of general form with more general reaction terms, such as the system

$$\frac{\partial u_1}{\partial t} - \sum_{j,l=1}^{n} u_{jl}^{(1)}(x, t) D_{jl} u_1 + \sum_{j=1}^{n} b_j^{(1)}(x, t) D_j u_1$$

$$+ c^{(1)}(x, t) u_1 = f_1(x, t, u_1, u_2), \qquad (x, t) \in Q_T,$$

$$\frac{\partial u_2}{\partial t} - \sum_{j,l=1}^{n} a_{jl}^{(2)}(x, t) D_{jl} u_2 + \sum_{j=1}^{n} b_j^{(2)}(x, t) D_j u_2$$

$$+ c^{(2)}(x, t) u_2 = f_2(x, t, u_1, u_2), \qquad (x, t) \in Q_T,$$

$$u_1(x, t) = g_1(x, t), \qquad (x, t) \in \partial_p Q_T,$$

$$u_2(x, t) = g_2(x, t), \qquad (x, t) \in \partial_p Q_T,$$

where $\Omega \subset \mathbb{R}^n$ is a bounded domain with $\partial \Omega \in C^{2,\alpha}$ for some $0 < \alpha < 1$, $f_i \in C^{\alpha, \alpha/2, \alpha}(\overline{Q}_T \times \mathbb{R}^2)$, $g_i \in C^{2+\alpha, 1+\alpha/2}(\overline{Q}_T)$, $a_{jl}^{(i)}, b_j^{(i)}, c^{(i)} \in C^{\alpha, \alpha/2}(\overline{Q}_T)$,

$a_{jl}^{(i)} = a_{lj}^{(i)}$ and there exist positive constants $\lambda^{(i)}$, $\Lambda^{(i)}$, such that

$$\lambda^{(i)}|\xi|^2 \leq \sum_{j,l=1}^n a_{jl}^{(i)}(x,t)\xi_j\xi_l \leq \Lambda^{(i)}|\xi|^2, \quad \forall \xi \in \mathbb{R}^n, \ (x,t) \in Q_T$$

for each $i = 1, 2$.

At the end of this section, we point out that the monotone method may be used to coupled elliptic systems and also to parabolic and elliptic systems with an arbitrary finite number of equations, see more details in [Pao (1992)].

Exercises

1. Prove Remarks 12.1.1 and 12.1.2.

2. Prove Lemmas 12.2.5–12.2.7.

3. Prove Theorems 12.2.2–12.2.4.

4. Apply the monotone method to general coupled uniformly parabolic systems.

5. Apply the monotone method to elliptic equations and coupled systems.

6. Establish the theory of monotone method for elliptic and parabolic systems with an arbitrary finite number of equations.

Chapter 13

Degenerate Equations

The last chapter of this book is devoted to elliptic and parabolic equations with degeneracy. We first consider linear equations and then discuss some kinds of quasilinear equations.

13.1 Linear Equations

Let $\Omega \subset \mathbb{R}^n$ be a bounded domain. Consider the equation

$$Lu \equiv -a_{ij}(x)D_{ij}u + b_i(x)D_iu + c(x)u = f(x), \quad x \in \Omega, \qquad (13.1.1)$$

where a_{ij} $(i, j = 1, \cdots, n)$, b_i $(i = 1, \cdots, n)$, c and f are functions in Ω with suitable regularity, $a_{ij} = a_{ji}$ and the matrix

$$A = \begin{pmatrix} a_{11} & \cdots & a_{1n} \\ \vdots & & \\ a_{n1} & \cdots & a_{nn} \end{pmatrix}$$

is nonnegative definite on $\overline{\Omega}$, denoted by $A \geq 0$, i.e.

$$a_{ij}(x)\xi_i\xi_j \geq 0, \quad \forall \xi = (\xi_1, \cdots, \xi_n) \in \mathbb{R}^n, \quad x \in \overline{\Omega}. \qquad (13.1.2)$$

Here, as before, repeated indices imply a summation from 1 up to n.

If A is positive definite, i.e. $A > 0$, then (13.1.1) is elliptic; otherwise, (13.1.1) is said to be degenerate. In case $b_n > 0$ and

$$\begin{pmatrix} a_{11} & \cdots & a_{1(n-1)} \\ \vdots & & \\ a_{(n-1)1} & \cdots & a_{(n-1)(n-1)} \end{pmatrix} > 0, \quad a_{jn} = a_{nj} = 0 \ (j = 1, \cdots, n),$$

(13.1.1) is a parabolic equation; this means that, parabolic equations are degenerate elliptic equations. If $b_n > 0$ and

$$\begin{pmatrix} a_{11} & \cdots & a_{1(n-1)} \\ & \vdots & \\ a_{(n-1)1} & \cdots & a_{(n-1)(n-1)} \end{pmatrix} \geq 0, \quad a_{jn} = a_{nj} = 0 \, (j = 1, \cdots, n),$$

then (13.1.1) is a degenerate parabolic equation. If

$$\begin{pmatrix} a_{11} & \cdots & a_{1m} \\ & \vdots & \\ a_{m1} & \cdots & a_{mm} \end{pmatrix} > 0, \quad 0 \leq m \leq n - 2,$$

$$a_{ij} = a_{ji} = 0 \, (i = m+1, \cdots, n; j = 1, \cdots, n),$$

then (13.1.1) is called an ultraparabolic equation. In the extreme case $a_{ij} = 0 \, (i, j = 1, \cdots, n)$, (13.1.1) degenerates into a first order equation.

Sometimes, it is convenient to write (13.1.1) in divergence form

$$Lu \equiv -D_j(a_{ij}(x)D_i u) + \beta_i(x)D_i u + c(x)u = f(x), \quad x \in \Omega \qquad (13.1.3)$$

where

$$\beta_i(x) = b_i(x) + D_j a_{ij}(x), \quad x \in \overline{\Omega} \quad (i = 1, \cdots, n).$$

13.1.1 *Formulation of the first boundary value problem*

Different from elliptic equations without degeneracy, to pose the first boundary value problem for degenerate elliptic equations, in general, we are not permitted to prescribe the boundary value on the whole boundary $\Sigma = \partial\Omega$.

Suppose that $\Sigma = \partial\Omega$ is piecewise smooth. Denote

$$\Sigma^0 = \{x \in \Sigma; a_{ij}(x)\nu_i(x)\nu_j(x) = 0\}$$

and

$$\beta(x) = \beta_i(x)\nu_i(x), \quad x \in \Sigma,$$

where $\vec{\nu} = (\nu_1, \cdots, \nu_n)$ is the unit normal vector inward to Σ; $\beta(x)$ is called the Fichera function. Divide Σ^0 as follows

$$\Sigma^0 = \Sigma_0 \cup \Sigma_1 \cup \Sigma_2$$

and denote

$$\Sigma_3 = \Sigma \backslash \Sigma^0,$$

where

$$\Sigma_0 = \{x \in \Sigma^0; \beta(x) = 0\},$$
$$\Sigma_1 = \{x \in \Sigma^0; \beta(x) < 0\},$$
$$\Sigma_2 = \{x \in \Sigma^0; \beta(x) > 0\}.$$

Then

$$\Sigma_3 = \{x \in \Sigma; a_{ij}(x)\nu_i(x)\nu_j(x) > 0\} \cup E,$$

where E is a possible subset of measure zero on Σ. Since $\Sigma = \partial\Omega$ is piecewise smooth, there might be a subset of measure zero on Σ, at any point of which, no normal exists.

The first boundary value problem for (13.1.1) or (13.1.3) is then formulated as follows

$$Lu = f, \quad x \in \Omega, \tag{13.1.4}$$

$$u\Big|_{\Sigma_2 \cup \Sigma_3} = g, \tag{13.1.5}$$

where g is a given function.

Let us observe some special examples.

If (13.1.1) is elliptic, then Σ^0 is an empty set, $\Sigma = \Sigma_3$ and we need to prescribe the boundary value on the whole boundary Σ.

For equations of the form

$$Lu \equiv -D_j(a_{ij}D_iu) + c(x)u = f(x), \quad x \in \Omega, \tag{13.1.6}$$

we have $\beta(x) = 0$, $\Sigma_0 = \Sigma^0$ and Σ_2 is empty. Thus only the boundary value on Σ_3 needs to be given.

Now we consider the parabolic equation

$$\frac{\partial u}{\partial t} - a_{ij}(x,t)D_{ij}u + b_i(x,t)D_iu + c(x,t)u = f(x,t),$$
$$(x,t) \in Q_T = \Omega \times (0,T), \tag{13.1.7}$$

where a_{ij} $(i,j = 1,\cdots,n)$, b_i $(i = 1,\cdots,n)$, c and f are functions on $\overline{Q}_T$ with suitable regularity and $a_{ij} = a_{ji}$ satisfy the condition

$$a_{ij}\xi_i\xi_j > 0, \quad \xi = (\xi_1,\cdots,\xi_n) \in \mathbb{R}^n, \ \xi \neq 0, \quad (x,t) \in \overline{Q}_T. \tag{13.1.8}$$

Denote $t = x_{n+1}$, $a_{(n+1)i} = a_{i(n+1)} = 0 \, (i = 1, \cdots, n+1)$, $b_{n+1} = 1$. Then (13.1.7) can be expressed as (13.1.1) or (13.1.3) with $i, j = 1, \cdots, n+1$, $x = (x_1, \cdots, x_n, x_{n+1})$ and $\Omega \times (0, T)$ in place of Ω. In the present case,

$$\Sigma = \partial Q_T = \left(\Omega \times \{t = x_{n+1} = 0\} \right) \cup \left(\Omega \times \{t = x_{n+1} = T\} \right) \cup \left(\partial \Omega \times (0, T) \right)$$

and the Fichera function is

$$\beta = \sum_{i=1}^{n+1} \beta_i \nu_i = \sum_{i=1}^{n} \beta_i \nu_i + \nu_{n+1}.$$

On the lower bottom $\Omega \times \{t = x_{n+1} = 0\}$, we have $\sum_{i,j=1}^{n+1} a_{ij} \nu_i \nu_j = 0$, $\beta = 1 > 0$ and hence $\Omega \times \{t = x_{n+1} = 0\} \subset \Sigma_2$, where the boundary value needs to be given. However, on the upper bottom $\Omega \times \{t = x_{n+1} = T\}$, $\sum_{i,j=1}^{n+1} a_{ij} \nu_i \nu_j = 0$, $\beta = -1 < 0$, which means that $\Omega \times \{t = x_{n+1} = T\} \subset \Sigma_1$, where we should not give the boundary value. Since a_{ij} satisfy (13.1.8), on the lateral boundary $\partial \Omega \times (0, T)$, $\sum_{i,j=1}^{n+1} a_{ij} \nu_i \nu_j = \sum_{i,j=1}^{n} a_{ij} \nu_i \nu_j > 0$, i.e. $\partial \Omega \times (0, T) = \Sigma_3$, where the boundary value needs to be given. Prescribing the boundary value on the lower bottom and the lateral boundary is just the usual formulation of the first boundary value problem for parabolic equations.

It is natural to ask why we formulate the first boundary value problem for (13.1.1) in the above manner. The basic idea is to search such a boundary value condition which can ensure the uniqueness and existence of the solution. A proper condition should first ensure that the homogeneous equation has only zero solution satisfying the homogeneous boundary value condition.

Let us first consider the special equation (13.1.6). Suppose that $u \in C^2(\overline{\Omega})$ is a solution of

$$Lu \equiv -D_j(a_{ij} D_i u) + cu = 0, \tag{13.1.9}$$

satisfying the homogeneous boundary value condition. Multiply both sides of (13.1.9) by u and integrate over Ω. After integrating by parts we obtain

$$0 = \int_\Omega u L u \, dx = - \int_\Omega u D_j(a_{ij} D_i u) dx + \int_\Omega cu^2 dx$$

$$= \int_\Sigma u a_{ij} D_i u \nu_j d\sigma + \int_\Omega a_{ij} D_i u D_j u dx + \int_\Omega cu^2 dx$$

$$= \int_{\Sigma^0} u a_{ij} D_i u \nu_j d\sigma + \int_{\Sigma_3} u a_{ij} D_i u \nu_j d\sigma$$

$$+ \int_\Omega a_{ij} D_i u D_j u dx + \int_\Omega cu^2 dx.$$

Since $a_{ij}\xi_i\xi_j \geq 0$ for all $\xi \in \mathbb{R}^n$ and $a_{ij}\nu_i\nu_j = 0$ on Σ^0, i.e. $a_{ij}\xi_i\xi_j$ achieves its minimum at $\xi = (\nu_1, \cdots, \nu_n)$, we have $a_{ij}\nu_j = 0 \, (i = 1, \cdots, n)$ on Σ^0 and

$$\int_{\Sigma^0} u a_{ij} D_i u \nu_j d\sigma = 0.$$

So, if $u\big|_{\Sigma_3} = 0$, then

$$\int_{\Sigma_3} u a_{ij} D_i u \nu_j d\sigma = 0$$

and we are led to

$$\int_\Omega a_{ij} D_i u D_j u dx + \int_\Omega cu^2 dx = 0,$$

from which we see that if (13.1.9) satisfies the structure condition

$$c(x) \geq c_0 > 0, \quad x \in \Omega, \tag{13.1.10}$$

then we finally derive $u \equiv 0$. As we have seen in the study of elliptic equations, the condition (13.1.10) seems to be reasonable.

Now we turn to the general equation (13.1.1) or (13.1.3). A similar derivation gives

$$0 = \int_\Omega u L u dx = -\int_\Omega u D_j(a_{ij} D_i u) dx + \int_\Omega u\beta_i D_i u dx + \int_\Omega cu^2 dx$$

$$= \int_\Sigma u a_{ij} D_i u \nu_j d\sigma + \int_\Omega a_{ij} D_i u D_j u dx$$

$$- \frac{1}{2} \int_\Sigma \beta u^2 d\sigma - \int_\Omega \left(\frac{1}{2} D_i \beta_i - c\right) u^2 dx. \tag{13.1.11}$$

Since $a_{ij}\nu_j = 0 \, (i = 1, \cdots, n)$ on Σ^0, we have

$$\int_\Sigma u a_{ij} D_i u \nu_j d\sigma = 0$$

if $u\big|_{\Sigma_3} = 0$. Also, if $u\big|_{\Sigma_3} = 0$, then $\int_\Sigma \beta u^2 d\sigma = \int_{\Sigma^0} \beta u^2 d\sigma$. It is natural to divide $\int_{\Sigma^0} \beta u^2 d\sigma$ into three parts:

$$\int_{\Sigma^0} \beta u^2 d\sigma = \int_{\Sigma_0} \beta u^2 dx + \int_{\Sigma_1} \beta u^2 dx + \int_{\Sigma_2} \beta u^2 dx.$$

By the definition of Σ_0, Σ_1, Σ_2, we have

$$\int_{\Sigma_0} \beta u^2 d\sigma = 0, \quad \int_{\Sigma_1} \beta u^2 d\sigma \leq 0, \quad \int_{\Sigma_2} \beta u^2 d\sigma \geq 0.$$

Only $\dfrac{1}{2}\displaystyle\int_{\Sigma_2} \beta u^2 d\sigma$ plays a negative role to our purpose. If $u\big|_{\Sigma_2} = 0$, then $\int_{\Sigma_2} \beta u^2 d\sigma = 0$ and from (13.1.11) we obtain

$$\int_\Omega a_{ij} D_i u D_j u dx - \frac{1}{2}\int_{\Sigma_1} \beta u^2 d\sigma - \int_\Omega \left(\frac{1}{2}D_i\beta_i - c\right) u^2 dx = 0, \quad (13.1.12)$$

from which we find that if the structure condition

$$-\left(\frac{1}{2}D_i\beta_i(x) - c(x)\right) \geq c_0 > 0, \quad x \in \Omega \qquad (13.1.13)$$

is assumed, then (13.1.12) implies $u \equiv 0$.

Summing up, we arrive at the following conclusion: in order that the first boundary value problem for (13.1.1) admits only one (classical) solution, it suffices to prescribe the boundary value on $\Sigma_2 \cup \Sigma_3$, provided the structure condition (13.1.13) is assumed. In fact, from the above derivation, we may obtain the following conclusion: for any function $u \in C^2(\overline{\Omega})$ satisfying $u\big|_{\Sigma_2\cup\Sigma_3} = 0$, there holds

$$\int_\Omega u^2 dx \leq \frac{1}{c_0}\int_\Omega u L u dx,$$

which implies

$$\int_\Omega u^2 dx \leq \frac{1}{c_0}\int_\Omega (Lu)^2 dx,$$

provided condition (13.1.13) is assumed.

13.1.2 *Solvability of the problem in a space similar to H^1*

In what follows, we confine ourselves to the consideration of the problem for (13.1.1) with the homogeneous boundary value condition

$$u\Big|_{\Sigma_2 \cup \Sigma_3} = 0. \tag{13.1.14}$$

Suppose that $u \in C^2(\overline{\Omega})$ is a solution of (13.1.1), (13.1.14). Multiply both sides of (13.1.3) by $v \in W = \left\{ v \in C^1(\overline{\Omega}); v\Big|_{\Sigma_3} = 0 \right\}$ and integrate over Ω. After integrating by parts, we obtain

$$\int_\Omega v f dx = B(u, v), \tag{13.1.15}$$

where

$$B(u, v) = \int_\Omega \left[a_{ij} D_i u D_j v - \beta_i u D_i v - (D_i \beta_i - c) uv \right] dx - \int_{\Sigma_1} \beta uv d\sigma.$$

Conversely, it is not difficult to verify that if $u \in C^2(\overline{\Omega}) \cap W$ satisfies (13.1.15), then u is a (classical) solution of (13.1.1), (13.1.14).

Let

$$(u, v)_H = \int_\Omega (a_{ij} D_i u D_j v + uv) dx - \int_{\Sigma_1} \beta uv d\sigma$$

and denote by H the completion of W endowed with the norm $\|\cdot\| = (\cdot, \cdot)_H^{1/2}$. It is easy to verify that H is a Hilbert space. H is not equivalent to H^1 unless $\beta \equiv 0$ on Σ_1.

Definition 13.1.1 Let $f \in L^2(\Omega)$. A function $u \in H$ is said to be a weak solution of (13.1.1), (13.1.14) in H if (13.1.15) holds for any $v \in W = \left\{ v \in C^1(\overline{\Omega}); v\Big|_{\Sigma_3} = 0 \right\}$.

Theorem 13.1.1 *Under the condition (13.1.13), the boundary value problem (13.1.1), (13.1.14) admits a unique weak solution in H.*

Proof. It is easy to see that, for any $u, v \in W$,

$$|B(u, v)| \le C \left(\int_\Omega (|Dv|^2 + v^2) dx + \int_{\Sigma_1} v^2 d\sigma \right)^{1/2} \|u\|_H.$$

Hence $B(u, v)$ can be uniquely extended to $H \times W$. Since for $v \in W$,

$$-\int_\Omega \beta_i v D_i v dx = \frac{1}{2} \int_{\Sigma_1} \beta v^2 d\sigma + \frac{1}{2} \int_{\Sigma_2} \beta v^2 d\sigma + \frac{1}{2} \int_\Omega D_i \beta_i v^2 dx$$

$$\geq \frac{1}{2} \int_{\Sigma_1} \beta v^2 d\sigma + \frac{1}{2} \int_{\Omega} D_i \beta_i v^2 dx,$$

we have

$$B(v, v) \geq \int_{\Omega} a_{ij} D_i v D_j v dx - \int_{\Omega} \left(\frac{1}{2} D_i \beta_i - c\right) v^2 dx - \frac{1}{2} \int_{\Sigma_1} \beta v^2 d\sigma.$$

Using (13.1.13), we obtain

$$B(v, v) \geq \delta \|v\|_H^2, \quad \forall v \in W$$

for some constant $\delta > 0$, which means that $B(u, v)$ is coercive. Thus, by a modified Lax-Milgram's theorem (§3.1.2), for any bounded linear functional $F(v)$ in H, there exists a unique $u \in H$, such that

$$F(v) = B(u, v), \quad \forall v \in W.$$

Clearly $\int_{\Omega} f v dx$ is a bounded linear functional in H. Thus there exists a unique $u \in H$ such that (13.1.15) holds for any $v \in W$. In other words, (13.1.1), (13.1.14) admits a unique weak solution in H. □

13.1.3 *Solvability of the problem in $L^p(\Omega)$*

Now we proceed to introduce another kind of weak solutions.

Suppose $u \in C^2(\overline{\Omega})$ is a solution of (13.1.1), (13.1.14). Multiply both sides of (13.1.13) by $v \in V = \left\{v \in C^2(\overline{\Omega}); v\big|_{\Sigma_1 \cup \Sigma_3} = 0\right\}$ and integrate over Ω. After integrating by parts twice, we are led to

$$\int_{\Omega} v f dx = \int_{\Omega} u L^* v dx, \tag{13.1.16}$$

where $L^* v$ is the conjugate of Lu, namely,

$$L^* v = -D_i(a_{ij} D_j v) - \beta_i D_i v + c^* v, \quad c^* = -D_i \beta_i + c, \quad x \in \Omega.$$

Conversely, if $u \in C^2(\overline{\Omega})$ and (13.1.16) holds for any $v \in V$, then u is a solution of (13.1.1), (13.1.14).

Definition 13.1.2 A function $u \in L^p(\Omega)$ is said to be a weak solution of (13.1.1), (13.1.14) in $L^p(\Omega)$, if (13.1.16) holds for any $v \in V = \Big\{v \in C^2(\overline{\Omega}); v\big|_{\Sigma_1 \cup \Sigma_3} = 0\Big\}$.

It is easy to see that weak solutions in H are weak solutions in $L^2(\Omega)$, but not weak solutions in $L^p(\Omega)$ $(p > 2)$.

To prove the existence of weak solutions in $L^p(\Omega)$, we need to establish some a priori estimates.

Proposition 13.1.1 *Suppose that $c > 0$, $c^* > 0$ on $\overline{\Omega}$. Then*

  i) For any $u \in C^2(\overline{\Omega})$ with $u\big|_{\Sigma_2 \cup \Sigma_3} = 0$ and $p \geq 1$,

$$\|u\|_{L^p(\Omega)} \leq \frac{p}{\min\limits_{\Omega}(c^* + (p-1)c)}\|Lu\|_{L^p(\Omega)}; \tag{13.1.17}$$

  ii) For any $v \in C^2(\overline{\Omega})$ with $v\big|_{\Sigma_1 \cup \Sigma_3} = 0$ and $q \geq 1$,

$$\|v\|_{L^q(\Omega)} \leq \frac{q}{\min\limits_{\Omega}(c + (p-1)c^*)}\|L^*v\|_{L^q(\Omega)}. \tag{13.1.18}$$

Proof. We merely prove (13.1.17). Multiply Lu by $(u^2 + \delta)^{(p-2)/2}u$ $(\delta > 0)$,

$$(u^2 + \delta)^{(p-2)/2}uLu = -(u^2 + \delta)^{(p-2)/2}uD_j(a_{ij}D_iu)$$
$$+ (u^2 + \delta)^{(p-2)/2}\beta_iD_iu + (u^2 + \delta)^{(p-2)/2}cu^2.$$

Substituting

$$(u^2 + \delta)^{(p-2)/2}uD_j(a_{ij}D_iu) = D_j\Big((u^2 + \delta)^{(p-2)/2}ua_{ij}D_iu\Big)$$
$$- ((p-1)u^2 + \delta)(u^2 + \delta)^{(p-4)/2}a_{ij}D_iuD_ju,$$

$$(u^2 + \delta)^{(p-2)/2}u\beta_iD_iu = D_i\Big(\frac{1}{p}(u^2 + \delta)^{p/2}\beta_i\Big) - \frac{1}{p}(u^2 + \delta)^{p/2}D_i\beta_i$$

into the above formula and integrating, we obtain

$$\int_\Omega (u^2 + \delta)^{(p-2)/2}uLudx$$
$$= \int_\Sigma (u^2 + \delta)^{(p-2)/2}ua_{ij}D_iu\nu_jd\sigma$$
$$+ \int_\Omega ((p-1)u^2 + \delta)(u^2 + \delta)^{(p-4)/2}a_{ij}D_iuD_judx$$
$$- \frac{1}{p}\int_\Sigma (u^2 + \delta)^{p/2}\beta_i\nu_id\sigma - \frac{1}{p}\int_\Omega (u^2 + \delta)^{p/2}D_i\beta_idx$$
$$+ \int_\Omega (u^2 + \delta)^{p/2}cu^2dx.$$

Since $a_{ij}\nu_j\big|_{\Sigma^0} = 0\,(i = 1,\cdots,n)$, $u\big|_{\Sigma_3} = 0$, the first integral of the right side vanishes. In addition,

$$-\frac{1}{p}\int_\Sigma (u^2+\delta)^{p/2}\beta_i\nu_i d\sigma = -\frac{1}{p}\int_{\Sigma_1}(u^2+\delta)^{p/2}\beta d\sigma - \frac{1}{p}\int_{\Sigma_2\cup\Sigma_3}\delta^{p/2}\beta d\sigma.$$

Thus

$$\int_\Omega (u^2+\delta)^{(p-2)/2}uLudx$$
$$= \int_\Omega \big((p-1)u^2+\delta\big)(u^2+\delta)^{(p-4)/2}a_{ij}D_iuD_judx$$
$$\quad -\frac{1}{p}\int_\Omega (u^2+\delta)^{p/2}D_i\beta_idx + \int_\Omega (u^2+\delta)^{(p-2)/2}cu^2dx$$
$$\quad -\frac{1}{p}\int_{\Sigma_1}(u^2+\delta)^{p/2}\beta d\sigma - \frac{1}{p}\int_{\Sigma_2\cup\Sigma_3}\delta^{p/2}\beta d\sigma,$$

from which, noting that the first and fourth terms of the right side are nonnegative, we obtain

$$\int_\Omega (u^2+\delta)^{(p-2)/2}uLudx$$
$$\geq \frac{1}{p}\int_\Omega (u^2+\delta)^{p/2}(c^*-c)dx + \int_\Omega (u^2+\delta)^{(p-2)/2}cu^2dx$$
$$\quad -\frac{1}{p}\int_{\Sigma_2\cup\Sigma_3}\delta^{p/2}\beta d\sigma.$$

Letting $\delta \to 0^+$ then gives

$$\frac{1}{p}\int_\Omega \big(c^*+(p-1)c\big)|u|^pdx \leq \frac{1}{p}\int_\Omega |u|^{p-1}Ludx \leq \frac{1}{p}\int_\Omega |u|^{p-1}|Lu|dx,$$

from which, (13.1.17) follows by using the assumption $c^* > 0$, $c > 0$ and Hölder's inequality. □

Theorem 13.1.2 *Suppose that $c^* > 0$, $c > 0$ on $\overline{\Omega}$. Then for any $f \in L^p(\Omega)$ with $p > 1$, (13.1.1), (13.1.14) admits a weak solution in $L^p(\Omega)$.*

Proof. From Hölder's inequality and (13.1.18), we have

$$\left|\int_\Omega fvdx\right| \leq \|f\|_{L^p(\Omega)}\|v\|_{L^q(\Omega)} \leq K_q\|L^*v\|_{L^q(\Omega)}\|f\|_{L^p(\Omega)}, \quad \forall v \in V,$$

where $1/p+1/q = 1$. So $\int_\Omega fvdx$ is a bounded linear functional in $\{L^*v; v \in V\} \subset L^q(\Omega)$. Let $\tilde{L}^q(\Omega)$ be the completion of $\{L^*v; v \in V\}$ in $L^q(\Omega)$. Then we can first extend $\int_\Omega fv dx$ to be a bounded linear functional in $\tilde{L}^q(\Omega)$ and then use Hahn-Banach's theorem to further extend it to be a bounded linear functional $l(w)$ in $L^q(\Omega)$. Thus, by Riesz's representation theorem, it can be expressed as $\int_\Omega uw dx$ with some $u \in L^p(\Omega)$, in particular,

$$\int_\Omega fv dx = l(L^*v) = \int_\Omega uL^*v dx.$$

$\square$

It is to be noted that, since the extension of $\int_\Omega fv dx$ to $L^q(\Omega)$ is not unique, we can not assert the uniqueness of the weak solution in $L^p(\Omega)$ from the proof of Theorem 13.1.2.

13.1.4 *Method of elliptic regularization*

A frequently applied approach in treating equations with degeneracy is the elliptic regularization. The basic idea is to consider the regularized equation

$$L_\varepsilon u \equiv -\varepsilon\Delta u + Lu = -\varepsilon\Delta u - a_{ij}D_{ij}u + b_i D_i u + cu = f, \quad x \in \Omega \quad (13.1.19)$$

with $\varepsilon > 0$ and hope to obtain the required weak solution of (13.1.1) as the limit of the solution of (13.1.19). (13.1.19) is an elliptic equation; we can solve it by means of those methods and theories presented in previous chapters.

Let u_ε be a solution of (13.1.19), (13.1.14). Under the condition $c \geq c_0 > 0$ on $\overline{\Omega}$, by the maximum principle, we have

$$|u_\varepsilon| \leq \sup \frac{|f|}{c_0},$$

which implies the existence of a subsequence $\varepsilon_k \to 0$ and a function u, such that

$$u_{\varepsilon_k} \rightharpoonup u \,(\varepsilon_k \to 0) \quad \text{in } L^2(\Omega), \quad \text{as } k \to \infty,$$

where $\rightharpoonup$ denotes the weak convergence.

We hope that the function u thus obtained is a solution, at least a weak solution of (13.1.1), (13.1.14), i.e. for any $v \in V = \left\{ v \in C^2(\overline{\Omega}); v\big|_{\Sigma_1 \cup \Sigma_3} = 0 \right\}$, (13.1.16) holds.

From (13.1.19), we have

$$
\int_\Omega f v dx = - \varepsilon \int_\Omega \Delta u_\varepsilon v dx + \int_\Omega L u_\varepsilon v dx
$$

$$
= \varepsilon \int_{\Sigma_0 \cup \Sigma_2} \frac{\partial u_\varepsilon}{\partial \vec{\nu}} v d\sigma + \varepsilon \int_\Omega D_i u_\varepsilon D_i v dx + \int_\Omega u_\varepsilon L^* v dx
$$

$$
= \varepsilon \int_{\Sigma_0 \cup \Sigma_2} \frac{\partial u_\varepsilon}{\partial \vec{\nu}} v d\sigma - \varepsilon \int_\Sigma u_\varepsilon \frac{\partial v}{\partial \vec{\nu}} d\sigma - \varepsilon \int_\Omega u_\varepsilon \Delta v dx
$$

$$
+ \int_\Omega u_\varepsilon L^* v dx. \tag{13.1.20}
$$

By the uniform boundedness and the weak convergence of u_{ε_k}, we may assert

$$
\varepsilon \int_\Sigma u_\varepsilon \frac{\partial v}{\partial \vec{\nu}} d\sigma \to 0, \quad \varepsilon \int_\Omega u_\varepsilon \Delta v dx \to 0,
$$

$$
\int_\Omega u_\varepsilon L^* v dx \to \int_\Omega u L^* v dx \quad (\varepsilon = \varepsilon_k \to 0).
$$

If, in addition, we can prove

$$
\varepsilon \int_{\Sigma_0 \cup \Sigma_2} \frac{\partial u_\varepsilon}{\partial \vec{\nu}} v d\sigma \to 0 \quad (\varepsilon = \varepsilon_k \to 0), \tag{13.1.21}
$$

then (13.1.16) follows from (13.1.20) by letting $\varepsilon = \varepsilon_k \to 0$ and u really is a weak solution in $L^2(\Omega)$ of (13.1.1), (13.1.14).

In order (13.1.21) holds, it suffices to establish the following key estimate:

$$
|D_i u_\varepsilon| \le M \varepsilon^{-1/2}, \quad \forall x \in \Sigma_0 \cup \Sigma_2 \quad (i = 1, \cdots, n).
$$

For the proof, we refer to [Oleĭnik and Radkevič (1973)].

13.1.5 *Uniqueness of weak solutions in $L^p(\Omega)$ and regularity*

It has been proved that the weak solution in $L^p(\Omega)$ with $p \ge 3$ of (13.1.1), (13.1.14) is unique, but it is not the case when $1 \le p < 3$ (see [Oleĭnik and Radkevič (1973)]). Here we merely sketch the method of proof, which is based on Holmgren's idea.

What we have to do is to prove that, if $u \in L^p(\Omega)$ satisfies the identity

$$\int_\Omega u L^* v dx = 0, \quad \forall v \in V = \left\{ v \in C^2(\overline{\Omega}); v \big|_{\Sigma_1 \cup \Sigma_3} = 0 \right\}, \qquad (13.1.22)$$

then $u = 0$ a.e. in Ω.

If for any $\varphi \in C_0^\infty(\Omega)$, the conjugate problem

$$L^* v = \varphi \text{ in } \Omega, \quad v \big|_{\Sigma_1 \cup \Sigma_3} = 0 \qquad (13.1.23)$$

had a solution $v \in C^2(\overline{\Omega})$, then we would have

$$\int_\Omega u \varphi dx = 0, \quad \forall \varphi \in C_0^\infty(\Omega)$$

and $u = 0$ a.e. in Ω would follow immediately.

However, it is difficult, even impossible, to prove the existence of classical solutions of the problem for the conjugate equation which is also degenerate. In view of this, instead of (13.1.23), one naturally considers its regularized problem

$$-\varepsilon \Delta v + L^* v = \varphi \text{ in } \Omega, \quad v \big|_\Sigma = 0 \quad (\varepsilon > 0). \qquad (13.1.24)$$

If $v \in C^2(\overline{\Omega})$ is a classical solution of (13.1.24), then, by assumption (13.1.22), we have

$$\int_\Omega u \varphi dx = -\varepsilon \int_\Omega u \Delta v dx + \int_\Omega u L^* v dx = -\varepsilon \int_\Omega u \Delta v dx.$$

To our purpose, it suffices to prove

$$\varepsilon \int_\Omega u \Delta v dx \to 0 \quad (\varepsilon \to 0). \qquad (13.1.25)$$

Under certain conditions, one can establish the estimate

$$\int_\Omega (\Delta v)^2 dx \le \frac{M}{\varepsilon^2} \qquad (13.1.26)$$

with some constant $M > 0$ (see [Oleĭnik and Radkevič (1973)]). (13.1.25) is an immediate consequence of this estimate. To verify this fact, we express $\varepsilon \int_\Omega u \Delta v dx$ as

$$\varepsilon \int_\Omega u \Delta v dx = \varepsilon \int_\Omega (u - u') \Delta v dx + \varepsilon \int_\Omega u' \Delta v dx$$

$$= \varepsilon \int_\Omega (u - u') \Delta v \, dx + \varepsilon \int_\Omega \Delta u' v \, dx,$$

where $u' \in C_0^\infty(\Omega)$. For any given $\delta > 0$, using (13.1.26), we may choose $u' \in C_0^\infty(\Omega)$ such that $\int_\Omega (u - u')^2 \, dx$ is so small to make

$$\left| \varepsilon \int_\Omega (u - u') \Delta v \, dx \right| \leq \varepsilon \left(\int_\Omega (u - u')^2 \, dx \right)^{1/2} \left(\int_\Omega (\Delta v)^2 \, dx \right)^{1/2} < \frac{\delta}{2}.$$

For fixed u', using the uniform boundedness of v in ε (following from the maximum principle), we have

$$\left| \varepsilon \int_\Omega \Delta u' v \, dx \right| < \frac{\delta}{2},$$

when $\varepsilon > 0$ is small enough.

Another important problem for degenerate equations is the regularity of weak solutions. Many authors have studied the global regularity of weak solutions by means of elliptic regularization. Uniform estimates in $C^k(\Omega)$ of solutions $\{u_\varepsilon\}$ of the regularized problems have been established under certain conditions. Based on these estimates, weak solutions of the original problem are proved to be functions in $C^k(\Omega)$. Also uniform estimates in some Sobolev space have been established under certain conditions and hence weak solutions are proved to be functions in this space.

The study of local regularity is related to the subellipticity of equations. A linear differential operator with C^∞ coefficients defined in Ω is said to be a subelliptic operator, if for any distribution u and any domain $\Omega' \subset\subset \Omega$, $Pu \in C^\infty(\Omega')$ implies $u \in C^\infty(\Omega')$.

It has been proved that any subelliptic operator possesses nonnegative or nonpositive characteristic form. Various conditions have been discovered for linear degenerate elliptic equations of second order to be subelliptic.

13.2　A Class of Special Quasilinear Degenerate Parabolic Equations – Filtration Equations

In this section, we proceed to discuss quasilinear equations. As seen in the previous section, the study of linear degenerate elliptic equations is more difficult than that for equations without degeneracy. Much more difficulty would be caused by the quasilinearity of equations. We do not attempt to present the argument for general degenerate elliptic equations. We merely introduce theory and methods for some special kinds of such equations. In this section, we are concerned with a certain kind of typical quasilinear degenerate parabolic equations, called filtration equations:

$$\frac{\partial u}{\partial t} = \Delta A(u), \tag{13.2.1}$$

where $A(s) \in C^1[0, +\infty)$ satisfies

$$A(0) = A'(0) = 0, \quad A'(s) > 0 \text{ for } s > 0. \tag{13.2.2}$$

Equation (13.2.1) is parabolic when $u > 0$. However, it degenerates when $u = 0$. If we do not restrict ourselves to the study of nonnegative solutions, then, instead, we assume that $A(s) \in C^1(-\infty, +\infty)$ satisfies

$$A(0) = A'(0) = 0, \quad A'(s) > 0 \text{ for } s \neq 0.$$

An important example of (13.2.1) is the Newtonian filtration equation

$$\frac{\partial u}{\partial t} = \Delta u^m \tag{13.2.3}$$

with $m > 1$, which corresponds to the slow diffusion. If we do not restrict ourselves to the study of nonnegative solutions, then (13.2.3) should be written as

$$\frac{\partial u}{\partial t} = \Delta(|u|^{m-1}u).$$

In this section, we merely consider the Cauchy problem for (13.2.1) with the initial condition

$$u(x, 0) = u_0(x), \quad x \in \mathbb{R}^n, \tag{13.2.4}$$

where $u_0(x)$ is a nonnegative and locally integrable function.

13.2.1 Definition of weak solutions

Let $Q_T = \mathbb{R}^n \times (0, T)$ and G be a subdomain of Q_T.

Definition 13.2.1 A nonnegative function u is said to be a weak solution of (13.2.1) in G, if $u, A(u) \in L^1_{\text{loc}}(G)$ and u satisfies

$$\iint_G \left(u \frac{\partial \varphi}{\partial t} + A(u) \Delta \varphi \right) dx dt = 0$$

for any $\varphi \in C_0^\infty(G)$.

Definition 13.2.2   A nonnegative function u is said to be a weak solution of (13.2.1), (13.2.4) in Q_T, if $u, A(u) \in L^1_{loc}(Q_T)$ and u satisfies

$$\iint_{Q_T} \left(u\frac{\partial \varphi}{\partial t} + A(u)\Delta\varphi \right) dxdt + \int_{\mathbb{R}^n} u_0(x)\varphi(x,0)dx = 0 \qquad (13.2.5)$$

for any $\varphi \in C^\infty(\overline{Q}_T)$, which vanishes when $|x|$ is large enough or $t = T$.

Remark 13.2.1   If $\dfrac{\partial A(u)}{\partial x_i} \in L^1_{loc}(Q_T)\, (i = 1, \cdots, n)$, then (13.2.5) can be transformed to the form

$$\iint_{Q_T} \left(u\frac{\partial \varphi}{\partial t} - \nabla A(u) \cdot \nabla\varphi \right) dxdt + \int_{\mathbb{R}^n} u_0(x)\varphi(x,0)dx = 0. \qquad (13.2.6)$$

If both u and $\dfrac{\partial A(u)}{\partial x_i}\, (i = 1, \cdots, n)$ are bounded, then, by approximation, it is easy to see that (13.2.6) holds for any $\varphi \in W^{1,\infty}(Q_T)$, vanishing when $|x|$ is large enough or $t = T$.

Remark 13.2.2   If u is a weak solution of (13.2.1), (13.2.4) in Q_T, then, for any $\tau \in (0, T)$, there holds

$$\iint_{Q_\tau} \left(u\frac{\partial \varphi}{\partial t} + A(u)\Delta\varphi \right) dxdt - \int_{\mathbb{R}^n} u(x,\tau)\varphi(x,\tau)dx$$

$$+ \int_{\mathbb{R}^n} u_0(x)\varphi(x,0)dx = 0 \qquad (13.2.7)$$

for any $\varphi \in C^\infty(\overline{Q}_T)$, vanishing when $|x|$ is large enough, where $Q_\tau = \mathbb{R}^n \times (0, \tau)$.

To prove, we choose $\varphi\eta_\varepsilon$ as a test function in (13.2.5), where $\eta_\varepsilon \in C^\infty[0,T]$ such that $\eta_\varepsilon(t) = 1$ for $t \in [0, \tau - \varepsilon]$, $\eta_\varepsilon(t) = 0$ for $t \in [\tau, T]$, $|\eta'_\varepsilon(t)| \leq \dfrac{C}{\varepsilon}$. Then we have

$$\iint_{Q_\tau} \eta_\varepsilon \left(u\frac{\partial \varphi}{\partial t} + A(u)\Delta\varphi \right) dxdt + \iint_{Q_\tau} u\varphi\eta'_\varepsilon dxdt + \int_{\mathbb{R}^n} u_0(x)\varphi(x,0)dx = 0,$$

from which, letting $\varepsilon \to 0$ and noticing that

$$\left| \iint_{Q_\tau} u\varphi\eta'_\varepsilon dxdt - \int_{\mathbb{R}^n} u(x,\tau)\varphi(x,\tau)dx \right|$$

$$= \left| \iint_{Q_\tau} \eta'_\varepsilon(t)\Big(u(x,t)\varphi(x,t) - u(x,\tau)\varphi(x,\tau) \Big) dxdt \right|$$

$$\leq \frac{C}{\varepsilon} \int_{\tau-\varepsilon}^{\tau} \left| \int_{\mathbb{R}^n} \big(u(x,t)\varphi(x,t) - u(x,\tau)\varphi(x,\tau) \big) dx \right| dt$$

$$\longrightarrow 0 \quad (\varepsilon \to 0),$$

we obtain (13.2.7).

Conversely, if for any $\tau \in (0,T)$ and any $\varphi \in C^\infty(\overline{Q}_T)$ vanishing when $|x|$ is large enough, (13.2.7) holds, then obviously, (13.2.5) holds for any $\varphi \in C^\infty(\overline{Q}_T)$, vanishing when $|x|$ is large enough or $t = T$.

13.2.2 *Uniqueness of weak solutions for one dimensional equations*

We first discuss the uniqueness of weak solutions of the Cauchy problem for (13.2.1) in one spatial dimension:

$$\frac{\partial u}{\partial t} = \frac{\partial^2 A(u)}{\partial x^2}, \qquad (x,t) \in Q_T, \qquad (13.2.8)$$

$$u(x,0) = u_0(x), \qquad x \in \mathbb{R}^n, \qquad (13.2.9)$$

where $A(s) \in C^1[0,+\infty)$ satisfies (13.2.2) and $0 \leq u_0(x) \in L^1_{\mathrm{loc}}(\mathbb{R})$.

Theorem 13.2.1 *The Cauchy problem (13.2.8), (13.2.9) has at most one weak solution u which is bounded together with the weak derivative $\dfrac{\partial A(u)}{\partial x}$.*

Proof. Let u_1, u_2 be weak solutions of (13.2.8), (13.2.9), which are bounded together with $\dfrac{\partial A(u_1)}{\partial x}, \dfrac{\partial A(u_2)}{\partial x}$. Then u_1, u_2 satisfy (13.2.6) and hence

$$\iint_{Q_T} \frac{\partial \varphi}{\partial t}(u_1 - u_2)\,dxdt = \iint_{Q_T} \left(\frac{\partial A(u_1)}{\partial x} - \frac{\partial A(u_2)}{\partial x} \right) dxdt \qquad (13.2.10)$$

for any $\varphi \in W^{1,\infty}(Q_T)$ which vanishes when $|x|$ is large enough or $t = T$. If

$$\varphi(x,t) = \int_T^t \Big(A(u_1(x,\tau)) - A(u_2(x,\tau)) \Big) d\tau \qquad (13.2.11)$$

could be chosen as a test function, then from (13.2.8) we would have

$$\iint_{Q_T} \big(A(u_1) - A(u_2) \big)(u_1 - u_2)\,dxdt$$

$$= \iint_{Q_T} \int_T^t \left(\frac{\partial A(u_1(x,\tau))}{\partial x} - \frac{\partial A(u_2(x,\tau))}{\partial x} \right) d\tau$$

$$\cdot \left(\frac{\partial A(u_1(x,t))}{\partial x} - \frac{\partial A(u_2(x,t))}{\partial x} \right) dx dt$$

$$= \frac{1}{2} \iint_{Q_T} \frac{\partial}{\partial t} \left(\int_T^t \left(\frac{\partial A(u_1)}{\partial x} - \frac{\partial A(u_2)}{\partial x} \right) d\tau \right)^2 dx dt$$

$$= -\frac{1}{2} \int_{-\infty}^{\infty} \left(\int_T^0 \left(\frac{\partial A(u_1)}{\partial x} - \frac{\partial A(u_2)}{\partial x} \right) d\tau \right)^2 dx$$

$$\leq 0$$

and hence $u_1 = u_2$ a.e. in $\mathbb{R}$ due to the condition (13.2.2).

However the function φ defined by (13.2.11) can not play the role of a test function because, in general, it does not vanish for large $|x|$, although we have $\varphi \in W^{1,\infty}(Q_T)$ and $\varphi(x,T) = 0$.

In view of this, it is natural to cut-off φ, i.e. to use the following function instead of φ:

$$\varphi_k(x,t) = \alpha_k(x)\varphi(x,t) = \alpha_k(x) \int_T^t \left(A(u_1) - A(u_2) \right) dx d\tau,$$

where $\alpha_k(x)$ is a smooth function such that $\alpha_k(x) = 1$ when $|x| \leq k - 1$, $\alpha_k(x) = 0$ when $|x| \geq k$, $0 \leq \alpha_k(x) \leq 1$ when $k - 1 \leq |x| \leq k$ and $\alpha'_k(x)$ is bounded uniformly in k.

Substituting $\varphi = \varphi_k$ into (13.2.10) yields

$$I_{1k}$$

$$= \iint_{Q_T} \alpha_k(x) \left(A(u_1) - A(u_2) \right)(u_1 - u_2) dx dt$$

$$= -\frac{1}{2} \int_{-\infty}^{\infty} \alpha_k(x) \left(\int_T^0 \left(\frac{\partial A(u_1)}{\partial x} - \frac{\partial A(u_2)}{\partial x} \right) d\tau \right)^2 dx$$

$$+ \iint_{Q_T} \alpha'_k(x) \left(\int_T^t \left(A(u_1) - A(u_2) \right) d\tau \right) \left(\frac{\partial A(u_1)}{\partial x} - \frac{\partial A(u_2)}{\partial x} \right) dx dt$$

$$\leq \iint_{Q_T^k} \alpha'_k(x) \left(\int_T^t \left(A(u_1) - A(u_2) \right) d\tau \right) \left(\frac{\partial A(u_1)}{\partial x} - \frac{\partial A(u_2)}{\partial x} \right) dx dt$$

$$= I_{2k}, \tag{13.2.12}$$

where

$$Q_T^k = \{ (x,t) \in Q_T; k - 1 \leq |x| \leq k, 0 < t < T \}.$$

Since $u_i, \dfrac{\partial A(u_i)}{\partial x}$ $(i = 1, 2)$ are bounded and $\alpha'_k(x)$ is bounded uniformly in k, I_{2k} is bounded, so is I_{1k}. By definition, $\alpha_k(x)$ increases with k, so does I_{1k}. Thus $\lim\limits_{k \to \infty} I_{1k}$ exists and we have

$$\lim_{k \to \infty} I_{1k} = \iint_{Q_T} \left(A(u_1) - A(u_2) \right)(u_1 - u_2) dx dt.$$

It is easy to verify that

$$\lim_{k \to \infty} I_{2k} = 0. \tag{13.2.13}$$

In fact, from the boundedness of $u_i, \dfrac{\partial A(u_i)}{\partial x}$ $(i = 1, 2)$ and the uniform boundedness of $\alpha'_k(x)$, we have

$$
\begin{aligned}
|I_{2k}| &\leq C \iint_{Q_T^k} |A(u_1) - A(u_2)| dx dt \\
&\leq C \Big(\iint_{Q_T^k} \left(A(u_1) - A(u_2) \right)^2 dx dt \Big)^{1/2} \\
&\leq C \Big(\iint_{Q_T^k} \left(A(u_1) - A(u_2) \right)(u_1 - u_2) dx dt \Big)^{1/2},
\end{aligned}
$$

where C is a constant independent of k. The finiteness of the integral $\iint_{Q_T} \left(A(u_1) - A(u_2) \right)(u_1 - u_2) dx dt$ implies

$$\lim_{k \to \infty} \iint_{Q_T^k} \left(A(u_1) - A(u_2) \right)(u_1 - u_2) dx dt = 0$$

and hence (13.2.13) holds. Combining (13.2.13) with (13.2.12) we finally obtain

$$\iint_{Q_T} \left(A(u_1) - A(u_2) \right)(u_1 - u_2) dx dt = \lim_{k \to \infty} I_{1k} = 0,$$

which implies that $u_1 = u_2$ a.e. in Q_T. $\qquad\square$

13.2.3 Existence of weak solutions for one dimensional equations

Now we proceed to discuss the existence of weak solutions of the Cauchy problem (13.2.8), (13.2.9).

Theorem 13.2.2   *Assume that $u_0(x) \geq 0$ is a continuous and bounded function in $\mathbb{R}$, $A(u_0(x))$ satisfies the Lipschitz condition, $A(s)$ is appropriately smooth and $\lim\limits_{s \to +\infty} A(s) = +\infty$. Then for any $T > 0$, the Cauchy problem (13.2.8), (13.2.9) admits a continuous, nonnegative and bounded weak solution u in Q_T such that $\dfrac{\partial A(u)}{\partial x}$ is bounded. Moreover, the solution u is classical in $\{(x,t) \in Q_T; u(x,t) > 0\}$.*

Proof.   Denote $v_0 = A(u_0)$ and choose a sequence of smooth functions $\{v_{0k}(x)\}$ converging to $v_0(x)$ uniformly as $k \to \infty$ and satisfying

$$0 < v_{0k}(x) \leq M, \quad \left|\frac{d}{dx}v_{0k}(x)\right| \leq K_0, \quad x \in \mathbb{R} \quad (k = 1, 2, \cdots)$$

with some constants M and K_0. Construct a sequence of smooth functions $\{w_k(x)\}$ such that

$$w_k(x) = \begin{cases} v_{0k}(x), & \text{when } |x| \leq k - 2, \\ M, & \text{when } |x| \geq k - 1 \end{cases}$$

and

$$0 < w_k(x) \leq M, \quad \left|\frac{d}{dx}w_k(x)\right| \leq N = \max\{K_0, M\}, \quad x \in \mathbb{R} \quad (k = 1, 2, \cdots).$$

Denote $v = A(u)$, $\Phi(v) = A^{-1}(v)$. Then (13.2.8) becomes

$$\Phi'(v)\frac{\partial v}{\partial t} = \frac{\partial^2 v}{\partial x^2}. \tag{13.2.14}$$

Consider the initial-boundary value problem for (13.2.14) with conditions

$$v(x, 0) = w_k(x), \quad x \in (-k, k), \qquad v(\pm k, t) = M, \quad t \in (0, T). \tag{13.2.15}$$

Since the initial-boundary value is positive, we may apply the standard theory of parabolic equations to assert that (13.2.14), (13.2.15) admits a classical solution $v_k(x, t)$ in $G_k = (-k, k) \times (0, T)$.

The maximum principle for classical solutions shows that

$$0 < \min_{\mathbb{R}} w_k(x) \leq v_k(x, t) \leq M. \tag{13.2.16}$$

A crucial step is to prove

$$\left|\frac{\partial v_k}{\partial x}\right| \leq N, \quad \text{in } G_k. \tag{13.2.17}$$

For this purpose, we first use the maximum principle to $P_k = \dfrac{\partial v_k}{\partial x}$ which satisfies

$$\Phi'(v_k)\frac{\partial P_k}{\partial t} = \frac{\partial^2 P_k}{\partial x^2} - \frac{1}{\Phi'(v_k)}\frac{\partial}{\partial x}\Phi'(v_k)\frac{\partial P_k}{\partial x}, \quad \text{in } G_k$$

and then obtain

$$\max_{\overline{G}_k}|P_k| = \max_{\overline{G}_k}\left|\frac{\partial v_k}{\partial x}\right| \leq \max_{\Gamma_k}|P_k| = \max_{\Gamma_k}\left|\frac{\partial v_k}{\partial x}\right|,$$

where $\Gamma_k = \partial_p G_k$ is the parabolic boundary of G_k. Since

$$\left.\frac{\partial v_k}{\partial x}\right|_{t=0} = |w_k'(x)| \leq N, \quad x \in [-k, k],$$

to prove (13.2.17), it suffices to show

$$\left.\left|\frac{\partial v_k}{\partial x}\right|\right|_{x=\pm k} \leq N, \quad t \in [0, T]. \tag{13.2.18}$$

Notice that v_k achieves its maximum M on the lateral $\{k\} \times [0, T]$. Hence

$$\left.\frac{\partial v_k}{\partial x}\right|_{x=k} \geq 0, \quad t \in [0, T]. \tag{13.2.19}$$

Consider the auxiliary function

$$z_k(x, t) = v_k(x, t) - M(x - k + 1), \quad (x, t) \in \overline{G}_k,$$

which satisfies

$$\Phi'(v_k)\frac{\partial z_k}{\partial t} = \frac{\partial^2 z_k}{\partial x^2}, \quad (x, t) \in D_k = (k - 1, k) \times (0, T),$$
$$z_k(x, 0) = w_k(x) - M(x - k + 1) = M(k - x), \quad x \in (k - 1, k),$$
$$z_k(k, t) = 0, \quad z_k(k - 1, t) = v_k(k - 1, t) > 0, \quad t \in (0, T).$$

Since z_k achieves its minimum $\min_{\overline{D}_k} z_k$ on $\{k\} \times [0, T]$, we have

$$\left.\frac{\partial z_k}{\partial x}\right|_{x=k} = \left.\frac{\partial v_k}{\partial x}\right|_{x=k} - M \leq 0, \quad t \in [0, T],$$

which combined with (13.2.19) yields

$$0 \leq \left.\frac{\partial v_k}{\partial x}\right|_{x=k} \leq M \leq N, \quad t \in [0, T].$$

Another part of (13.2.18) can be proved similarly.

The estimate (13.2.17) implies the uniform Lipschitz continuity of v_k in x: for any $(x,t),(y,t) \in \overline{Q}_T$ and sufficiently large k such that $(x,t),(y,t) \in \overline{G}_k$,

$$|v_k(x,t) - v_k(y,t)| \leq N|x - y|. \qquad (13.2.20)$$

Denote $u_k = A^{-1}(v_k)$. The existence of the inverse function $A^{-1}(s)$ for $s \in [0, +\infty)$ follows from the assumptions $A'(s) > 0$ and $\lim_{s \to +\infty} A(s) = +\infty$. From (13.2.14) we have

$$\frac{\partial u_k}{\partial t} = \frac{\partial^2 v_k}{\partial x^2}. \qquad (13.2.21)$$

For any $(x,t),(y,s) \in \overline{Q}_T$, choose k large enough such that $(x,t),(y,s) \in \overline{G}_k$, $x + |\Delta t|^{1/2} \in [-k, k]$ with $\Delta t = t - s$. Integrating (13.2.21) over $(x, x + |\Delta t|^{1/2}) \times (s,t)$ yields

$$\int_x^{x+|\Delta t|^{1/2}} \Big(u_k(z,t) - u_k(z,s)\Big)dz$$

$$= \int_s^t \int_x^{x+|\Delta t|^{1/2}} \frac{\partial u_k}{\partial t}(z,\tau)dzd\tau$$

$$= \int_s^t \int_x^{x+|\Delta t|^{1/2}} \frac{\partial^2 v_k}{\partial x^2}(z,\tau)dzd\tau$$

$$= \int_s^t \Big(\frac{\partial v_k}{\partial x}(x + |\Delta t|^{1/2},\tau) - \frac{\partial v_k}{\partial x}(x,\tau)\Big)d\tau.$$

Hence from (13.2.17) we obtain

$$\left|\int_x^{x+|\Delta t|^{1/2}} \Big(u_k(z,t) - u_k(z,s)\Big)dz\right| \leq 2N|\Delta t|.$$

Using the mean value theorem for integrals, we see that there exists $x^* \in [x, x + |\Delta t|^{1/2}]$ such that

$$\int_x^{x+|\Delta t|^{1/2}} \Big(u_k(z,t) - u_k(z,s)\Big)dz = \Big(u_k(x^*,t) - u_k(x^*,s)\Big)|\Delta t|^{1/2}.$$

Thus

$$|u_k(x^*,t) - u_k(x^*,s)| \leq 2N|\Delta t|^{1/2}$$

and hence for some constant C

$$|v_k(x^*,t) - v_k(x^*,s)| = |A(u_k(x^*,t)) - A(u_k(x^*,s))|$$

$$= |A'(\xi_k)||u_k(x^*, t) - u_k(x^*, s)|$$
$$\leq C|\Delta t|^{1/2}.$$

Combining this with (13.2.20) we obtain

$$|v_k(x, t) - v_k(y, s)| \leq |v_k(x, t) - v_k(x^*, t)| + |v_k(x^*, t) - v_k(x^*, s)|$$
$$+ |v_k(x^*, s) - v_k(y, s)|$$
$$\leq N(|x - x^*| + |x^* - y|) + C|\Delta t|^{1/2}$$
$$\leq (2N + C)(|x - y| + |\Delta t|^{1/2})$$

and assert the uniform continuity of $\{v_k\}$ in G_k with exponent $\{1, 1/2\}$. This together with (13.2.16), (13.2.18) implies that there exists a subsequence of $\{v_k\}$, supposed to be $\{v_k\}$ itself, such that $\{v_k\}$ converges to a certain continuous function v uniformly in any compact subset of $\overline{Q}_T$, and $\left\{\dfrac{\partial v_k}{\partial x}\right\}$ weak star converges to $\dfrac{\partial v}{\partial x}$ in any bounded subdomain of $\overline{Q}_T$. Furthermore it is easy to see that $\{u_k\}$ $(u_k = A^{-1}(v_k))$ converges to $u = A^{-1}v$ uniformly in any compact subset of $\overline{Q}_T$.

Obviously u and $\dfrac{\partial A(u)}{\partial x}$ are bounded. Given any test function φ, i.e. $\varphi \in C^\infty(\overline{Q}_T)$ such that $\varphi = 0$ for large $|x|$ or $t = T$. Let k be large enough such that supp$\varphi \subset G_k$. Multiplying (13.2.21) by φ, integrating over Q_T and integrating by parts, we obtain

$$\iint_{Q_T} \left(u_k \frac{\partial \varphi}{\partial t} - \frac{\partial A(u_k)}{\partial x} \cdot \frac{\partial \varphi}{\partial x}\right) dx dt + \int_{-\infty}^{\infty} \varphi(x, 0) A^{-1}(w_k(x)) dx = 0,$$

from which it follows by letting $k \to \infty$ and noticing that for large k, $w_k(x) = v_{0k}(x)$ converges to $v_0 = A(u_0)$ uniformly in any finite interval, that u satisfies (13.2.6), i.e. u is a weak solution of (13.2.8), (13.2.9).

Finally, we prove that u is classical in $\{(x, t) \in Q_T; u(x, t) > 0\}$. Let $(x_0, t_0) \in Q_T$, $u(x_0, t_0) > 0$. Then in some neighborhood $U \subset Q_T$ of (x_0, t_0), we have

$$u(x, t) \geq \alpha_0 > 0$$

with some constant α_0. Hence for any $(x, t) \in U$ and large k,

$$u_k(x, t) \geq \frac{\alpha_0}{2} > 0.$$

This means that for large k, u_k satisfies

$$\frac{\partial u_k}{\partial t} = \frac{\partial}{\partial x}\left(a(x,t)\frac{\partial u_k}{\partial t}\right), \quad (x,t) \in U$$

with $a(x,t) = A'(u_k)$, which is uniformly parabolic in U. From the standard theory of parabolic equations, it follows that for large k, u_k is uniformly bounded and equicontinuous in $C^2(U)$. Thus $u \in C^2(U)$ and u satisfies (13.2.8) in the classical sense.   □

13.2.4   *Uniqueness of weak solutions for higher dimensional equations*

Now we turn to the higher dimensional case. We first study the uniqueness of weak solutions. The same method as the proof of Theorem 13.2.1 can be used to prove that (13.2.1), (13.2.4) admits at most one weak solution u which is bounded together with $\nabla A(u)$. However, it is difficult to prove the existence of such weak solutions in higher dimensional case; one can not obtain such solutions even under rather restrictive conditions on $A(u)$ and $u_0(x)$.

In what follows, we present a uniqueness theorem for equation (13.2.3), which is valid also for equations (13.2.1) satisfying (13.2.2).

Theorem 13.2.3   *Suppose $0 \le u_0 \in L^1(\mathbb{R}^n) \cap L^\infty(\mathbb{R}^n)$. Then the Cauchy problem (13.2.3), (13.2.4) admits at most one weak solution in $L^1(Q_T) \cap L^\infty(Q_T)$.*

Proof.   From Remark 13.2.1, we have

$$\int_{\mathbb{R}^n} u_i(x,\tau)\varphi(x,\tau)dx - \int_{\mathbb{R}^n} u_0(x,\tau)\varphi(x,0)dx$$

$$= \iint_{Q_\tau}\left(u_i\frac{\partial\varphi}{\partial t} + u_i^m \Delta\varphi\right)dxdt \quad (i = 1,2)$$

for any $\tau \in (0,T)$ and $\varphi \in C^\infty(\overline{Q}_T)$, vanishing when $|x|$ is large enough.

Let $u = u_1 - u_2$. Then

$$\int_{\mathbb{R}^n} u(x,\tau)\varphi(x,\tau)dx = \iint_{Q_\tau}\left(u\frac{\partial\varphi}{\partial t} + (u_1^m - u_2^m)\Delta\varphi\right)dxdt$$

$$= \iint_{Q_\tau} u\left(\frac{\partial\varphi}{\partial t} + a\Delta\varphi\right)dxdt, \quad\quad (13.2.22)$$

where

$$a(x,t) = \begin{cases} \dfrac{u_1^m(x,t) - u_2^m(x,t)}{u_1(x,t) - u_2(x,t)}, & \text{if } u_1(x,t) \neq u_2(x,t), \\ mu_1^{m-1}(x,t), & \text{if } u_1(x,t) = u_2(x,t), \end{cases} \quad (x,t) \in Q_\tau.$$

If for any function $g \in C_0^\infty(\mathbb{R}^n)$, the problem

$$\begin{cases} \dfrac{\partial\varphi}{\partial t} + a\Delta\varphi = 0, & (x,t) \in Q_\tau, \\ \varphi(x,\tau) = g(x), & x \in \mathbb{R}^n \end{cases} \quad (13.2.23)$$

had a solution $\varphi \in C^\infty(\overline{Q}_T)$, vanishing when $|x|$ is large enough, then from (13.2.22) we would have

$$\int_{\mathbb{R}^n} u(x,\tau)g(x)dx = 0, \quad (13.2.24)$$

which would imply $u(x,\tau) = 0$ a.e. for $x \in \mathbb{R}^n$ and this is just what we want to prove.

However, since the coefficient a in (13.2.23) is merely a locally integrable function, (13.2.23) does not admit any smooth solution in general and even if (13.2.23) does admit, the solution can not have compact support in x in general. In view of this, we replace a by

$$a_k(x,t) = \rho_k(x,t) * a(x,t) + \frac{1}{k}, \quad (x,t) \in Q_\tau,$$

where ρ_k is a mollifier in $\mathbb{R}^{n+1}$, and consider the boundary value problem

$$\begin{cases} \dfrac{\partial\varphi}{\partial t} + a_k\Delta\varphi = 0, & |x| < R, 0 < t < \tau, \\ \varphi(x,t) = 0, & |x| = R, 0 < t < \tau, \\ \varphi(x,\tau) = g(x), & |x| < R, \end{cases} \quad (13.2.25)$$

where $R > R_0 + 1$ such that supp $g(x) \subset B_{R_0} = \{x \in \mathbb{R}^n; |x| < R_0\}$. We choose ρ_k such that

$$\int_0^\tau \int_{B_R} (a - \rho_k * a)^2 dx dt \leq \frac{1}{k^2}. \quad (13.2.26)$$

Let φ_k be a solution of (13.2.25) and extend it to the whole Q_τ by setting $\varphi_k = 0$ outside $\overline{B}_R \times [0,\tau]$. Since the extended function $\varphi_k = 0$ may not necessarily be a sufficiently smooth function in Q_τ, we use a function

$\xi_R \in C_0^\infty(\mathbb{R}^n)$ with the following properties to "cut-off" φ_k: $0 \le \xi_R(x) \le 1$, $\xi_R(x) = 1$ for $|x| < R - 1$, $\xi_R(x) = 0$ for $|x| > R - 1/2$,

$$|\nabla \xi_R(x)| + |\Delta \xi_R(x)| \le C. \tag{13.2.27}$$

Here and below, as we did before, we use C to denote a universal constant independent of R and k, which may take different values on different occasions. Choosing $\varphi = \xi_R \varphi_k$ in (13.2.22) gives

$$\int_{\mathbb{R}^n} u(x, \tau) g(x) \xi_R(x) dx = \iint_{Q_\tau} (u_1^m - u_2^m)(2\nabla \xi_R \cdot \nabla \varphi_k + \varphi_k \Delta \xi_R) dx dt$$

$$+ \iint_{Q_\tau} u \xi_R (a - a_k) \Delta \varphi_k dx dt$$

$$= I_k + J_k. \tag{13.2.28}$$

Now we are ready to estimate I_k and J_k. Multiplying the equation in (13.2.25) by $\Delta \varphi_k$, integrating over $B_R \times (t, \tau)$ and integrating by parts, we obtain for any $0 < t < \tau$

$$\frac{1}{2} \int_{B_R} |\nabla \varphi_k(x, t)|^2 dx + \int_t^\tau \int_{B_R} a_k (\Delta \varphi_k)^2 dx ds = \frac{1}{2} \int_{B_R} |\nabla g|^2 dx,$$

from which it follows, in particular

$$\int_0^\tau \int_{B_R} |\nabla \varphi_k|^2 dx ds \le C, \tag{13.2.29}$$

$$\int_0^\tau \int_{B_R} a_k (\Delta \varphi_k)^2 dx ds \le C. \tag{13.2.30}$$

Using (13.2.27), (13.2.29) and noticing that $u_i \in L^\infty(Q_T)\,(i = 1, 2)$ and φ_k is uniformly bounded, we obtain

$$|I_k| \le C \int_0^\tau \int_{B_R \setminus B_{R-1}} (u_1^m + u_2^m)(|\nabla \varphi_k| + 1) dx dt$$

$$\le C \int_0^\tau \int_{B_R \setminus B_{R-1}} (u_1 + u_2) dx dt. \tag{13.2.31}$$

Using (13.2.30) and noticing that $u_i \in L^\infty(Q_T)\,(i = 1, 2)$ and φ_k is uniformly bounded, we obtain

$$|J_k| \le C \Big(\int_0^\tau \int_{B_R} \frac{(a - a_k)^2}{a_k} dx dt \Big)^{1/2} \Big(\int_0^\tau \int_{B_R} a_k (\Delta \varphi_k)^2 dx dt \Big)^{1/2}$$

$$\leq C\Big(\int_0^\tau \int_{B_R} \frac{(a - a_k)^2}{a_k} dxdt\Big)^{1/2}.$$

Using (13.2.26), we further obtain

$$|J_k| \leq C\sqrt{k}\Big(\int_0^\tau \int_{B_R} \Big(a - \rho_k * a - \frac{1}{k}\Big)^2 dxdt\Big)^{1/2} \leq \frac{C}{\sqrt{k}}. \qquad (13.2.32)$$

Combining (13.2.31), (13.2.32) with (13.2.28), we finally arrive at

$$\int_{\mathbb{R}^n} u(x, \tau)g(x)\xi_R(x)dx \leq |I_k| + |J_k|$$

$$\leq C\int_0^\tau \int_{B_R \backslash B_{R-1}} (u_1 + u_2)dxdt + \frac{C}{\sqrt{k}}. \qquad (13.2.33)$$

Let $k \to \infty$ and then $R \to \infty$. Since $u_i \in L^1(Q_T)$ $(i = 1, 2)$, the right side of (13.2.33) tends to zero and hence (13.2.24) holds. □

Remark 13.2.3 *It is to be noted that requiring weak solutions to belong to $L^1(Q_T) \cap L^\infty(Q_T)$ is too restrictive, which means that they must be "small" at infinity and thus even the nonzero constant solutions are excluded. Fortunately, those weak solutions determined by initial data with compact support satisfy such condition.*

13.2.5 Existence of weak solutions for higher dimensional equations

Now we discuss the existence of weak solutions of the Cauchy problem (13.2.3), (13.2.4).

Theorem 13.2.4 *Assume that $u_0 \in L^1(\mathbb{R}^n) \cap L^\infty(Q_T)$ and $u_0(x) \geq 0$. Then the Cauchy problem (13.2.3), (13.2.4) admits a weak solution $u \in L^1(Q_T) \cap L^\infty(Q_T)$.*

Proof. As we did in the proof of Theorem 13.2.2, the basic idea is to regularize the initial value and then to establish some estimates for the approximate solutions to obtain the desired compactness.

First we choose a sequence of positive numbers R_k and η_k such that

$$R_k \to +\infty, \quad \eta_k R_k^n \to 0, \quad \text{as } k \to \infty, \qquad (13.2.34)$$

and then construct $u_{0k} \in C_0^\infty(B_{R_k})$ such that

$$\|u_{0k}\|_{L^\infty(\mathbb{R}^n)} \leq \|u_0\|_{L^\infty(\mathbb{R}^n)}, \qquad (13.2.35)$$

$$\|u_{0k} - u_0\|_{L^1(\mathbb{R}^n)} \to 0, \quad \text{as } k \to \infty. \tag{13.2.36}$$

To this purpose, we may define $v_{0k} = u_0$ for $|x| \leq B_{R_k-1}$, $v_{0k} = 0$ elsewhere and then mollify it.

Consider the initial-boundary value problem

$$\begin{cases} \dfrac{\partial u_k}{\partial t} = \Delta u_k^m, & (x,t) \in B_{R_k} \times (0,T), \\ u(x,t) = \eta_k, & (x,t) \in \partial B_{R_k} \times (0,T), \\ u(x,0) = u_{0k}(x) + \eta_k, & x \in B_{R_k}. \end{cases} \tag{13.2.37}$$

According to the classical theory, (13.2.37) admits a smooth solution u_k. The maximum principle and (13.2.35) imply that

$$\eta_k \leq u_k \leq \|u_0\|_{L^\infty(\mathbb{R}^n)} + \eta_k. \tag{13.2.38}$$

Multiplying the equation in (13.2.37) by $pu_k^{p-1}(1 \leq p < +\infty)$ gives

$$\begin{aligned} \frac{\partial u_k^p}{\partial t} &= pu_k^{p-1}\Delta u_k^m \\ &= p\,\mathrm{div}(u_k^{p-1}\nabla u_k^m) - p\nabla u_k^{p-1}\cdot\nabla u_k^m \\ &= p\,\mathrm{div}(u_k^{p-1}\nabla u_k^m) - mp(p-1)u_k^{m+p-3}|\nabla u_k|^2. \end{aligned}$$

Integrating both sides of the above equality over $B_{R_k} \times (0,t)\,(0 < t \leq T)$ and noticing that

$$\frac{\partial u_k}{\partial \mu} \leq 0, \quad \text{on } \partial B_R \times (0,T),$$

where μ is the normal vector outward to $\partial B_{R_k} \times (0,T)$, we deduce

$$\int_{B_{R_k}} u_k^p(x,t)dx - \int_{B_{R_k}} (u_{0k}(x) + \eta_k)^p dx$$

$$= p\int_0^t \int_{\partial B_{R_k}} u_k^{p-1}\frac{\partial u_k^m}{\partial \mu}d\sigma d\tau - mp(p-1)\int_0^t \int_{B_{R_k}} u_k^{m+p-3}|\nabla u_k|^2 dx d\tau$$

$$\leq -mp(p-1)\int_0^t \int_{B_{R_k}} u_k^{m+p-3}|\nabla u_k|^2 dx d\tau$$

or

$$\int_{B_{R_k}} u_k^p(x,t)dx + mp(p-1)\int_0^t \int_{B_{R_k}} u_k^{m+p-3}|\nabla u_k|^2 dx d\tau$$

$$\leq \int_{B_{R_k}} (u_{0k}(x) + \eta_k)^p dx, \quad 0 < t \leq T.$$

Using (13.2.34), (13.2.36), we see that the right side of the above inequality tends to $\int_{\mathbb{R}^n} u_0^p dx$ as $k \to \infty$, which implies that for any fixed $p \in [1, \infty)$, the left side is bounded. This combined with (13.2.38) gives, in particular

$$\int_{B_{R_k}} u_k(x, t) dx \le C, \quad 0 \le t \le T \tag{13.2.39}$$

and

$$\int_0^T \int_{B_{R_k}} |\nabla u_k^m| dx dt \le m^2 \int_0^T \int_{B_{R_k}} u_k^{2(m-1)} |\nabla u_k|^2 dx dt \le C. \tag{13.2.40}$$

Multiply the equation in (13.2.37) by $m u_k^{m-1} \dfrac{\partial u_k}{\partial t}$,

$$\frac{4m}{(m+1)^2} \left(\frac{\partial u_k^{(m+1)/2}}{\partial t} \right)^2 = \text{div}\left(\frac{\partial u_k^m}{\partial t} \nabla u_k^m \right) - \frac{1}{2} \frac{\partial}{\partial t} |\nabla u_k^m|^2$$

and integrate the resulting equality over $B_{R_k} \times (t, T)$ $(0 \le t < T)$. Noticing that

$$\frac{\partial u_k}{\partial t} = 0, \quad \text{on } \partial B_{R_k} \times (0, T),$$

we obtain

$$\frac{4m}{(m+1)^2} \int_t^T \int_{B_{R_k}} \left(\frac{\partial u_k^{(m+1)/2}}{\partial t} \right)^2 dx d\tau$$

$$= -\frac{1}{2} \int_{B_{R_k}} |\nabla u_k^m(x, T)|^2 dx + \frac{1}{2} \int_{B_{R_k}} |\nabla u_k^m(x, t)|^2 dx$$

$$\le \frac{1}{2} \int_{B_{R_k}} |\nabla u_k^m(x, t)|^2 dx, \quad 0 \le t < T.$$

Integrating with respect to t over $(0, T)$ we further obtain

$$\frac{4m}{(m+1)^2} \int_0^T \int_{B_{R_k}} t \left(\frac{\partial u_k^{(m+1)/2}}{\partial t} \right)^2 dx dt$$

$$= \frac{4m}{(m+1)^2} \int_0^T \int_t^T \int_{B_{R_k}} \left(\frac{\partial u_k^{(m+1)/2}}{\partial t} \right)^2 dx d\tau dt$$

$$\le \frac{1}{2} \int_0^T \int_{B_{R_k}} |\nabla u_k^m(x, t)|^2 dx dt.$$

Using this and (13.2.40) and the uniform boundedness of u_k, we finally derive

$$\int_0^T \int_{B_{R_k}} t\left(\frac{\partial u_k^m}{\partial t}\right)^2 dxdt = m^2 \int_0^T \int_{B_{R_k}} tu_k^{2(m-1)}\left(\frac{\partial u_k}{\partial t}\right)^2 dxdt$$

$$= \frac{4m}{(m+1)^2}\int_0^T \int_{B_{R_k}} tu_k^{m-1}\left(\frac{\partial u_k^{(m+1)/2}}{\partial t}\right)^2 dxdt$$

$$\leq C. \tag{13.2.41}$$

The estimates (13.2.39)–(13.2.41) and Kolmogrow's theorem imply that for any $\delta \in (0,T)$, $R > 0$, $\{u_k^m\}$ is strongly compact in $L^2(B_R \times (\delta, T))$ and hence there exists a subsequence of $\{u_k^m\}$, supposed to be $\{u_k^m\}$ itself, which converges almost everywhere to a certain function v in Q_T

$$u_k^m \to v \text{ a.e. in } Q_T, \quad \text{as } k \to \infty,$$

i.e.

$$u_k \to u = v^{1/m} \text{ a.e. in } Q_T, \quad \text{as } k \to \infty.$$

(13.2.38), (13.2.39) imply that $u \in L^1(Q_T) \cap L^\infty(Q_T)$.

Given any $\varphi \in C^\infty(\overline{Q}_T)$, vanishing when $|x|$ is large enough or $t = T$. We have, for large k,

$$\iint \left(u_k\frac{\partial \varphi}{\partial t} + u_k^m \Delta\varphi\right)dxdt + \int_{\mathbb{R}^n}(u_{0k}(x) + \eta_k)\varphi(x,0)dx = 0.$$

Here we regard u_k as zero outside $\overline{B}_{R_k} \times [0,T]$. Letting $k \to \infty$, we see that u is a weak solution of problem (13.2.3), (13.2.4). $\qquad\square$

Remark 13.2.4 *Different from the one dimensional case, the above existence theorem does not provide a continuous solution.*

13.3   General Quasilinear Degenerate Parabolic Equations

The most general quasilinear degenerate parabolic equations, written in divergence form, are as follows

$$\frac{\partial u}{\partial t} = \frac{\partial}{\partial x_j}\left(a_{ij}(x,t,u)\frac{\partial u}{\partial x_i}\right) + \frac{\partial}{\partial x_i}b_i(x,t,u) + c(x,t,u),$$

where $a_{ij} = a_{ji}$ and

$$a_{ij}\xi_i\xi_j \geq 0, \quad \forall\xi = (\xi_1,\cdots,\xi_n) \in \mathbb{R}^n.$$

Here, as before, repeated indices imply a summation from 1 up to n. However, in this section, we merely consider the one dimensional case and, for simplicity, we will present the arguments merely for equations of the form

$$\frac{\partial u}{\partial t} = \frac{\partial^2 A(u)}{\partial x^2} + \frac{\partial B(u)}{\partial x}, \tag{13.3.1}$$

where $A(s)$, $B(s) \in C^1(\mathbb{R})$ and $A(s)$ satisfies

$$A(0) = 0, \quad A'(s) \geq 0 \text{ for } s \in \mathbb{R}.$$

We will illustrate the theory and methods by discussing the first initial-boundary value problem for (13.3.1), i.e. the problem with conditions

$$\begin{aligned}
A\big(u(0,t)\big) &= A\big(u(1,t)\big) = 0, & t &\in (0,T), & (13.3.2) \\
u(x,0) &= u_0(x), & x &\in (0,1). & (13.3.3)
\end{aligned}$$

Equation (13.3.1) degenerates whenever $u = 0$. If the set $E = \{s \in \mathbb{R}; A'(s) = 0\}$ does not contain any interior point, then (13.3.1) is said to be weakly degenerate; otherwise, (13.3.1) is said to be strongly degenerate.

13.3.1 Uniqueness of weak solutions for weakly degenerate equations

Denote $Q_T = (0,1) \times (0,T)$.

Definition 13.3.1 A function $u \in L^1(Q_T)$ is said to be a weak solution of the first initial-boundary value problem (13.3.1)–(13.3.3), if $A(u)$, $B(u) \in L^1(Q_T)$ and the integral equality

$$\iint_{Q_T} \left(u\frac{\partial \varphi}{\partial t} + A(u)\frac{\partial^2 \varphi}{\partial x^2} - B(u)\frac{\partial \varphi}{\partial x} \right) dxdt + \int_0^1 u_0(x)\varphi(x,0)dx = 0$$

holds for any function $\varphi \in C^\infty(\overline{Q}_T)$ with $\varphi(0,t) = \varphi(1,t) = \varphi(x,T) = 0$.

It is easy to verify that if $u \in C^1(\overline{Q}_T) \cap C^2(Q_T)$ is a classical solution of (13.3.1)–(13.3.3), then u is a weak solution of the problem; conversely, if $u \in C^1(\overline{Q}_T) \cap C^2(Q_T)$ is a weak solution of (13.3.1)–(13.3.3), then u is a classical solution of the problem. To verify the latter part, it should be noted that $A(u(x,t))\big|_{x=0,1} = 0$ implies $u(x,t)\big|_{x=0,1} = 0$ because of the strict monotonicity of $A(s)$, which follows from the assumption that the set $E = \{s \in \mathbb{R}; A'(s) = 0\}$ contains no interior point.

Theorem 13.3.1 *Assume that $u_0(x) \in L^\infty(0,1)$, $A(s)$, $B(s) \in C^1(\mathbb{R})$ and (13.3.1) is weakly degenerate. Then the first initial-boundary value problem (13.3.1)–(13.3.3) admits at most one weak solution in $L^\infty(Q_T)$.*

We will prove the theorem by means of Holmgren's approach. A crucial step is to establish the L^1-estimate for the derivatives of the solutions of the adjoint equation. The proof of our theorem will be completed by using this estimate together with some L^2-type estimates for the same solutions.

Let $u_1, u_2 \in L^\infty(Q_T)$ be weak solutions of (13.3.1)–(13.3.3). By definition, we have

$$\iint_{Q_T} (u_1 - u_2)\Big(\frac{\partial \varphi}{\partial t} + \tilde{A}\frac{\partial^2 \varphi}{\partial x^2} - \tilde{B}\frac{\partial \varphi}{\partial x}\Big)dxdt = 0$$

for any $\varphi \in C^\infty(\overline{Q}_T)$ with $\varphi(0,t) = \varphi(1,t) = \varphi(x,T) = 0$, where

$$\tilde{A} = \tilde{A}(u_1, u_2) = \int_0^1 A'(\theta u_1 + (1-\theta)u_2)d\theta, \qquad (x,t) \in Q_T,$$

$$\tilde{B} = \tilde{B}(u_1, u_2) = \int_0^1 B'(\theta u_1 + (1-\theta)u_2)d\theta, \qquad (x,t) \in Q_T.$$

If for any $f \in C_0^\infty(Q_T)$, the problem

$$\begin{cases} \dfrac{\partial \varphi}{\partial t} + \tilde{A}\dfrac{\partial^2 \varphi}{\partial x^2} - \tilde{B}\dfrac{\partial \varphi}{\partial x} = f, & (x,t) \in Q_T, \\ \varphi(0,t) = \varphi(1,t) = 0, & t \in (0,T), \\ \varphi(x,T) = 0, & x \in (0,1) \end{cases} \tag{13.3.4}$$

had a solution φ in $C^\infty(\overline{Q}_T)$, then we would have

$$\iint_{Q_T} (u_1 - u_2)f dxdt = 0$$

and $u_1 = u_2$ a.e. in Q_T would follow from the arbitrariness of f. However, since $\tilde{A}$ and $\tilde{B}$ are merely bounded and measurable, it is difficult to discuss the solvability of problem (13.3.4). Even if we have established the existence of solutions of the problem, the solutions are not smooth in general. In view of this situation, we consider some approximation of (13.3.4).

For sufficiently small $\eta > 0, \delta > 0$, let

$$\lambda_\eta^\delta = \begin{cases} (\eta + \tilde{A})^{-1/2}\tilde{B}, & \text{if } |u_1 - u_2| \geq \delta, \\ 0, & \text{if } |u_1 - u_2| < \delta. \end{cases}$$

Since $A(s)$ is strictly increasing and $u_1, u_2 \in L^\infty(Q_T)$, there must be constants $L(\delta) > 0, K(\delta) > 0$ depending on δ but independent of η, such that

$$\tilde{A} = \frac{A(u_1) - A(u_2)}{u_1 - u_2} \geq L(\delta), \quad |\lambda_\eta^\delta| \leq K(\delta), \quad \text{if } |u_1 - u_2| \geq \delta.$$

Let $\tilde{A}_\varepsilon$ and $\lambda_{\eta,\varepsilon}^\delta$ be C^∞ approximations of $\tilde{A}$ and λ_η^δ respectively, such that

$$\lim_{\varepsilon \to 0} \tilde{A}_\varepsilon = \tilde{A}, \quad \lim_{\varepsilon \to 0} \lambda_{\eta,\varepsilon}^\delta = \lambda_\eta^\delta, \quad \text{a.e. in } Q_T,$$
$$\tilde{A}_\varepsilon \leq C, \quad |\lambda_{\eta,\varepsilon}^\delta| \leq K(\delta), \quad \text{in } Q_T,$$

where C is a constant independent of ε.

Denote

$$\tilde{B}_{\eta,\varepsilon}^\delta = \lambda_{\eta,\varepsilon}^\delta (\eta + \tilde{A}_\varepsilon)^{1/2}.$$

For given $f \in C_0^\infty(Q_T)$, consider the approximate problem of (13.3.4)

$$\frac{\partial \varphi}{\partial t} + (\eta + \tilde{A}_\varepsilon)\frac{\partial^2 \varphi}{x^2} - \tilde{B}_{\eta,\varepsilon}^\delta \frac{\partial \varphi}{\partial x} = f, \qquad (x,t) \in Q_T, \qquad (13.3.5)$$

$$\varphi(0,t) = \varphi(1,t) = 0, \qquad\qquad t \in (0,T), \qquad (13.3.6)$$

$$\varphi(x,T) = 0, \qquad\qquad x \in (0,1). \qquad (13.3.7)$$

The existence of solutions in $C^\infty(\overline{Q}_T)$ follows from the standard theory of parabolic equations.

Lemma 13.3.1 *The solution φ of (13.3.5)–(13.3.7) satisfies*

$$\sup_{Q_T} |\varphi(x,t)| \leq C, \qquad (13.3.8)$$

$$\iint_{Q_T} (\eta + \tilde{A}_\varepsilon)\left(\frac{\partial^2 \varphi}{\partial x^2}\right)^2 dx dt \leq K(\delta)\eta^{-1}, \qquad (13.3.9)$$

$$\iint_{Q_T} \left(\frac{\partial \varphi}{\partial x}\right)^2 dx dt \leq K(\delta)\eta^{-1}. \qquad (13.3.10)$$

Here and in the sequel, we use C to denote a universal constant, independent of δ, η and ε, and $K(\delta)$ a constant depending only on δ, which may take different values on different occasions.

Proof. (13.3.8) follows from the maximum principle. To prove (13.3.9) and (13.3.10), we multiply (13.3.5) by $\frac{\partial^2 \varphi}{\partial x^2}$ and integrate over Q_T. Integrating by parts and using (13.3.6), (13.3.7) yield

$$\frac{1}{2}\int_0^1 \left(\frac{\partial\varphi(x,0)}{\partial x}\right)^2 dx + \iint_{Q_T}(\eta+\tilde{A}_\varepsilon)\left(\frac{\partial^2\varphi}{\partial x^2}\right)^2 dxdt$$

$$- \iint_{Q_T}\tilde{B}_{\eta,\varepsilon}^\delta \frac{\partial\varphi}{\partial x}\cdot\frac{\partial^2\varphi}{\partial x^2}dxdt = \iint_{Q_T}f\frac{\partial^2\varphi}{\partial x^2}dxdt.$$

Using Young's inequality and noticing that $|\lambda_{\eta,\varepsilon}^\delta| \le K(\delta)$, we obtain

$$\iint_{Q_T}(\eta+\tilde{A}_\varepsilon)\left(\frac{\partial^2\varphi}{\partial x^2}\right)^2 dxdt$$

$$\le \iint_{Q_T}\lambda_{\eta,\varepsilon}^\delta(\eta+\tilde{A}_\varepsilon)^{1/2}\frac{\partial\varphi}{\partial x}\cdot\frac{\partial^2\varphi}{\partial x^2}dxdt + \iint_{Q_T}f\frac{\partial^2\varphi}{\partial x^2}dxdt$$

$$\le \frac{1}{4}\iint_{Q_T}(\eta+\tilde{A}_\varepsilon)\left(\frac{\partial^2\varphi}{\partial x^2}\right)^2 dxdt$$

$$+ K(\delta)\iint_{Q_T}\left(\frac{\partial\varphi}{\partial x}\right)^2 dxdt + C\eta^{-1}. \tag{13.3.11}$$

Integrating by parts and using (13.3.6) and Young's inequality give

$$\iint_{Q_T}\left(\frac{\partial\varphi}{\partial x}\right)^2 dxdt = -\iint_{Q_T}\varphi\frac{\partial^2\varphi}{\partial x^2}dxdt$$

$$\le \alpha\iint_{Q_T}(\eta+\tilde{A}_\varepsilon)\left(\frac{\partial^2\varphi}{\partial x^2}\right)^2 dxdt + C\alpha^{-1}\eta^{-1}. \tag{13.3.12}$$

for any $\alpha > 0$. Substituting this into (13.3.11) and choosing $\alpha > 0$ (depending only on δ) small enough, we derive (13.3.9). (13.3.10) follows from (13.3.9) and (13.3.12). $\qquad\square$

Lemma 13.3.2   *The solution φ of (13.3.5)–(13.3.7) satisfies*

$$\sup_{t\in(0,T)}\int_0^1\left|\frac{\partial\varphi(x,t)}{\partial x}\right|dx \le C. \tag{13.3.13}$$

Proof.   For small $\beta > 0$, let

$$\text{sgn}_\beta s = \begin{cases} 1, & \text{if } s \ge \beta, \\ \dfrac{s}{\beta}, & \text{if } |s| < \beta, \\ -1, & \text{if } s \le -\beta, \end{cases}$$

$$I_\beta(s) = \int_0^s \text{sgn}_\beta\theta d\theta.$$

Differentiate (13.3.5) with respect to x, multiply the resulting equality by $\text{sgn}_\beta \dfrac{\partial \varphi}{\partial x}$ and integrate over $S_t \equiv (0, 1) \times (t, T)$ $(0 \le t < T)$. Then we obtain

$$\iint_{S_t} \frac{\partial}{\partial t} I_\beta \left(\frac{\partial \varphi}{\partial x} \right) dx d\tau + \iint_{S_t} \frac{\partial}{\partial x} \left[(\eta + \tilde{A}_\varepsilon) \frac{\partial^2 \varphi}{\partial x^2} \right] \text{sgn}_\beta \left(\frac{\partial \varphi}{\partial x} \right) dx d\tau$$
$$- \iint_{S_t} \frac{\partial}{\partial x} \left[\tilde{B}^\delta_{\eta,\varepsilon} \frac{\partial \varphi}{\partial x} \right] \text{sgn}_\beta \left(\frac{\partial \varphi}{\partial x} \right) dx d\tau = \iint_{S_t} \text{sgn}_\beta \left(\frac{\partial \varphi}{\partial x} \right) \frac{\partial f}{\partial x} dx d\tau.$$

Hence, integrating by parts and using (13.3.7) yield

$$\int_0^1 I_\beta \left(\frac{\partial \varphi(x, t)}{\partial x} \right) dx$$
$$= - \iint_{S_t} (\eta + \tilde{A}_\varepsilon) \left(\frac{\partial^2 \varphi}{\partial x^2} \right)^2 \text{sgn}'_\beta \left(\frac{\partial \varphi}{\partial x} \right) dx d\tau$$
$$+ \iint_{S_t} \tilde{B}^\delta_{\eta,\varepsilon} \frac{\partial \varphi}{\partial x} \cdot \frac{\partial^2 \varphi}{\partial x^2} \text{sgn}'_\beta \left(\frac{\partial \varphi}{\partial x} \right) dx d\tau - \iint_{S_t} \text{sgn}_\beta \left(\frac{\partial \varphi}{\partial x} \right) \frac{\partial f}{\partial x} dx d\tau$$
$$+ \int_t^T \left((\eta + \tilde{A}_\varepsilon) \frac{\partial^2 \varphi}{\partial x^2} - \tilde{B}^\delta_{\eta,\varepsilon} \frac{\partial \varphi}{\partial x} \right) \text{sgn}_\beta \left(\frac{\partial \varphi}{\partial x} \right) \Big|_{x=0}^{x=1} d\tau. \qquad (13.3.14)$$

The first term of the right side is nonnegative. The third term is bounded. In addition, using (13.3.5), (13.3.6) and the fact $f \in C_0^\infty(Q_T)$, we see that

$$\left((\eta + \tilde{A}_\varepsilon) \frac{\partial^2 \varphi}{\partial x^2} - \tilde{B}^\delta_{\eta,\varepsilon} \frac{\partial \varphi}{\partial x} \right) \text{sgn}_\beta \left(\frac{\partial \varphi}{\partial x} \right) \Big|_{x=0}^{x=1} = \left(f - \frac{\partial \varphi}{\partial t} \right) \text{sgn}_\beta \left(\frac{\partial \varphi}{\partial x} \right) \Big|_{x=0}^{x=1} = 0,$$

which shows that the last term of the right side of (13.3.14) is equal to zero. Thus we obtain

$$\int_0^1 I_\beta \left(\frac{\partial \varphi(x, t)}{\partial x} \right) dx \le C(\eta, \varepsilon, \delta) \int_t^T d\tau \int_{\left\{ x \in [0,1]; \, \left| \frac{\partial \varphi}{\partial x} \right| \le \beta \right\}} \left| \frac{\partial^2 \varphi}{\partial x^2} \right| dx + C,$$

from which (13.3.13) follows by letting $\beta \to 0$ and using a known result (see [Saks (1964)], 131–133) to conclude that

$$\int_{\left\{ x \in [0,1]; \, \left| \frac{\partial \varphi}{\partial x} \right| \le \beta \right\}} \left| \frac{\partial^2 \varphi}{\partial x^2} \right| dx \to 0, \qquad \text{as } \beta \to 0.$$

$\square$

Proof of Theorem 13.3.1. Given $f \in C_0^\infty(Q_T)$. Let φ be a solution of (13.3.5)–(13.3.7). Then

$$\iint_{Q_T} (u_1 - u_2) f \, dx dt = \iint_{Q_T} (u_1 - u_2) \left(\frac{\partial \varphi}{\partial t} + (\eta + \tilde{A}_\varepsilon) \frac{\partial^2 \varphi}{\partial x^2} - \tilde{B}^\delta_{\eta,\varepsilon} \frac{\partial \varphi}{\partial x} \right) dx dt.$$

By the definition of weak solutions, we have

$$\iint_{Q_T} (u_1 - u_2)\Big(\frac{\partial \varphi}{\partial t} + \tilde{A}\frac{\partial^2 \varphi}{\partial x^2} - \tilde{B}\frac{\partial \varphi}{\partial x}\Big)\,dxdt = 0.$$

Hence

$$\iint_{Q_T} (u_1 - u_2)f\,dxdt$$

$$= \iint_{Q_T} (u_1 - u_2)\eta\frac{\partial^2 \varphi}{\partial x^2}\,dxdt + \iint_{Q_T} (u_1 - u_2)(\tilde{A}_\varepsilon - \tilde{A})\frac{\partial^2 \varphi}{\partial x^2}\,dxdt$$

$$- \iint_{Q_T} (u_1 - u_2)(\tilde{B}^\delta_{\eta,\varepsilon} - \tilde{B})\frac{\partial \varphi}{\partial x}\,dxdt \qquad (13.3.15)$$

Now we proceed to estimate each term of the right side of (13.3.15). First, from Lemma 13.3.1,

$$\Big|\iint_{Q_T} (u_1 - u_2)(\tilde{A}_\varepsilon - \tilde{A})\frac{\partial^2 \varphi}{\partial x^2}\,dxdt\Big|$$

$$\leq C\Big(\iint_{Q_T} (\tilde{A}_\varepsilon - \tilde{A})^2\,dxdt\Big)^{1/2}\Big(\iint_{Q_T} \Big(\frac{\partial^2 \varphi}{\partial x^2}\Big)^2\,dxdt\Big)^{1/2}$$

$$\leq K(\delta)\eta^{-1}\Big(\iint_{Q_T} (\tilde{A}_\varepsilon - \tilde{A})^2\,dxdt\Big)^{1/2}.$$

Hence

$$\lim_{\varepsilon \to 0} \iint_{Q_T} (u_1 - u_2)(\tilde{A}_\varepsilon - \tilde{A})\frac{\partial^2 \varphi}{\partial x^2}\,dxdt = 0. \qquad (13.3.16)$$

Denote

$$G_\delta = \{(x,t) \in Q_T; |u_1 - u_2| < \delta\},$$
$$F_\delta = \{(x,t) \in Q_T; |u_1 - u_2| \geq \delta\}.$$

Using Lemma 13.3.1 and Lemma 13.3.2, we have

$$\Big|\iint_{G_\delta} (u_1 - u_2)(\tilde{B}^\delta_{\eta,\varepsilon} - \tilde{B})\frac{\partial \varphi}{\partial x}\,dxdt\Big|$$

$$\leq \Big|\iint_{G_\delta} (u_1 - u_2)\tilde{B}^\delta_{\eta,\varepsilon}\frac{\partial \varphi}{\partial x}\,dxdt\Big| + \Big|\iint_{G_\delta} (u_1 - u_2)\tilde{B}\frac{\partial \varphi}{\partial x}\,dxdt\Big|$$

$$\leq \delta\Big(\iint_{G_\delta} (\tilde{\lambda}^\delta_{\eta,\varepsilon})^2\,dxdt\Big)^{1/2}\Big(\iint_{G_\delta} \Big(\frac{\partial \varphi}{\partial x}\Big)^2\,dxdt\Big)^{1/2} + C\delta\iint_{G_\delta} \Big|\frac{\partial \varphi}{\partial x}\Big|\,dxdt$$

$$\leq K(\delta)\eta^{-1/2}\Big(\iint_{G_\delta}(\tilde{\lambda}^\delta_{\eta,\varepsilon})^2\,dxdt\Big)^{1/2}+C\delta.$$

Since $\lim\limits_{\varepsilon\to 0}\tilde{\lambda}^\delta_{\eta,\varepsilon}=0$ a.e. in G_δ, it follows that

$$\overline{\lim_{\varepsilon\to 0}}\Big|\iint_{G_\delta}(u_1-u_2)(\tilde{B}^\delta_{\eta,\varepsilon}-\tilde{B})\frac{\partial\varphi}{\partial x}\,dxdt\Big|\leq C\delta.$$

Using Lemma 13.3.1, we have

$$\Big|\iint_{F_\delta}(u_1-u_2)(\tilde{B}^\delta_{\eta,\varepsilon}-\tilde{B})\frac{\partial\varphi}{\partial x}\,dxdt\Big|$$

$$\leq C\Big(\iint_{F_\delta}(\tilde{B}^\delta_{\eta,\varepsilon}-\tilde{B})^2\,dxdt\Big)^{1/2}\Big(\iint_{F_\delta}\Big(\frac{\partial\varphi}{\partial x}\Big)^2\,dxdt\Big)^{1/2}$$

$$\leq K(\delta)\eta^{-1/2}\Big(\iint_{F_\delta}(\tilde{B}^\delta_{\eta,\varepsilon}-\tilde{B})^2\,dxdt\Big)^{1/2},$$

from which, noticing that $\lim\limits_{\varepsilon\to 0}\tilde{B}^\delta_{\eta,\varepsilon}=\tilde{B}$ in F_δ, we infer

$$\lim_{\varepsilon\to 0}\iint_{F_\delta}(u_1-u_2)(\tilde{B}^\delta_{\eta,\varepsilon}-\tilde{B})\frac{\partial\varphi}{\partial x}\,dxdt=0.$$

Therefore

$$\overline{\lim_{\varepsilon\to 0}}\Big|\iint_{Q_T}(u_1-u_2)(\tilde{B}^\delta_{\eta,\varepsilon}-\tilde{B})\frac{\partial\varphi}{\partial x}\,dxdt\Big|\leq C\delta. \tag{13.3.17}$$

It remains to treat the first term of the right side of (13.3.15). For any $\gamma>0$,

$$\Big|\iint_{Q_T}(u_1-u_2)\frac{\partial^2\varphi}{\partial x^2}\,dxdt\Big|$$

$$\leq\iint_{F_\gamma}\Big|(u_1-u_2)\frac{\partial^2\varphi}{\partial x^2}\Big|\,dxdt+\iint_{G_\gamma}\Big|(u_1-u_2)\frac{\partial^2\varphi}{\partial x^2}\Big|\,dxdt$$

$$\leq C\sup_{F_\gamma}\tilde{A}^{-1/2}\iint_{F_\gamma}\tilde{A}^{1/2}\Big|\frac{\partial^2\varphi}{\partial x^2}\Big|\,dxdt+\gamma\iint_{G_\gamma}\Big|\frac{\partial^2\varphi}{\partial x^2}\Big|\,dxdt$$

$$\leq C\sup_{F_\gamma}\tilde{A}^{-1/2}\iint_{F_\gamma}\Big|(\tilde{A}^{1/2}_\varepsilon-\tilde{A}^{1/2})\frac{\partial^2\varphi}{\partial x^2}\Big|\,dxdt$$

$$+C\sup_{F_\gamma}\tilde{A}^{-1/2}\iint_{F_\gamma}\tilde{A}^{1/2}_\varepsilon\Big|\frac{\partial^2\varphi}{\partial x^2}\Big|\,dxdt+\gamma\iint_{G_\gamma}\Big|\frac{\partial^2\varphi}{\partial x^2}\Big|\,dxdt$$

$$\leq C\sup_{F_\gamma}\tilde{A}^{-1/2}\Big(\iint_{F_\gamma}(\tilde{A}^{1/2}_\varepsilon-\tilde{A}^{1/2})^2\,dxdt\Big)^{1/2}\Big(\iint_{Q_T}\Big(\frac{\partial^2\varphi}{\partial x^2}\Big)^2\,dxdt\Big)^{1/2}$$

$$+ C \sup_{F_\gamma} \tilde{A}^{-1/2} \Big(\iint_{Q_T} \tilde{A}_\varepsilon \Big(\frac{\partial^2 \varphi}{\partial x^2} \Big)^2 dxdt \Big)^{1/2}$$

$$+ C\gamma \Big(\iint_{Q_T} \Big(\frac{\partial^2 \varphi}{\partial x^2} \Big)^2 dxdt \Big)^{1/2}$$

$$\leq CL(\gamma)^{-1/2} \Big(\iint_{F_\gamma} (\tilde{A}_\varepsilon^{1/2} - \tilde{A}^{1/2})^2 dxdt \Big)^{1/2} \Big(\iint_{Q_T} \Big(\frac{\partial^2 \varphi}{\partial x^2} \Big)^2 dxdt \Big)^{1/2}$$

$$+ CL(\gamma)^{-1/2} \Big(\iint_{Q_T} \tilde{A}_\varepsilon \Big(\frac{\partial^2 \varphi}{\partial x^2} \Big)^2 dxdt \Big)^{1/2}$$

$$+ C\gamma \Big(\iint_{Q_T} \Big(\frac{\partial^2 \varphi}{\partial x^2} \Big)^2 dxdt \Big)^{1/2}$$

$$\leq CL(\gamma)^{-1/2} K(\delta)^{1/2} \Big(\iint_{F_\gamma} (\tilde{A}_\varepsilon^{1/2} - \tilde{A}^{1/2})^2 dxdt \Big)^{1/2}$$

$$+ CL(\gamma)^{-1/2} K(\delta)^{1/2} \eta^{-1/2} + C\gamma K(\delta)^{1/2} \eta^{-1},$$

where we have used (13.3.9) again and the fact that $\tilde{A} \geq L(\gamma) > 0$ in F_γ. Since $\lim_{\varepsilon \to 0} \tilde{A}_\varepsilon = \tilde{A}$ a.e. in Q_T, letting $\varepsilon \to 0$ in the last inequality gives

$$\overline{\lim_{\varepsilon \to 0}} \Big| \iint_{Q_T} (u_1 - u_2) \frac{\partial^2 \varphi}{\partial x^2} dxdt \Big| \leq CL(\gamma)^{-1/2} K(\delta)^{1/2} \eta^{-1/2} + C\gamma K(\delta)^{1/2} \eta^{-1}.$$

Now we choose $\gamma = \delta K(\delta)^{-1/2}$. Then we are led to

$$\overline{\lim_{\varepsilon \to 0}} \Big| \iint_{Q_T} (u_1 - u_2) \eta \frac{\partial^2 \varphi}{\partial x^2} dxdt \Big|$$

$$\leq CL(\delta K(\delta)^{-1/2})^{-1/2} K(\delta)^{1/2} \eta^{1/2} + C\delta. \tag{13.3.18}$$

Combining (13.3.16)–(13.3.18) with (13.3.15) we finally obtain

$$\Big| \iint_{Q_T} (u_1 - u_2) f dxdt \Big| \leq CL(\delta K(\delta)^{-1/2})^{-1/2} K(\delta)^{1/2} \eta^{1/2} + C\delta,$$

from which, it follows by first letting $\eta \to 0$ and then letting $\delta \to 0$, that

$$\iint_{Q_T} (u_1 - u_2) f dxdt = 0.$$

The proof of our theorem is complete.

13.3.2 Existence of weak solutions for weakly degenerate equations

Theorem 13.3.2 *Assume that u_0 is Lipschitz continuous in $[0,1]$ with $u_0(0) = u_0(1) = 0$, $A(s), B(s)$ are appropriately smooth, $\lim\limits_{s \to \pm\infty} A(s) = \pm\infty$ and (13.3.1) is weakly degenerate. Then the first initial-boundary value problem (13.3.1)–(13.3.3) admits a continuous weak solution.*

To prove the theorem, we consider the following regularized problem

$$\frac{\partial u_\varepsilon}{\partial t} = \frac{\partial^2 A_\varepsilon(u_\varepsilon)}{\partial x^2} + \frac{\partial B(u_\varepsilon)}{\partial x}, \qquad (x,t) \in Q_T, \qquad (13.3.19)$$

$$u_\varepsilon(0,t) = u_\varepsilon(1,t) = 0, \qquad t \in (0,T), \qquad (13.3.20)$$

$$u_\varepsilon(x,0) = u_{0\varepsilon}(x), \qquad x \in (0,1), \qquad (13.3.21)$$

where

$$A_\varepsilon(s) = \varepsilon s + A(s), \quad s \in \mathbb{R} \quad (\varepsilon > 0)$$

and $u_{0\varepsilon}$ is a smooth function approximating u_0 uniformly with

$$u_{0\varepsilon}(0) = u_{0\varepsilon}(1) = u_{0\varepsilon}'(0) = u_{0\varepsilon}'(1) = u_{0\varepsilon}''(0) = u_{0\varepsilon}''(1) = 0$$

and $|u_{0\varepsilon}'|$ uniformly bounded.

Let u_ε be a smooth solution of this problem, whose existence follows from the classical theory of parabolic equations. We need some estimates to ensure the compactness of $\{u_\varepsilon\}$.

First, the maximum principle implies that

$$\sup_{Q_T} |u_\varepsilon(x,t)| \le M \qquad (13.3.22)$$

with constant M independent of ε.

Next, we have

Lemma 13.3.3 *Let u_ε be a solution of problem (13.3.19)–(13.3.21). Then*

$$\left| \frac{\partial A_\varepsilon(u_\varepsilon)}{\partial x}(x,t) \right|_{x=0,1} \le C, \quad t \in [0,T] \qquad (13.3.23)$$

with the constant C independent of ε.

Proof. Let

$$w_\varepsilon(x,t) = \int_0^{u_\varepsilon(x,t)} \lambda_\varepsilon(s)ds, \quad (x,t) \in Q_T, \qquad (13.3.24)$$

where

$$\lambda_\varepsilon(s) = \frac{A'_\varepsilon(s)}{\theta(s)}, \quad s \in \mathbb{R} \tag{13.3.25}$$

and $\theta(s)$ is an auxiliary function of the form $\theta(s) = \alpha + s$ with an arbitrary constant α greater than M, the constant in (13.3.22). For example, we may choose $\alpha = M + 1$. Then

$$\frac{\partial w_\varepsilon}{\partial x} = A'_\varepsilon(u_\varepsilon)\frac{\partial u_\varepsilon}{\partial x}/\theta(u_\varepsilon) = \frac{\partial A_\varepsilon(u_\varepsilon)}{\partial x}/\theta(u_\varepsilon), \qquad (x,t) \in Q_T,$$

$$\frac{\partial w_\varepsilon}{\partial t} = A'_\varepsilon(u_\varepsilon)\frac{\partial u_\varepsilon}{\partial t}/\theta(u_\varepsilon) = \frac{\partial A_\varepsilon(u_\varepsilon)}{\partial t}/\theta(u_\varepsilon), \qquad (x,t) \in Q_T.$$

Using (13.3.19), one can easily check that w_ε satisfies

$$\frac{\partial w_\varepsilon}{\partial t} - A'_\varepsilon(u_\varepsilon)\frac{\partial^2 w_\varepsilon}{\partial x^2} - \left(\frac{\partial w_\varepsilon}{\partial x}\right)^2 - B'(u_\varepsilon)\frac{\partial w_\varepsilon}{\partial x} = 0, \quad (x,t) \in Q_T,$$

in which we have a term $\left(\dfrac{\partial w_\varepsilon}{\partial x}\right)^2$; as will be seen bellow, this term plays an important role in our proof. If we simply set $w_\varepsilon = \dfrac{\partial A_\varepsilon(u_\varepsilon)}{\partial x}$, then in the equation for w_ε, this term disappears. This is just why we introduce the auxiliary function $\theta(x)$.

Define an operator H as follows:

$$H[w] \equiv \frac{\partial w}{\partial t} - A'_\varepsilon(u_\varepsilon)\frac{\partial^2 w}{\partial x^2} - \left(\frac{\partial w}{\partial x}\right)^2 - B'(u_\varepsilon)\frac{\partial w}{\partial x}, \quad (x,t) \in Q_T.$$

Then $H[w_\varepsilon] = 0$. Let

$$v_\varepsilon(x,t) = K(x-1) - w_\varepsilon(x,t), \quad (x,t) \in Q_T$$

with constant K to be determined. By a simple calculation, using (13.3.22), we see that, for sufficiently large $K > 0$,

$$H[v_\varepsilon] = -2\left(\frac{\partial w_\varepsilon}{\partial x} - \frac{K}{2}\right)^2 - \frac{K^2}{2} - KB'(u_\varepsilon) - H[w_\varepsilon]$$

$$\leq -\frac{K^2}{2} - KB'(u_\varepsilon) < 0, \quad \text{in } Q_T.$$

From this it follows that v_ε can not achieve its maximum at any point inside Q_T. In addition, since from (13.3.20), (13.3.21) and the uniform boundedness of $u'_{0\varepsilon}$, we have, for large K,

$$v_\varepsilon(0,t) = -K < 0, \quad v_\varepsilon(1,t) = 0, \quad t \in [0,T],$$

$$\frac{\partial v_\varepsilon}{\partial x}\bigg|_{t=0} = K - \frac{\partial w_\varepsilon}{\partial x}\bigg|_{t=0} = K - \frac{A'_\varepsilon(u_{0\varepsilon}(x))u'_{0\varepsilon}(x)}{\theta(u_{0\varepsilon}(x))} > 0, \quad x \in [0,1],$$

we can assert that the maximum of v_ε must be zero and must achieve at $x = 1$. Thus

$$\frac{\partial v_\varepsilon}{\partial x}\bigg|_{x=1} \geq 0, \quad t \in [0,T]$$

and hence

$$\frac{\partial A_\varepsilon(u_\varepsilon)}{\partial x}\bigg|_{x=1} = (\alpha + u_\varepsilon)\frac{\partial w_\varepsilon}{\partial x}\bigg|_{x=1} \leq (\alpha + u_\varepsilon)\bigg|_{x=1} K \leq C, \quad t \in [0,T].$$

Similarly, we can prove that

$$\frac{\partial A_\varepsilon(u_\varepsilon)}{\partial x}\bigg|_{x=1} \geq -C, \quad t \in [0,T].$$

Therefore the conclusion (13.3.23) for $x = 1$ is proved. Similarly, we can prove another part of (13.3.23). $\qquad\Box$

Lemma 13.3.4 *Let u_ε be a solution of problem (13.3.19)–(13.3.21). Then*

$$\left|\frac{\partial A_\varepsilon(u_\varepsilon)}{\partial x}(x,t)\right| \leq C, \quad (x,t) \in Q_T \qquad (13.3.26)$$

with the constant C independent of ε.

Proof. Define w_ε and $\lambda_\varepsilon(s)$ as in (13.3.24) and (13.3.25) with $\theta(s)$ to be determined. The first requirement is that $\theta(s)$ has positive lower bound on $|s| \leq M$ (M is the constant in (13.3.22)). Then from (13.3.19) we see that w_ε satisfies

$$\frac{\partial w_\varepsilon}{\partial t} - A'_\varepsilon(u_\varepsilon)\frac{\partial^2 w_\varepsilon}{\partial x^2} - \theta'(u_\varepsilon)\left(\frac{\partial w_\varepsilon}{\partial x}\right)^2 - B'(u_\varepsilon)\frac{\partial w_\varepsilon}{\partial x} = 0, \quad (x,t) \in Q_T$$

and $v_\varepsilon = \dfrac{\partial w_\varepsilon}{\partial x}$ satisfies

$$\frac{\partial v_\varepsilon}{\partial t} - A'_\varepsilon(u_\varepsilon)\frac{\partial^2 v_\varepsilon}{\partial x^2} - \left(2\theta'(u_\varepsilon)v_\varepsilon + B'(u_\varepsilon) + A''_\varepsilon(u_\varepsilon)\frac{\partial u_\varepsilon}{\partial x}\right)\frac{\partial v_\varepsilon}{\partial x}$$
$$- \frac{\theta''(u_\varepsilon)}{\lambda_\varepsilon(u_\varepsilon)}v_\varepsilon^3 - \frac{B''(u_\varepsilon)}{\lambda_\varepsilon(u_\varepsilon)}v_\varepsilon^2 = 0, \quad (x,t) \in Q_T.$$

Multiplying the above equality by v_ε gives

$$-\frac{1}{2}\frac{\partial v_\varepsilon^2}{\partial t} - A'_\varepsilon(u_\varepsilon)v_\varepsilon\frac{\partial^2 v_\varepsilon}{\partial x^2} - \frac{1}{2}\left(2\theta'(u_\varepsilon)v_\varepsilon + B'(u_\varepsilon) + A''_\varepsilon(u_\varepsilon)\frac{\partial u_\varepsilon}{\partial x}\right)\frac{\partial v_\varepsilon^2}{\partial x}$$

$$- \frac{\theta''(u_\varepsilon)}{\lambda_\varepsilon(u_\varepsilon)} v_\varepsilon^4 - \frac{B''(u_\varepsilon)}{\lambda_\varepsilon(u_\varepsilon)} v_\varepsilon^3 = 0, \quad (x,t) \in Q_T. \qquad (13.3.27)$$

If v_ε^2 achieves its maximum at some point on the parabolic boundary, then, by Lemma 13.3.3, (13.3.26) holds clearly. Suppose that v_ε^2 achieves the maximum at some point (x_0, t_0) not on the parabolic boundary. Then at (x_0, t_0), the sum of the first three terms of the left side of (13.3.27) is nonnegative and hence

$$- \frac{\theta''(u_\varepsilon)}{\lambda_\varepsilon(u_\varepsilon)} v_\varepsilon^4 - \frac{B''(u_\varepsilon)}{\lambda_\varepsilon(u_\varepsilon)} v_\varepsilon^3 \leq 0, \quad (x,t) \in Q_T,$$

namely

$$-\theta''(u_\varepsilon) v_\varepsilon^2 - B''(u_\varepsilon) v_\varepsilon \leq 0, \quad (x,t) \in Q_T,$$

from which it follows by using Young's inequality that for $\delta > 0$,

$$-\theta''(u_\varepsilon) v_\varepsilon^2 \leq \delta v_\varepsilon^2 + \frac{1}{4\delta}(B''(u_\varepsilon))^2, \quad (x,t) \in Q_T,$$

namely

$$(-\theta''(u_\varepsilon) - \delta) v_\varepsilon^2 \leq \frac{1}{4\delta}(B''(u_\varepsilon))^2, \quad (x,t) \in Q_T.$$

If $\theta(s)$ is chosen such that $\theta''(s)$ has negative upper bound on $|s| \leq M$, then we can choose $\delta > 0$ so small that

$$v_\varepsilon^2(x,t) \leq C, \quad (x,t) \in Q_T$$

with constant C independent of ε. This inequality implies (13.3.26), if $\theta(s)$ is required to have positive lower bound on $|s| \leq M$. The choice of such functions $\theta(s)$ is quite free, for example, we may choose

$$\theta(s) = 1 + (M - s)(M + s), \quad s \in \mathbb{R}.$$

This completes the proof of our lemma. $\qquad\qquad\qquad\qquad\square$

Lemma 13.3.5 *Let u_ε be a solution of problem (13.3.19)–(13.3.21). Then for any $(x,t), (y,s) \in Q_T$,*

$$|A_\varepsilon(u_\varepsilon(x,t)) - A_\varepsilon(u_\varepsilon(y,s))| \leq C(|x - y| + |t - s|^{1/2}) \qquad (13.3.28)$$

where the constant C is independent of ε.

Proof. Since Lemma 13.3.4 implies that

$$|A_\varepsilon(u_\varepsilon(x,t)) - A_\varepsilon(u_\varepsilon(y,t))| \le C|x-y|, \quad (x,t), (y,t) \in Q_T, \quad (13.3.29)$$

it remains to further prove

$$|A_\varepsilon(u_\varepsilon(x,t)) - A_\varepsilon(u_\varepsilon(x,s))| \le C|t-s|^{1/2}, \quad (x,t), (x,s) \in Q_T. \quad (13.3.30)$$

Suppose, for example, $\Delta t = t - s > 0$. Given $\alpha \in (0,1)$ arbitrarily and denote $d = (\Delta t)^\alpha$. We may choose $d \le 1/2$; otherwise, (13.3.30) follows immediately from the uniform boundedness of $\{u_\varepsilon\}$.

In case $x+d \le 1$, we integrate (13.3.19) over $(x, x+d) \times (s, t)$. Integrating by parts then gives

$$\int_x^{x+d} (u_\varepsilon(\xi, t) - u_\varepsilon(\xi, s)) d\xi = \int_s^t \left(\frac{\partial A_\varepsilon(u_\varepsilon)}{\partial x} + B(u_\varepsilon) \right) \Big|_x^{x+d} dt. \quad (13.3.31)$$

Using the mean value theorem for integrals, we see that

$$\int_x^{x+d} (u_\varepsilon(\xi, t) - u_\varepsilon(\xi, s)) d\xi = d(u_\varepsilon(x^*, t) - u_\varepsilon(x^*, s))$$

for some $x^* \in [x, x+d]$. Combining this with (13.3.31) and using (13.3.22) and Lemma 13.3.4, we obtain

$$|u_\varepsilon(x^*, t) - u_\varepsilon(x^*, s)| \le C(\Delta t)^{1-\alpha}.$$

This, together with (13.3.29) gives

$$\begin{aligned}
&|A_\varepsilon(u_\varepsilon(x,t)) - A_\varepsilon(u_\varepsilon(y,s))| \\
&\le |A_\varepsilon(u_\varepsilon(x,t)) - A_\varepsilon(u_\varepsilon(x^*,t))| + |A_\varepsilon(u_\varepsilon(x^*,t)) - A_\varepsilon(u_\varepsilon(x^*,s))| \\
&\quad + |A_\varepsilon(u_\varepsilon(x^*,s)) - A_\varepsilon(u_\varepsilon(x,s))| \\
&\le C(\Delta t)^\alpha + C(\Delta t)^{1-\alpha} + C(\Delta t)^\alpha \\
&= C(2(\Delta t)^\alpha + (\Delta t)^{1-\alpha}),
\end{aligned}$$

which implies (13.3.30), if we take $\alpha = 1/2$.

If $x + d > 1$, then since $d \le 1/2$, we have $x > 1 - d \ge 1/2$ and can obtain the same conclusion by integrating (13.3.19) over $(x - d, x) \times (s, t)$. The proof is complete. $\qquad\square$

Proof of Theorem 13.3.2. Denote

$$w_\varepsilon(x, t) = A_\varepsilon(u_\varepsilon(x, t)), \quad (x, t) \in \overline{Q}_T.$$

Lemma 13.3.5 and (13.3.22) imply the uniform boundedness and equicontinuity of $\{w_\varepsilon\}$ in Q_T. Hence there exists a subsequence, still denoted by $\{w_\varepsilon\}$, and a function $w \in C^{1,1/2}(\overline{Q}_T)$, such that

$$\lim_{\varepsilon \to 0} w_\varepsilon(x,t) = w(x,t), \quad \text{uniformly in } Q_T.$$

Let $\psi(s)$ be the inverse function of $A(s)$, whose existence for $s \in \mathbb{R}$ follows from the strict monotonicity of $A(s)$ and the assumption $\lim_{s \to \pm\infty} A(s) = \pm\infty$. Then

$$u(x,t) = \lim_{\varepsilon \to 0} u_\varepsilon(x,t) = \lim_{\varepsilon \to 0} \psi(w_\varepsilon(x,t) - \varepsilon u_\varepsilon(x,t)), \quad (x,t) \in \overline{Q}_T$$

exists and $u \in C(\overline{Q}_T)$. To prove that u is a weak solution of problem (13.3.1)–(13.3.3), notice that, from (13.3.19)–(13.3.21), for any $\varphi \in C^\infty(\overline{Q}_T)$ with $\varphi(0,t) = \varphi(1,t) = \varphi(x,T) = 0$, we can obtain

$$\iint_{Q_T} \left(u_\varepsilon \frac{\partial\varphi}{\partial t} + w_\varepsilon \frac{\partial^2\varphi}{\partial x^2} - B(u_\varepsilon)\frac{\partial\varphi}{\partial x} \right) dxdt + \int_0^1 u_{0\varepsilon}(x)\varphi(x,0)dx = 0$$

and hence, by letting $\varepsilon \to 0$,

$$\iint_{Q_T} \left(u \frac{\partial\varphi}{\partial t} + w \frac{\partial^2\varphi}{\partial x^2} - B(u)\frac{\partial\varphi}{\partial x} \right) dxdt + \int_0^1 u_0(x)\varphi(x,0)dx = 0.$$

Since $w = A(u)$, by definition, u is a weak solution of (13.3.1)–(13.3.3). Theorem 13.3.2 is proved.

Theorem 13.3.3 *If in addition to the assumptions of Theorem 13.3.2, suppose that*

$$|A(s_1) - A(s_2)| \geq \lambda|s_1 - s_2|^m \tag{13.3.32}$$

for some constants $m > 1$ and $\lambda > 0$, then the weak solution u of problem (13.3.1)–(13.3.3) given in Theorem 13.3.2 is Hölder continuous, precisely, $u \in C^{1/m,1/(m+1)}(\overline{Q}_T)$.

Proof. In the proof of Theorem 13.3.2, in fact, we have reached $A(u(x,t)) \in C^{1,1/2}(\overline{Q}_T)$, which follows from (13.3.28) by letting $\varepsilon \to 0$. Thus, using the assumption (13.3.32), we obtain

$$|u(x,t) - u(y,s)| \leq \lambda^{-1/m}|A(u(x,t)) - A(u(y,s))|^{1/m}$$
$$\leq C(|x-y| + |t-s|^{1/2})^{1/m}$$
$$\leq C(|x-y|^{1/m} + |t-s|^{1/(2m)}), \quad (x,t),(y,s) \in Q_T,$$

i.e. $u \in C^{1/m, 1/(2m)}(\overline{Q}_T)$. We further prove that $u \in C^{1/m, 1/(m+1)}(\overline{Q}_T)$. First, using (13.3.32) and Lemma 13.3.5 gives

$$
\begin{aligned}
|u_\varepsilon(x,t) - u_\varepsilon(y,t)| \le & \lambda^{-1/m} |A(u_\varepsilon(x,t)) - A(u_\varepsilon(y,t))|^{1/m} \\
\le & C |A_\varepsilon(u_\varepsilon(x,t)) - A_\varepsilon(u_\varepsilon(y,t))|^{1/m} \\
& + C\varepsilon^{1/m} |u_\varepsilon(x,t) - u_\varepsilon(y,t)|^{1/m} \\
\le & C|x - y|^{1/m} + C\varepsilon^{1/m}.
\end{aligned}
\tag{13.3.33}
$$

Next, for any given $\alpha \in (0,1)$, by an argument similar to the proof of Lemma 13.3.5, we can assert that for any $x \in (0,1)$, there exists $x^* \in (0,1)$ with $|x - x^*| \le d = (\Delta t)^\alpha$ (suppose $\Delta t = t - s > 0$), such that

$$
|u_\varepsilon(x^*, t) - u_\varepsilon(x^*, s)| \le C(\Delta t)^{1-\alpha}.
$$

Combining this with (13.3.33) gives

$$
\begin{aligned}
|u_\varepsilon(x,t) - u_\varepsilon(x,s)| \le & |u_\varepsilon(x,t) - u_\varepsilon(x^*,t)| \\
& + |u_\varepsilon(x^*,t) - u_\varepsilon(x^*,s)| + |u_\varepsilon(x^*,s) - u_\varepsilon(x,s)| \\
\le & C((\Delta t)^{\alpha/m} + \varepsilon^{\alpha/m} + (\Delta t)^{1-\alpha}).
\end{aligned}
\tag{13.3.34}
$$

Letting $\varepsilon \to 0$ in (13.3.33), (13.3.34) and choosing $\alpha = \dfrac{m}{m+1}$, we are led to

$$
\begin{aligned}
|u(x,t) - u(y,t)| & \le C|x - y|^{1/m}, & (x,t), (y,t) \in Q_T, \\
|u(x,t) - u(x,s)| & \le C|t - s|^{1/(m+1)}, & (x,t), (x,s) \in Q_T,
\end{aligned}
$$

which imply that $u \in C^{1/m, 1/(m+1)}(\overline{Q}_T)$. $\qquad \square$

13.3.3 *A remark on quasilinear parabolic equations with strong degeneracy*

Now we turn to the strong degenerate equation (13.3.1), i.e. equation (13.3.1) with $A'(s) \ge 0$ and $E = \{s \in \mathbb{R}; A'(s) = 0\}$ containing interior points. Problems for such equation are much more difficult to study than those for equation with weak degeneracy. The root of difficulty is that the solutions of such equation might be discontinuous. This can be exposed in the following consideration. Suppose $E \supset [a, b]$ $(a < b)$. Then for $u \in [a, b]$, (13.3.1) degenerates into the first order conversation law

$$
\frac{\partial u}{\partial t} = \frac{\partial B(u)}{\partial x},
\tag{13.3.35}
$$

whose solutions, as is well known, might have discontinuity, even if the initial-boundary value is smooth enough.

The first problem is how to define solutions with discontinuity for (13.3.1). Motivated by the theory of shock waves, a meaningful discontinuous solution u of (13.3.35) should satisfy the so-called entropy condition

$$(\bar{u} - k)\gamma_t \leq (\bar{B}(u) - B(k))\gamma_x, \quad \forall k \in \mathbb{R} \tag{13.3.36}$$

at the points of discontinuity in addition to the integral identity

$$\iint_{Q_T} \left(u \frac{\partial \varphi}{\partial t} - B(u) \frac{\partial \varphi}{\partial x} \right) dx dt = 0, \quad \forall \varphi \in C_0^\infty(Q_T). \tag{13.3.37}$$

Here $\bar{u} = \dfrac{1}{2}(u^+ + u^-)$, $u^\pm$ are the approximate limits of u at the points of discontinuity and (γ_t, γ_x) is the unit normal vector to the line of discontinuity. It is not difficult to verify that (13.3.36) and (13.3.37) imply the following integral inequality

$$\iint_{Q_T} \operatorname{sgn}(u - k)\left((u - k) \frac{\partial \varphi}{\partial t} - (B(u) - B(k)) \frac{\partial \varphi}{\partial x} \right) dx dt \geq 0,$$

$$\forall 0 \leq \varphi \in C_0^\infty(Q_T), \forall k \in \mathbb{R}. \tag{13.3.38}$$

In fact, at least for piecewise continuous functions u, (13.3.36), (13.3.37) are equivalent to (13.3.38). It was Kruzhkov who first defined weak solutions of (13.3.35) in this way and proved the existence and uniqueness of weak solutions of the Cauchy problem for (13.3.35).

Inspired by Kruzhkov's idea, Vol'pert and Hudjaev defined weak solutions of strongly degenerate equation (13.3.1) as follows.

Definition 13.3.2   A function $u \in L^\infty(Q_T)$ is said to be a weak solution of (13.3.1), if $\dfrac{\partial A(u)}{\partial x} \in L^1_{\text{loc}}(Q_T)$ and

$$\iint_{Q_T} \operatorname{sgn}(u - k)\left((u - k) \frac{\partial \varphi}{\partial t} - \frac{\partial A(u)}{\partial x} \cdot \frac{\partial \varphi}{\partial x} - (B(u) - B(k)) \frac{\partial \varphi}{\partial x} \right) dx dt \geq 0,$$

$$\forall 0 \leq \varphi \in C_0^\infty(Q_T), \forall k \in \mathbb{R}.$$

The existence and uniqueness of the weak solution thus defined has been proved in $BV(Q_T)$ for both the initial-boundary value problem and the Cauchy problem.

By $BV(Q_T)$, it is meant the set of all functions of locally bounded variation, i.e. a subset of $L^1_{\text{loc}}(Q_T)$, in which the weak derivatives of each function are Radon measures on Q_T.

The existence can be proved by means of parabolic regularization; the basic idea is the same as the proof of Theorem 13.3.2 but it is different in techniques. The proof of uniqueness is rather difficult, which is based on a deep study of functions in $BV(Q_T)$ and a complicated derivation (see [Wu, Zhao, Yin and Li (2001)]).

Bibliography

Adams R. A., Sobolev Spaces, Academic Press, New York-London, 1975.

Chen Yazhe, Parabolic Equations of Second Order, Beijing University Press, China, 2003.

Chen Yazhe and Wu Lancheng, Elliptic Equations and Systems of Second Order, Science Press, China, 1997.

Cui Zhiyong, Jin Dejun and Lu Xiguan, Introduction to Linear Partial Differential Equations, Jilin University Press, 1991.

Evans L. C., Weak Convergence Methods for Nonlinear Partial Differential Equations, CBMS 74, American Mathematical Society, Providence, RI, 1990.

Friedman A., Partial Differential Equations of Parabolic Type, Pentice-Hall, Inc., 1964.

Gilbarg D. and Trudinger N. S., Elliptic Partial Differential Equations of Second Order, Springer-Verlag, Heidelberg-New York, 1977.

Gu Liankun, Parabolic Equations of Second Order, Xiamen University Press, 1995.

Jiang Zejian and Sun Shanli, Functional Analysis, Higher Education Press, 1994.

Ladyženskaja O. A., Solonnikov V. A. and Ural'ceva N. N., Linear and Quasilinear Equations of Parabolic Type, Transl. Math. Mono., 23, American Mathematical Society, Providence, RI, 1968.

Ladyženskaja O. A. and Ural'ceva N. N., Linear and Quasilinear Elliptic Equations, English Transl., Academic Press, New York, 1968.

Lieberman G. M., Second Order Parabolic Differential Equations, World Scientific Publishing Co., Inc., River Edge, NJ, 1996.

Maz'ja V. G., Sobolev Spaces, English Transl., Springer-Verlag, Berlin-Heidelberg, 1985.

Oleĭnik O. A. and Radkevič E. V., Second Order Differential Equations with Nonnegative Characteristic Form, American Mathematical Society, Rhode Island and Plenum Press, New York, 1973.

Pao C. V., Nonlinear Parabolic and Elliptic Equations, Plenum Press, New York, 1992.

Saks S., Theory of the Integral, English Transl., Dover Publications, Inc., New York, 1964.

Wu Zhuoqun, Zhao Junning, Yin Jingxue and Li Huilai, Nonlinear Diffusion Equations, World Scientific Publishing Co., Singapore, 2001.

Zhong Chengkui, Fan Xianling and Chen Wenyuan, Introduction to Nonlinear Functional Analysis, Lanzhou University Press, 1998.

Index

$C^k(\Omega)$, 2
$C_0^k(\Omega)$, 2
$C^{k,\alpha}(\Omega)$, 7
$C^{k,\alpha}(\overline{\Omega})$, 7
$H^k(\Omega)$, 15
L^2 theory, 39, 60, 71, 94, 241, 256
L^∞ norm estimate, 233, 264
L^p estimate, 255, 260, 264, 266, 271
L^p norm estimate, 264, 273
$V(Q_T)$, 24
$V_2(Q_T)$, 24
$W_p^{2k,k}(Q_T)$, 24
$W^{k,p}(\Omega)$, 15
$W_0^{k,p}(\Omega)$, 15
BMO(Q_0), 259
$\overset{\bullet}{W}{}_p^{2k,k}(Q_T)$, 24
$\overset{\circ}{W}{}_p^{2k,k}(Q_T)$, 24
t-anisotropic Campanato space, 197
t-anisotropic Poincaré's inequality,
   28, 204
t-anisotropic Sobolev space, 24
t-anisotropic embedding theorem, 26

relatively strongly compact, 16

a modified Lax-Milgram's theorem,
   73, 94
Aleksandrov's maximum principle,
   264, 273
Arzela-Ascoli's theorem, 241

barrier function, 291, 293
barrier function technique, 242, 282
Bernstein approach, 282, 296, 297
bilinear form, 64
boundary gradient estimate, 282, 287,
   289
boundary Hölder's estimate, 287, 289
boundary Hölder's estimate for
   gradients, 308

Caccioppoli's inequality, 168, 200,
   205, 208
Campanato space, 160, 198, 256
Cauchy problem, 369
Cauchy's inequality, 1
Cauchy's inequality with ε, 2
classical Harnack's inequality, 133
coercive, 64
compact continuous field, 316
compact continuous mapping, 315
compact embedding theorem, 20
comparison principle, 233, 235, 238
completely continuous field, 316
completely continuous mapping, 315
conjugate operator, 48
continuity module, 261
contraction mapping principle, 246
coupled elliptic system, 354
coupled parabolic system, 336
critical point, 314
critical value, 314
cut-off function, 5

De Giorgi iteration, 105, 110, 115, 282
degenerate equation, 355
diagonal process, 244
difference operator, 47
Dirichlet problem, 39
discontinuous solution, 400
domain additivity, 315, 317
domain of (A)-type, 160

Ehrling–Nirenberg–Gagliardo's
  interpolation inequality, 17, 260
elliptic regularization, 365
embedding theorem, 19, 204
energy method, 71, 94
entropy condition, 400
estimate near the boundary, 122
existence and uniqueness of the
  classical solution, 240, 248–251
existence and uniqueness of the
  strong solution, 264, 272
existence and uniqueness of the weak
  solution, 46, 62, 75
existence of classical solutions, 233
exterior ball property, 242

Fichera function, 356
filtration equation, 368
fixed point method, 277
Fredholm's alternative theorem, 67

Galerkin's method, 85, 101
general elliptic equations, 60, 187
general linear elliptic equations, 248,
  260
general linear parabolic equations,
  251, 271
general parabolic equations, 94, 231
general quasilinear degenerate
  parabolic equations, 384
global estimate, 193
global gradient estimate, 282, 296
global Hölder's estimate for gradients,
  310
global regularity, 56, 92

Hölder space, 7

Hölder's estimate, 143, 155, 282, 301
Hölder's estimate for gradients, 282,
  307
Hölder's inequality, 2
Harnack's inequality, 131, 141, 145,
  156, 282, 284
heat equation, 71, 111, 123, 199, 250,
  266
heat equation with strong nonlinear
  source, 317
Hilbert-Schmidt's theorem, 85
homogeneous heat equation, 111, 123,
  145

in the sense of distributions, 41
integral characteristic of Hölder
  continuous functions, 161, 198
interior estimate, 178, 200
interior Hölder's estimate, 155, 284
interior Hölder's estimate for
  gradients, 307
interior regularity, 50
interpolation inequality, 17, 200
invariance of compact homotopy, 317
invariance of homotopy, 315
inverse Hölder's inequality, 125
inverse Poincaré's inequality, 124
iteration lemma, 177, 205, 207, 213,
  301

Kronecker's existence theorem, 315,
  317

Laplace's equation, 105, 118, 131
Lax-Milgram's theorem, 64
Leray-Schauder degree, 315
Leray-Schauder's fixed point theorem,
  277
linear elliptic equation, 240, 255, 264
linear parabolic equation, 249, 272
Lipschitz space, 8
local boundedness estimate, 116, 118,
  120, 121, 123, 126
local flatting, 6
lower sequence, 325

maximum estimate, 282
maximum principle, 233, 237, 241, 282
mean value formula, 131
method of continuity, 246
method of solidifying coefficients, 188, 263
Minkowski's inequality, 2
mixed quasimonotone, 337
mollifier, 4
monotone method, 323
monotone sequence, 339
more general quasilinear elliptic equations, 310
more general quasilinear equations, 310
more general quasilinear parabolic equations, 311
Morrey's theorem, 282, 302
Moser iteration, 105, 121, 123, 137, 282, 285

near bottom estimate, 211
near boundary estimate, 181, 191
near lateral boundary estimate, 219
nonhomogeneous heat equation, 112, 126
normality, 314, 316

one-sided Lipschitz condition, 324
ordered supersolution and subsolution, 324, 338

partition of unity, 6
Poincaré's inequality, 21, 28
Poisson's equation, 39, 47, 107, 120, 122, 178, 181, 240, 255, 287
property of segment, 16
property of uniform inner cone, 17

quasilinear degenerate parabolic equation, 368
quasilinear elliptic equation, 277
quasilinear parabolic equation, 280
quasimonotone, 337
quasimonotone nondecreasing, 337
quasimonotone nonincreasing, 337

regular point, 313
regular value, 314
regularity near the boundary, 53
regularity of weak solutions, 47, 50, 89
relatively weakly compact, 16
rescaling, 23, 204
Riesz's representation theorem, 41, 61
Rothe's method, 79, 96

Schauder's estimate, 159, 187, 197, 199, 231, 233, 264
Schwarz's inequality, 2
sector, 324, 339
semi-difference method, 79
shock wave, 400
sign rule, 297
smoothing operator, 3, 58
Sobolev conjugate exponent, 20
Stampacchia's interpolation theorem, 259
strong solution, 255, 263, 266, 272
strongly compact, 16
strongly degenerate, 385
subelliptic operator, 368
subellipticity, 368
subsolution, 324, 338
supersolution, 324, 338
support, 3

test function, 51, 54, 72, 73, 370–372, 377
the first boundary value problem, 356
the first initial-boundary value problem, 71
topological degree, 313
topological degree method, 313
trace of functions in $H^1(\Omega)$, 29

uniform exterior ball property, 242
uniform parabolicity, 94, 231
uniformly elliptic, 60
upper sequence, 325

weak derivative, 14
weak Harnack's inequality, 154, 284
weak maximum principle, 105, 107,
  111, 112
weak solution, 40, 60, 361, 362, 369,
  370, 385
weak subsolution, 116, 123

weak supersolution, 116, 123
weakly compact, 16
weakly degenerate, 385

Young's inequality, 1
Young's inequality with ε, 2